W9-ATY-414

Introduction to
Telecommunications
Second Edition

Anu A. Gokhale
Illinois State University
Normal, IL

Australia Canada Mexico Singapore Spain United Kingdom United States

THOMSON

DELMAR LEARNING

Introduction to Telecommunications, 2E

by Anu A. Gokhale

Vice President, Technology and Trades SBU:
Alar Elken

Editorial Director:
Sandy Clark

Acquisitions Editor:
Stephen Helba

Development Editor:
Dawn Daugherty

Marketing Manager:
David Garza

Channel Manager:
Fair Huntoon

Marketing Coordinator:
Casey Bruno

Production Director:
Mary Ellen Black

Production Editor:
Toni Hansen

Art/Design Coordinator:
Francis Hogan

Senior Editorial Assistant:
Dawn Daugherty

Production Services:
TIPS Technical Publishing, Inc.

Printed in the United States of America
1 2 3 4 5 XX 06 05 04

For more information contact Thomson Delmar Learning
Executive Woods
5 Maxwell Drive, PO Box 8007
Clifton Park, NY 12065-8007
Or find us on the World Wide Web at
www.delmarlearning.com

Library of Congress Cataloging-in-Publication Data
Gokhale, Anu A.
Introduction to Telecommunications/ Anu A. Gokhale.—2nd ed.
p. cm.
Includes bibliographical references index.
ISBN 1-4018-5648-9
1. Telecommunication. I. Title.

TK5101 .G596 2004
621.382—dc22

2004008777

This book is dedicated to my husband, Ashok,
for his continued encouragement, support, and sacrifices,
and to our daughter, Ashwini,
for her patience and understanding.

Contents

PREFACE

Introduction to Telecommunications, Second Edition, has been upgraded to include the latest developments in telecommunications with an introduction to emerging technologies. Specifically, here are just a few examples of technical topics that have been added: data coding method for Gigabit Ethernet, Bluetooth, IPv6, protocols such as H.323 used for multimedia transmission over IP, and transform coding technique for data compression. On the telecom business side, new topics include the process of calculating ROI (return on investment) and implementing a business continuance strategy and its enabling storage technology. In light of these additional topics, sections within chapters have been reorganized in a logical, progressively dependent order.

A new feature in this second edition is that most chapters have an accompanying Case Study—a decisional scenario where students are given brief descriptions of relevant facts and parameters of action, which provides students greater insights into real-world applications. The questions based on the case study pose possible outcomes, and then students individually, or as part of a team, have to make and support a decision. Case studies make material come alive and bolster students' enthusiasm and interest in courses, while preparing them to enter the workforce with real-world perspectives.

This text is intended for a course that forms a basis for a career in telecommunications, one of the most crucial technologies of the information age. The book has been developed for the purpose of helping university, community college, or technical institute students gain a comprehensive understanding of telecommunications technologies, their applications, and their implications for business. As each year passes, telecommunications grow in scope. It influences each one of us as an individual, and society as a whole. This book encourages readers to think about systems in ways that will continue to serve them well as technology continues to evolve.

The material presented here is based on the telecommunications and electronics courses that I have taught in the last ten years. Even after years of teaching telecommunications classes, I was not satisfied with the scope, clarity, and readability of available textbooks. Some texts are aimed at advanced computer or electrical engineering students, while others cover only the business aspects of telecommunications. I sought to create a simple, easy-to-read and interesting textbook that not only provides a thorough up-to-date presentation of the latest telecommunications technologies, but also examines its business aspects.

Outstanding Features

The text uses real-world examples and analogies to explain technical concepts. The reader is also presented with a discussion of practical, management-oriented applications of those concepts. The goal of this book is to integrate technical and business aspects to equip the reader with a global understanding of telecommunications. Current and emerging technologies are covered, as well as traditional material. The extensive glossary and acronyms included in the back of the book are designed to help the professional deal quickly with issues at hand. The book's comprehensive coverage will make it a useful reference tool.

Pedagogy

Each chapter opens with a list of Key Terms and Objectives to which students can refer while reading the material. Each section covers essential concepts before discussing one or more applications or examples. The figures are particularly helpful learning tools, as they provide graphical representations of technical material. Each chapter has several solved examples, so that the reader learns to analyze problems and evaluate solutions. The content is reinforced by a Summary, and the Review Questions draw attention to the main concepts presented in the chapter.

Organization

This book follows a logical progression as outlined here.

Chapter 1, An Overview of Telecommunications, walks the reader through the evolution of the industry, from the technological and business perspective, with the ultimate focus being the user or customer. It introduces a variety of terms commonly used in telecommunications and provides a foundation for following chapters.

Chapter 2, Electronics for Telecommunications, discusses relevant electronics concepts and their applications in telecommunications. Included are sections on communications system parameters, such as noise, and an in-depth analysis of different modulation and multiplexing schemes.

Chapter 3, Transmission Media, deals with copper and fiber-optic cables. Cable construction, specifications, applications, and transmission impairments are among the topics covered in this chapter. It concludes with a detailed discussion of cabling architecture, including installation and facilities hardware.

Chapter 4, Voice Communications, begins with the way a voice call is handled through the telephone set, followed by the switched network, line, and trunk signaling. Included are intelligent network services, such as automatic call distribution, and business telephone systems, such as private branch exchanges and Centrex. It concludes with a statistical discussion of designing a network to achieve a balance between the cost and quality of service to the user.

Chapter 5, Wireless Communications, provides a review of wireless technologies beginning with cellular mobile telephone system and analog and digital wireless access schemes. In addition to discussing microwave, radio, and infrared LANs, the chapter covers satellite communications and international wireless systems.

Chapter 6, Data Communications, discusses fundamentals of data communications including character codes, data coding methods, error detection and correction, data link protocols, and the Open Systems Interconnect model. It focuses on local area network, a mainstream technology that is at the heart of every network, from a small business to a large corporation. There are separate sections on Peer-to-Peer, Client/Server, Ethernet, Token Ring, and Fiber Distributed Data Interface. This chapter addresses the role of internetworking devices in a network.

Chapter 7, Wide Area Network and Broadband Technologies, takes a comparative look at widely used technologies such as Synchronous Optical Network, Frame Relay, Switched Multimegabit Data Service, Asynchronous Transfer Mode, and Integrated Services Digital Network. Residential or small business access technologies such as Digital Subscriber Line, Cable Modems, Fixed Wireless, VSAT, and Passive Optical Network are also discussed.

Chapter 8, Internet and Converged Networks, offers an in-depth discussion of the Transmission Control Protocol/Internet Protocol model, and evaluates the role of Virtual Private Networks, Intranets, and Extranets. The reader gets an insight about how voice is carried over data networks, relevant multimedia specifications and protocols, and data compression schemes.

Chapter 9, Network Management, covers a wide variety of issues that are part of the responsibilities of a network manager. Policy management, evaluation of network hardware and software, network administration and maintenance, and network services are the highlights of this chapter. A significant portion is devoted to different storage strategies and network security, vital concerns among both users and network managers.

Chapter 10, Telecom Policy and Business Contracts, deals with the Telecom policy, its intent, and its implications. The chapter analyzes the business aspects of telecommunications and how technology must be merged with business goals to achieve organizational success. A discussion of the elements of electronic commerce, Service Level Agreement,

Return on Investment, and Total Cost of Ownership provides valuable information that will help the reader become an effective telecom manager.

Introduction to Telecommunications' goal is to give readers a comprehensive knowledge base from which to accomplish the objectives on their way to a successful career in telecommunications. Readers will gain insights into meaningful business applications of telecommunications technologies and the technical concepts will provide a foundation for future learning.

Supplements

Instructor's Manual—Provides PowerPoint slides, multiple choice questions and answers, as well as critical-thinking exercises for classroom instruction.

Online Companion—Visit *Introduction to Telecommunications'* Online Companion at *www.electronictech.com* for up-to-date information on evolving telecommunications technologies.

Acknowledgments

The author and Thomson Delmar Learning wish to thank the following individuals who reviewed the manuscript and provided suggestions that have been incorporated into this second edition of Introduction to Telecommunications:

Scott Rosen
Santa Rosa Community College
Santa Rosa, California

Don Schrum
Vatterott College
St. Louis, Missouri

Michael Simpkins
Texas State Technical College
Marshall, Texas

Thomas Whalen
Mississippi Gulf Coast Community College
Gautier, Mississippi

About The Author

Anu Gokhale has completed over fourteen years of university teaching and is currently a Professor at Illinois State University. Originally from India, she has a master's degree in physics–electronics from the College of William & Mary and a doctorate in industrial technology from Iowa State University. She has experience as a consultant and trainer in industries, has published in several refereed journals, and pursues multi-year projects funded by agencies like the National Science Foundation. She is actively involved with the Institute of Electrical and Electronics Engineers (IEEE) and currently serves as Secretary of the Central Illinois Section.

1

AN OVERVIEW OF TELECOMMUNICATIONS

KEY TERMS

Telecommunication
Telephony
Symbol
Local Exchange Carrier (LEC)
Inter Exchange Carrier (IXC)
Equal Access
Local Access and Transport Area (LATA)
Network
Topology
Public Network
Private Network
Virtual Private Network (VPN)
Circuit Switching
Bursty Traffic
Message Switching
Packet Switching
Reliability
Cell Switching
Distributed Computing
Scalability
Centralized Computing
Redundant Array of Independent Disks (RAID)
Uninterruptible Power Supply (UPS)
Standards
Open Systems Interconnection (OSI) Model

OBJECTIVES

Upon completion of this chapter, you should be able to:

- Discuss the meaning of the term telecommunication and how its implied meaning has changed with time
- Outline the history of *telecommunications* technologies
- Summarize the evolution of the telecommunications industry
- Discuss network classification and characteristics
- Identify the role of national and international organizations in establishing and implementing telecommunications standards
- Analyze the challenges of telecommunications technologies
- Describe career opportunities for telecommunications professionals

INTRODUCTION

Communication has always been an integral part of our lives. Family relations, education, government, business, and other organizational activities are all totally dependent on communications. It is such a commonplace activity that we take it for granted. Yet, without communications most modern human activity would come to a stop and cease to exist. To a great extent, the success of almost every human activity is highly dependent on how the available communications methods and techniques are effectively utilized. The purpose of this book is to provide a firm foundation of the concepts involved in modern communications systems. This book effectively integrates business with technology to give the reader a broad perspective on the continuously evolving world of telecommunications. The general background and terminology introduced in this chapter will be revisited later in greater detail.

WHAT IS TELECOMMUNICATION?

The word **telecommunication** has its roots in two words: *Tele* in Greek meaning "distant" and *communicatio* in Latin meaning "connection." Telecommunication is the distant transfer of meaningful information from one location (the sender, transmitter, or source) to a second location (the receiver or destination). Today, the term *telecommunication* is used in a very broad sense to imply transfer of information over cable (copper or fiber) or wireless media and includes all of the hardware and software necessary for its transmission and reception.

A first important step in the route toward a modern information society and the information superhighway is the ability to represent information in digital form as binary digits or bits. These bits are then stored electronically and transmitted either as electrical or light pulses over a physical network or by broadcast signals between sites. An important advantage of digital communication lies in its versatility. Almost any form of information—audio, video, or data—can be encoded into bits, transmitted, and then decoded back into the desired final form at the receiver. As a result, it is almost always possible to establish a communications system that will transfer the exact types of information needed.

The term **telephony** is limited to the transmission of sound over wire or wireless. It connotes voice or spoken and heard information and it usually assumes a temporarily dedicated point-to-point connection rather than broadcast connection. Not long ago, telecommunication implied communication by wire, but with the use of radio waves to transmit information, the distinction between telephony and telecommunication has become difficult to make. With the arrival of computers and the transmittal of digital information over telephone systems, voice messages can be sent as connectionless packets. Digitization allows text, images, sound, and graphics to be stored, edited, manipulated, and interacted within the same format, and this in turn has led to the development of multimedia applications.

HISTORY OF TELECOMMUNICATIONS

A timeline of the major developments in telecommunications during the 19th century is shown in Figure 1–1. The developments have provided opportunities that go far beyond the vision of telephony on which this industry was built. This section has been divided into a history of telecommunications technologies and a history of the telecommunications industry. The reader will get an insight about how technological developments interact with business and government regulations, with the ultimate focus being the user or customer.

Year	Major Development
1837	Samuel Morse invents the telegraph
1858	Transoceanic telegraph cable is laid
1876	Alexander Graham Bell invents the telephone
1885	Incorporation of the American Telephone and Telegraph company (AT&T)
1888	Heinrich Hertz discovers the electromagnetic wave
1895	Marconi begins experimenting with wireless telegraph

Figure 1–1 Timeline of the major developments in telecommunications from 1800 to 1900.

History of Telecommunications Technologies

The information age began with the telegraph, which was invented by Samuel F.B. Morse in 1837. This was the first instrument to transform information into electrical form and transmit it reliably over long distances. The telegraph was followed by Alexander Graham Bell's invention of the telephone in 1876. The magneto-telephone was one of the first telephones on which both transmission and reception were done with the same instrument. After Heinrich Hertz discovered electromagnetic waves in 1888, Guglielmo Marconi invented the radio—the first wireless electronic communications system—in 1901. Industrialization in the twentieth century made life faster and more complex. To cope with these demands, engineers worked to find new means of calculating, sorting, and processing information, which led to the invention of the computer.

Telegraph

The earliest form of electrical communication, the original Morse telegraph of 1837 did not use a key and sounder. Instead, it was a device designed to print patterns at a distance. These represented the more familiar dots (short beeps) and dashes (long beeps) of the Morse code, shown in Figure 1–2. In 1844, Morse developed a key and sounder for his first commercial *telegraph*. At the transmitting end a telegrapher closed a switch or

A	•–	N	–•	1	•––––
B	–•••	O	–––	2	••–––
C	–•–•	P	•––•	3	•••––
D	–••	Q	––•–	4	••••–
E	•	R	•–•	5	•••••
F	••–•	S	•••	6	–••••
G	––•	T	–	7	––•••
H	••••	U	••–	8	–––••
I	••	V	•••–	9	––––•
J	•–––	W	•––	0	–––––
K	–•–	X	–••–	.	•–•–•–
L	•–••	Y	–•––	,	––••––
M	––	Z	––••	?	••––••

Figure 1–2 Morse code.

telegraph key in a certain pattern of short and long closures to represent a letter of the alphabet. The electrical energy on the wire was sent in the same pattern of short and long bursts. At the receiving end, this energy was converted into a pattern of sound clicks that was decoded by a telegrapher. The code used by both transmitter and receiver is the Morse code. With the advent of the electric telegraph and the laying of the transoceanic cable in 1858, a person's range of communication expanded to thousands of miles, the message delivery time dropped to seconds, and the information rate was maintained in the range of 5 to 100 words per minute.

Telephone

Invented by Bell and his assistant, Thomas A. Watson, the *telephone* marked a significant development in the history of electrical communications systems. In the earliest magneto-telephone of 1876, depicted in Figure 1–3, the speaker's voice was converted into electrical energy patterns that could be sent over reasonably long distances over wires to a receiver, which would convert these energy patterns back into the original sound waves for the listener. This system provided many of the long-range communications capabilities of the telegraph, but also had the convenience of speaking and hearing directly so that everyone could use the system. Its rate of information transfer was limited only by the rate of human speech. Telecommunication includes the telephony technology associ-

Figure 1–3 Alexander Graham Bell's magneto-telephone.

(photo courtesy of the Smithsonian Institution *Neg # 74-2496*)

ated with the electronic transmission of voice, fax, or other information between distant parties using systems historically associated with the telephone.

Radio

The first commercial wireless voice transmitting system utilizing electromagnetic waves, the *radio*, was built in the United States in 1906. Hertz discovered the electromagnetic wave in 1888, and in 1895, Marconi began experimenting with wireless telegraphy. Once man learned to encode and decode the human voice in a form that could be superimposed onto electromagnetic waves and transmitted to receivers, this communication approach was used directly with human speech. Now the human voice was transmitted to remote locations thousands of miles away, picked up by receivers, and converted to speech by speakers. This development opened new opportunities for wireless communications. A wartime ban on nonmilitary broadcasting delayed the acceptance of radio; the first regular commercial radio broadcast began in 1920. Thereafter, hundreds of amateur stations sprang up. World War II was a stimulus to wireless communications leading to the widespread development of cellular, mobile, satellite, and consumer radio and television systems.

Computer

Computers have revolutionized the way we live and work. The key developments that have brought us to our present state of computing include the development of numbers, the introduction of mechanical aids to calculation, the evolution of electronics, and the impact of electronics on computing. Although no one person may be credited with the invention of the *computer*, we will begin to track its history with an American mathematician and physicist, John Vincent Atanasoff, who designed the first electronic computer in early 1939. The marriage of computers and communications in 1941 was a major milestone that had synergistic effects on both technologies as they developed. In that year, a message recorded in telegraph code on punched paper tape was converted to a code used to represent the message data on punched cards read by a computer. Any signal must change to convey information, and these changes are coded into a **symbol**; a symbol is the smallest indivisible part of a signal. The *signaling rate* is given by the number of symbols per second, where the symbols may be binary or multilevel; examples are given in Figure 1–4. The *data rate* is defined as the number of bits per second.

Symbol Type	Examples
Alphabetic	26 letters of the alphabet
Numerical	digits 0 to 9
Binary	0 and 1
Morse code	dots, dashes, and spaces

Figure 1–4 Examples of symbols.

The modern computer era commenced with the first large-scale automatic digital computer, commonly referred to as Mark I, developed by Howard Aiken between 1939 and 1944. Perhaps one of the most important milestones in the history of electronics was the invention of the transistor in 1948 by John Bardeen, Walter Houser Brattain, and William Bradford Shockley, all of whom worked for Bell Telephone Labs at the time. The invention of the Integrated Circuit (IC) by Fairchild and Texas Instruments in 1961 marked another turning point for the computing industry. It became possible to develop miniaturized devices, such as amplifiers and microprocessors, which had low power requirements. The ICs are at the heart of all telecommunications equipment. The desktop Personal Computer (PC) made its market debut in the early 1970s after Intel developed the microprocessor in 1971. There has been a burgeoning growth in computer applications since the Internet and desktop computers came together in early 1980s.

History of the Telecommunications Industry

After its incorporation in 1885, the American Telephone and Telegraph (AT&T) company dominated the telecommunications market. Until recently, the combined Bell system was both the predominant **Local Exchange Carrier** (LEC) and the long distance carrier. AT&T owned the world's largest telecommunications manufacturing facility and the premier telecommunications research laboratory. Universal telephone service became available to practically all Americans, and the American switched circuit telephone network became the best in the world. As a result of AT&T's burgeoning growth and market dominance in the 1950s and 1960s, the company became a subject of recurrent Department of Justice antitrust actions.

In the late 1960s, Microwave Communications, Inc. (MCI) began constructing a microwave network between Chicago and St. Louis. MCI took its interconnection request to the courts and prevailed, though it nearly drove the company into bankruptcy. In 1976, the Federal Communications Commission (FCC) opened long-distance telephone service to competition from other long-distance carriers, also called **Inter Exchange Carriers (IXCs)**. Users had to dial an additional seven-digit number to access these IXCs versus just dialing "1" to access AT&T. Line-side access or trunk-side access, as shown in Figure 1–5, characterized LEC services at the local switching office. Four-wire trunk-side access was available to only AT&T, while all other IXCs had two-wire line-side access. The line-side access represented by Feature Group A does not support *Automatic Number Identification (ANI)*, which is the capability of a local switching office to automatically identify the calling station and is usually used for accounting and billing information. The Feature Group characteristics are summarized in Figure 1–6.

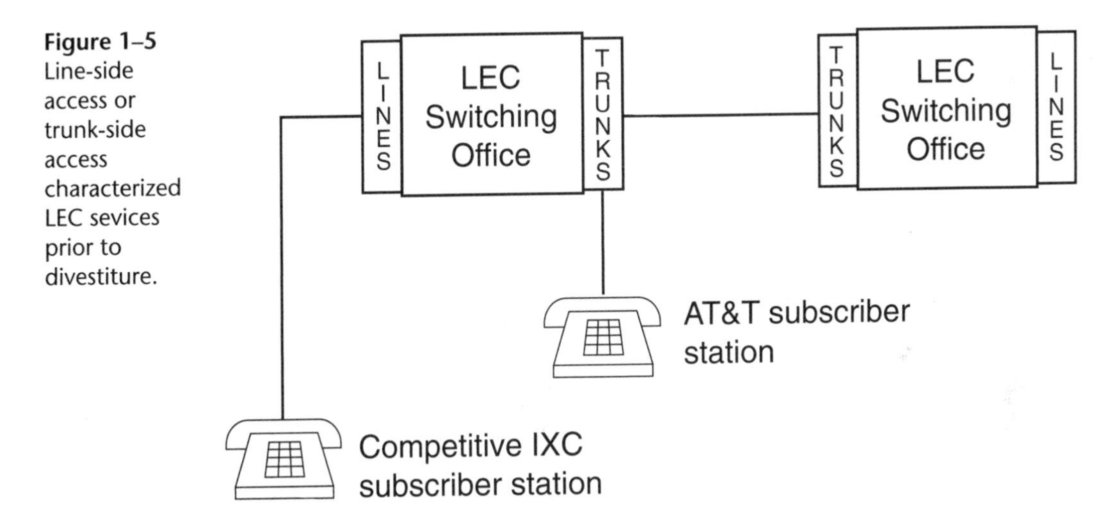

Figure 1–5 Line-side access or trunk-side access characterized LEC sevices prior to divestiture.

The AT&T monopoly prompted the U.S. Justice Department to file an antitrust lawsuit against the company in early 1974. The outcome was a restructuring agreement that was signed in 1982 and went into effect January 1, 1984. The divestiture or breakup of AT&T resulted in the formation of seven Regional Bell Operating Companies (RBOCs), also called *Baby Bells*. AT&T retained its long distance network and the capability to sell business telephone systems, but gave up the ownership of the local telephone companies, which then became part of the newly formed RBOCs.

Each of the seven RBOCs shown in Figure 1–7 had different BOCs in its geographical area. Over the years, federal and state lawmakers heavily regulated practically all aspects of the business operations of the RBOCs. Within this structure, monopoly telephone companies essentially agreed to provide local services at heavily regulated prices in return for the governmental guarantee that they would be the only market provider and would

Feature Group	Characteristics
A	✦ Two-wire line-side access ✦ No Automatic Number Identification (ANI) ✦ Poor quality ✦ Not used anymore
B	✦ Four-wire trunk-side access ✦ Supports partial ANI ✦ High quality ✦ Not used anymore
C	✦ Four-wire trunk-side access ✦ Supports ANI ✦ High quality ✦ Available only to AT&T ✦ Used by LECs prior to the divestiture agreement ✦ Not used anymore
D	✦ Four-wire trunk-side access ✦ Supports ANI ✦ High quality ✦ Represents equal access ✦ Provided by LECs to all IXCs

Figure 1–6 LEC access services categorized by feature group.

have the opportunity to earn a reasonable profit. As part of the decree, these providers of local telecommunications services, also known as LECs, had to provide equal access to all competing long-distance carriers. In 1999, the number of RBOCs has shrunk from seven to four as SBC Communications bought Pacific Telesis and Ameritech, and Bell Atlantic absorbed NYNEX.

Equal Access

Equal Access meant that all IXCs had connections that were identical to AT&T's connection to the local telephone network. The LECs were required to provide four-wire trunk-side access to all competing IXCs; they therefore had to upgrade their equipment from Feature Group C to Feature Group D. The *Point of Presence (POP)* is where the LEC and IXCs are interconnected, which is also known as a *Point of Interface (POI)*. When a user originates a long-distance call, the LEC's switching equipment must decide which IXC the user wants to handle the call. Each user pre-subscribes to a preferred IXC, and the pre-selected IXC is known as the Primary Interexchange Carrier (PIC). Callers can reach other IXCs by dialing a carrier access code, 101XXXX, where XXXX is a number assigned to each IXC. Thus, we have so many 101XXXX options available today.

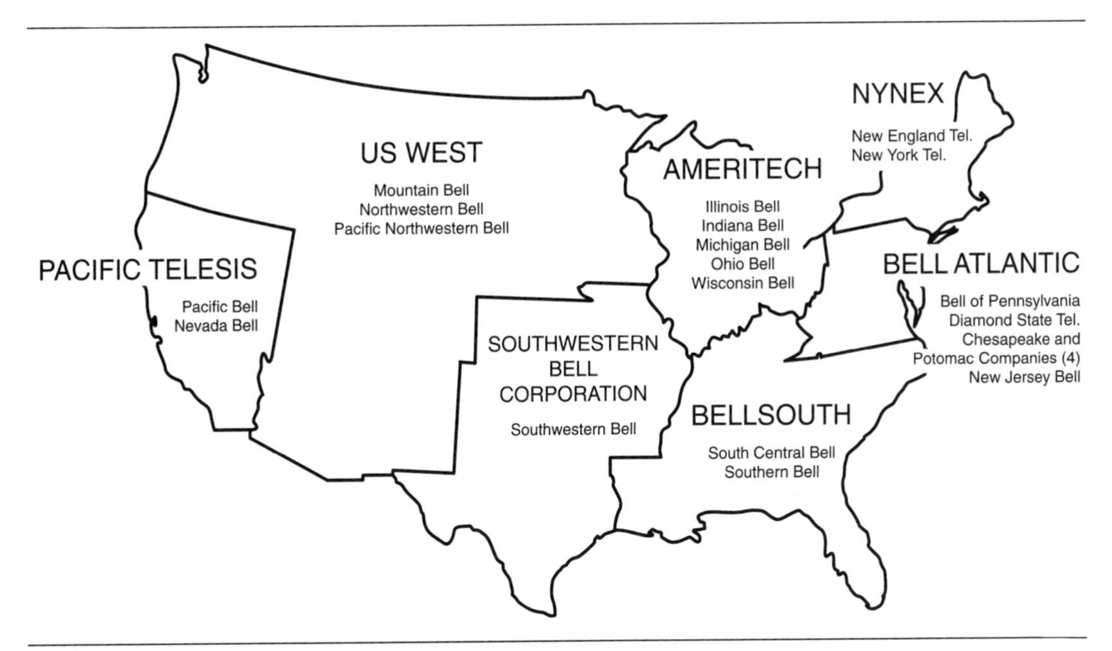

Figure 1–7 Seven RBOCs formed as a result of the divestiture agreement.

Local Access and Transport Area (LATA)

The **Local Access and Transport Area (LATA)** concept was another significant outcome of the divestiture agreement of 1984. The LATA was a predetermined area used to govern who could carry calls in what area. There were two main types of calling using the LATA concept: IntraLATA transport belonged to the LECs, and InterLATA transport belonged to the IXCs or long distance carriers, as shown in Figure 1–8. Most companies used common terms to describe the various categories by which they marketed their services. They included IntraLATA, Intrastate, Interstate, inbound toll-free and calling card services. IntraLATA calls, sometimes known as local long distance, were calls that were outside the local calling area but inside the LATA and were carried by the LEC. Intrastate calls were calls made within the state but outside the LATA. Interstate calls were calls made from one state to another. Both Intrastate and Interstate are part of InterLATA and require a long distance carrier. LEC/IXC facilities and services used to complete InterLATA calls are illustrated in Figure 1–9.

<table>
<tr><th colspan="2">Intra-LATA</th><th colspan="2">Inter-LATA</th></tr>
<tr><td colspan="2">LEC</td><td>LEC</td><td>IXC</td></tr>
<tr><td>Access</td><td>Transport</td><td>Access</td><td>Transport</td></tr>
</table>

Figure 1–8 IntraLATA and InterLATA services.

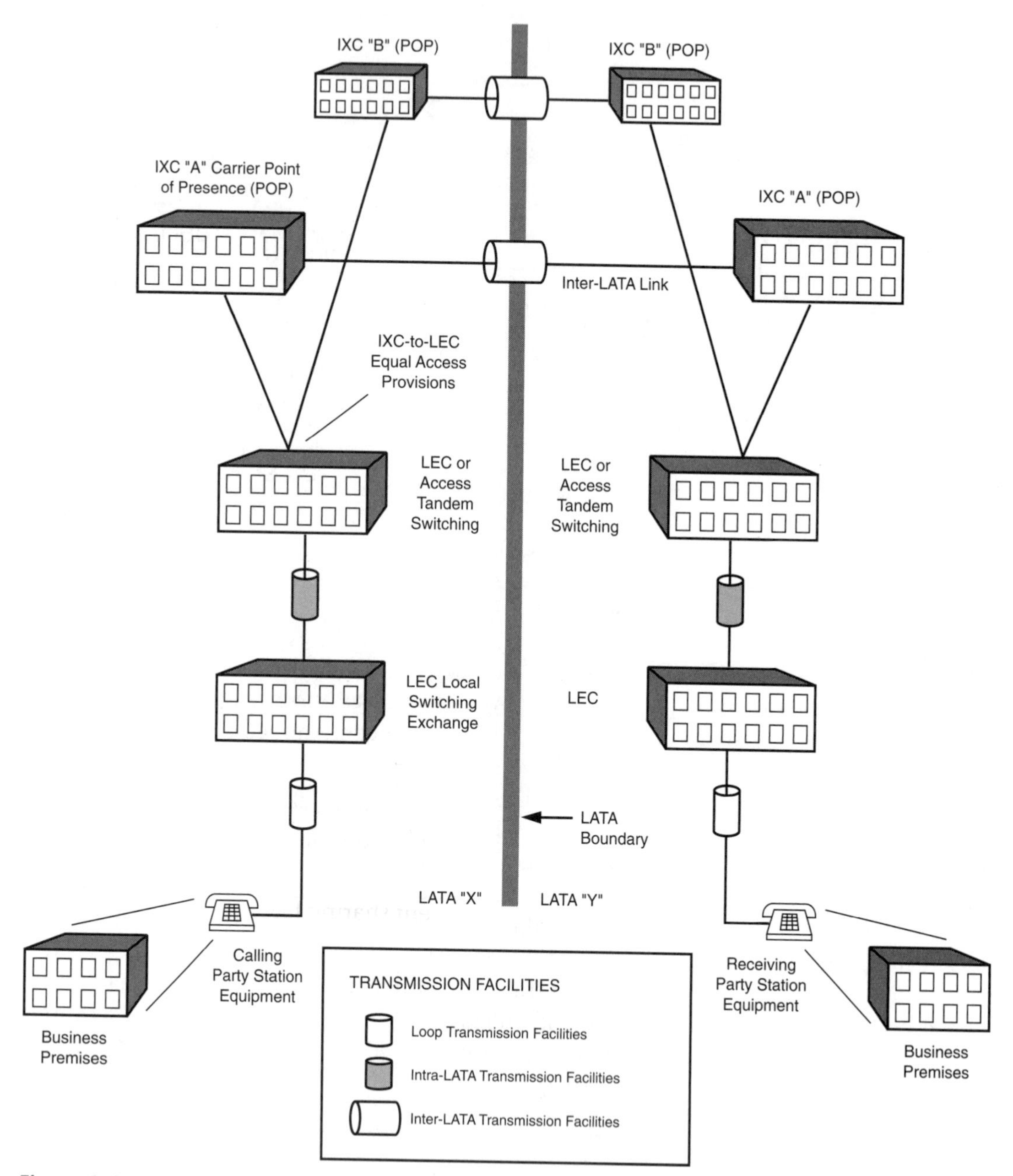

Figure 1–9 LEC/IXC facilities configuration.

Telecommunications Act of 1996

Regarded as the first major reform to the 1936 telecommunications legislation that established the Federal Communications Commission (FCC), the Telecommunications Act of 1996 deregulated local phone markets with the intent to make telecommunications services (an industry known for its bureaucracy) competitive. Until its passage, practically all LECs operated as local franchised monopolies. The act was aimed at deregulating the market and increasing competition among service providers.

TELECOMMUNICATIONS NETWORKS

In information technology (IT), a **network** is a series of points or nodes interconnected by communication paths. The connection points are known as network nodes or switching exchanges. Networks can interconnect with other networks and can therefore contain subnetworks. Every network has a *backbone,* which is a larger transmission line that carries data gathered from smaller lines that interconnect with it. Traditionally, the telephone network was the largest network of computers interconnecting networks owned by different carriers. The *Public Switched Telephone Network (PSTN)* still remains the lifeline of most communications. The advent of data communications and a need to interconnect computers resulted in an emergence of data networks.

Data networks increase an organization's efficiency, productivity, and profitability by combining the geographically dispersed resources—the skills of different people and the power of different hardware and software. Networking computers provides the following benefits:

- Powerful, Flexible Collaboration: Networks enable users to instantaneously and effortlessly view, change, and exchange information. Electronic collaboration frees people from spending considerable time, effort, and money traveling or communicating by less effective means.
- Cost-effective Sharing of Equipment: Equipment sharing has significant benefits. It enables a company to buy equipment with features that one would not otherwise be able to afford and to ensure that this equipment is used to its full potential. Networks enable users to share resources such as printers, modems, facsimile machines, data storage devices such as hard disks and CD-ROM drives, data backup devices such as tape drives, and all networkable software.
- Software Management: In a networked environment, software installation and update is easier and more efficient since the software is loaded only on the host system, such as a mainframe or minicomputer, and authorized personnel can have immediate access. In addition, networks make it easier to track software licenses since a central host houses the software. In contrast, it can be very expensive and time-consuming to install, update, and keep track of software on every individual machine.

- Freedom to Choose the Right Tool: In a networked environment, users may choose to work on the type of computer best suited for their job, without placing restrictions on their file-sharing capabilities.
- Flexible Use of Computing Power: One of the most powerful things a network can do is use the processing power of two or more computers. This can be done in two ways: remote login or distributed parallel processing. In remote login, a user working on his or her own computer can simultaneously log into or use the processing power of another computer that may be sitting idle. In distributed parallel processing, computers are networked to run programs that are too big to run on individual microcomputers.
- Secure Management of Sensitive Information: Sophisticated networks have extremely powerful security features that enable flexible control over user access to information and equipment.
- Easy, Effective Worldwide Communication: By implementing a complete suite of networking products, users are able to connect computing equipment at different, widely dispersed geographic locations into one cohesive network so that they are able to almost instantaneously pass critical data to multiple locations anywhere in the world.

INTERNET

Let us trace the history of the Internet, which is a network of data networks. The term "Internet" was first used in 1982 but its history dates back to 1969. It is a global network of computers linked mainly via the telephone system and the academic, research, and commercial computing network. Large networks have sharing and exchange arrangements with other large networks so that even larger networks are created. In an Internet, a *backbone* is a set of paths that local or regional networks connect to for long-distance interconnection. The first prototype of the Internet was ARPANET, funded in 1969 by the Defense Advanced Research Projects Agency (DARPA) of the Department of Defense. One important characteristic of ARPANET and other networks funded by DARPA was the commitment to a standard communications protocol suite, the Transmission Control Protocol/Internet Protocol (TCP/IP), which permits transmission of information among systems of different kinds. Each network's host, whether it is a local, regional, national, or international network, still shares the common TCP/IP protocol suite to connect to the Internet.

In 1978, a UNIX-to-UNIX copy program resulted in the formation of worldwide UNIX-based communications networks. The USENET (User's Network) was developed in 1979, followed by the CSNET (Computer Science Network) and BITNET (Because It's Time Network) in 1981. These can be described as the first major networks to be based solely on interest and willingness to connect rather than disciplinary specialty, mainframe type, or funding source. Some of the standard options available on CSNET and BITNET were electronic mail and file transfer services. In 1989, CREN (Corporation for Research and Education Networking) represented the merging of BITNET and CSNET. In the mid-1980s the National Science Foundation (NSF) established a number of supercomputer centers. A high-speed communications network known as the NSFNET (NSF Network) linked the centers and provided users with electronic access to the data stored on the computers. The NSFNET is the most prominent of the Internet backbones.

The Internet is a superhighway information network limited only by the rate at which the network components can transmit and handle data. The World Wide Web (WWW) became a functioning part of the Internet only in 1990, but the growth in the number of computer hosts connected to the Internet since then has been exponential. The point-and-click Graphical User Interface (GUI) of the WWW allows access to a global network of computers by millions of people who have no formal training in computer technology. In 1991, as a result of the extraordinary economic and social importance of an adequate information infrastructure, the federal government enacted legislation designed to rationalize and upgrade the Internet. It is this upgraded, harmonized network that is the National Research and Education Network (NREN). NREN is a high-speed backbone network designed to provide U.S. academic and research institutions with supercomputer resources. Figure 1–10 provides an outline of the evolution of the Internet.

Year	Major Development
1969	ARPANET was funded by the DARPA commitment to a standard communications protocol, the TCP/IP
1978	Development of the Unix-to-Unix copy program
1981	Development of CSNET and BITNET based soley on interest and willingness to connect
1982	The term *Internet* is coined
1986	Establishment of NSFNET, a network of supercomputers
1989	CSNET and BITNET merge to form CREN
1990	World Wide Web (WWW) becomes a functioning part of the Internet
1991	Federal government upgrades the Internet to a high-speed backbone network, the NREN
mid-1990s	Emergence of Intranets, which are corporate networks based on Internet standards

Figure 1–10 Outline of the evolution of the Internet.

CLASSIFICATION OF DATA NETWORKS

Networks can be characterized in several different ways and classified by:

- Spatial distance, such as *Local Area Network (LAN), Metropolitan Area Network (MAN),* and *Wide Area Network (WAN);*
- Topology or general configurations of networks, such as the ring, bus, star, tree, mesh, hybrid, and others;
- Network ownership, such as public, private, or virtual private;
- Type of switching technology such as circuit, message, packet, or cell switching;
- Type of computing model, such as centralized or distributed computing; and
- Type of information it carries such as voice, data, or both kinds of signals.

Classification by Spatial Distance

The geographic expanse of a network is a very important characteristic that may determine other factors, such as speed and ownership. The most common designations are the LAN and the WAN, with the MAN being a less common designation. WAN technology connects sites that are in diverse locations, while LAN technology connects machines within a site. Let us take an example of a university campus. A single department or college has its own LAN. These departmental or college LANs are then connected to the university LAN or MAN. The university LAN or MAN is connected to the WAN via *leased lines*, which are private lines that provide a permanent pathway between two communicating stations. Another example is an enterprise or a corporate network. It may consist of multiple LANs that may be interconnected over a distance to form a WAN. Figure 1–11 provides an overview of the characteristics of a LAN, MAN, and WAN.

Wide Area Network (WAN)

A WAN usually refers to a network that covers a large geographical area and uses common carrier circuits to connect intermediate nodes. The WAN for a multinational company may be global, whereas a WAN for a small company may cover only few cities. A major factor that distinguishes a WAN is that it utilizes circuits leased from telephone companies or common carriers. This restricts the communications facilities and transmission speeds to those normally provided by such companies. Transmission rates typically range from kbps (kilobits per second) to Mbps (Mega or Million bits per second), with 56 kbps, 64 kbps, 2 Mbps, 34 Mbps, and 45 Mbps being most common. WAN transmission technologies discussed in Chapters 7 and 8 include data communications protocols such as TCP/IP, Systems Network Architecture (SNA), Asynchronous Transfer Mode (ATM), X.25, Frame Relay, and others.

	LAN	MAN	WAN
Typical Geographic Expanse	Less than 5 km	5 to 50 km	More than 50 km
Ownership	Private	Private/Public	Private/Public
Transmission Rate	Mbps to Gbps	kbps to Mbps	kbps to Mbps
Typical Applications	✦ Industrial plants ✦ Business offices ✦ College campuses ✦ Single departments	✦ Frequently provides a shared connection to other networks using a link to a WAN ✦ City networks	✦ Connects offices in different cities using leased lines

Figure 1–11 Characteristics of a LAN, MAN, and WAN.

Metropolitan Area Network (MAN)

A MAN typically covers an area of between 5 and 50 km in diameter, about the size of a city, and acts as a high-speed network to allow sharing of regional resources (similar to a large LAN). A MAN (like a WAN) is not generally owned by a single organization, but rather a consortium of users or by a single network provider who sells the service to the users. The level of service provided to each user must therefore be negotiated with the MAN operator, and some performance guarantees are normally specified. Its primary customers are companies that need a lot of high-speed digital service within a relatively small geographic area. A MAN is also frequently used to provide a shared connection to other networks using a link to a WAN.

Local Area Network (LAN)

A LAN is the most common type of data network. Typical installations are in industrial plants, office buildings, college or university campuses, or similar locations. LANs are installed by organizations that want their own high-quality, high-speed communication links where data transmission speeds range from 10 to 1000 Mbps. LANs allow users to share computer-related resources within an organization and may be used to provide a shared access to remote users through a router connected to a MAN or a WAN. Intermediate node devices such as repeaters, bridges, switches, and routers allow LANs to be connected together to form larger LANs. At the local level, a *backbone* is a line or set of lines that LANs connect to for a WAN connection, or within a LAN to span distances efficiently—for example, between buildings. The discussion of LANs is extensive and deals with many other topics; for this reason, Chapter 6 is dedicated to complete coverage of data communications in LANs.

Classification by Topology

A **topology** (derived from the Greek word "*topos* meaning "place") is a description of any kind of location in terms of its physical layout. In the context of communication networks, a topology pictorially describes the configuration or arrangement of a network, including its nodes and connecting lines. The ring, bus, and star are the three basic network topologies. Different topologies are depicted in Figure 1–12.

Ring

Ring is a network topology or circuit arrangement in which each device is attached along the same signal path to two other devices and forms a path in the shape of a ring. Each device in the ring has a unique address. To avoid collisions, information flow is unidirectional, and a controlling device intercepts and manages the flow to and from every station on the ring by granting a token or permission to send or receive. The advantages of the ring network are that it is easy and inexpensive to install, and even if one connection is down, the network will still work. Its disadvantages are that the network must be shut down for reconfiguration, and it is difficult to troubleshoot. The token ring, Fiber Distributed Data Interface (FDDI), and Synchronous Optical Network (SONET) are examples of ring networks.

Bus

Bus is a term that is used in two somewhat different contexts. In the context of a computer, a bus is the data path on the computer's motherboard that interconnects the microprocessor with attachments to the motherboard in expansion slots, such as disk drives and graphics adapters. In a network, a bus topology is a circuit arrangement in which all devices are directly attached to a line and all signals pass through each of the devices. Each device has a unique identity and can recognize those signals intended for it. The advantages of a bus network are that it is inexpensive, simple, and easy to configure, connect to, and expand. Its major disadvantage is that if the backbone goes down, the whole network goes down. Also, increasing the number of users will cause the network to become slower, and performance may be unpredictable under heavy load conditions. The network must be shut down to add any new users, and troubleshooting can be very time consuming. The 10Base2 Thin Ethernet, explained in Chapter 3, is typically implemented as a bus.

Star

Star is a network in which all computers are connected to a central node, called a hub, which rebroadcasts all transmissions received from any peripheral node to all peripheral nodes on the network, including the originating node. Thus, all peripheral nodes may communicate with all others by transmitting to, and receiving from, the central node only. The advantages of a star network are that it is simple and robust, it is faster than

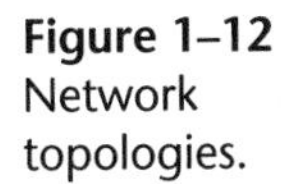

Figure 1–12 Network topologies.

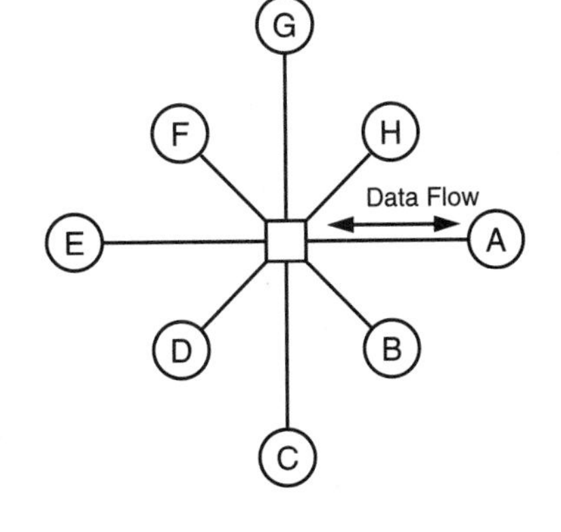

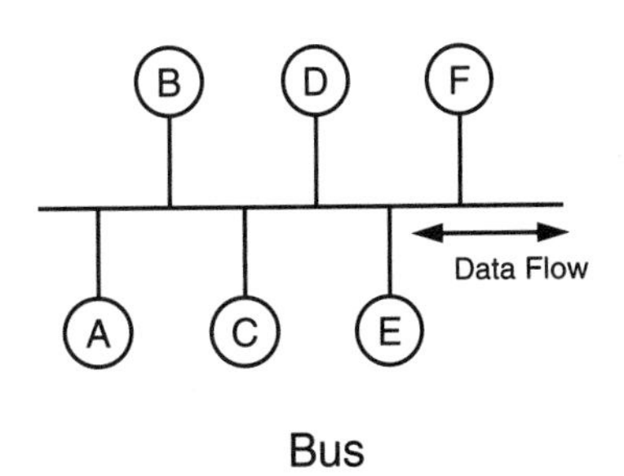

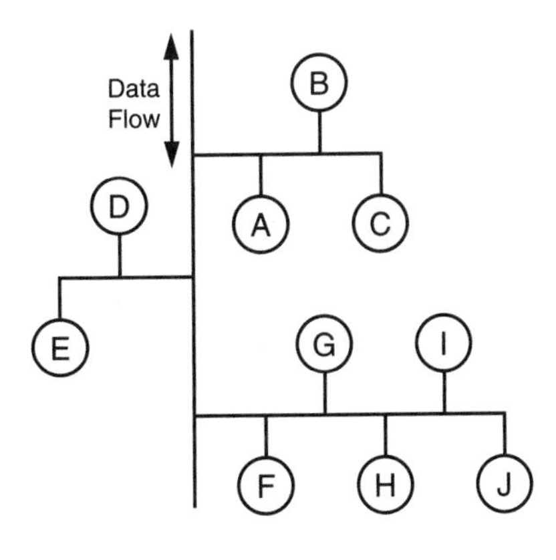

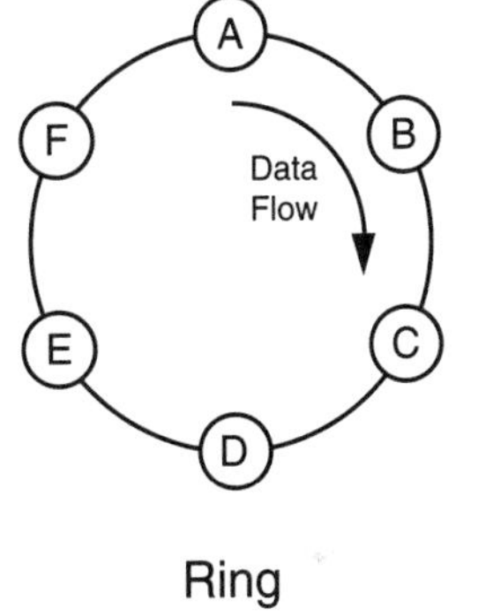

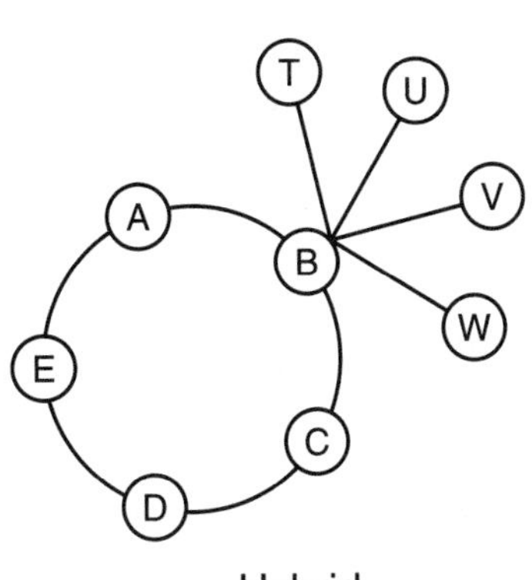

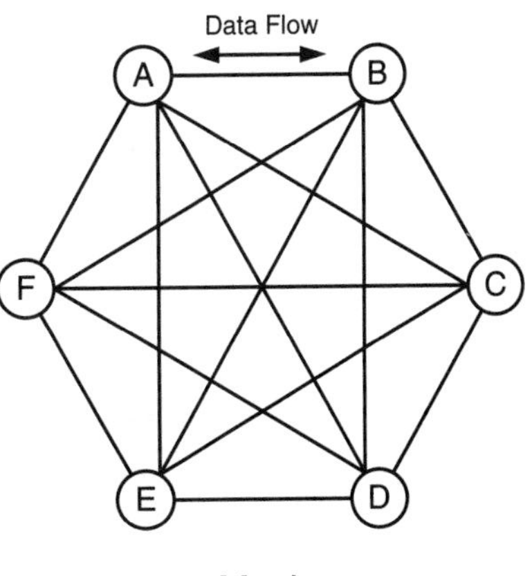

ring or bus, has greater stability, is easy to set up, reconfigure, and troubleshoot, has low configuration costs, and provides for a centralized administration and security control. If a transmission line linking a peripheral node to the central node fails, it will result in the isolation of that peripheral node, but the remaining network is not affected. The disadvantage is that if one of the hubs fails or a hub cable fails it will shut down that segment of the network. Also, a hub has limited ports, and an increase in the number of users may involve additional network expenses. The 10BaseT, 100BaseT, and 1000BaseT Ethernet, explained in Chapter 3, are implemented in a star topology.

Tree

Tree, also known as hierarchical network, is a network topology that from a purely topologic viewpoint resembles an interconnection of star networks. The individual peripheral nodes are required to transmit to and receive from one other node only, toward a central node, and are not required to act as repeaters or regenerators. The function of the central node may be distributed. The top node in the structure is called a root node.

Mesh

Mesh topology is similar to a hierarchical structure except that there are more interconnections between nodes at different levels, or even at the same level. At a minimum, there are at least two nodes with two or more paths between them. In a fully interconnected mesh each node is connected to every other node, although this is cost prohibitive and therefore seldom implemented. The PSTN is a classic example of mesh topology with multiple interconnections making the network virtually failsafe. LAN backbone switches may be connected using a mesh topology.

Hybrid

Hybrid network is a combination of two or more basic network topologies. Instances where two basic network topologies are connected together and retain the basic network character cannot be classified as a hybrid network. For example, a tree network connected to a tree network is still a tree network. Therefore, a hybrid network is created only when two different basic network topologies are connected and the resulting network topology fails to meet any one of the basic topology definitions. For example, ring and star networks connected together exhibit hybrid network topologies.

Classification by Ownership

Networks can also be classified according to their ownership. The two broad categories are public networks and private networks. Virtual private network (VPN) is a newly emerged third category.

Public Network

A **public network** refers to a network owned by a common carrier for use by its customers. The term is usually applied to the PSTN, but it could also mean packet switched public data networks. The public data network is typically operated by a telecommunications administration or a recognized private operating agency for the specific purpose of providing data transmission services for the public. The advantage of a public network is that it provides services or access to locations that a company might not otherwise be able to afford. As the capital and operational costs are shared by a number of users, the common carrier can achieve good utilization of its network and provide high-quality service at a reasonable cost.

Private Network

A **private network** is built for exclusive use by a single organization. When traffic among a company's business locations is sufficiently high, it may be cheaper to shift the internal traffic from public switched networks to a private switched network. It can be designed to address specific communications requirements of the organization as it is built around particular traffic patterns. Also, it gives the company full control of the network's operation and potentially superior security. At times, the flexibility and autonomous operation may be bought at a higher cost. The State Farm insurance company has one of the largest private networks in the world.

Virtual Private Network (VPN)

Virtual Private Networks (VPNs) are gaining popularity because they combine the advantages of both private networks and public networks. VPNs are encrypted tunnels through a shared private or public network that forward data over the shared media rather than over dedicated leased lines. The operation of a VPN is very similar to that of a telephone connection over a public telephone network. In a telephone call, there is a dedicated connection between two parties for the entire duration of the call. Similarly, a VPN is characterized by dedicated connections set up between sites on a public network and controlled by software and protocols during the connection. After the session of data transmission is terminated, the connection between the sites is abandoned. A VPN allows sharing of the Internet's structure of routers, switches, and transmission lines, while providing security for the users. The cost factor is a compelling argument for replacing a private network with a VPN, because sharing leased lines in a public network such as the Internet can cut monthly recurring costs by an order of magnitude. However, using public networks for highly sensitive corporate data, such as financial information, can pose security problems.

Classification by Switching Technology

Another broad way of classifying networks is by the technology used in switching circuits. The cost and the required quality of transmission dictate the technology implemented. For example, voice or video is not very tolerant of delays, as opposed to data. Therefore, voice circuits mostly employ circuit-switching or cell-switching techniques, whereas packet switching is most efficient for data communications. The following paragraphs provide an overview but we will study these techniques in more detail in later chapters. Figure 1–13 identifies the strengths and weaknesses of different switching technologies.

Circuit Switching

Circuit switching systems, sometimes called connection-oriented networks, are ideal for communications that require information to be transmitted in real-time. Voice services have been traditionally supported via circuit-based techniques. For over a century, the telecommunications infrastructure developed around this technology. It has two major disadvantages. The first is that an entire communication channel must remain dedicated to two users regardless of whether they actually need the full channel capacity for the entire time. This is especially inefficient for data communications characterized by **bursty traffic,** where there are peak periods of data transmission followed by periods in which no transmission takes place. The second disadvantage is that a constant connection for the entire time during which a transmission traverses the channel gives an intruder time to pick up on a sequential cohesive message.

Message Switching

Message switching, also known as a *store-and-forward system,* accepts a message from a user, stores it, and forwards it to its destination according to the priority set by the sender. Its primary advantage is that the sender and receiver do not need to be online simultaneously. The storage time may be minimal so that forwarding is almost instantaneous. If the receiving device is unavailable, or if the switching device is waiting for more favorable rates, the messages may be stored for longer periods. In any case, the network queues messages and releases the originating device. Its two disadvantages are longer response time as compared with circuit or packet switching and the added cost of storage facilities in the switching device. An example is a domestic or international Telex.

Packet Switching

Packet switching permits data or digital information to proceed over virtual telecommunications paths that use shared facilities and are in use only when information is actually being sent. It is made possible by breaking information streams into individual packets—which are blocks of data characters delimited by header and trailer records—and routing them using addressing information contained within the packet. In contrast to a circuit-

Switching Technology	Strengths	Weaknesses
Circuit	✦ Ideal for real-time applications such as voice ✦ Guaranteed quality of service	✦ Inefficient use of channel capacity ✦ Susceptible to eavesdropping ✦ Inappropriate for data communication
Message	✦ Sender and receiver do not need to be online simultaneously	✦ Longer response time ✦ Added cost of storage facilities
Packet	✦ Efficient use of network facilities ✦ Most appropriate for data communication	✦ Real-time applications such as voice and video conferencing may suffer from poor qualilty of service
Cell	✦ Viable technology for real-time applications ✦ Capable of providing measures for quality of service	✦ Inefficient transfer of IP packets

Figure 1–13 Strengths and weaknesses of different switching technologies.

switched network with dedicated connections between stations, a packet-switched network establishes virtual connections between stations. **Reliability** of a network specifies the ability of a packet to reach its destination.

In a *permanent virtual circuit,* the routing between stations is fixed and packets always take the same route. In a *switched virtual circuit,* the routing is determined with each packet. Individual packets from a single message may travel over different networks as they seek the most efficient route to their destination. Network nodes are controlled by software with algorithms that determine the route. At the receiving station, packets may arrive out of sequence, but the control information allows them to be reassembled in proper order. This technology permits massive amounts of data to be transmitted rapidly and efficiently without tying up a specific circuit or path for any extended length of time. Packet switching technology is primarily digital and designed for data communication. Most WAN protocols, including TCP/IP, X.25, and Frame Relay use packet-switching techniques.

Cell Switching

Cell switching is a relatively new technique that has gained rapid popularity. It combines aspects of both circuit and packet switching to produce networks with low latency and high throughput. The fast processing of fixed length cells maintains a constant rate data channel. ATM is currently the most prominent cell-switched technology; digital voice, data, and video information can simultaneously travel over a single ATM network.

Classification by Computing Model

There are two basic types of computing models: centralized computing and distributed computing. In the past, centralized computing was the mainstay of corporate data communications. However, the increased availability of microprocessor-based desktop computers gave rise to distributed computing. Now, much of the processing load is offloaded from the mainframe and performed by the desktop.

Distributed Computing

Distributed computing spreads users across several smaller systems, and thus limits the disruption that will be caused if one of the systems goes down. A *client/server* setup is a classic example of a distributed network. The client part is any other network device or process that makes requests to use server resources and services. If one server went out of service, only users connected to that server would be affected by the outage; the rest of the network would continue to function normally. This distributed design is therefore inherently superior to centralized designs in which even a single mainframe failure can bring down the whole network.

The n-tier application structure implies a client/server program model, where n stands for a positive integer, and the application program is distributed among n separate computers. Its most common form is a three-tier application in which user interface programming is in the user's computer, business logic is in a more centralized computer, and needed data is in a computer that manages a database. In a two-tier application, business logic and database management functions are merged in a single computer. Where there are more than three tiers involved, the additional tiers in the application are usually associated with the business logic tier. In addition to the advantages of distributing programming and data throughout a network, n-tier applications have the advantage that any one tier can be updated independently of the other tiers. Communication between the program tiers uses special program interfaces such as those provided by the Common Object Request Broker Architecture (CORBA).

A distributed network has the following attributes:

- Flexibility; in other words, easily customizable because one can use equipment from several vendors to build or expand a network without losing the initial investment in hardware
- Low centralized computer costs, but higher end-user equipment and network management costs
- Fault-tolerance, since even a catastrophic server failure can still be a manageable event

- Scalability, since distributed systems use public networks as a sort of expansion bus to link the smaller systems together
- Ability to be implemented in both LAN and WAN technologies

Scalability is the ability to smoothly increase the power and/or number of users in an environment without major redesigns, at a reasonable cost. Distributed processing provides a structure that can be upgraded in phases to support newer technologies as well as an increasing number of users, so as to ensure high user satisfaction. Distributed networks make it possible for companies to build enterprise networks using modular, low-cost components and to build fault tolerant server arrays for large offices.

Centralized Computing

Centralized computing involves accessing a central computer, called the mainframe, which does all processing associated with most tasks. Initially, input to the computer was performed using interactive dumb terminals. Later, smart terminals provided for batched input to the mainframe. Batch terminals help to reduce network costs by taking advantage of switching networks. Centralized computing is often found in retail chains where stores download sales information to the mainframe at the end of the day. A centralized network has the following attributes:

- Lack of flexibility and customization
- High centralized computer costs, but lower end-user equipment and network management costs
- Suitability for mission-critical information
- Ability to be implemented in WAN technologies

Thin-client architecture is a newer implementation of the older centralized computing model. In this network, the level of computing power on each desktop may vary between end users. In many cases, administrators enable or disable certain functions, depending on the needs of the particular user, while retaining centralized control. A common profile of the worker for whom a thin client desktop, also called a network computer, makes a good match is one who frequently uses a remote database and relies on a limited number of applications.

In successful thin-client architectures, commands flow from the client to the server, and only a small amount of data flows back to the client. This is ideal for terminal-like applications, for example, locating a hotel reservation. In this case, there is no need to download the entire set of data to read just one entry from a reservations database. On the other hand, thick clients are highly efficient for some applications. For example, it would be quite cumbersome to edit a document in a thin-client architecture, where the document is downloaded one paragraph at a time. The objective is to balance the transfer of data from server to client and the transfer of processing from the client to the server. A security benefit of this strategy is that all potentially sensitive data resides on the server, so there is none on the harder-to-secure client workstations. Servers can be configured with varying degrees of security measures.

Thin-client architecture also gives agencies a bit of fault tolerance; if the server is properly protected with a **Redundant Array of Independent Disks (RAID)** and an **Uninterruptible Power Supply (UPS)**, data will not be lost as a result of hard disk failures or power outages. RAID is a way of storing the same data in different places, thus, redundantly on multiple hard disks. By placing data on multiple disks, Input/Output (I/O) operations can overlap in a balanced way, thereby improving performance and fault tolerance. A UPS is a device whose battery kicks in after sensing a loss of power from the primary source and allows a computer to keep running for at least a short time. Software is available that automatically saves any data being worked on when the UPS becomes activated. The UPS also provides protection from power surges by intercepting the surge so that it does not damage the computer. If the terminals lose power, users simply log back on when power is restored and resume working in their applications where they left off.

Classification by Type of Information

All information can be classified into three basic types: data, audio or voice, and video. The term *data communications* is used to describe digital transmission of information. *Voice communications* primarily refers to telephone communications. *Video communications* include one-way transmissions such as Cable TV (CATV), and two-way transmissions such as videoconferencing. Communications have evolved from dedicated networks for voice, data, and video to converged data/voice/video networks. In the past, data communications were limited to text and numeric data. However, with current developments in technology, any information that can be reduced to 0s and 1s is data. The telecommunications industry is no longer dominated by telephony; data traffic and Internet are now taking over with converged communications networks becoming a reality.

TELECOMMUNICATIONS STANDARDS

The broad goal of setting standards for the telecommunications industry is connectivity, compatibility, and open networking of communications and computer systems from multiple vendors. **Standard***s* are documented agreements containing technical specifications or other precise criteria to be used consistently as rules, guidelines, or definitions of characteristics to ensure that the products, processes, and services are fit for their purpose. A standard provides benefits to users, as well as the industry. It enables users to buy components in a competitive open market. At the same time, a standard provides manufacturers with a system that accommodates current products and offers a template for future product design. Adoption of the standards by any country, whether it is a member of the organization or not, is entirely voluntary.

In the United States and internationally, many organizations and associations are involved in the standards process; the field of players is vast and sometimes not closely coordinated. In the United States the complex infrastructure includes political bodies at both the state and national levels, most notably the U.S. Congress. It also includes regulatory bodies at both the state and national levels, most notably the FCC. In addition, the infrastructure includes standards bodies at the regional and, importantly, at the national and international levels because an international standard facilitates trade and global competition. The national, regional, and international standards-setting process is a vital element of the infrastructure that delivers information technology to meet societal demands for new products and services. In recent years, there has been significant growth in industry consortia aimed at facilitating the marketplace introduction of products and services that comply with new standards. The political and regulatory bodies impact a marketplace system that is vital to matching information technology solutions to the needs of end users. The most prominent organizations are shown in Figure 1–14. The following paragraphs provide a description of the role played by these. The standards adopted by these organizations are presented throughout this book.

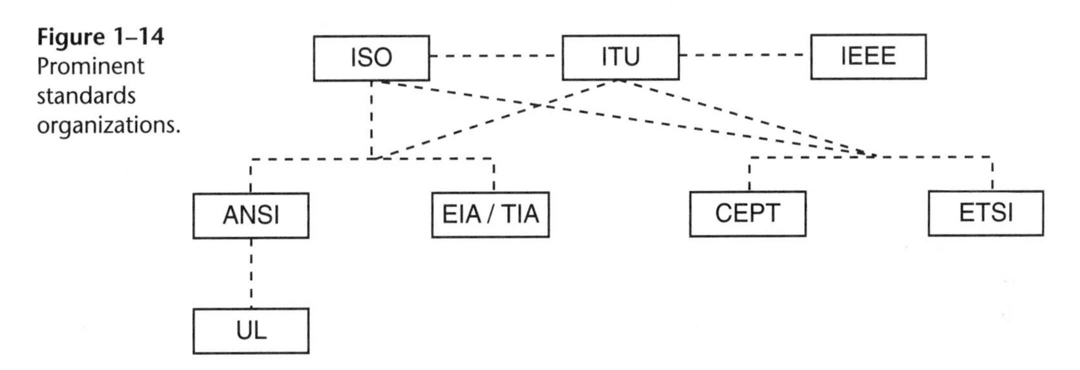

Figure 1–14 Prominent standards organizations.

International Standards Organizations

The existence of non-harmonized standards for similar technologies in different countries or regions can contribute to technical barriers to international trade. An agreement on world standards helps rationalize the international trading process. Users have more confidence in products and services that conform to international standards. Assurance of conformity can be provided by manufacturers' declarations or by audits carried out by independent bodies, which has led to the establishment of numerous international standards organizations.

International Standards Organization (ISO)

A non-governmental organization established in 1947, the *International Standards Organization (ISO)* is the most prominent worldwide federation of national standards bodies from some 130 countries (one from each country). Its mission is to promote the development of standardization and related activities in the world with a view toward facilitating the international exchange of goods and services and developing cooperation in the spheres of intellectual, scientific, technological, and economic activity. The ISO's work results in international agreements that are published as international standards. The technical work of the ISO is highly decentralized and is carried out in a hierarchy of technical committees, subcommittees, and working groups. In these committees, qualified representatives of industry, research institutes, government authorities, consumer bodies, and national or international organizations from all over the world come together as equal partners in the resolution of global standardization problems. One example is the role of the ISO in the development of the **Open Systems Interconnection Model**, or OSI model.

Back in the 1970s computer networking was completely vendor-developed and proprietary. ISO became concerned about the compatibility and consistency within networks and in 1977 developed the OSI model to provide standardized guidelines for creating and operating these network systems. The OSI model, depicted in Figure 1–15, is one of the most widely-used networking models for data communications. The model was established to enable interconnectivity between systems despite differences in underlying architectures, and to modularize components used in network communications. Typically, only the lower layers are implemented in hardware, with the higher layers being implemented in software. Each layer performs a specific task, but all of the layers are necessary for communications to occur. Sometimes, each task is performed by a separate piece of hardware or software; at other times, a single program may perform the functions of several layers. Overall consistency and application remain as long as the guidelines for each layer are adhered to by the manufacturers, facilitating multivendor equipment interoperability. You can use the OSI reference model to visualize how information travels throughout a network; a detailed description is provided in Chapter 6 of this book.

The scope of the ISO is not limited to any particular branch; it covers all technical fields except electrical and electronic engineering, which is a responsibility of the International Electrotechnical Commission (IEC). Founded in 1906, the IEC is the international standards and conformity assessment body that prepares and publishes international standards for all electrical, electronic and related technologies. U.S. participation, through the U.S. National Committee (USNC), is strong in the IEC. In the field of information technology, a joint ISO/IEC technical committee does the work.

International Telecommunication Union (ITU)

Headquartered in Geneva, Switzerland, the *International Telecommunication Union (ITU)* is an international organization within which governments and the private sector coordinate global telecommunications networks and services. The ITU-T, Telecommunication Standardization Sector, was created in 1993 within the framework of the ITU, replacing the former International Radio Consultative Committee (CCIR) and the International Tele-

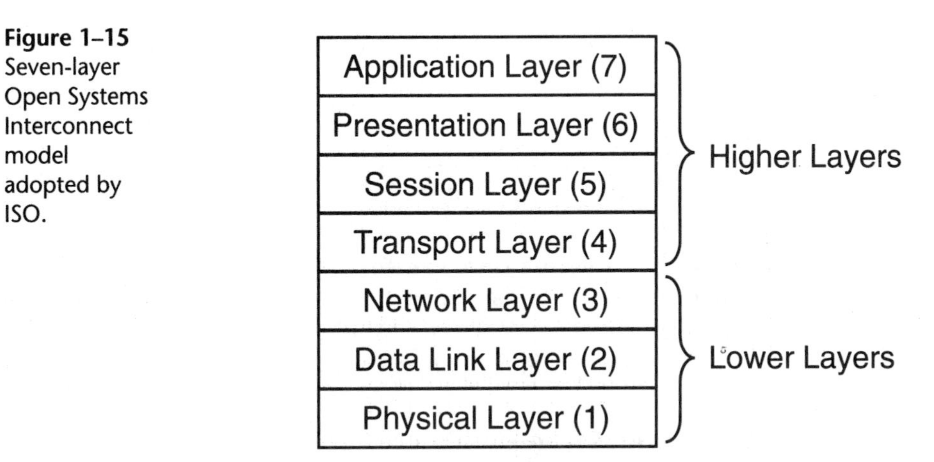

Figure 1–15 Seven-layer Open Systems Interconnect model adopted by ISO.

phone and Telegraph Consultative Committee (CCITT) whose origins are over 100 years old. The ITU-T, which is one of the three sectors of the ITU, studies technical, operating, and tariff questions and adopts recommendations on them with a view toward standardizing telecommunications on a worldwide basis. The ITU is composed of study groups that work in four-year time increments. After a four-year session, the study groups present their work to plenary assembly for approval. Plenary assemblies coincide with leap years. The Telecommunication Standardization Bureau (TSB) provides support for the work of the ITU-T Sector and diffuses the information worldwide. As an example, the International Mobile Telecommunication 2000 standard for wireless communications developed by the ITU has been adopted worldwide in next-generation mobile communications systems.

Institute of Electrical and Electronics Engineers (IEEE)

A worldwide technical, professional, and educational organization, the *Institute of Electrical and Electronics Engineers (IEEE)*, promotes networking, information sharing, and leadership through its technical publishing, conferences, and consensus-based standards activities. The IEEE is a catalyst for technological innovation and a leading authority in technical areas ranging from computer engineering, biomedical technology, and telecommunications to electric power, aerospace, and consumer electronics. The predecessors of IEEE, the AIEE (American Institute of Electrical Engineers) and the IRE (Institute of Radio Engineers) date to 1884. The AIEE laid the foundations for all work on electrical industry standards in the United States. The establishment of IRE in 1912 was prompted by the interests and needs of those specializing in the expanding field of radio and wireless communications. Many of the original members of the IRE were members of the AIEE, and both organizations continued to have members in common until they merged to form the IEEE in 1963. The IEEE continues to expand as information technologies grow in importance and as the career and technical needs of members broaden. The IEEE 802.x standards for local area networks are some of the most widely implemented data communications standards.

National Regulatory and Standards Organizations

In the United States, the need for standards and the need for technical progress sometimes conflict because standards often are not set until the technology has been proven in practice. But the only way to prove a technology is through extensive use. As a result, when it is time to set a standard, a large base of installed equipment is already in use. Competing manufacturers are represented on the standards-setting bodies to preclude the adoption of proprietary standards. Sometimes, organizations collaborate to produce standards that are adopted by the entire industry, such as the ANSI/EIA/TIA 568 cabling standard developed in accordance with the NEC (National Electrical Code). Government agencies such as the FCC play a very important role in regulating the industry.

Federal Communications Commission (FCC)

An independent United States government agency, the *Federal Communications Commission (FCC)* is directly responsible to Congress. The FCC was established by the Communications Act of 1934 and is charged with regulating interstate and international communications by radio, television, wire, satellite, and cable. The FCC's jurisdiction covers the 50 states, the District of Columbia, and U.S. possessions. Wire and radio communication facilities that aid the national defense form one of the basic requirements of the Communications Act. The FCC provides leadership to create new opportunities for competitive technologies and services for the American public. In particular, it focuses on consumer protection to ensure that consumers are empowered and treated fairly in an environment marked by greater competition and convergence of technology and industry sectors.

American National Standards Institute (ANSI)

Founded in 1918 by five engineering societies and three government agencies, the *American National Standards Institute (ANSI)* remains a private, nonprofit, voluntary standardization organization supported by a diverse constituency. The Institute represents the interests of its nearly 1,400 corporate, organization, government agency, institutional, and international members. ANSI was a founding member of the ISO and plays an active role in its governance. Through ANSI, the United States has immediate access to the ISO and the IEC standards development processes. As a sole U.S. representative and dues-paying member of the ISO, ANSI promotes international use of U.S. standards, advocates U.S. policy and technical positions in international and regional standards organizations, and encourages the adoption of international standards as national standards. The Underwriters Laboratories, Inc. (UL) and others are all ANSI accredited standards developers. They have registered standards under the Continuous Maintenance option.

Telecommunications Industry Association (TIA)

Accredited by the ANSI to develop voluntary industry standards for a wide variety of telecommunications products, the *Telecommunications Industry Association (TIA)*'s Standards and Technology Department is composed of five divisions that sponsor over 70 standards-setting formulating groups. The committees and subcommittees sponsored by the five divisions are Fiber Optics, User Premises Equipment, Network Equipment, Wireless Communications, and Satellite Communications. Within TIA, representatives from manufacturers, service providers, and end-users (including the government) serve on the formulating groups involved in standards setting. To ensure representation for the positions of U.S. telecommunications equipment producers in the international arena, TIA also participates in international standards setting activities, such as the ITU and the Inter-American Telecommunication Commission (CITEL).

European Standards Organizations

European organizations were a result of the integration movements in Western Europe in the 1950s. The efforts to introduce broad regional cooperation in the field of posts and telecommunications resulted in the formation of regional standards bodies.

European Conference of Postal and Telecommunications Administrations (CEPT)

Established in 1959, the *European Conference of Postal and Telecommunications Administrations (CEPT)* now covers almost the entire geographical area of Europe with its 43 members. CEPT's activities include cooperation on commercial, operational, regulatory, and technical standardization issues. In 1988, CEPT decided to create the European Telecommunications Standards Institute (ETSI), into which all its telecommunication standardization activities were transferred. The new CEPT, which deals exclusively with sovereign/regulatory matters, has established two committees on telecommunications issues: the European Radio-communication Committee (ERC), and the European Committee for Regulatory Telecommunications Affairs (ECTRA).

European Telecommunications Standards Institute (ETSI)

A non-profit organization, the *European Telecommunications Standards Institute (ETSI)*'s mission is to determine and produce the telecommunications standards. In Europe, telecommunications standardization is an important step towards building a harmonized economic market. The European Commission has set an ambitious pace for achieving a unified single market and the members of the European Free Trade Association and other CEPT countries strongly support this goal. The role and purpose of ETSI is defined in part as follows:

- Establishing a European forum for discussions on sovereign and regulatory issues in the field of post and telecommunications issues
- Providing mutual assistance among members with regard to the settlement of sovereign/regulatory issues

- Strengthening and fostering cooperation among European countries and promoting and facilitating relations between European regulators
- Influencing, through common positions, developments within ITU in accordance with European goals
- Creating a single Europe on posts and telecommunications sectors

De facto Standards

Large companies have enough market power to set proprietary standards that others must follow to be compatible. Examples are Microsoft's Windows and IBM's SNA for WANs. The voice networks in the United States were largely designed in accordance with AT&T proprietary standards. Although in some cases international standards organizations have adopted proprietary standards, in other cases they are in conflict. For example, ITU's Signaling System Number 7 (SS7) is incompatible with AT&T's Common Channel Interoffice Signaling (CCIS) protocol that was used in long-distance switching equipment.

Open Computing

Standards have allowed heterogeneous systems to communicate with each other and exchange data. These capabilities drove cost down and productivity up, while increasing both speed to market and business agility. Just as open standards were critical to the emergence of the Internet, open source software will play increasingly critical roles. Open source—whose better known implementations include the Linux operating system and the Apache web server—ships with complete source code and can be freely copied and redistributed. As much a philosophy as a method of licensing software, open source technology is inexpensive and provides the building blocks to support in-house IT development that is far less reliant on proprietary file formats and network protocols. Open computing provides businesses with a way to treat technology components as discrete modules that can be mixed and matched.

CHALLENGES OF TELECOMMUNICATION TECHNOLOGIES

Electronic communication has enabled people to interact in a timely fashion on a global level in social, economic, and scientific areas. The range and immediacy of electronic communications are two of the most obvious reasons why this type of communication is so important. The objective of the telecommunications system is to interconnect users, whether they are people or systems communicating over data, voice, or video circuits. Networks of many organizations have isolated islands of automation. The telecommunications engineer/manager is challenged to connect these islands. Linking engineering, production, business functions, and management into one computerized information system can reduce cost while improving product quality, productivity, and customer satisfaction—thereby making the companies more competitive.

This book explores telecommunications in the broadest way possible: in the context of powerful interrelated thrusts in information technology; in competition; and in globalization. The long-standing goal of the telecommunications industry has been to provide voice, data, and images in any combination, anywhere, at any time, with convenience and economy. This objective will be made possible by highly intelligent, high-capacity multimedia networks accessible by a multitude of advanced multifunction terminals. The various types of information terminals in the hands of people will act as gateways to the intelligence stored in switched networks around the world. Moreover, we will see communications and entertainment blend into integrated or converged communications networks.

CAREERS IN TELECOMMUNICATIONS

Market-driven companies have realized that one of the keys to owning a market segment is the effective use of information that already resides within the enterprise. Information is regarded as both a valuable business asset and a foundation for an enterprise's competitive advantage. These organizations are transforming themselves into information-driven enterprises in which consistent and comprehensive information about customers, markets, competitors, products, and technologies acts as a catalyst that drives all processes and activities. They are reinventing themselves over and over again through the most dynamic, robust technology available. The companies taking on the challenges of marketing their products in a global economy are opening their doors to a growing number of IT professionals.

Most companies view IT—a broad term covering all aspects of processing and managing information—as a tool to implement solutions. Therefore, IT professionals have a wide variety of career opportunities in diverse fields. For example, the life sciences information technology field includes bioinformatics, computational hardware, database storage, and software. In an industrial environment, IT functions include converging network deployments, wireless applications, data modeling, production and process control, security, and the impact of e-commerce. IT professionals design, develop, support, and manage computer software, hardware, and networks—including Internet applications.

Telecom technicians and engineers design, interface, install, test, and troubleshoot network systems and the advanced electronic devices of today and the future. Some specific examples are LAN and WAN design and installation, network cable installation and certification, and troubleshooting telecom equipment (telephone switches and networking hardware). Over the last decade, wireless communications has seen a rise in opportunities—communications antenna designer, microwave and satellite technicians, and wireless systems designers and installers. There will always be a demand for both research and development and technical support in all the diverse fields that encompass telecommunications.

Traditional job titles such as programmer and systems analyst once defined where people fit into the IT world. But these titles may be losing their luster in an era when skills and experience seem to outweigh titles in determining rank and pay. Even the hierarchy of job titles is breaking down. For example, one version of the IT hierarchy lists these jobs in ascending order: programmer analyst, senior systems programmer, senior systems analyst, project manager, network administrator, and computer operations manager. But the salary does not necessarily go in that order. Titles are likely to get more confusing in the future because the roles people are playing are diversifying.

There was a time when responsibility was easily defined. Now it is a matrixed world, and we are all working cross-functionally. This brings us to the question: What do companies look for when hiring new employees? The new infrastructure includes electronic messaging, office productivity tools, enterprise resource planning, and Internet technologies. Industry requires a full range of technological skills, from mainframe to client/server to Web-based development with the latest in e-commerce and object-based design. The ability to work with leading clients on critical business issues continues to be a key factor in an increasingly global operation. Recruiters are generally looking for a blend of business knowledge and technical expertise, as they want people who can use technology to solve business problems. Prospective employees must understand how technologies interact and how they support business transactions. Therefore, this book is well balanced to provide the reader with technical knowledge and applications, as well as business aspects of telecommunications technologies.

SUMMARY

Communication is necessary for human development, and society's progress goes hand-in-hand with the ability to communicate. In our personal lives, we have always had a need to share our thoughts and experiences. In business, the goal of all communications applications is increased productivity. Traditionally, telecommunications referred to voice communication by wire. Today, it implies transmission of any type of information such as data, voice, video, or image by wire or wireless. Distance, location, time, and volume are traditional barriers to the movement of information, but high-speed communications are breaking them down at an unprecedented rate. Next generation networks will be more heterogeneous and versatile and, at the same time, they will be readily available to a significantly wider segment of the world's population than they are today.

Circuit switching is ideal when data must be transmitted quickly and must arrive in the same order in which it is sent. This is the case with most real-time data, such as voice communications and live audio and video. Packet switching is more efficient and robust for data that can withstand some delays in transmission, such as e-mail messages and Web pages. Cell switching attempts to combine the best of both worlds—the guaranteed delivery of circuit-switched networks, and the robustness and efficiency of packet-switching networks.

The interconnected yet heterogeneous nature of a global telecommunications network requires standards so that the devices can seamlessly communicate with one another. ISO, IEEE, ITU, EIA, TIA, ANSI and CEPT are some of the notable standards organizations. Before the widespread use of the Internet, the normal evolution for a business was to start small, serving customers in one geographic area, then expand regionally, then nationally, and finally enter the international business market. Today, a Web site gives a company with a few employees instant international exposure and access to a global customer base, which also brings worldwide competition. This revolution has resulted in a vast new range of challenges and opportunities for telecommunications professionals.

REVIEW QUESTIONS

1. Explain the term "telecommunication" and how its implied meaning has changed over time.

2. Outline major developments in telecommunications technologies.

3. Track the history of the telecommunications industry.

4. Define the following terms:
   - A. LATA
   - B. Equal Access
   - C. Backbone
   - D. Leased Lines
   - E. Public Network
   - F. Private Network
   - G. Virtual Private Network
   - H. Circuit Switching
   - I. Message Switching
   - J. Bursty Traffic
   - K. Packet Switching
   - L. Centralized Computing
   - M. Distributed Computing
   - N. Client/Server

5. Discuss the evolution of the Internet.

6. Analyze the characteristics of WANs, MANs, and LANs.

7. Describe the following network configurations:
   - A. Ring
   - B. Bus
   - C. Star
   - D. Tree
   - E. Mesh
   - F. Hybrid

8. Evaluate the importance of standards in the field of telecommunications.

9. Identify international, regional, and national telecommunications organizations or regulating agencies and explain the role played by each.

10. Discuss career opportunities for telecommunications professionals and the challenges faced by the industry.

2

ELECTRONICS FOR TELECOMMUNICATIONS

KEY TERMS

Bandwidth
Broadband
Baseband
Asynchronous
Synchronous
Efficiency of Transmission
Overheads
Simplex, Half-Duplex, and Full-Duplex
Serial
Parallel
Universal Asynchronous Receiver Transmitter (UART)
Analog
Digital
Codec
Local Loop
Modem
Noise
Signal-to-Noise Ratio (SNR)
Bit Error Rate (BER)
Modulation
Time Domain
Frequency Domain
Frequency Shift Keying (FSK)
Phase Shift Keying (PSK)
Quadrature Amplitude Modulation (QAM)
Sampling
Multiplexing

OBJECTIVES

Upon completion of this chapter, you should be able to:

- Identify the basic components of a communications system
- Discuss different communications system parameters
- Analyze different modulation techniques
- Explain different multiplexing schemes
- Evaluate real-life applications of different modulation and multiplexing technologies

INTRODUCTION

Electronics began with pioneer work in two closely related fields: electricity and magnetism. The *electromagnetic (E/M) spectrum,* which includes all oscillating signals from 30 Hz at the low-frequency end to several hundred GHz at the high-frequency end, plays a major role in telecommunications. The radio waves provide a wireless path for information transmission, while wavelengths in the near-infrared region are used in fiber-optic communications. Figure 2–1 provides the names given to different frequency ranges in the E/M spectrum. The FCC has jurisdiction over the use of this spectrum for communications in the United States.

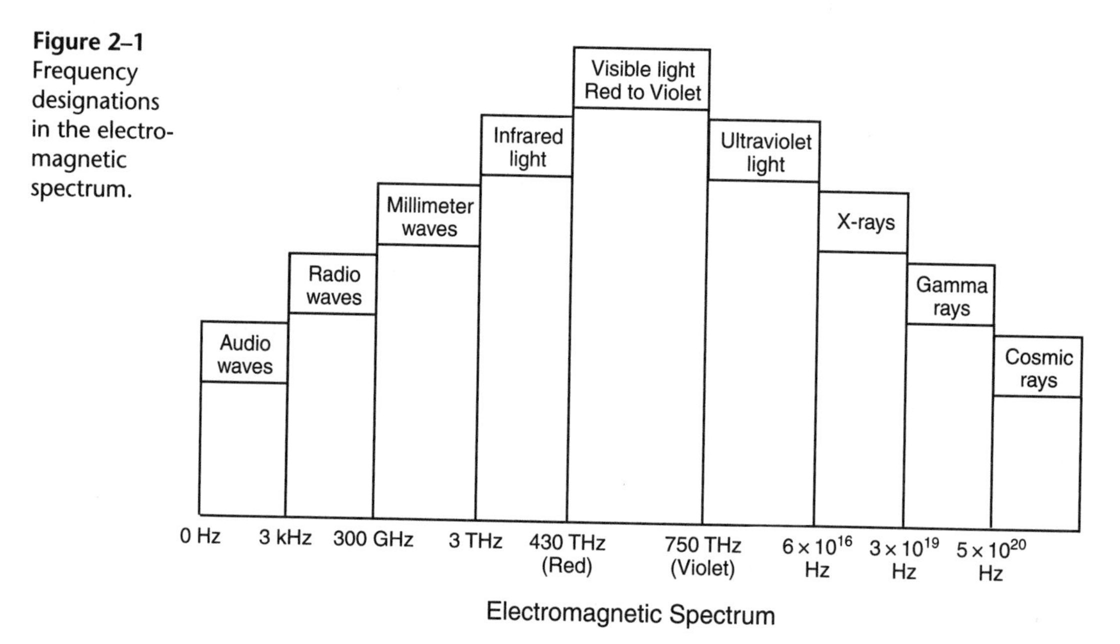

Figure 2–1 Frequency designations in the electromagnetic spectrum.

A typical block diagram of an electronic communications system is shown in Figure 2–2. Electronic communication uses electrical energy to transmit the information to be communicated. Information can be defined as any physical pattern that is meaningful to both sender and receiver. The source of the information can be either a person or a machine. The original form of the information can be a written document, a sound pattern such as human speech, or a light pattern such as a picture. The transmitter converts the information from its original form to some kind of signal, usually an electrical or electromagnetic signal, so that it can travel through a channel, such as cables, or through space, to a receiver. The receiver converts the electrical signal back to its original form so that it can be understood by a person or a machine. In this chapter, we will study communications system parameters, relevant electricity/electronics concepts, and different modulation and multiplexing techniques.

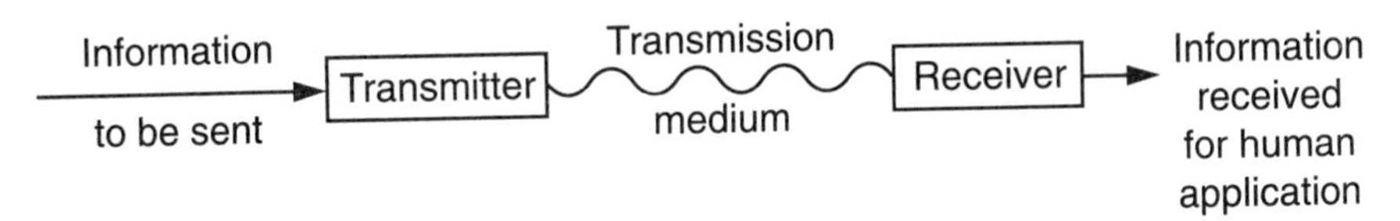

Figure 2–2 Block diagram of an electronic communications system.

COMMUNICATIONS SYSTEM PARAMETERS

The cost of a system interacts with and relates to each of the requirements listed in the following sections. Obviously, the user always wants the most performance at the least cost, with good reliability and convenience. This is measured in terms of *price to performance* ratio. The type of information to be transmitted and bandwidth requirement are prime system parameters that determine network design and architecture. The other requirements fall behind them.

Type of Information

Each type of information—data, voice, and video—has specific transmission system requirements. The major requirement is that voice and video communications require a constant rate of information transfer and cannot tolerate any delays, which is in direct contrast with bursty data communications that transfer information at a variable rate and on demand. Networks have traditionally been separated by the type of information because of these significant differences in traffic characteristics. Networks have evolved; for example, the PSTN that was originally designed for voice carries data too. The next-generation public network is a packet-based infrastructure that integrates data, voice, and video communications.

Bandwidth

Bandwidth (BW) is the range of frequencies that can be transmitted with minimal distortion. The BW is equal to the rate of information transfer, which is the amount of information that is communicated from the source to the destination in a fixed amount of time, typically one second. BW is also a measure of the transmission capacity of the communications medium. There is a general rule that relates BW and information capacity. *Hartley's Law*, which states that the amount of information that can be transmitted in a given time is directly proportional to bandwidth, is represented by Equation 2–1.

$$I = ktBW \tag{2–1}$$

where I = amount of information that can be transmitted

k = a constant that depends on the type of modulation

t = transmission time in seconds

BW = channel bandwidth

From the above equation, it is clear that the greater the channel bandwidth, the greater the amount of information you can transmit in a given time. You can still transmit the same amount of information over a narrower channel except that it will take longer. As you progress through this book, you will see that bandwidth has started to drive the evolution of computing. High-bandwidth applications include Web browsing, e-commerce, audio and video streaming, real-time document sharing, videoconferencing, on-line gaming, and digital TV. As the movement for transmission of data, voice, and video traffic over the same networks continues to gain momentum, the demand for bandwidth keeps growing.

For digital devices, the bandwidth is expressed in bits per second (bps). In most cases, the bandwidth is the same as channel frequency, so 100 MHz is analogous to 100 Mbps. For analog devices, bandwidth is expressed in cycles per second, or Hertz (Hz), and the minimum required channel BW is determined by the difference between upper- and lower-frequency limits of the signal, as indicated in Figure 2–3. For example, since most human speech falls in the frequency range of 200–3000 Hz, the minimum bandwidth requirement is 2800 Hz, but 4000 Hz is allotted.

Figure 2–3
Concept of bandwidth.

$$
\begin{aligned}
BW &= f_2 - f_1 \\
&= 3000 - 200 \\
&= 2800 \text{ Hz}
\end{aligned}
$$

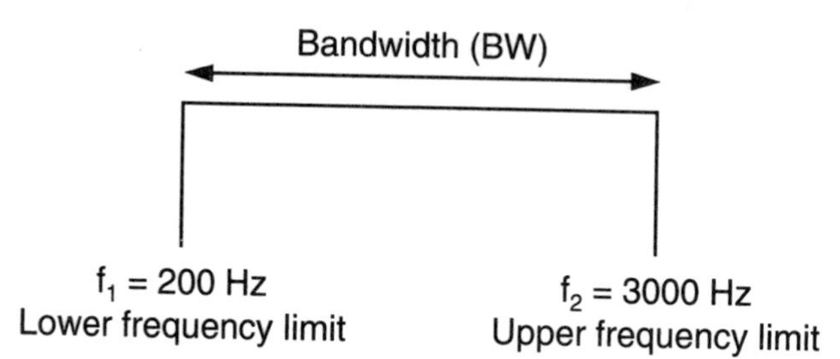

Broadband versus Baseband

There are two types of transmission systems: *broadband* and *baseband.* The term **broadband**, which originated in the CATV industry, involves the simultaneous transmission of multiple channels over a single line. The channel allocation is based on different multiplexing schemes that we will study later. **Baseband** refers to the original frequency range of a signal before it is modulated into a higher and more efficient frequency range, but the term is more commonly used to indicate digital transmission of a single channel at a time. It offers advantages such as low cost and ease of installation as well as maintenance, and most importantly, high transmission rates. Most data communications use baseband transmission; however, the push is toward broadband communication that integrates voice, data, and video over a single line.

Synchronous versus Asynchronous

Communications are designated as *synchronous* or *asynchronous* depending on how the timing and framing information is transmitted. The framing for **asynchronous** communication is based on a single character, while that for **synchronous** communication is based on a much bigger block of data. Synchronous signals require a coherent clock signal called a data clock between the transmitter and receiver for correct data interpretation. The clock recovery circuit in the receiver extracts the data clock signal frequency from the stream of incoming data and data synchronization is achieved. Also, there are a special series of bits called synchronization (SYN) characters that are transmitted at the beginning of every data block to achieve synchronization. Each data block represents hundreds or even thousands of data characters. Typically, two 8-bit SYN codes signal the start of a transmission. At the end of the block is a special code (ETX) signaling the end of the transmission. One or more error codes usually follow. Thus, such systems are more expensive and complex but extremely efficient, since all the bits transmitted are message bits except the bits in the synchronization and error detection characters.

Asynchronous transmission incorporates the use of framing bits—start and stop bits—to signal the beginning and end of each data character because the data clock signals at the transmitter and receiver are not synchronized—although they must operate at the same frequency. It is more cost-effective but inefficient compared with synchronous transmission. For every character that is transmitted, the asynchronous transmission system adds a start bit and a stop bit, and some also add a parity bit for error-detection. **Efficiency of transmission** is the ratio of the actual message bits to the total number of bits, including message and control bits, as shown in Equation 2–2. In any transmission, the synchronization, error detection, or any other bits that are not messages are collectively referred to as **overheads**, represented in Equation 2–3. The higher the overheads, the lower the efficiency of transmission, as shown in Equation 2–4.

$$\text{Efficiency} = \frac{M}{M+C} \times 100\% \qquad (2\text{–}2)$$

$$\text{Overhead} = \left(1 - \frac{M}{M+C}\right) \times 100\% \qquad (2\text{–}3)$$

where M = Number of message bits
C = Number of control bits

In other words,

$$\text{Efficiency}\ \% = 100 - \text{Overhead}\ \% \qquad (2\text{–}4)$$

Example 2–1

Problem

Find the efficiency and overhead for an asynchronous transmission of a single 7-bit ASCII (American Standard Code for Information Interchange) character with one start bit, one stop bit, and one parity bit per character.

Solution

$$\text{Efficiency} = \frac{7}{7+3} \times 100\%$$
$$= 70\%$$
$$\text{Overhead }\% = 100 - \text{Efficiency }\%$$
$$= 30\%$$

Simplex, Half-Duplex, and Full-Duplex

Simplex refers to communications in only one direction from the transmitter to the receiver. There is no acknowledgement of reception from the receiver, so errors cannot be conveyed to the transmitter. **Half-duplex** refers to two-way communications but in only one direction at a time. **Full-duplex** refers to simultaneous two-way transmission. For example, a radio is a simplex device, a walkie-talkie is a half-duplex device, and certain computer video cards are full-duplex devices. Similarly, radio or TV broadcast is a simplex system, transfer of inventory data from a warehouse to an accounting office is a half-duplex system, and videoconferencing represents a full-duplex application.

Serial versus Parallel

Serial transmission refers to the method of transmitting the bits (0s and 1s) one after another along a single path. It is slow, cost-effective, has relatively few errors, and is practical for long distances. **Parallel** transmission is described as transmitting a group of bits at a single instant in time, which requires multiple paths. For example, to transfer a byte (8-bit data word), parallel transmission requires eight separate wires or communications channels. It is fast (higher data transfer rate) but expensive, and it is practical only for short distances. Most transmission lines are serial, whereas information transfer within computers and communications devices is in parallel. Therefore, there must be techniques for converting between parallel and serial, and vice versa. Such data conversions are usually accomplished by a **Universal Asynchronous Receiver Transmitter (UART)**.

Figure 2–4 is a general block diagram of a UART. At the transmit section, parallel data from the computer, usually in 8-bit words, is put on an internal data bus. Before being transmitted, the data is stored first in a buffer storage register and then sent to a shift register. A shift register is a sequential logic circuit made up of a number of flip-flops connected

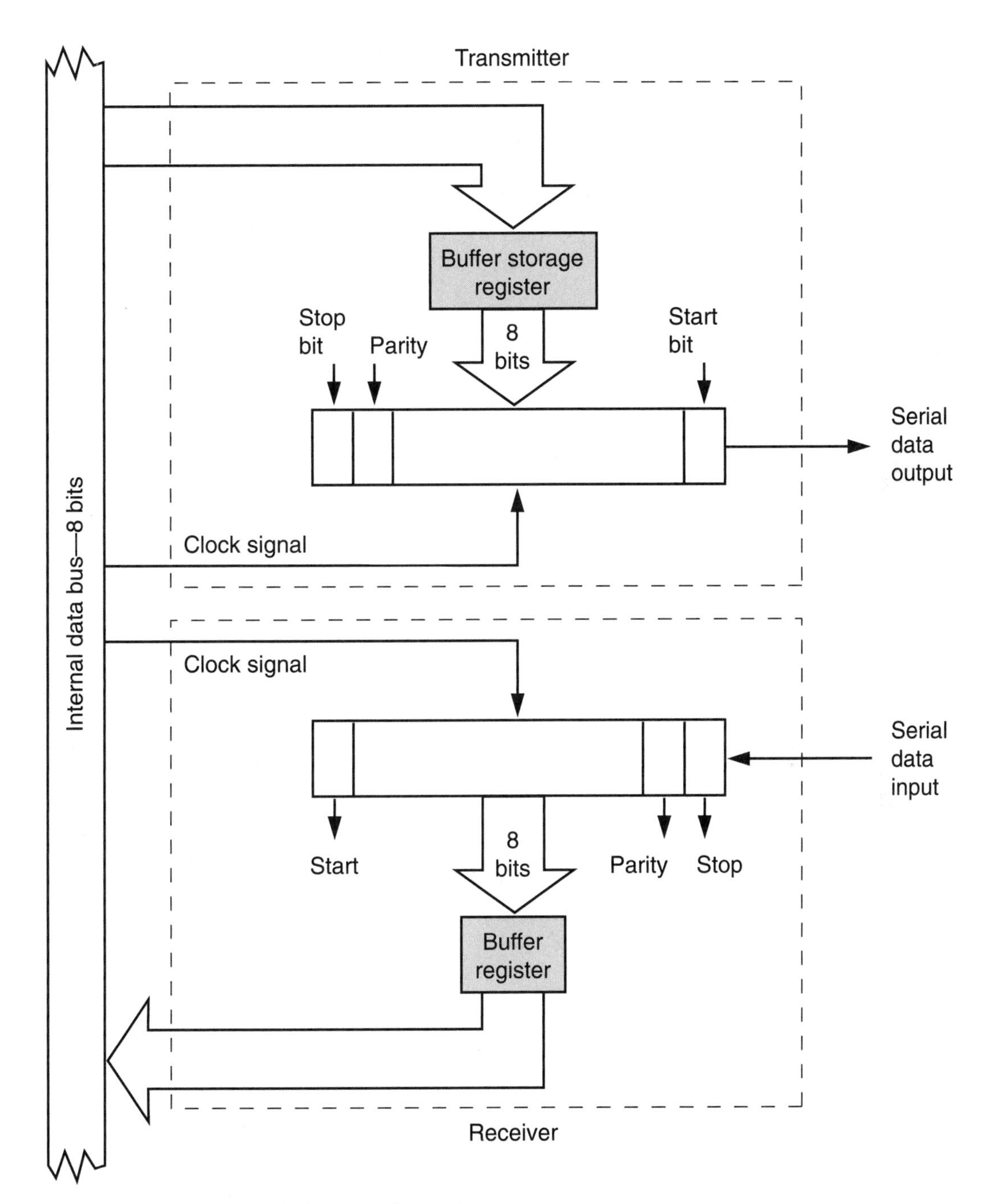

Figure 2–4 General block diagram of a UART.

in cascade, as shown in Figure 2–5. A clock signal shifts the data out serially, one bit at a time. The internal circuitry adds start and stop bits and a parity bit. The start and stop bits signal the beginning and end of the word, and the parity bit is used to detect error. The resulting serial data is transmitted one bit at a time to a serial interface.

At the receive section of the UART, serial data is shifted into a shift register where the start, stop, and parity bits are stripped off. The remaining data is transferred to a buffer storage register and then onto the internal data bus. The clock and control logic circuits in the UART control all internal shifting and data transfer operations under the direction of control signals from the computer. All this circuitry is typically contained within a single Integrated Circuit (IC).

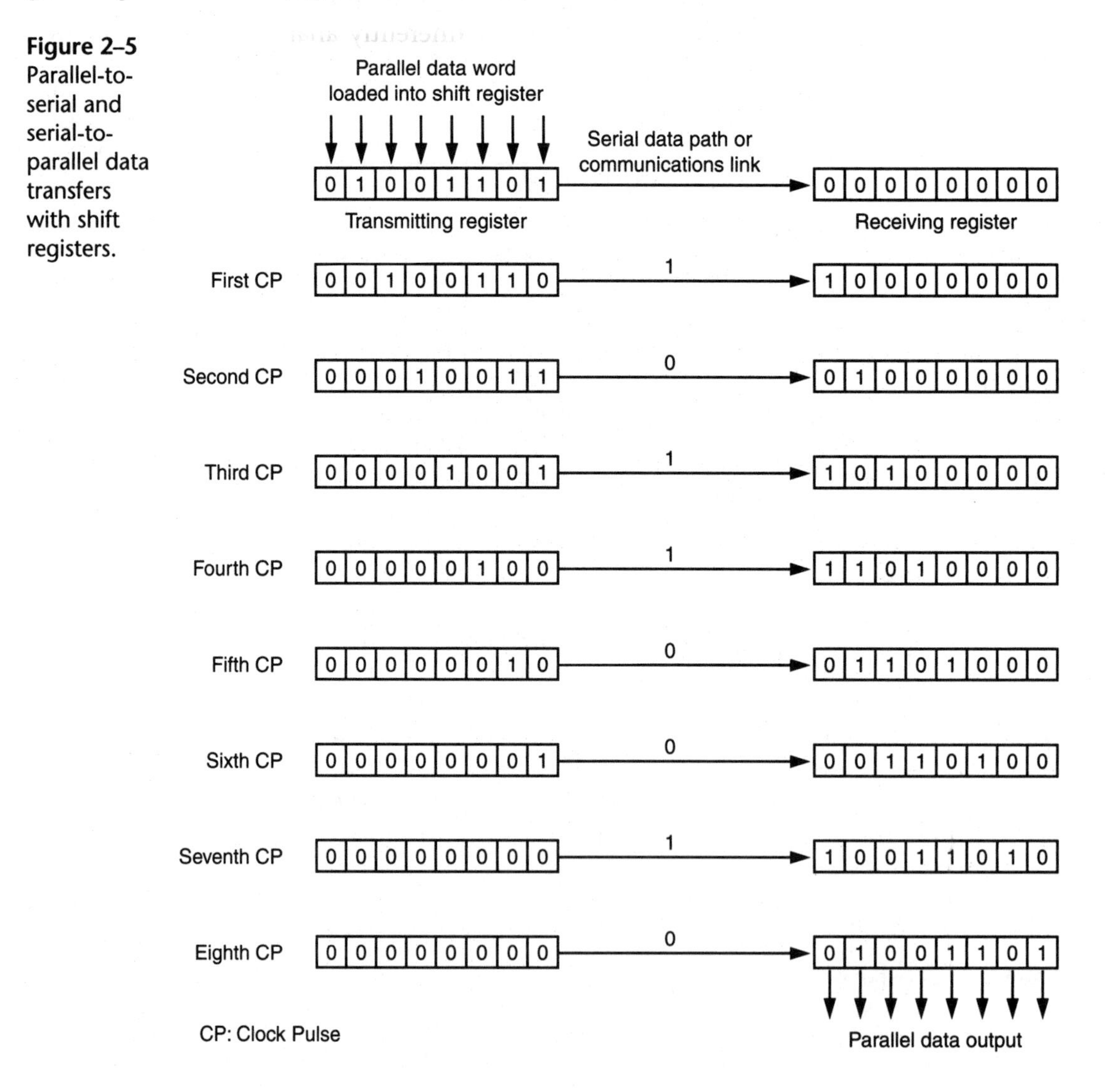

Figure 2–5 Parallel-to-serial and serial-to-parallel data transfers with shift registers.

Analog versus Digital

Information that needs to be communicated may be in analog or digital form. **Analog** signals are continuously varying quantities, while **digital** signals are discrete quantities, most commonly binary (On or Off, High or Low, 1 or 0), as shown in Figure 2–6. Voices, images, and temperature readings from a sensor are all examples of analog data. In digital transmission, as all information is reduced to a stream of 0s and 1s, you can use a single network for voice, data, and video. Digital circuits are cheaper, more accurate, more reliable, have fewer transmission errors, and are easier to maintain than analog circuits. A vast infrastructure exists for analog signaling, and much of it can be adapted to transport digital signals as well. The public telephone network, cable TV infrastructure, and practically every form of wireless communication are inherently analog transmission media that have been adapted for digital transmissions.

Figure 2–6 Analog and digital signals.

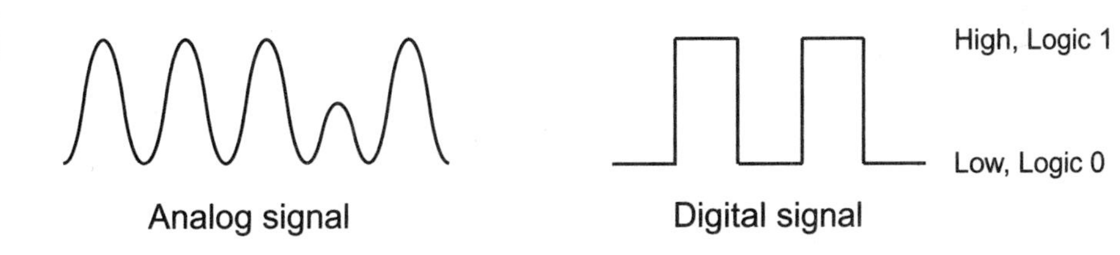

Analog data can be encoded as an analog signal, for example, cassette tape player, and audio as well as video components of a TV program. Digital data is regularly represented by digital signals, for example, e-mail. Also, analog data is commonly encoded with digital signals. When you scan an image or capture a sound on the computer, you are converting analog data to digital signals. This analog-to-digital conversion is usually accomplished with a special device or process referred to as a **codec**, which is short for coder-decoder. The conversion process is explained later in this chapter.

Digital transmission has replaced analog in most parts of the PSTN except the telephone **local loop**, which is a pair of copper wires that runs from a telephone to a local switching station. Although voice is the primary signal carried by the local loop, this network is now widely used to carry digital information, or data, as well. There are two primary problems in transmitting digital data over the telephone network:

1. If a binary signal were applied directly to the telephone network, it simply would not pass. The reason is that binary signals are usually switched DC pulses, that is, the 1s and 0s are represented by pulses of a single polarity, usually positive; and the transformers, capacitive coupling, and other AC circuitry virtually ensure that no DC signals get through. The telephone line is designed to carry only AC analog signals that are usually of a specific frequency range: 300–3000 Hz is most common.

2. Binary data is usually transmitted at high speeds and this high-speed data would essentially be filtered out by the system with its limited bandwidth. A *filter* is a tuned device that passes certain desirable frequencies and rejects the others. Figure 2–7 provides a graphic representation of different types of filters.

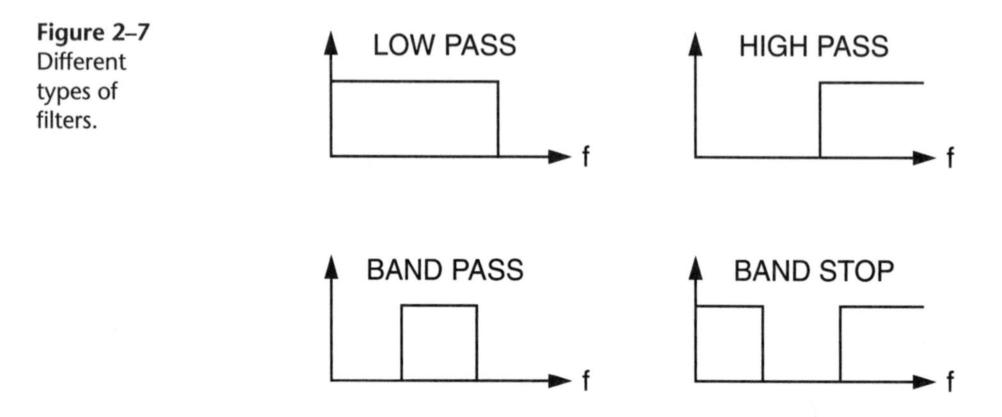

Figure 2–7 Different types of filters.

So the question is: How do we transmit data over the local loop? The answer is by using a **modem (MOdulator/DEModulator)**, which converts digital signals that it receives from a serial interface of a computer into analog signals for transmission over the telephone local loop, and vice versa. One can connect a computer over a telephone line to a remote server by using a modem.

Figure 2–8 shows block diagrams for two different types of signals, analog and digital, transmitted over different channels. In Figure 2–8 (a), an analog signal is sent over a single channel with no modulation. A typical example would be an ordinary public-address system, with a microphone, an amplifier, and a speaker, using twisted-pair wire as a channel. Figure 2–8 (b) shows analog transmission using modulation and demodulation, of which broadcast radio and television are good examples. Figure 2–8 (c) and (d) start with a digital source such as a data file from a computer. In (c), the channel can handle the digital signal directly, but in (d), the channel is analog so an intermediate step is the modulation-demodulation process accomplished by a modem. Examples include a radio channel and data transmission over an ordinary telephone connection. Lastly, Figure 2–8 (e) and (f) show an analog signal that is digitized at the transmitter and converted back to analog form at the receiver. The difference between these two systems is that in (e), the transmission is digital, while in (f), with the transmission channel being analog, modulation and demodulation are required.

Let us consider the scenario of transmitting information between a computer and a telephone line, which is depicted in Figure 2–9. First, a UART chip or IC, which resides in the Central Processing Unit (CPU) of a computer, performs parallel-to-serial and serial-to-parallel data transfers, thereby providing an interface between a computer and a modem. The modem performs digital-to-analog and analog-to-digital conversion, and it interfaces directly with an analog, serial, telephone line. The different modulation schemes utilized by modems are discussed later in this chapter.

Figure 2–8 Analog and digital transmissions.

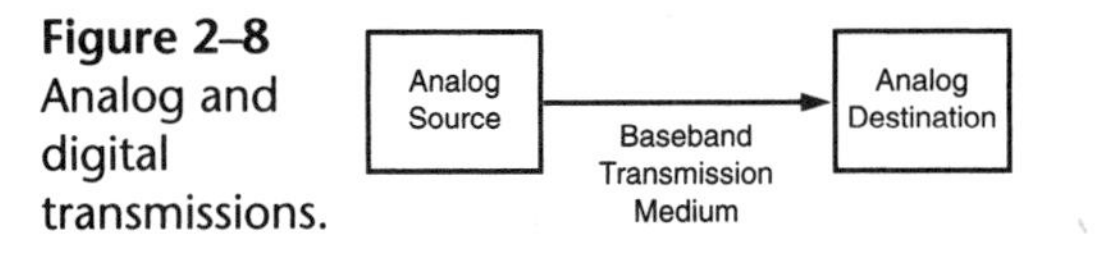

(a) Analog Signal and Baseband Transmission

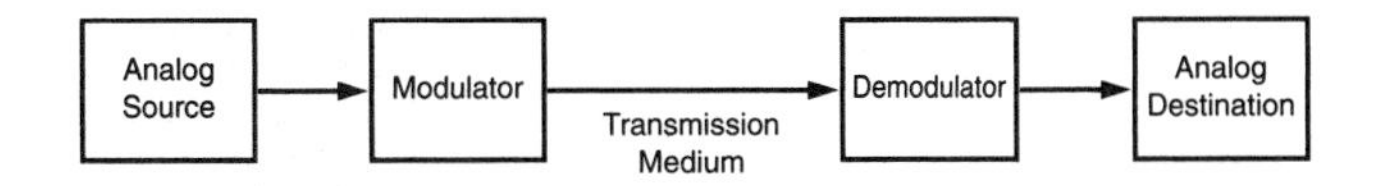

(b) Analog Transmission Using Modulation and Demodulation

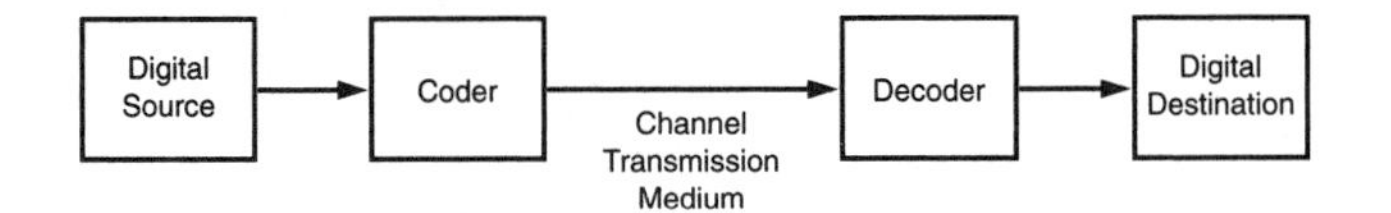

(c) Digital Signal Transmitted on Digital Channel

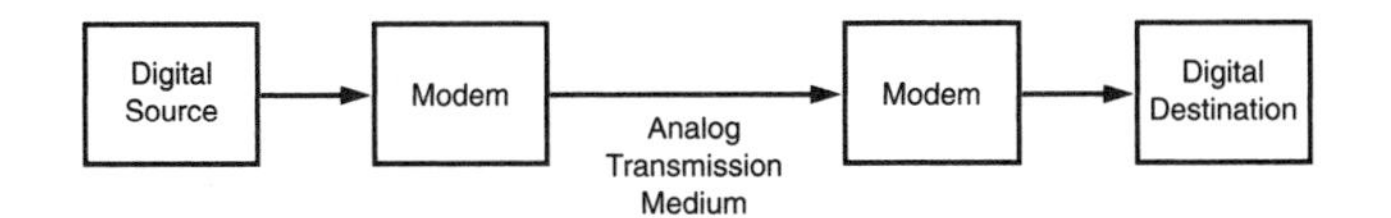

(d) Digital Signal Transmitted by Modem

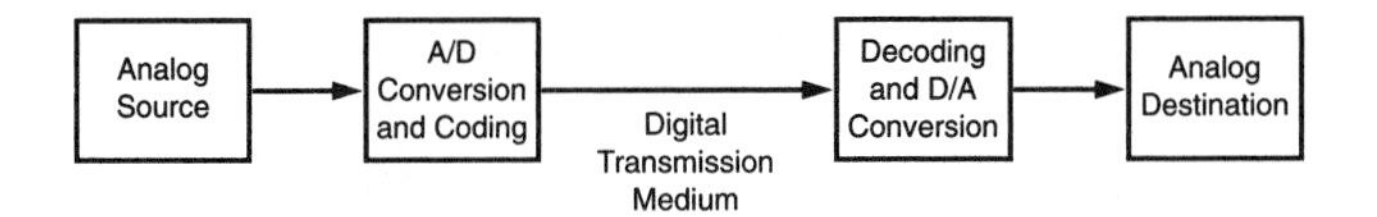

(e) Analog Signal Transmitted Digitally

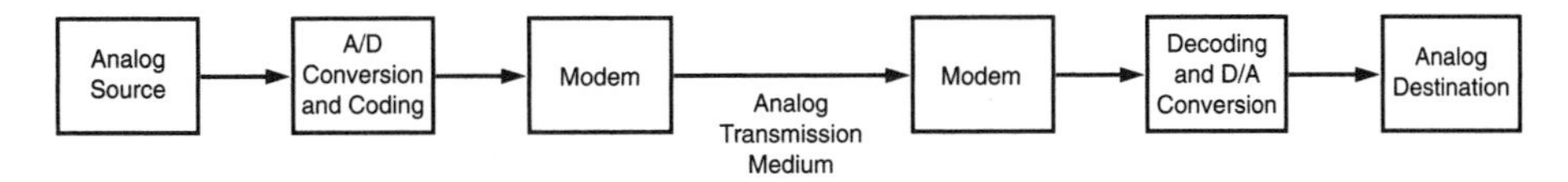

(f) Analog Signal Digitized and Transmitted by Modem

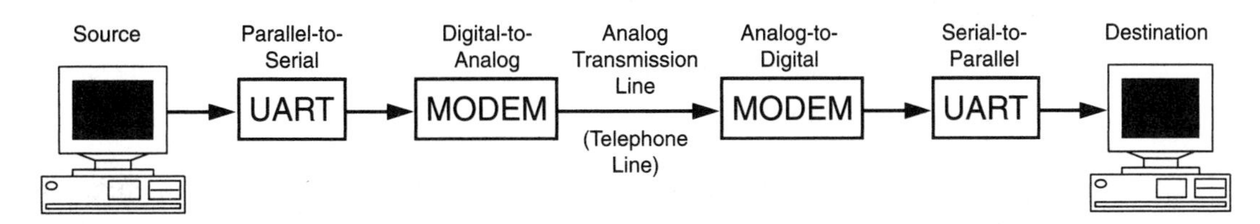

Figure 2–9 Interfacing a computer with a telephone line.

Noise

The three factors, bandwidth, power, and noise determine the theoretical capacity of a channel. Consisting of undesired, usually random, variations that interfere with the desired signals and inhibit communication, **noise** originates both in the channel and in the communication equipment. Although it cannot be eliminated completely, its effects can be reduced by various means. It is helpful to divide noise into two types: *internal noise,* which originates within the communication equipment, and *external noise,* which is a property of the channel.

External noise consists of man-made noise, atmospheric, and space noise. Man-made noise is generated by equipment that produces sparks, such as automobile engines and electric motors with brushes. Also, any equipment with fast rise-time voltage or current can generate interference, like light dimmers and computers. A typical solution for a computer, for instance, involves shielding and grounding the case and all connecting cables and installing a low-pass filter on the power line where it enters the enclosure.

Atmospheric noise is often called *static* because lightning, which is a static electricity discharge, is its principal source. Since it occurs in short, intense bursts with relatively long periods of time between bursts, it is often possible to improve communication by simply disabling the receiver for the duration of the burst. This technique is called *noise blanking.* Space noise is mostly solar noise, which can be a serious problem with satellite reception when the satellite is in line between the antenna and the sun. It is more important at higher frequencies because most of the space noise at lower frequencies is absorbed by the upper atmosphere. On the other hand, atmospheric noise dominates at lower frequencies.

Internal noise is generated in all electronic equipment—both passive components like resistors and cables, and active devices like diodes and transistors. *Thermal noise* is produced by the random motion of electrons in a conductor due to heat. It is an equal mixture of noise of all frequencies, and is sometimes called *white noise,* by analogy with white light, which is an equal mixture of all colors. The term *noise* is often used alone to refer to this type of noise, which is found everywhere in electronic circuitry. The noise power in a conductor is a function of its temperature, as shown by Equation 2–5:

$$P_N = kTBW \tag{2–5}$$

where P_N = internal noise power in Watts

k = Boltzmann's constant, 1.38×10^{-23} Joules/Kelvin (J/K)

T = absolute temperature in Kelvin (K)

BW = operating bandwidth in Hertz

The temperature in degrees Kelvin can be found by adding 273 to the Celsius temperature. The previous equation shows that noise power is directly proportional to bandwidth, which means that high bandwidth communications are associated with higher noise. The only way to reduce noise is to decrease the temperature or the bandwidth of a circuit, or both. Amplifiers used with very low signal levels are often cooled artificially to reduce noise. The technique is called *cryogenics* and may involve, for example, cooling the first stage of a receiver for radio astronomy by immersing it in liquid nitrogen. The other method of noise reduction, bandwidth reduction, will be referred to many times throughout this book. Using a bandwidth greater than required for a given application is simply an invitation to problems with noise.

Shot noise has a power spectrum that resembles that for thermal noise by having equal energy in every hertz of bandwidth, at frequencies from DC into the GHz region. It is created by random variations in current flow in active devices such as transistors and semiconductor diodes. *Excess noise,* also called *flicker noise* or *pink noise,* varies inversely with frequency. It is rarely a problem in communication circuits, because it declines with increasing frequency and is usually insignificant above approximately one kHz.

The main reason for studying and calculating noise power or voltage is the effect that noise has on the desired signal. In analog systems, noise makes the signal unpleasant to watch or listen to, and in extreme cases, difficult to understand. Once noise and distortion are present, there is usually no way to remove them. In addition, the effects of these impairments are cumulative: noise will be added in the transmitter, the channel, the receiver and, if the communications system involves several trips through amplifiers and channels—as in a long-distance telephone system—the noise will gradually increase with increasing distance from the source. In digital transmission of analog signals, the conversion of infinitely variable analog signal to digital form introduces error. This will inevitably result in the loss of some information and the creation of a certain amount of noise and distortion.

In communications, it is not really the amount of noise that concerns us, but rather the amount of noise compared to the level of the desired signal. That is, it is the *ratio* of signal to noise power that is important, rather than the noise power alone. This **Signal-to-Noise Ratio (SNR)**, usually expressed in decibel (dB), is one of the most important specifications of any communication system. The decibel is a logarithmic unit used for comparisons of power levels or voltage levels. In order to understand the implication of dB, it is important to know that a sound level of zero dB corresponds to the threshold of hearing, which is the smallest sound that can be heard. A normal speech conversation would measure about 60 dB. The SNR is given by Equation 2–6:

$$\text{SNR (dB)} = 10\log_{10}\left(\frac{P_S}{P_N}\right) \qquad (2\text{–}6)$$

where P_S is the signal power

P_N is the noise power

The *Hartley-Shannon Theorem* (commonly referred to as *Shannon's Limit*) states that the maximum data rate for a communication channel is governed by a channel's bandwidth and SNR. Typical values of SNR range from about 10 dB for barely intelligible speech to

Example 2–2

Problem

A receiver has an input power of 42.2 mW while the noise power is 33.3 μW. Find the SNR for the receiver.

Solution

$$\text{SNR (dB)} = 10\log_{10}\left(\frac{P_S}{P_N}\right)$$

$$= 10\log_{10}\left(\frac{42.2}{.0333}\right)$$

$$= 31.03 \text{ dB}$$

90 dB or more for compact-disc audio systems. A SNR of zero dB would mean that the noise has the same power as the signal, which would be absolutely unacceptable for any transmission system. Another quantity that is used to determine the signal quality is the *noise figure (NF)* also called the *noise factor,* which is related to the *noise ratio (NR).* These can be computed by using Equations 2–7, 2–8, and 2–9.

$$\text{NR} = \frac{(\text{SNR})_{\text{input}}}{(\text{SNR})_{\text{output}}} \qquad (2\text{–}7)$$

where $(\text{SNR})_{\text{input}}$ is the signal-to-noise ratio at the input

$(\text{SNR})_{\text{output}}$ is the signal-to-noise ratio at the output

$$\text{NF} = 10 \log \text{NR} \qquad (2\text{–}8)$$

Therefore,

$$\text{NF (dB)} = \text{SNR}_{\text{input}}\text{ (dB)} - \text{SNR}_{\text{output}}\text{ (dB)} \qquad (2\text{–}9)$$

An amplifier or receiver will always have more noise at the output than at the input because the amplifier or receiver generates internal noise, which will be added to the signal. And even though the signal may be amplified, that noise will be amplified along with it. Since the SNR at the output will be less than the SNR at the input, the noise figure will always be greater than 1. A receiver that contributes zero noise to the signal would have a noise figure of 1, or 0 dB; but such a noise figure is not attainable in practice. The lower the noise figure, the better the amplifier.

Data and voice signals exhibit entirely different tolerances to noise. Data signals may be satisfactory in the presence of white noise, but the same can be bothersome to

Example 2–3

Problem

Suppose the SNR at the input of an amplifier is 25 dB and its NF is 10 dB. Find the SNR at the amplifier output.

Solution

$$NF\ (dB) = SNR_{input}\ (dB) - SNR_{output}\ (dB)$$

$$SNR_{output}\ (dB) = SNR_{input}\ (dB) - NF\ (dB)$$

$$= (25 - 10)\ dB$$

$$= 15\ dB$$

humans. On the other hand, impulse noise (clicks, pops, or sometimes frying noise) will destroy a data signal on a circuit but might be acceptable for speech communication.

Digital systems are not immune from noise and distortion, but it is possible to reduce their effect. Consider the simple digital signal shown in Figure 2–10. Suppose that a transmitter generates 1 V for binary one, and 0 V for a binary zero. The receiver examines the signal in the middle of the pulse, and has a decision threshold at 0.5 V; that is, it considers any signal with amplitude greater than 0.5 V to be a one, and any amplitude less than that to represent a zero. This is achieved mainly by a *quantizer* circuit at the receiver end, whose function is to determine whether the incoming digital signal has a voltage level corresponding to binary 0 or binary 1. The basic design concern is to minimize the impact of channel noise at the receiver.

Figure 2–10 (a) shows the signal as it emerges from the transmitter, and Figure 2–10 (b) shows it after its passage through a channel that adds noise and distorts the pulse. In spite of the noise and distortion, the receiver has no difficulty deciding correctly whether the signal is a zero or a one. Since the binary value of the pulse is the only information in the signal, the distortion has had no effect on the transmission of information.

The received signal of Figure 2–10 (b) could now be used to generate a new pulse train to send further down the channel. This receiver-transmitter combination, which is called a *repeater* and illustrated in Figure 2–10 (c), has not only avoided the addition of any distortion of its own, but has also removed the effects of noise and distortion that were added by the channel preceding the repeater. Unfortunately, since noise is random, it is possible for a noise pulse to have any amplitude, including one that will cause a transition to the wrong level. Extreme distortion of pulses can cause errors as demonstrated in Figure 2–11. Errors can never be eliminated completely, but by judicious choice of such parameters as signal levels and bit rates, it is possible to reduce the probability of error to a very small value. There are even techniques to detect and correct some of the errors.

Figure 2–10 Removal of noise and distortion from digital signal.

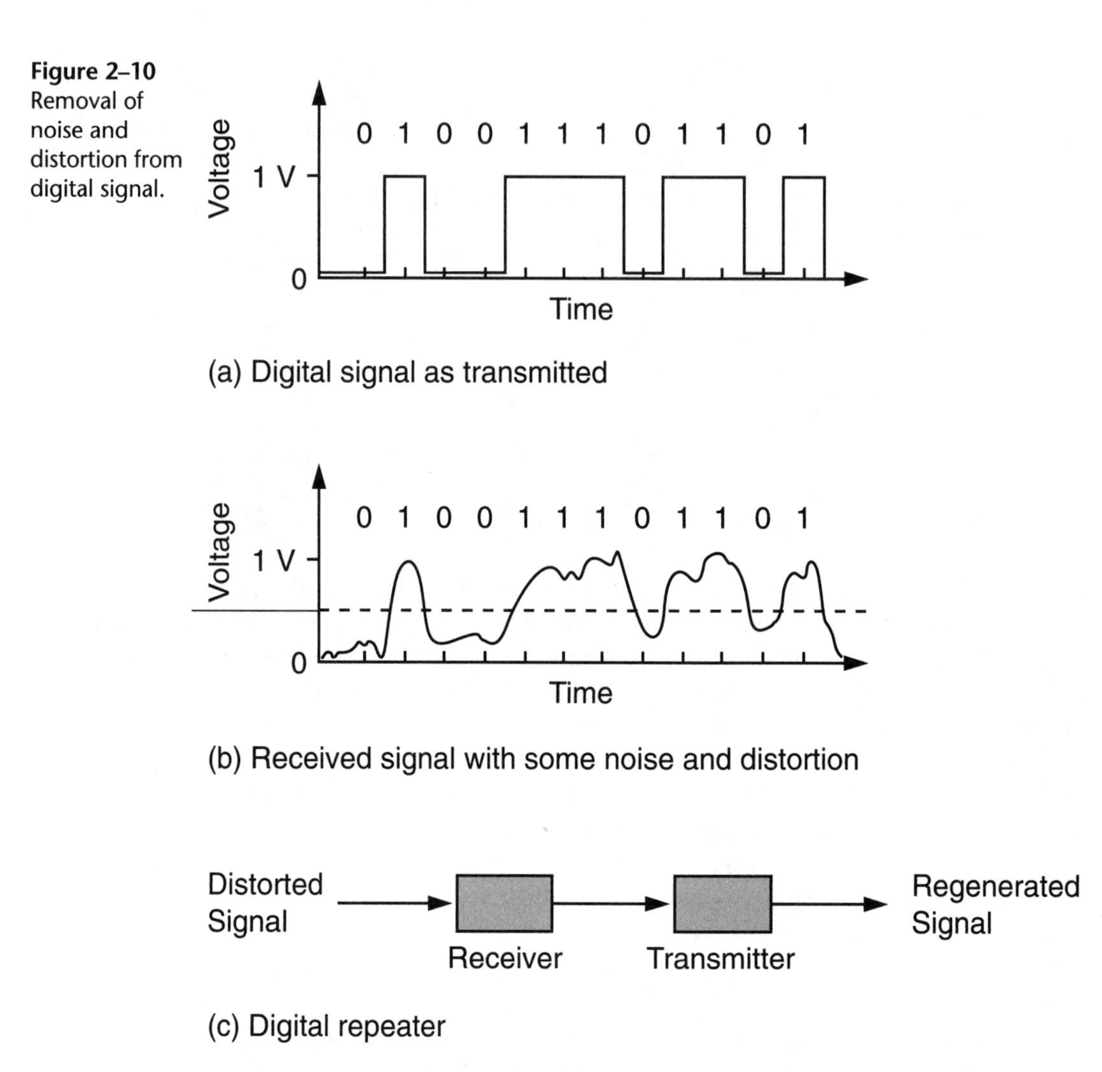

While signal-to-noise ratio is used as a performance measure for analog systems, the **Bit Error Rate (BER)** is a prime factor in a digital system. It is the number of bits in error expressed as a portion of transmitted bits. For example, a BER of 10^{-9} (which equals $1/10^9$) means one bit is in error for each one billion bits received.

MODULATION

Modulation is a means of controlling the characteristics of a signal in a desired way. The modulation is done at the transmitter, while an inverse process, called *demodulation* or detection, takes place at the receiver to restore the original baseband signal. There are many ways to modulate a signal, such as Amplitude Modulation (AM), Frequency Modulation

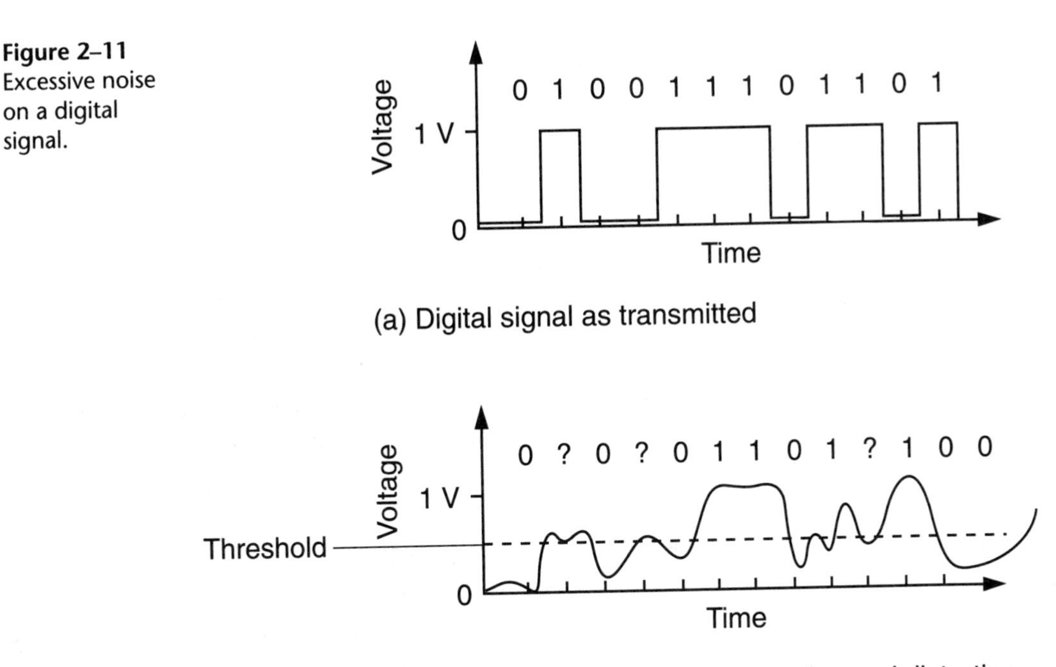

Figure 2–11 Excessive noise on a digital signal.

(FM), Phase Modulation (PM), and Pulse Modulation. Both AM and FM are used in radio broadcast. Pulse modulation is mainly used for analog-to-digital conversion. In modulation, the amplitude, frequency, or phase of a carrier wave is changed in accordance with the modulating signal in order to transmit information. The resultant is called a modulated wave. This concept is illustrated in Figure 2–12.

A carrier, which is usually a sine wave, is generated at a frequency much higher than the highest modulating signal frequency. Equation 2–10 is a general equation for a sine wave carrier:

$$e(t) = E_C \sin(\varpi_C t + \theta) \tag{2–10}$$

where $e(t)$ = instantaneous amplitude or voltage of the sine wave at time t

E_C = maximum amplitude or peak voltage

ϖ_C = frequency in radians per second

t = time in seconds

θ = phase angle in radians

In the mathematics concerning modulation, frequency is expressed in radians per second to make the equation simpler. Of course, frequency is usually given in Hertz rather than in radians per second when practical devices are being discussed, but it is easy to convert between the two systems using $\varpi = 2\Pi f$. In modulation, the instantaneous amplitude

Figure 2–12 Concept of modulation.

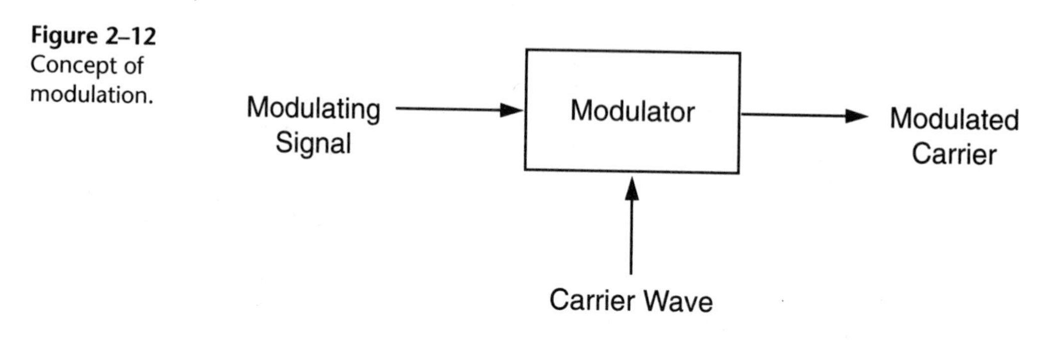

of the modulating signal is used to vary some parameter of the carrier. The parameters that can be changed are amplitude E_C, frequency ϖ_C, and phase θ. Combinations are also possible; for example, many schemes for transmitting digital information use both amplitude and phase modulation.

Fourier Theorem and Spectral Analysis

The sine wave, discovered by the French mathematician Baron Jean Baptiste Joseph Fourier during the early 19th century, is important because it is the fundamental waveform from which more complex waveforms can be created. The *Fourier theorem* states that any periodic function or waveform can be expressed as the sum of sine waves with frequencies at integer or harmonic multiples of the fundamental frequency of the waveform, with appropriate maximum amplitudes and phases. The theorem also specifies the procedure for analyzing a waveform to determine the amplitudes and phases of the sine waves that compromise it. Fourier's discovery, applied to a time-varying signal, can be expressed mathematically as follows (Equation 2–11):

$$f(t) = A_0 + A_1 \cos \varpi t + B_1 \sin \varpi t + A_2 \cos 2\varpi t + B_2 \sin 2\varpi t + A_3 \cos 3\varpi t + B_3 \sin 3\varpi t + \ldots \quad (2\text{–}11)$$

where $f(t)$ = any well-behaved function of time such as voltage $\upsilon(t)$ o1r current $i(t)$

A_n and B_n = real-number coefficients that can be positive, negative, or zero

ϖ = radian frequency of the fundamental

There are two general ways of looking at signals: the **time domain** and the **frequency domain**, which are two different representations of the same information. An oscilloscope displays signals in the time-domain and provides a graph of voltage with respect to time. Signals can also be described in the frequency domain, where amplitude or power is shown on one axis and frequency is displayed on the other. Amplitudes, when plotted graphically as a function of frequency, result in a plot or graph called the amplitude spectrum of the waveform or signal. A spectral representation of the square of the amplitude spectrum is called the power spectrum. A *Fourier analysis* or *spectrum analysis* done by a *spectrum analyzer* provides an amplitude spectrum of the signal.

As illustrated in Figure 2–13, a sine wave has energy only at its fundamental frequency for the frequency domain, so it can be shown as a straight line at that frequency. Frequency-domain representations are very useful in the study of communication systems; for instance, the bandwidth of a modulated signal can easily be found if the baseband signal can be represented in the frequency domain. An unmodulated sine-wave carrier would exist at only one frequency and so would have zero bandwidth. However, a modulated signal is no longer a single sine wave, and it will therefore occupy a greater bandwidth. The inverse of Fourier analysis is *Fourier synthesis*, which is a process of adding together the sine waves to recreate the complex waveform.

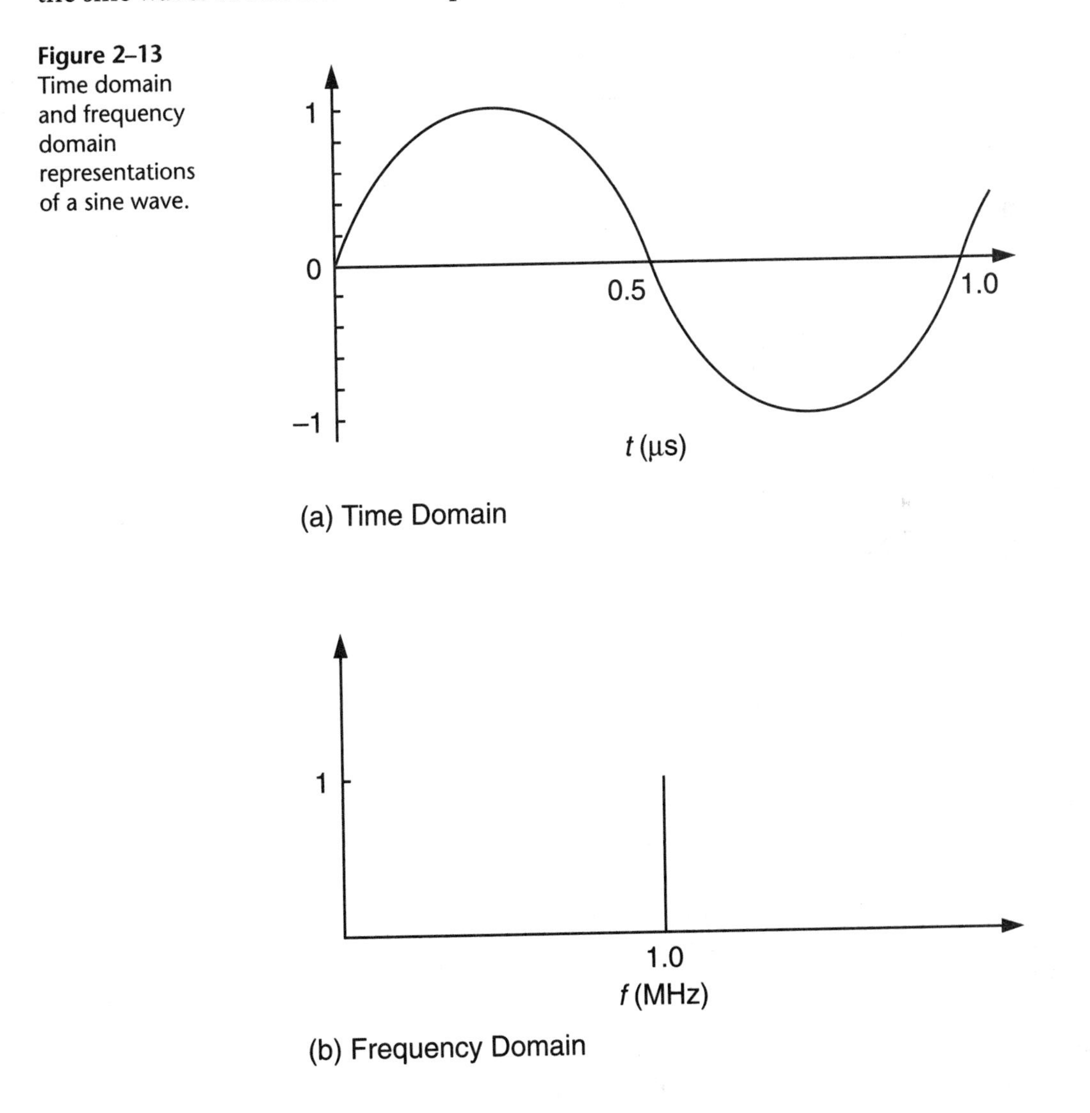

Figure 2–13 Time domain and frequency domain representations of a sine wave.

Amplitude Modulation (AM)

Amplitute Modulation (AM) is one of the oldest and simplest forms of modulation used for analog signals. In AM, an audio signal's varying voltage is applied to a carrier. Its amplitude changes in accordance with the modulating voice signal, while its frequency remains unchanged. This principle is shown in Figure 2–14.

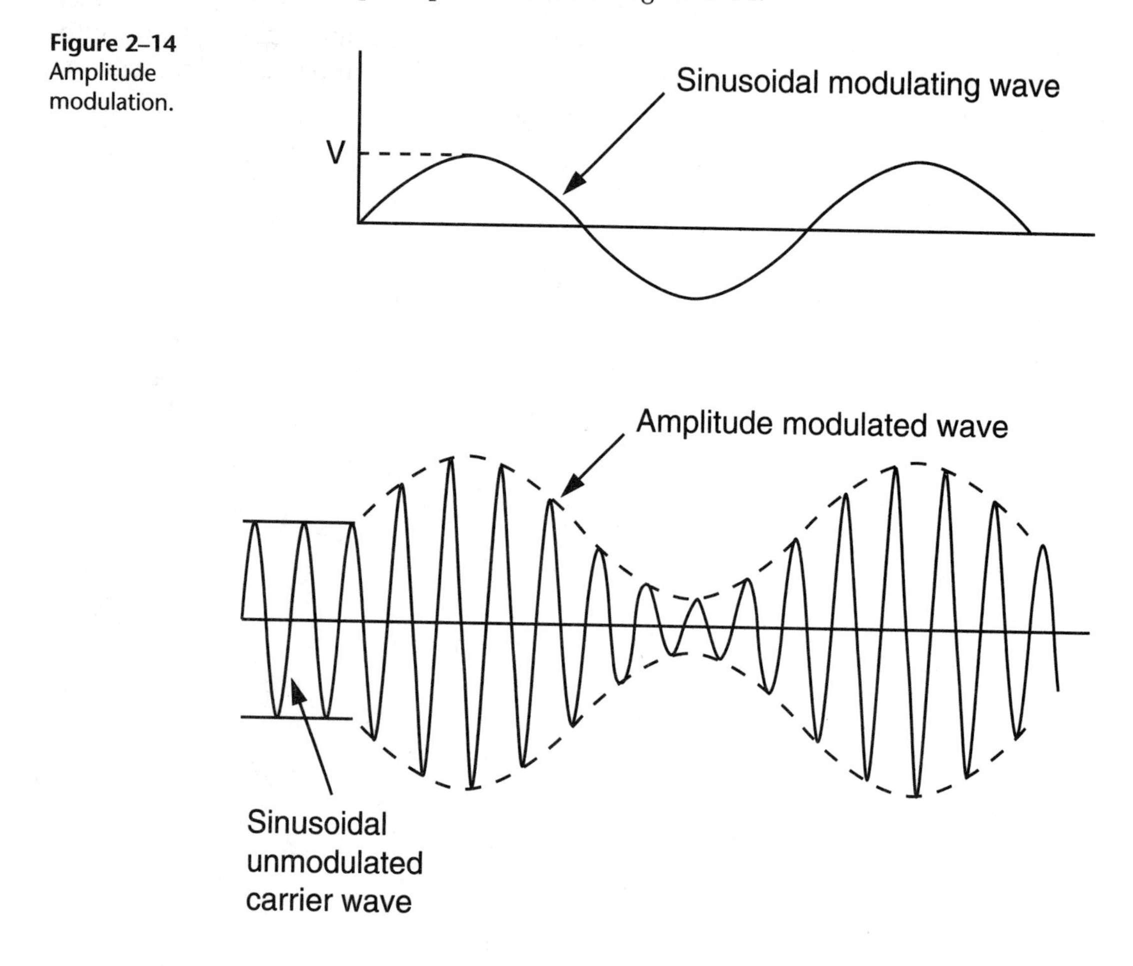

Figure 2–14 Amplitude modulation.

Frequency Modulation (FM)

Edwin Armstrong invented frequency modulation (FM) in 1935. In FM, frequency of the carrier changes in accordance with the amplitude of the input signal, but its amplitude remains unchanged as shown in Figure 2–15. This makes FM modulation more immune to noise than is AM and improves the overall signal-to-noise ratio of the communications system. Since the amplitude (voltage) stays the same, the output power of a FM signal is constant, unlike the varying AM power output. However, the amount of bandwidth necessary to trans-

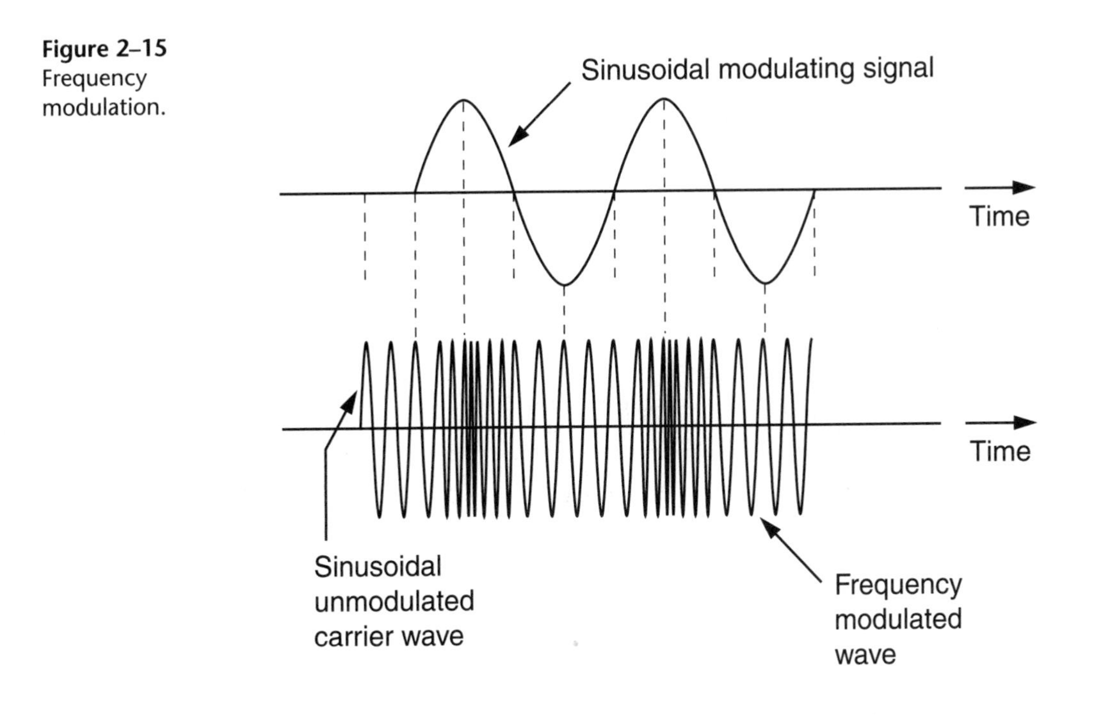

Figure 2–15
Frequency modulation.

mit a FM signal is greater than that necessary for AM—a limiting constraint for some systems. Also, the circuits used for FM are much more complex than those used for AM.

As an example, let us consider a carrier frequency, also called *center frequency*, of 1 MHz. Assume that because of FM modulation, the center frequency is made to deviate 75 kHz by the audio baseband signal. This change from center is the *frequency deviation*, which, in this example, is ±75 kHz or 150 kHz. The 75 kHz deviation is for the loudest audio signal with the greatest amplitude in the baseband modulating signal. The FM radio broadcast band is 88 to 108 MHz, with stations spaced every 200 kHz or 0.2 MHz. Examples of carrier frequencies are 92.1, 96.3, and 104.5 MHz. The 200 kHz spacing between carrier frequencies is needed to allow for a total swing of 150 kHz, with a guard band of 25 kHz on each side to prevent interference between adjacent stations.

Frequency Shift Keying

Frequency Shift Keying (FSK) is a popular implementation of FM for data applications and is used in low-speed modems. A carrier is switched between two frequencies—one for mark (logic 1) and the other for space (logic 0)—as indicated in Figure 2–16. There are always guard bands that reduce the effects of bleedover between adjacent channels, which is a condition more commonly referred to as *crosstalk*. For full-duplex operation, there are two pairs of mark and space frequencies. All these frequencies are well inside

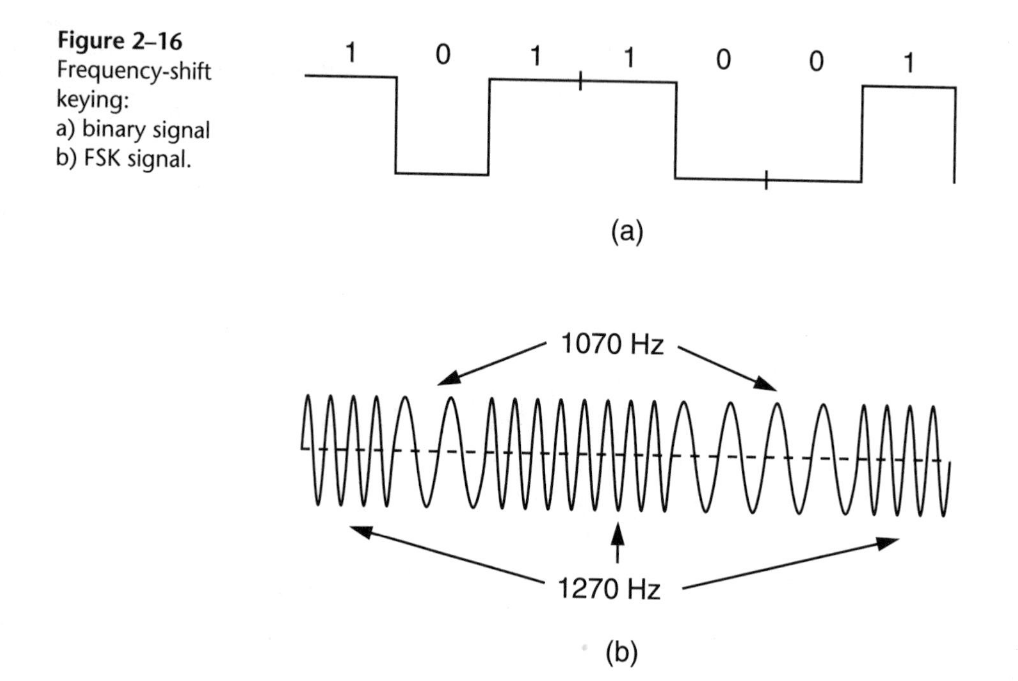

Figure 2–16 Frequency-shift keying:
a) binary signal
b) FSK signal.

the telephone pass-band, and they are sufficiently removed from each other to prevent crosstalk between the sidebands that are generated by modulation. FSK has applications for digital communications via high-frequency radio waves. Here, the system specifies the frequency shift between mark and space for a center frequency. So when a mark (logic 1) is transmitted, the center frequency may be lowered, for example, by 42.5 Hz, and when a space (logic 0) is transmitted, the center frequency may be raised by 42.5 Hz. Thus, if the center frequency is 425 Hz, a mark represents 382.5 Hz, while a space represents 467.5 Hz. This process is called FSK.

Phase Modulation (PM)

In PM, the amount of *phase-shift* of the carrier changes in accordance with the modulating signal; in effect, as the amount of phase-shift changes, the carrier frequency changes. Since PM results in FM, it is often referred to as indirect FM. Phase shift is a time difference between two sine waves of the same frequency. Figure 2–17 illustrates several examples of phase shift. Note that a phase shift of 180° represents the maximum difference and is also known as phase reversal. The advantage of using PM over FM is that the carrier can be optimized for frequency accuracy and stability. This type of modulation is easily adaptable to data or digital applications.

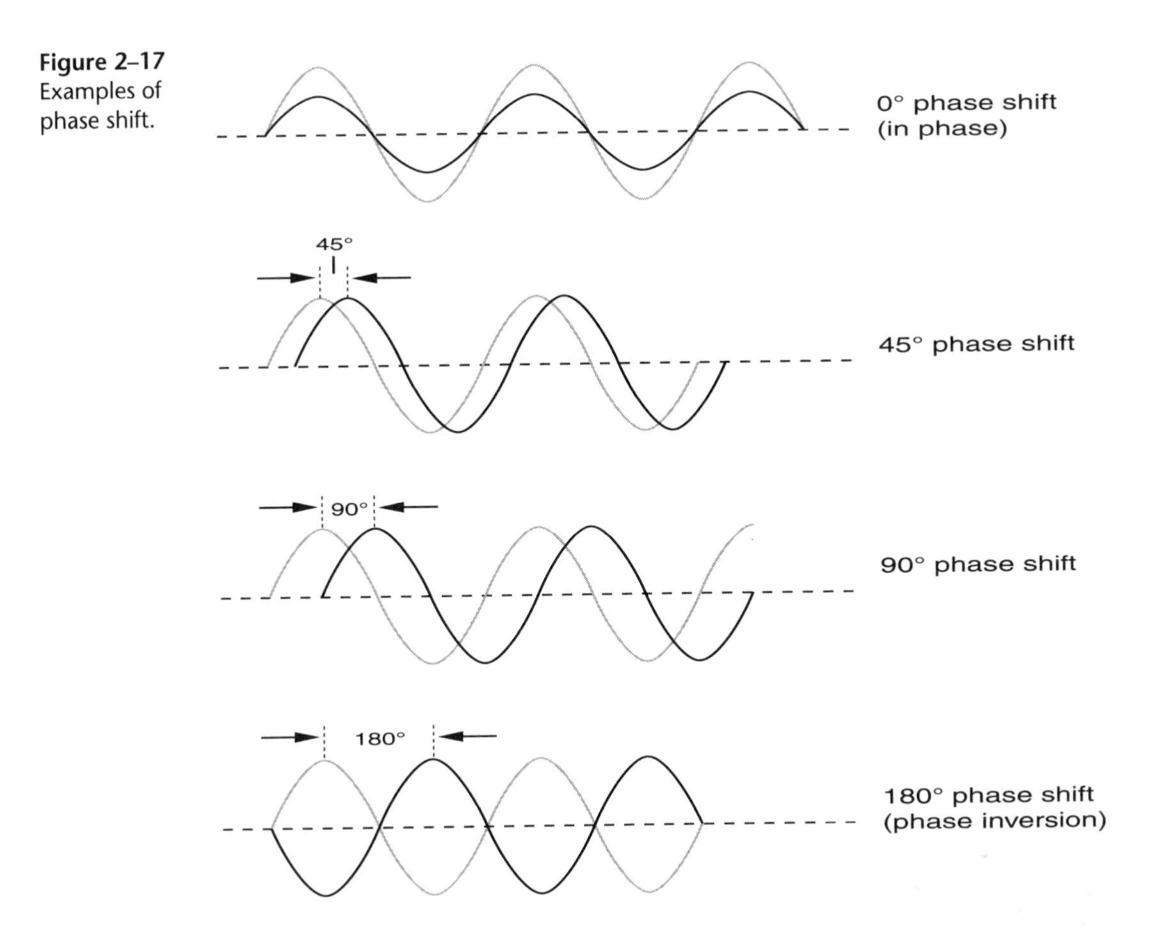

Figure 2–17 Examples of phase shift.

Phase Shift Keying (PSK)

Phase Shift Keying (PSK) is the most popular implementation of PM for data applications. In PSK the binary signal (0 or 1) to be transmitted changes the phase shift of a sine wave accordingly. Figure 2–18 illustrates the simplest form of PSK, known as binary PSK (BPSK). During the time that a binary 0 occurs, the carrier signal is transmitted with one phase, but when binary 1 occurs, the carrier signal is transmitted with 180° phase shift. The main problem with BPSK is that the speed of data transmission is limited in a given bandwidth. One way to increase the binary data rate while not increasing the bandwidth requirement for signal transmission is to encode more than one bit per phase change. Most PSK modems use Quadrature PSK (or 4–PSK), where each symbol represents two bits, as illustrated in Figure 2–19.

Baud rate is defined as the number of symbols (or signal transitions) transmitted in one second. Equation 2–12 gives the relationship between the baud rate and the bit rate.

$$\text{Bit rate} = \text{Baud rate} \times \text{Bits per Symbol} \quad (2\text{–}12)$$

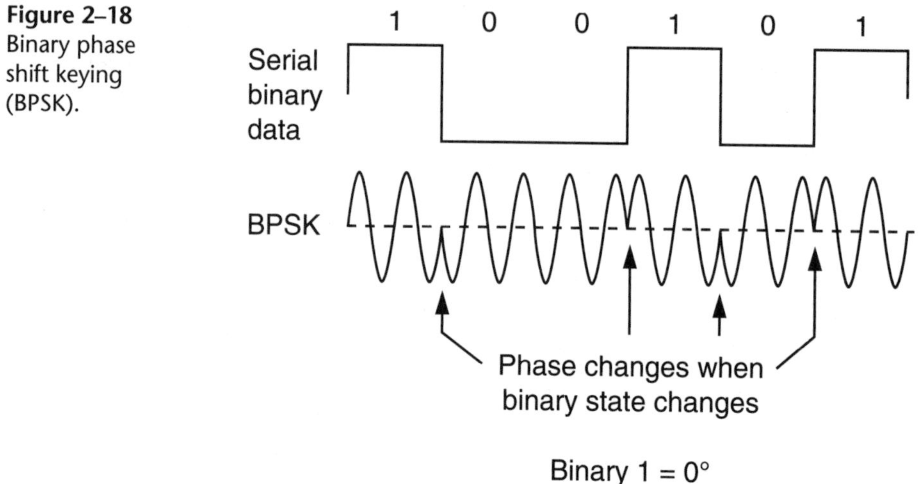

Figure 2–18 Binary phase shift keying (BPSK).

Figure 2–19 Quadrature PSK modulation.

Bit		Phase shift
0	0	45°
0	1	135°
1	0	225°
1	1	315°

135° = 01
45° = 00
225° = 10
315° = 11

Example 2–4

Problem

Find the transmission bit rate if the baud rate is 1200 and there are two bits per symbol or signal transition.

Solution

Bit rate = Baud rate × Bits per Symbol

Therefore, Bit rate = 1200 × 2

Bit Rate = 2400 bps

Quadrature Amplitude Modulation (QAM)

A Quadrature Amplitude Modulation (QAM) modem uses two amplitude-modulated carriers with a 90° phase angle between them. These are added to produce a signal with an amplitude and phase angle that can vary continuously. The number of amplitude-phase combinations could be infinite, but a practical limit is reached when the difference between adjacent combinations becomes too small to be detected reliably in the presence of noise and distortion. For example, the V.32bis modem has a modulation rate of 2400 baud and 14,400 bps, where each signal transition represents six data bits. The term *bis* comes from Latin, meaning *second;* in other words, the second and enhanced release of the standard. Third releases are designated *ter,* translated from Latin as *third.* The modulation techniques specified by the ITU for some of the common modems are outlined in Figure 2–20. The high-speed 56 kbps modems, such as V.90 modems, take advantage of the digital portion of the phone network to achieve higher speeds that were not possible with purely analog lines.

Type	Bit rate	Modulation
V.21	300	Frequency Shift Keying (FSK)
V.22	1200	Phase Shift Keying (PSK)
V.22bis	2400	Quadrature Amplitude Modulation (QAM)
V.32	9600	Quadrature Amplitude Modulation (QAM)
V.32bis	14400	Quadrature Amplitude Modulation (QAM)
V.34	28800	Quadrature Amplitude Modulation (QAM) plus other advanced techniques
V.90	$56000_{downstream}$ $33600_{upstream}$	Pulse Code Modulation (PCM) with differing upstream and downstream speeds

Figure 2–20 Modulation techniques for some of the common modems.

Pulse Modulation

Pulse modulation, which includes a variety of schemes, is used for both analog and digital signals. For analog signals, the process involves **sampling** where a snapshot (sample) of the waveform is taken for a brief instant of time, but at regular intervals. These instantaneous amplitudes are the sample values, or samples, of the signal waveform. The rate at which a signal is sampled is called the *sampling rate,* and it is expressed as the number of samples per second. The *sampling interval* is the time interval between each sample. The sampling rate is the reciprocal of the sampling interval.

In 1928 Henry Nyquist determined the optimum sampling rate. The *Nyquist sampling theorem* states that if a waveform is sampled at a rate at least twice the maximum frequency component in the waveform, then it is possible to reconstruct that waveform from the periodic samples without any distortion. Therefore, if the maximum frequency component in the signal is F_{max}, then the optimum sampling rate equals $2F_{max}$. The sampling rate is sometimes called the Nyquist rate or Nyquist frequency. If a signal has a maximum frequency component of 5 kHz, then the sampling rate is 10,000 kHz, which is the same as 10,000 samples per second. The sampling process converts an analog signal into a train of pulses of varying amplitude, but at a constant frequency.

Analog-to-Digital Conversion consists of three stages:

- The first stage is a low-pass filtering of the analog signal, called an anti-*aliasing* filter, to prevent any alias frequencies from appearing due to under-sampling of an unexpected high frequency. Aliasing, a penalty for a sampling rate that is too low, is a form of distortion in which the reconstructed original signal results in a lower-frequency signal.
- The second stage is the sampling of the analog signal at the Nyquist rate, the result of which is a series of pulses at the Nyquist sampling rate with amplitudes equal to the sample values. These pulses represent a Pulse Amplitude Modulated (PAM) signal.
- The third stage transforms these pulses into a digital signal. The amplitude of the pulses is quantized, and the quantized values are coded as binary numbers. The binary numbers become a stream of on-off pulses. A number of pulses together then represent a binary number. The process of encoding analog samples as a series of on-off pulses is referred to as *Pulse Code Modulation (PCM).*

Pulse Amplitude Modulation (PAM)

Pulse Amplitude Modulation (PAM) generates pulses whose amplitude variation corresponds to that of the modulating waveform, as shown in Figure 2–21 (b). Like AM, it is very sensitive to noise. While PAM was deployed in early AT&T Private Branch Exchanges, there are no practical implementations in use today. However, PAM is an important first step in PCM.

Pulse Position Modulation (PPM)

Pulse Position Modulation (PPM) is closely related to PWM. All pulses have the same amplitude and duration but their timing varies with the amplitude of the modulating signal, represented in Figure 2–21 (c). The random arrival rate of pulses makes this unsuitable for transmission.

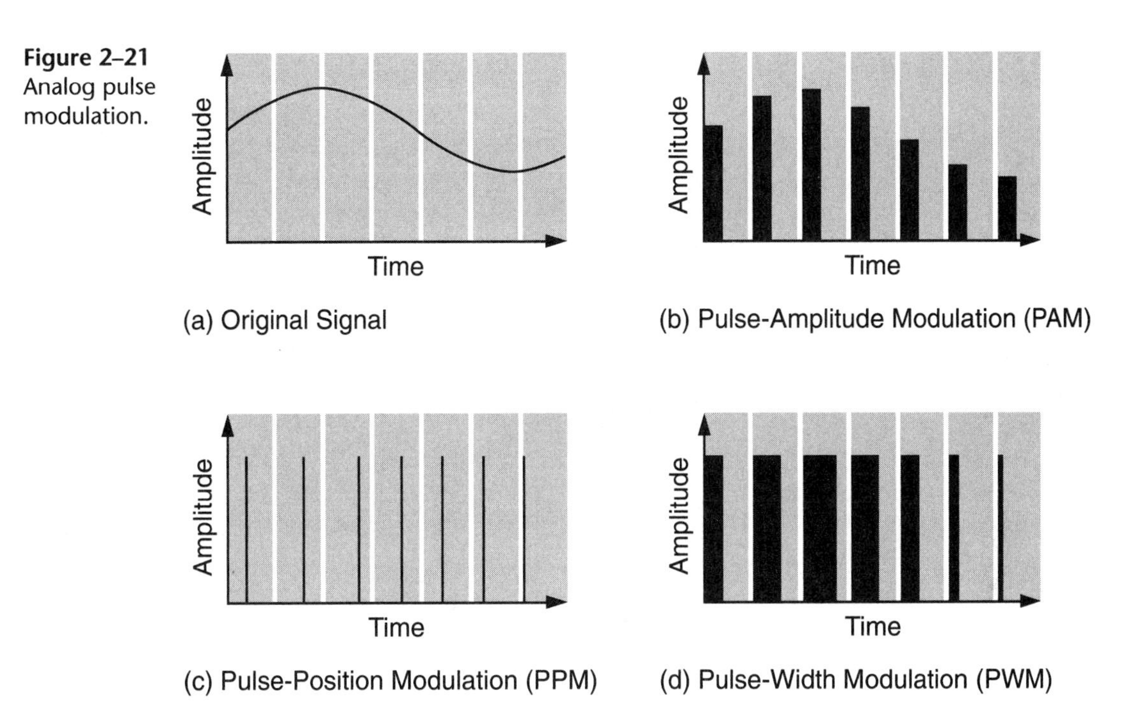

Figure 2–21 Analog pulse modulation.

Pulse Width Modulation (PWM)

The *Pulse Width Modulation (PWM)* technique generates pulses at a regular rate, whose length or width is controlled by the modulating signal's amplitude as depicted in Figure 2–21 (d). PWM is unsuitable for transmission because of the varying pulse-width.

Pulse Code Modulation (PCM)

Pulse Code Modulation (PCM) is the only technique that renders itself well to transmission. It is the most commonly used method of coding digital signals and is also used for transmitting telephone (analog) signals digitally. For analog signals, the amplitude of each sample of a signal is converted to a binary number. A common pattern for coding the transmitted information is by using a *character code* such as ASCII. A character code specifies a unique string of 0s and 1s to identify a character. The receiver detects either the presence (1) or absence (0) of a pulse. When it detects this pattern of 0s and 1s in a given period of time, it interprets the transmitted code by finding the corresponding character represented by it. The frequency range that can be represented through PCM modulation depends upon the sampling rate.

T-1 Carrier uses PCM as depicted in Figure 2–22. The allotted bandwidth per voice channel is 4 kHz. According to the Nyquist theorem, an analog signal must be sampled at twice its highest frequency to obtain an accurate digital representation of the information content of the signal. Therefore, the voice channel must be sampled at 8 kHz. A pulse code modulator samples the voice 8,000 times every second, converts each sample to an eight-bit digital word, and transmits it over a line interspersed with similar digital signals from 23 other channels. Each PCM voice channel operates at 64 kbps (8 bits/sample and 8000 samples/sec). Repeaters spaced at appropriate intervals regenerate the 24-channel signal with an aggregate of 1.536 Mbps (equals 24×64 kbps). With additional 8 kbps for synchronization, this technique results in a 24-channel 1.544 Mbps digital signal known as T-1. Each of the 24 channels can be used for either data or digital voice communications.

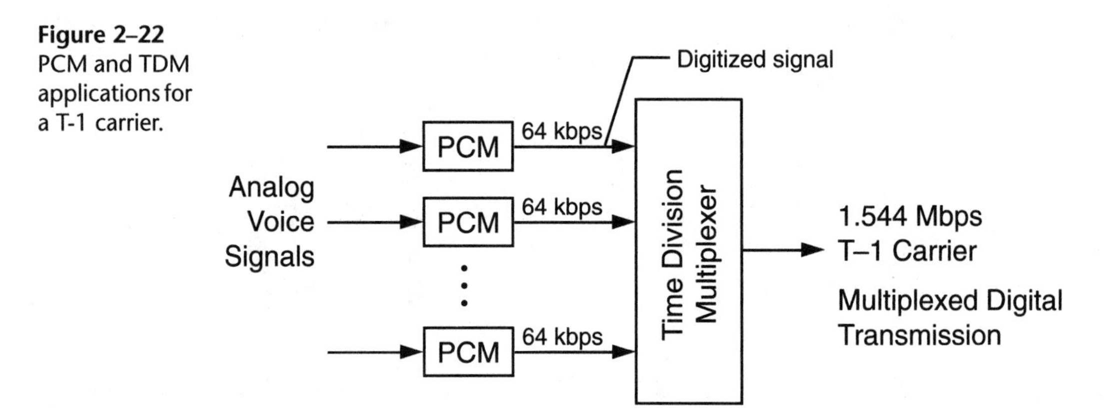

Figure 2–22 PCM and TDM applications for a T-1 carrier.

MULTIPLEXING

Multiplexing is the process in which two or more signals are combined for transmission over a single communications path. This concept is conveyed in Figure 2–23. Multiplexing has made communications very economical by transmitting thousands of independent signals over a single transmission line. There are three predominant ways to multiplex: Frequency Division Multiplexing (FDM), Time Division Multiplexing (TDM), and Wavelength Division Multiplexing (WDM). WDM is used exclusively in optical communications.

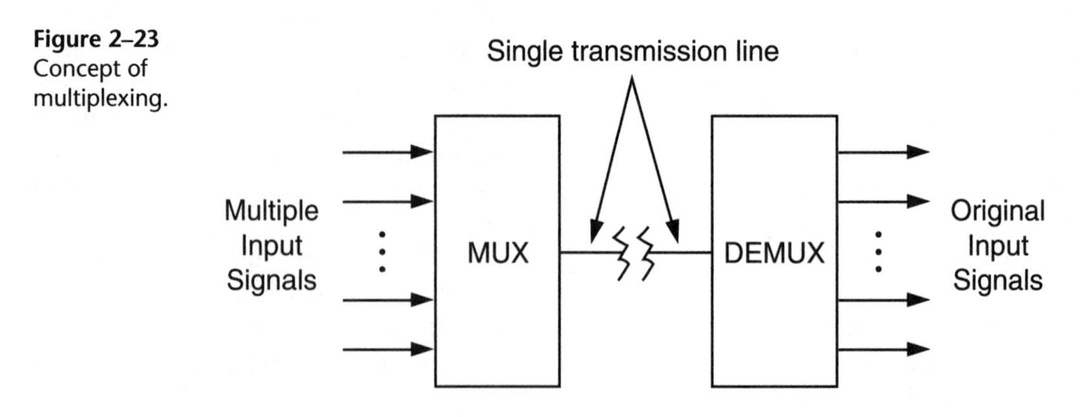

Figure 2–23 Concept of multiplexing.

Frequency Division Multiplexing (FDM)

Frequency Division Multiplexing (FDM) is predominantly used in analog communications. Figure 2–24 shows a general block diagram of an FDM system where each signal is assigned a different carrier frequency. The modulated carrier frequencies are combined for transmission over a single line by a multiplexer (MUX). There is always some unused frequency range between channels, known as guard band. At the receiving end of the communications link, a demultiplexer (DEMUX) separates the channels by their frequency and routes them to the proper end users. A two-way communications circuit requires a multiplexer/demultiplexer at each end of the long-distance, high-bandwidth cable. FDM was the first multiplexing scheme to enjoy wide-scale network deployment.

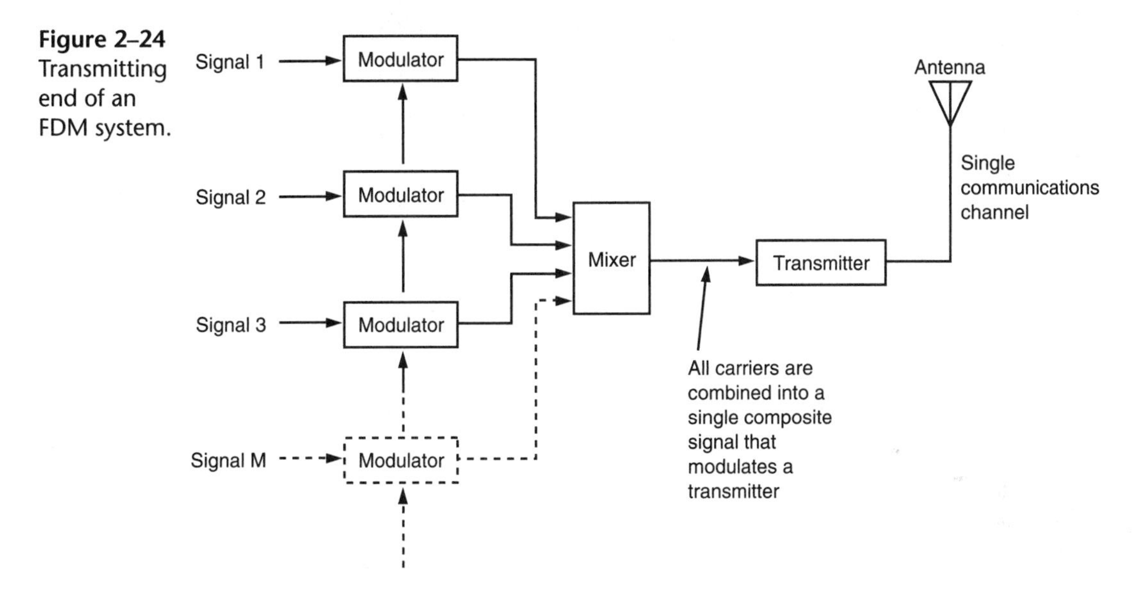

Figure 2–24 Transmitting end of an FDM system.

FDM is widely used in FM stereo broadcast. It preserves compatibility with monoreceivers and requires only a slight increase in BW. In stereo, two microphones are used to generate two separate audio signals, generally called the left L and right R. The two microphones pick up sound from a common source, such as a voice or band, but from different directions. This separation of the two microphones provides sufficient difference in the audio signals to provide more realistic reproduction of the original sound. The L and R are fed to a circuit where they are combined to form sum $L + R$ and difference $L - R$ signals. FDM techniques are used to transmit these two independent signals on a single channel.

To explore this concept further, consider how different voice channels can be placed on a single wire or cable using FDM. Each voice channel requires a maximum 4 kHz bandwidth and therefore modulates a different carrier frequency spaced 4 kHz apart. The 12 carrier frequencies are 60 kHz, 64 kHz, and so on, through 108 kHz, causing the 12 voice channels to occupy non-overlapping frequencies. The resulting separate bandwidths are summed so the channels can be stacked on top of each other in the frequency spectrum. As shown in Figure 2–25, twelve voice channels are combined into a *group*. Five groups form a *supergroup*, and ten supergroups form a *mastergroup*. This mastergroup can handle a total of 12 × 5 × 10 = 600 channels. Figure 2–26 provides the Bell System's hierarchy of FDM groups. FDM's disadvantages stem from analog circuitry, crosstalk and the difficulty of interfacing an FM transmitter with digital sources such as a computer; also, an FM channel remains idle when not in use.

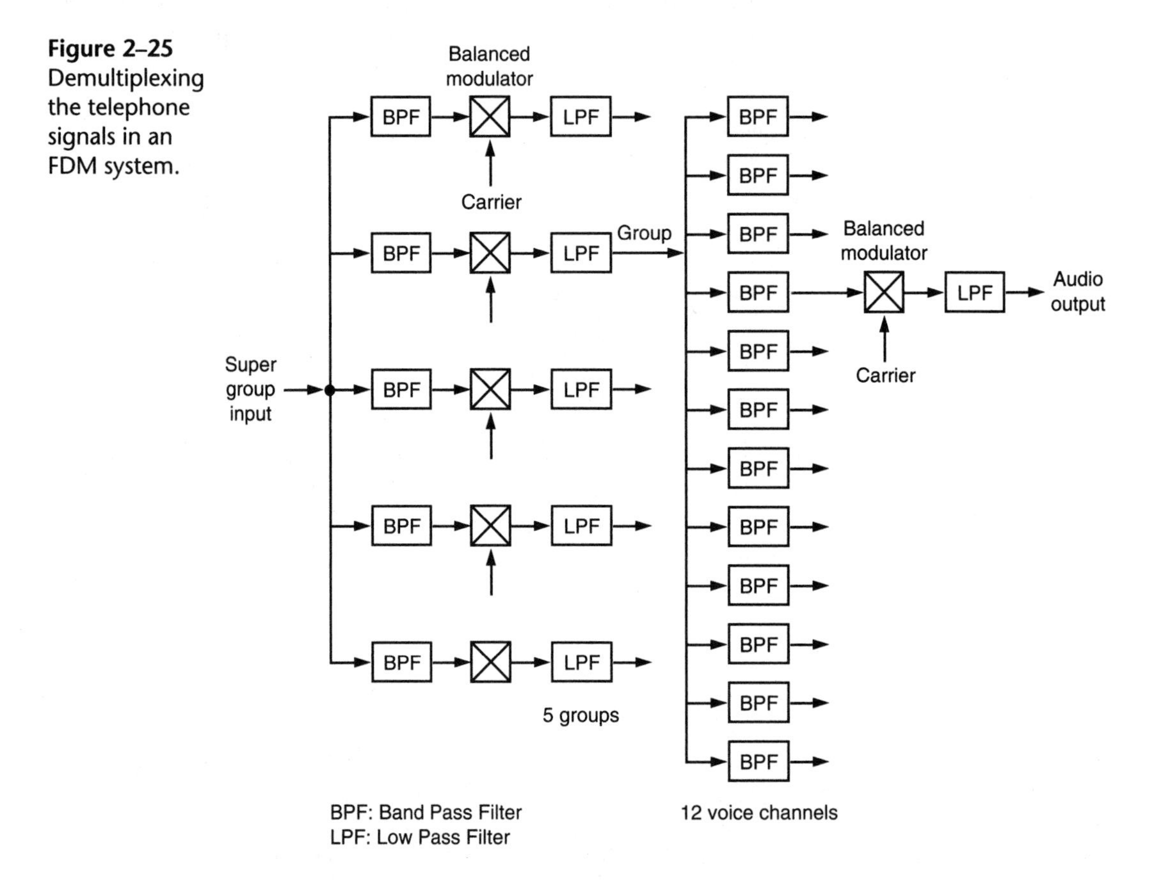

Figure 2–25 Demultiplexing the telephone signals in an FDM system.

Time Division Multiplexing (TDM)

While FDM has been used to great advantage in increasing system capacity, the use of TDM offers even greater system improvements. TDM is protocol insensitive and is capable of combining various protocols and different types of signals, such as voice and data, onto a single high-speed transmission link. It is more efficient than FDM, as there is no need for guard bands. In order to use TDM, the transmission must be digital in nature so an essential component of TDM is the process of sampling the analog signal in time. In

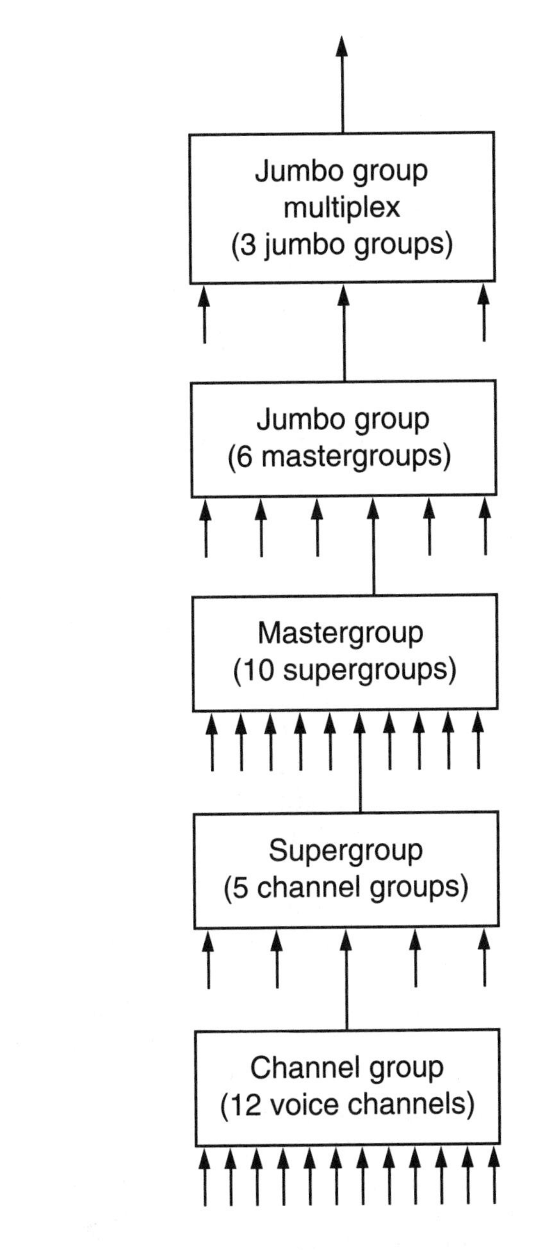

Figure 2–26 Hierarchy of the Bell System's FDM groups.

order to transmit telephone conversations, speech, which is an analog signal, is converted to a digital signal, transmitted, and then reconverted into analog at the receiving telephone. The main disadvantages of TDM are the greater complexity of digital systems and the greater transmission bandwidth required. Large-scale, low-cost ICs are reducing the difficulty and expense of constructing complex circuitry, and data-compression techniques are beginning to decrease the bandwidth penalty. In general, the advantages outweigh the disadvantages.

A T-1 Carrier uses TDM where each of the 24 channels is assigned an 8-bit time slot, as depicted in Figure 2–27. A framing bit is used to synchronize the system. For 24 channels, there are a total of 193 bits (24 × 8 + 1 framing bit) occurring 8,000 times a second, as shown in Figure 2–28. This gives a bit rate of 1.544 Mbps (193 × 8,000). Digital channels offer much more versatility and much higher speed than analog channels. Furthermore, the digital signal is much more immune to channel noise than is the analog signal.

Figure 2–27
T-1 frame.

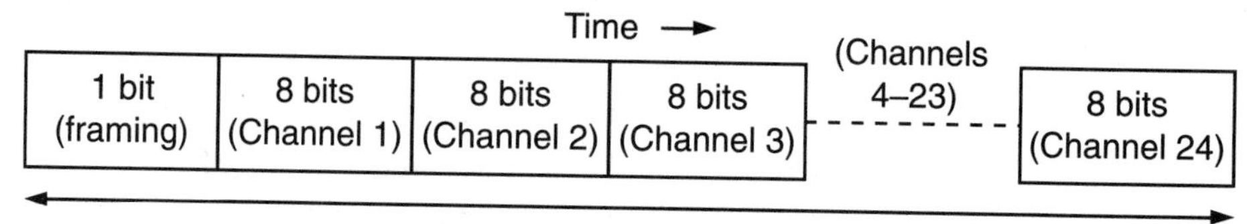

However, at 1.544 Mbps, T-1 lines simply do not have sufficient bandwidth to deal with the new demands being made on networks. Yet fiber-based T-3s at 45 Mbps bandwidth and 10 times the cost are overkill for many small and mid-sized businesses. Moreover, T-3 circuits are not easily available to many businesses, while T-1 lines are ubiquitous. The price, bandwidth, and availability gap between T-1 and T-3 has businesses and service providers searching for cost-effective ways to fulfill needs. *Inverse multiplexing* of T-1s benefits carriers and end users alike in bridging this bandwidth gap between 1.5 Mbps and 45 Mbps, which is a critical range for many wide area network applications.

Inverse multiplexing of T-1s is the process of distributing a serial data stream, bit by bit onto multiple T-1s, then reassembling the original data stream at the receiving end. The chief benefit of T-1 inverse multiplexing is that it uses the ubiquitous T-1 infrastructure to create clear data channels from 3 to 12 Mbps. The primary work of the inverse multiplexer is to assure that the bits are reassembled in the correct order. A very small portion of the T-1 payload is taken over for *metaframing,* which keeps the T-1s aligned in spite of minor timing differences and unequal circuit delays. Since there is no industry standard as of yet for bit-based T-1 inverse multiplexing, inverse multiplexers use proprietary metaframing techniques, which means that the devices at both ends of a data channel must be from the same vendor. Specifically, channels 1 through 8 of the T-3 are assigned to voice, channels 25 through 28 are assigned to Internet access, and channels 9 through 24 are available as spare capacity for voice and/or data.

There are basically three different TDM schemes: Conventional TDM, Statistical TDM (STDM), and Cell-Relay or ATM. STDM includes Conventional STDM, Frame Relay, and X.25 networking.

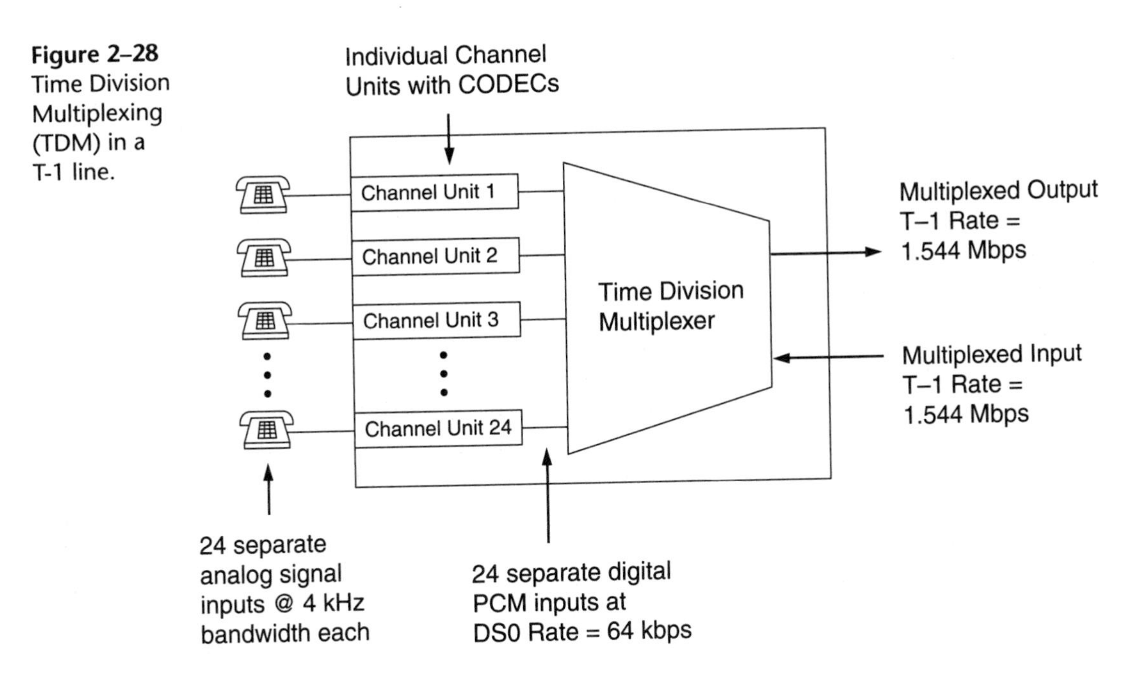

Figure 2–28 Time Division Multiplexing (TDM) in a T-1 line.

Conventional TDM

Conventional TDM systems usually employ either *bit-interleaved* or *byte-interleaved* multiplexing schemes. Clocking (bit timing) is critical in conventional TDM. All sources of I/O and clock frequencies must be derived from a central, traceable source for the greatest efficiency. In bit-interleaved TDM, a single data bit from an I/O port is output to the aggregate or the single communications channel, followed by a data bit from another I/O port, and so on, with the process repeating itself. A time-slice is reserved on the aggregate channel for each individual I/O port. Since the time-slices for each I/O port are known to both the transmitter and the receiver, the only requirement is for the transmitter and receiver to be in step. This is accomplished through the use of a synchronization channel between the two multiplexers. The synchronization channel transports a fixed pattern that the receiver uses to acquire synchronization. Total I/O bandwidth cannot exceed that of the aggregate minus the bandwidth requirements for the synchronization channel.

Bit-interleaved TDM is simple and efficient and requires little or no buffering of I/O data, but it does not fit in well with a microprocessor-driven, byte-based environment. In byte-interleaved multiplexing, complete words (bytes) from the I/O channels are placed sequentially, one after another, onto the high-speed aggregate channel. Otherwise, the process is identical to bit-interleaved multiplexing. Byte-interleaved systems were heavily deployed from the late 1970s to around 1985. In 1984, with the divestiture of AT&T and the launch of T-1 facilities and services, many companies jumped into the private networking market, pioneering a generation of intelligent TDM called STDM networks.

With Conventional TDM, the time slots are allocated on a constant basis. Thus, if a channel does not need to transmit data, the channel bandwidth goes unused during that time slot. This inefficiency is overcome by STDM techniques. The term *statistical* refers to the fact that the time slots are allocated on a need basis.

Statistical Time Division Multiplexing (STDM)

Statistical Time Division Multiplexing (STDM) allocates slices on demand, but it needs to know the address of the station, which is an additional overhead. A block diagram of a STDM application is shown in Figure 2–29. Its advantage is that there is no idle time, but a buffer is needed to handle simultaneous requests. In this scheme, the underlying assumption is that not all channels are transmitting all the time. A statistical multiplexer (stat mux) has an aggregate transmission BW that is less than the sum of channel BWs because the aggregate bandwidth is used only when there is actual data to be transported from I/O ports. The receiver knows the destination port for the data it receives because the transmitter sends not only the data but also an address. The address identifies the port the data is destined for. The stat mux assigns variable time slots every second depending upon the number of users and the amount of data transmitted by each. Frame Relay, X.25, and Switched MultiMegabit Data Service (SMDS) are all categorized as STDM systems, and are discussed in depth in Chapter 7.

STDM's biggest disadvantage is that it is I/O protocol sensitive. Therefore, a stat mux has difficulty supporting transparent I/O data and unusual protocols. To support these I/O data types, many STDM systems have provisions to support conventional TDM I/O traffic through the use of adjunct/integrated modules. This Conventional STDM was very popular in the late 1970s to mid 1980s and is still used, although the market for these units is dwindling. In Conventional STDM, as I/O traffic arrives at the stat mux, it is buffered and then inserted into frames. The receiving units remove the I/O traffic from the aggregate frames. Statistical multiplexers are ideally suited for the transport of asynchronous I/O data, as they can take advantage of the inherent latency in asynchronous communications. However, they can also multiplex synchronous protocols by *spoofing*, again taking advantage of the latency between blocks or frames. *Spoofing* refers to simulating a communications protocol by a program that is interjected into a normal sequence of processes for the purpose of adding some useful function.

Time Assignment Speech Interpolation (TASI) represents an analog STDM scheme. These systems were in limited use in the 1980s and were particularly adept at sharing voice circuits, specifically Private Branch Exchange (PBX) trunks. In normal telephone conversations, a majority of time is spent in a latent (idle) state. TASI trunks allocate snippets of voice from another channel during this idle time. As digital speech processing became more common, TASI systems called Digital Speech Interpolation (DSI) were created. These had analog inputs and digital outputs.

Both TASI and DSI systems suffer from some major drawbacks. First, users can notice a lot of voice clipping when a little bit of speech is lost while waiting for the TASI mux to

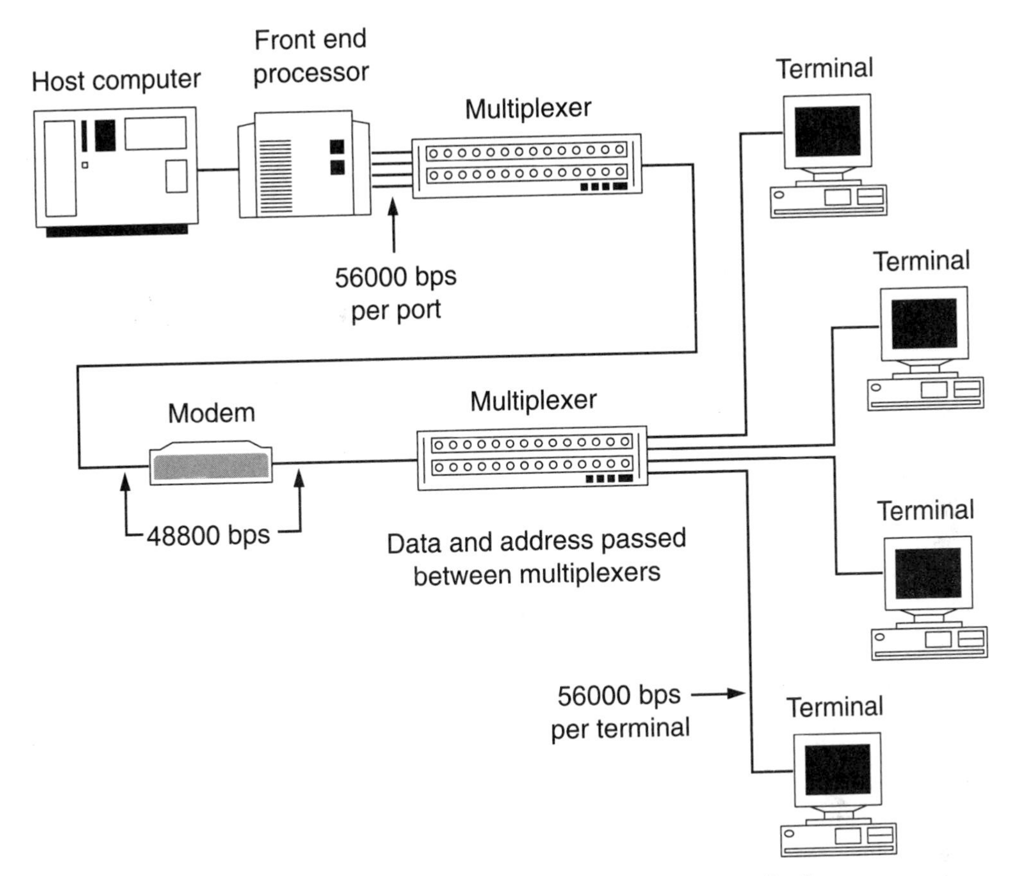

Figure 2–29 Block diagram of a Statistical Time Division Multiplexing application.

detect valid speech and allocate bandwidth. Clipping also occurs when there is insufficient bandwidth. In addition, TASI and DSI units are very susceptible to audio input levels and may have problems with the transport of voice-band data, for example, modem signals.

Wavelength Division Multiplexing (WDM)

Wavelength Division Multiplexing (WDM) is a cost-effective way to increase the capacity of fiber-optic communications. In its early form, WDM was capable of carrying signals over two widely-spaced wavelengths, and for a relatively short distance, as depicted in Figure 2–30. Each wavelength of light can transmit encoded information at an optimum data rate. Therefore, multiplexing distinct wavelengths of light leads to a significant increase in the total throughput.

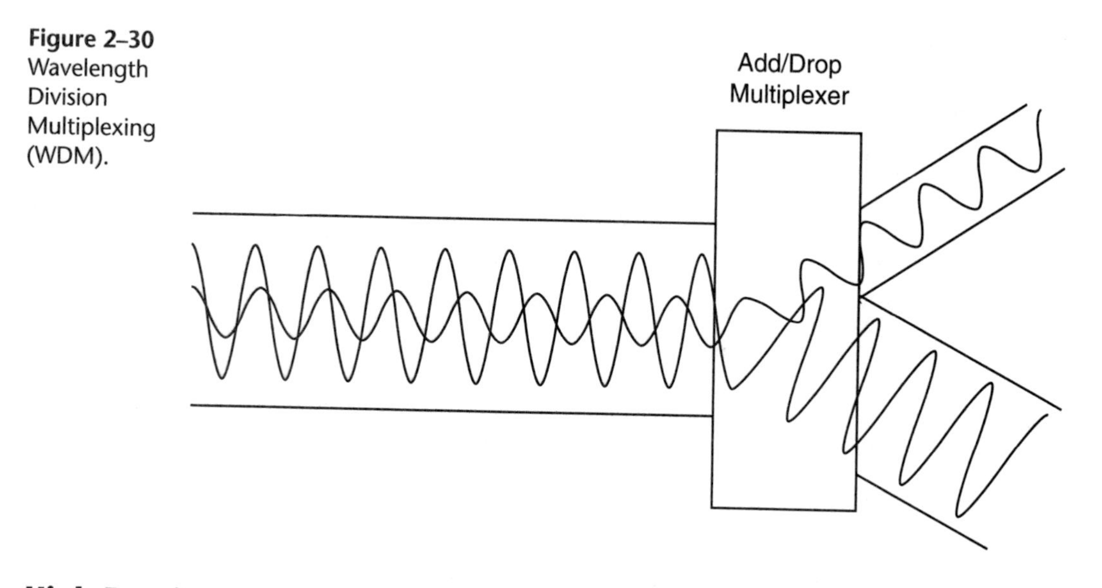

Figure 2–30 Wavelength Division Multiplexing (WDM).

High-Density or Dense WDM (DWDM)

Improvements in optical filters and narrowband lasers enabled *High-density* or *Dense WDM (DWDM)* to combine more than two wavelengths and multiplex the signals as separate channels across a single optical fiber. The invention of the flat-gain wideband optical amplifier dramatically increased the viability of DWDM systems by greatly extending the transmission distance. DWDM technology is typically found at the core of carrier networks. Optical fiber technology has undergone many improvements since the first lines were laid in the ground nearly 20 years ago. Rather than digging up and replacing these lines, telecommunications companies have searched for ways to maximize bandwidth and minimize dispersion in older fibers. The challenge of delivering greater bandwidth surged research efforts in managing wavelengths.

Until 1998, the predominant driver in DWDM deployment was long-distance transport with the network architecture being point-to-point DWDM. The goal was to send as many channels across a single fiber as far as possible. But DWDM is now migrating into MAN and LAN applications where the bandwidth demands exceed the physical limitations of the existing fiber and where the economics of installing DWDM systems are more attractive than upgrading the entire installed fiber plant. A key advantage of DWDM is that it is independent of the protocol and the bit-rate as well. DWDM-based networks can transmit data in IP, ATM, SONET, and Ethernet and handle bit rates between 100 Mbps and 2.5 Gbps. Thus, DWDM-based networks can carry different types of traffic at different speeds over the same optical channel.

A block diagram of a DWDM system is shown in Figure 2–31; the key elements are transmitters that consist of tunable semiconductor lasers, electro-optical modulating and multiplexing components for combining and separating signals, single-mode optical fibers, and receivers that include optical amplifiers and photodetectors. The system

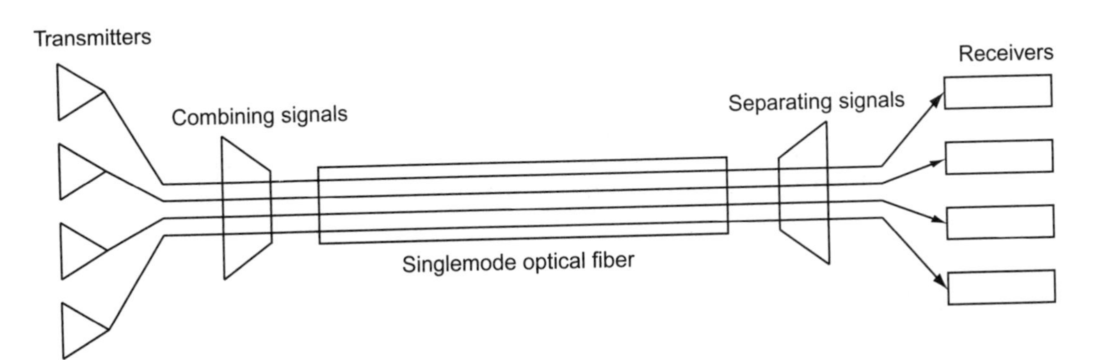

Figure 2–31 Block diagram of a DWDM system.

employs at least four multiplexing devices: one mux and one demux for each direction of traffic. Several devices may be cascaded to multiplex the desired number of channels. DWDM systems require that the laser wavelength be extremely stable. Optical amplifiers boost light signals and thereby extend their reach without converting them back to electrical form. Demonstrations have been made of ultra-wideband amplifiers that can boost light signals carrying over 100 channels (or wavelengths). A network using such an amplifier could easily handle a terabit of information. The add/drop optical multiplexers circumvent the need to demultiplex transmissions into electronic signals prior to rerouting or amplifying them. Furthermore, optical routers switch DWDM traffic wavelength by wavelength as it comes into optical hub sites, allowing carriers to establish meshed optical nets. The result is a more reliable network with virtually no downtime because the routers share network data and are smart enough to quickly route around failures. However, the drawback to DWDM is that it only functions point to point. So if a point fails, all calls on the path are lost until an alternative path can be set up and individual sessions are reestablished.

DWDM has been compared to a multilane highway for carrying data, in contrast to a single line in the case of a traditional TDM implementation. Rather than trying to pack more vehicles into one lane at increasingly higher speeds, DWDM makes full use of all the lanes. Perhaps more importantly, it also enables the various highway lanes to move at different speeds and to carry different types of information. But managing the abundant bandwidth DWDM affords is a growing challenge.

For example, a single-mode optical fiber with an attenuation of 0.2 dB per km at 1,550 nm is capable of accommodating a set of wavelengths each spaced apart by a few tenths of nm (50 GHz to 100 GHz). Thus, it has an estimated transmission capacity in the THz regime. This indicates that instead of using a single wavelength laser to transmit information along the optical fiber, we can use multiple wavelength lasers to transmit far more information along the same channel, thereby increasing the total capacity of optical transmission. The use of 48 distinct wavelength lasers, each modulated at 2.5 Gbps, represents an effective transmission rate of 48 × 2.5 Gbps, which is equal to 120 Gbps.

The use of 100 distinct wavelength lasers could increase the effective data throughput to Tbps. Future developments will be in different modulation technologies capable of achieving these speeds; this trend will move wide-area networking speeds from Mbps to Gbps and eventually to Tbps. The idea appears to be moving toward reality as many companies are providing advanced DWDM technologies that allow the service or trunk providers to upgrade their system capacity in accordance with the ever-increasing demand for information.

SUMMARY

Much of society's progress in social, economic, and scientific endeavors can be related to improvements in the ability to communicate. Communication is the transfer of information in the form of physical patterns from a source to the destination. Electronic communication uses physical patterns of electrical signals to transmit information rapidly and over long ranges from one point to another. When evaluating communications systems, basic design parameters such as rate of information transfer, system reliability, and cost must be considered. The information capacity of a channel is limited by its power, bandwidth, and noise. In the telecommunications industry, significant research and development efforts have focused on how to superimpose an increasing amount of information on a single transmission medium. These have resulted in different modulation and multiplexing techniques for efficient transfer of information.

REVIEW QUESTIONS

1. Identify the basic components of a communications system.

2. Define the term and give an example of a practical application for each:

   A. Bandwidth
   B. Baseband
   C. Broadband
   D. Analog Transmission
   E. Digital Transmission
   F. Serial Transmission
   G. Parallel Transmission
   H. Synchronous
   I. Asynchronous
   J. Simplex
   K. Half-Duplex
   L. Full-Duplex

3. State the Fourier theorem and distinguish between time domain and frequency domain representations of a signal.

4. Assess the value of representing signals in the frequency domain.

5. Determine the internal noise power in watts for a microwave amplifier that generates an equivalent noise temperature of 140° K at an operating bandwidth of 500 MHz.

6. A receiver produces a noise power of 200 mW with no signal. The output level increases to 5 W when a signal is applied. Calculate the signal-to-noise ratio as a power ratio and in decibels.

7. What is the implication of a SNR of zero dB?

8. Compute the maximum noise power at the input of a communications receiver in order to maintain a 40 dB SNR for an input signal power of 20 μW.

9. Calculate the BER if there were six bad bits in a total transmission of 10,000 bits.

10. Explain the principle of operation for each of the following techniques:

    A. FSK
    B. QPSK
    C. QAM

11. Explain the use of PCM and TDM in T-1 carriers.

12. Determine the efficiency of a T-1 carrier.

13. Discuss the current status of DWDM.

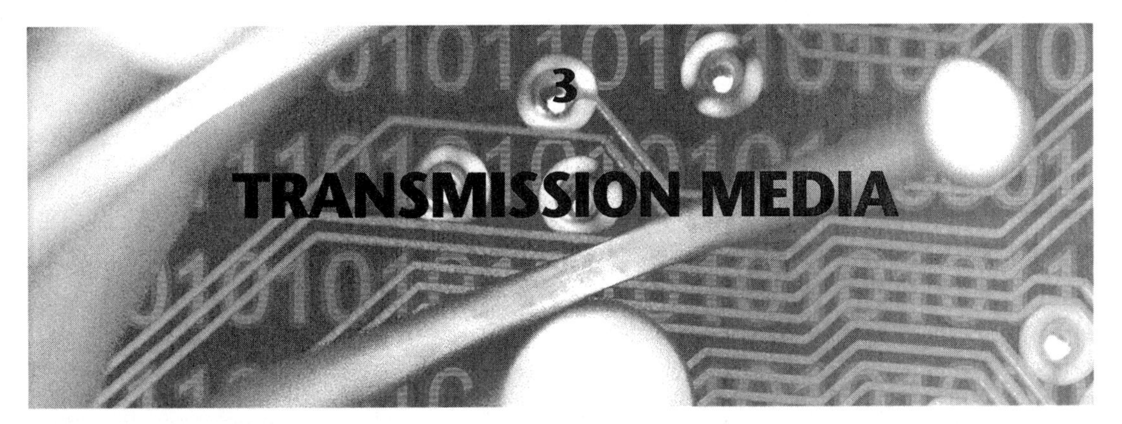

3 TRANSMISSION MEDIA

KEY TERMS

Coaxial Cable
Unshielded Twisted Pair (UTP) cable
Shielded Twisted Pair (STP) cable
Attenuation
Resistance (R)
Impedance (Z) Matching
Echo
Four-wire Terminating Sets
Crosstalk
Bend Radius
Electromagnetic Interference (EMI)
Fiber-optic Cable
Core
Cladding
Refractive Index
Snell's Law
Total Internal Reflection
Numerical Aperture (NA)
Dispersion
Structured Wiring

OBJECTIVES

Upon completion of this chapter, you should be able to:

- Develop an understanding of different transmission media
- Distinguish between wired and wireless communications and their applications
- Categorize different types of copper cables and analyze their applications
- Discuss the current status of Enhanced Category 5 and higher grade cable
- Explain the construction of an optical fiber and a fiber-optic cable
- Describe the principle of operation in the propagation of light through fiber
- Differentiate between different types of fiber-optic cables and their applications
- Analyze the characteristics of fiber-optic cable as compared to copper cable
- Examine different transmission impairments for copper and fiber-optic cables
- Describe the different components and standards for structured wiring
- Determine appropriate transmission media for different applications

INTRODUCTION

The physical path over which the information flows from transmitter to receiver is called the transmission medium or the channel. Transmission media can be classified into two major categories: wired and wireless. Wired includes different types of copper and fiber-optic cables, while wireless includes infrared, radio, microwave and satellite transmission. The performance specifications of cables are important when selecting a specific type of cable to determine its suitability for specific applications. The two major factors are construction and installation. Chapter 5 addresses wireless communications; in this chapter, we will focus on wired media.

There are several specifications that cover different aspects of cabling in North America. The IEEE 802 addresses local area network standards applicable for data communications. The ANSI/TIA/EIA 568 standard developed in conjunction with the Canadian Standards Association (CSA) deals with recommended methods and practices for installation and termination of telephony and networking cable. The 568 specifications are designed to be automatically in accordance with the National Electrical Code (NEC), which is an overall specification for all wiring in the United States. Although the ANSI/TIA/EIA compliance is not required by local building codes, any company planning a wiring system is well advised to follow the standard. In Europe, the CE (Conformitè Europèenne) mark means that a product complies with an applicable European directive. All regulated products placed for sale in the European market must display the CE marking.

COPPER CABLES

Copper wire is the most commonly used medium for communications circuits; the oldest installed cables were copper and it is still the most used material for connecting devices. The cost of a cable is a function of the cost of the materials and of the manufacturing process. Thus, cables with larger diameter, involving more copper conductor and more insulation, are more expensive than those with small diameter. The three main types of copper cables include *coaxial, Unshielded Twisted Pair (UTP),* and *Shielded Twisted Pair (STP).* Let us begin with a study of the construction and application of each of these cable types.

Coaxial Cable

Coaxial cable, depicted in Figure 3–1, is a two-conductor cable in which one conductor forms an electromagnetic shield around the other; the two conductors are separated by insulation. This is a constant impedance transmission cable. It is primarily used for CATV since it provides a bandwidth of nearly 1 GHz into the home. Coaxial cable is used for long distance, low attenuation, and low noise transmission of information. In fact, the research and test deployment of CATV-based Internet delivery systems is a growth industry. The telephone companies also resort to coaxial cable to transmit 140 Mbps data sig-

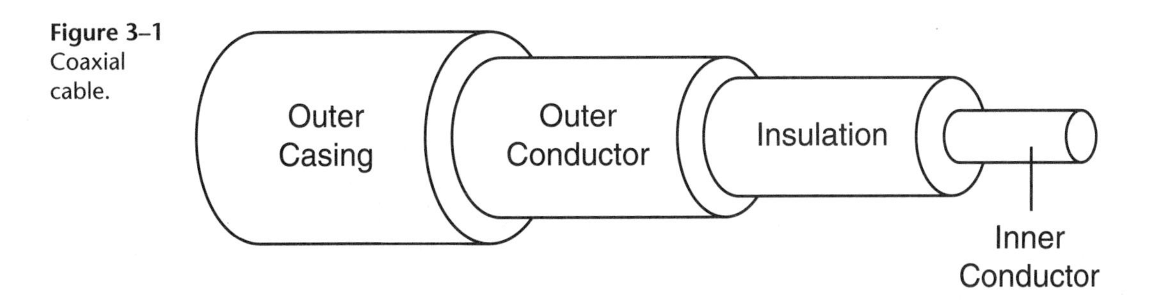

Figure 3–1 Coaxial cable.

nals between telephone switch buildings with a hop distance of up to 2 km. For data applications, it is classified into two categories: *thick coax* or *10Base5,* and *thin coax* or *10Base2.* In this designation, the *10* refers to the transmission speed of 10 Mbps, the *Base* refers to baseband signaling, and the *2* and *5* refer to the coaxial cable maximum segment length in meters. For instance, in 10Base2, the *2* refers to 200 m (185 m has been rounded off to 200 m).

10Base5

10Base5 interface, also known as *Thicknet,* is based upon the use of thick, inflexible, coaxial cable. The cable is firm because the center conductor is solid. It serves as a backbone transmission medium for the LAN. It is primarily used for facility-wide installations and is typically installed as a physical bus linking one *Telecommunications Closet (TC)* to another. A TC is an enclosure in which wiring is terminated; a building may have multiple telecommunications closets. Nowadays, in most LANs, the 10Base5 backbone has been replaced by fiber. When compared to thin coax, the thick coax is less susceptible to interference and can carry much more data.

10Base2

10Base2 interface, also known as *Thinnet,* is based upon the use of thin, flexible, less expensive coaxial cable. Unlike the thick coax, the center of the thin coax is stranded, which makes it relatively flexible. It is primarily used in office environments, and offers some advantages over the general-purpose UTP. Thin coax cabling provides greater distance, allows T-connectors implementing bus topology, offers higher noise immunity and does not involve crossovers. The biggest disadvantage is the difficulty of terminating coaxial cable, which has been the main driving force in UTP rapidly becoming the de facto standard for horizontal wiring. Another important reason is the advancements in manufacturing techniques with new categories of UTP increasing the bandwidth availability to the desktop.

Unshielded Twisted Pair (UTP)

Unshielded Twisted Pair (UTP), illustrated in Figure 3–2, is the copper media inherited from telephony that is being used for increasingly higher data rates. A *twisted pair* is a pair of copper wires with diameters of 0.4 to 0.8 mm that are twisted together and protected by a thin polyvinyl-chloride (PVC) or Teflon jacket. The amount of twist per inch for each cable pair has been scientifically determined and must be strictly observed because it serves a purpose. The twisting increases the electrical noise immunity and reduces crosstalk as well as the bit error rate (BER) of the data transmission. UTP is a very flexible, low-cost media and can be used for either voice or data communications. Its greatest disadvantage is the limited bandwidth, which restricts long-distance transmission with low error rates.

UTP can be made with a variety of materials, sizes of conductors, and numbers of pairs inside a single cable. All UTP cables come in both solid and stranded filament. Solid filament cables are more rigid and usually intended for trunk cabling. Stranded filament cables are more pliable and generally targeted for patch cables. Figure 3–3 shows part of the TIA/EIA 568 specifications that include transmission speeds and applications for different categories of UTP cable. The standard recommends a 22 or 24 AWG wire. Jacks and plugs conform to the Uniform Service Ordering Code (USOC, pronounced "you-sock") numbers, which were originally developed by the Bell System, and are endorsed by the FCC. A RJ-45 (ISO 8877) 8-pin connector is recommended for UTP cable. The plug is the male component crimped on the end of the cable, while the jack is the female component in a wall plate or patch panel.

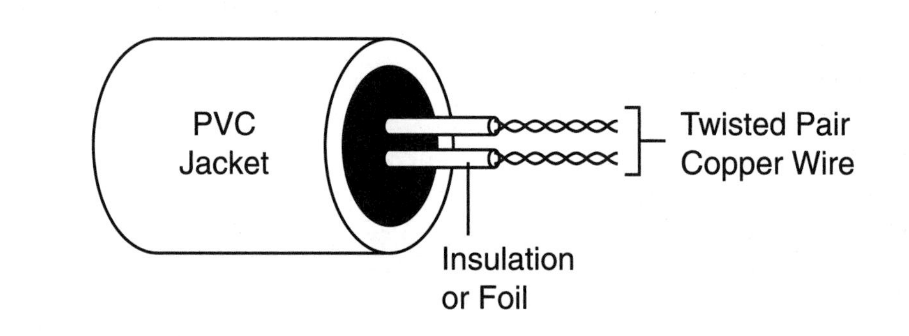

Figure 3–2 Unshielded Twisted Pair (UTP).

Categories of UTP Cable

Category 3 (CAT 3) twisted pair cable is used in implementing the popular 10BaseT interface, where the *T* represents twisted-pair cable. It is not the same as the regular *silver satin* phone cable because the pairs in the phone cable are not twisted. Although CAT 3 is used for voice and data communications, the market trend is to abandon CAT 3 in favor of installing CAT 5, especially for data.

Category 5 (CAT 5) cabling was standardized in 1995 by the TIA in the United States and by the ISO internationally. It consists of four pairs that are wrapped in a thermal

Category	Specified Data Rate	Application
CAT 1	Less than 1 Mbps	Telephone wiring (only audio signals, not for data)
CAT 2	4 Mbps	4 Mbps Token Ring
CAT 3	16 Mbps	10BaseT Ethernet
CAT 4	20 Mbps	16 Mbps Token Ring
CAT 5	100 Mbps	100BaseT Ethernet 155 Mbps ATM
CAT 5E	100 Mbps	100BaseT Ethernet 155 Mbps ATM
CAT 6	250 Mbps	1000BaseT Broadband
CAT 7	600 Mbps	Broadband—over 1 GHz bandwidth per pair

Figure 3–3 UTP cable: categories and their applications.

plastic insulator twisted around one another, and encased in a flame-retardant polymer. It has a maximum operating frequency of 100 MHz suitable for token ring, 100BaseT Ethernet, and 155 Mbps ATM. But this is slow when compared with the next generation of LAN protocols such as Gigabit Ethernet and high-speed ATM that push frequency requirements into the hundreds of MHz—for example, 350 MHz in the case of 622 Mbps ATM.

That fact has prompted vendors to roll out CAT 5E, CAT 6, and CAT 7 cable, which can handle frequencies in excess of 100 MHz. These components are mechanically and electrically backwards compatible with CAT 5. CAT 6 channel specifications for attenuation and NEXT are set up to 250 MHz, assuring at least double the channel bandwidth of CAT 5. For CAT 6 to support Gigabit Ethernet, proper cable installation is critical, unlike more tolerant environments like Fast Ethernet over CAT 5E. The bi-directional dual duplex transmission scheme employed by 1000BaseT actually requires each end of a channel to transmit on one conductor of each of the four pairs simultaneously. CAT 6 and CAT 7 cable is more suitable for current and future delay-sensitive applications. CAT 7 cable rated at 600 Mbps features individually shielded or screened twisted pairs (STP or ScTP) of wires, making it ideal for installation in locations with a high potential for electromagnetic interference (EMI).

T-1

T-1, sometimes referred to as a DS-1, consists of two pairs of UTP 19 AWG wire. It is a popular leased line option for businesses connecting to the Internet backbone since it provides a way of expanding networking capability and controlling costs. Its most common external use that is not part of the telephone network is to provide high-speed access from the customer's premises to the public network. A T-1 line supports 24 full-duplex channels, each of which is rated at 64 kbps, and can be configured to carry voice or data traffic. Most telephone companies allow businesses to buy one or more of these individual channels, known as fractional T-1 access. The fractional T-1 lines provide less bandwidth but are also less expensive. Typically, fractional T-1 lines are sold in increments of 56 kbps, where the extra 8 kbps per channel is the overhead used for data management.

Shielded Twisted Pair (STP)

Shielded Twisted Pair (STP), depicted in Figure 3–4, is a 150 Ω cable composed of two copper pairs. Each copper pair is wrapped in metal foil and then sheathed in an additional braided metal shield and an outer PVC jacket. The shielding absorbs radiation and reduces the EMI. As a result, STP can handle higher data speeds than UTP. The main drawback of STP is its high cost, although STP is less expensive than fiber-optic cabling. In addition, STP is bulkier than UTP, which poses problems for installations with crowded conduits. Foil Twisted Pair (FTP) or Screened Twisted Pair (ScTP) are variations of the original STP. They are thinner and less expensive, as they use a relatively thin overall outer shield. STP is used extensively by the telephone company for moving digitized information over distances of 2 km between repeaters, to span the distance of several miles between telephone company switching stations.

The IEEE 802.3 transmission medium characteristics for different types of cables discussed above are tabulated in Figure 3–5 .

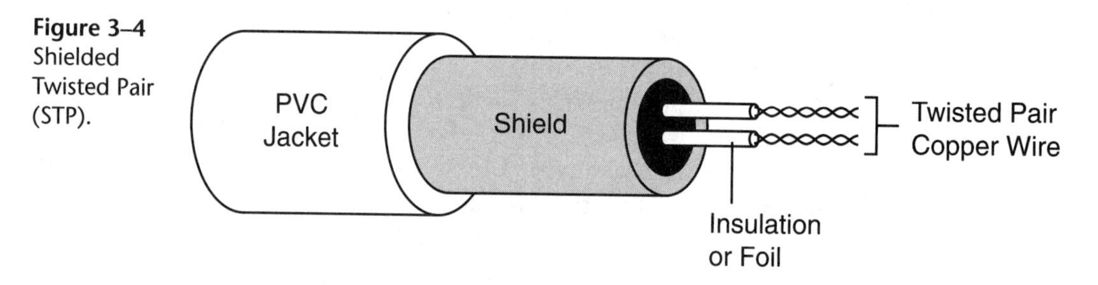

Figure 3–4 Shielded Twisted Pair (STP).

Attenuation in Copper Cables

All transmission impairments collectively result in undesired signals or noise that adversely affect the SNR that are also referred to as *attenuation*. **Attenuation** is the loss of power that occurs in a signal as it travels down a cable. It is commonly measured in dB and is given by Equation 3–1:

	10Base2	10Base5	10BaseT	100BaseTX	100BaseT4	100BaseFX
Data Rate (Mbps)	10	10	10	100	100	100
Signaling	Baseband	Baseband	Baseband	Baseband	Baseband	Baseband
Media	50 ohm Thin Coax	50 ohm Thick Coax	CAT 3 or higher grade UTP	2-pair CAT 5 UTP or Type 1 and 2 STP	4-pair CAT 3 or higher grade UTP	2-strands of 62.5/125 multimode fiber
Topology	Bus	Bus	Star	Star	Star	Star
Maximum Segment Length (m)	185	500	100	100	100	400
Maximum Network Span (m)	925	2500	500	200	200	400

Figure 3–5 IEEE 802.3 transmission medium characteristics.

$$\text{Attenuation (dB)} = 10\log_{10}\left(\frac{P_O}{P_I}\right) \tag{3–1}$$

where P_O is the output power

P_I is the input power

In any telecommunications circuit, a signal traveling on a cable becomes weaker the further it travels. At some point, the signal becomes too weak for the network hardware to interpret it reliably. Thus, there is a maximum cable run for every signal so that the signal at the far end is powerful enough to be detected by a receiver. For copper cables, attenuation varies with:

- Frequency
- Resistance
- Impedance Mismatch
- Crosstalk

As a general rule, attenuation increases with frequency. Ideally, all frequencies should undergo the same attenuation, but in reality, higher frequencies are attenuated more than lower ones, which results in attenuation distortion. Original local loop deployments targeted analog voice services in the 4 kHz region of the spectrum and ignored future utilization of higher-frequency bands. To overcome loss and extend reach, phone companies opted to reduce the series resistance of the line by using larger gauge wire. They also increased the series inductance of the line with loading coils and used analog electronic amplifiers to provide compensating gain to the transmission line. This places definite limits on the rate of data transmission.

Resistance (R)

The **Resistance (R)** of a cable depends upon the *specific resistance* or *resistivity* of the material, the length, and the cross-sectional area of the cable. The specific resistance, ρ, expressed in circular-mil ohms per foot, enables the resistance of different materials to be compared according to their natures, regardless of different areas or lengths. The specific resistance for different conductors is listed in Figure 3–6. Figure 3–7 lists the standard wire sizes specified using a system known as the American Wire Gauge (AWG). The gauge numbers specify the size of round wire in terms of its diameter and cross-sectional area and its resistance per foot at a temperature of 25°C. The cross-sectional area of round wire

Material	Description and Symbol	Specific Resistance (ρ) at 20° C cmil • Ω/ft	Temperature Coefficient per °C (α)	Melting Point (°C)
Aluminum	Element (Al)	17	0.0004	660
Constantan	Alloy, 55% Cu, 45% Ni	295	0 (average)	1210
Copper	Element (Cu)	10.4	0.004	1083
Gold	Element (Au)	14	0.004	1063
Iron	Element (Fe)	58	0.006	1535
Manganin	Alloy, 84% Cu, 12% Mn, 4% Ni	270	0 (average)	910
Nichrome	Alloy, 65% Ni, 23% Fe, 12% Cr	676	0.0002	1350
Nickel	Element (Ni)	52	0.005	1452
Silver	Element (Ag)	9.8	0.004	961
Steel	Alloy, 99.5% Fe, 0.5% C	100	0.003	1480
Tungsten	Element (W)	33.8	0.005	3370

Note: Listings approximate only, since precise values depend on exact composition of material.

Figure 3–6 Properties of conducting materials.

AWG	Diameter (mils)	Area (cmils)	Ohms per 1000 ft of Copper Wire at 25°C*	AWG	Diameter (mils)	Area (cmils)	Ohms per 1000 ft of Copper Wire at 25°C*
1	289.3	83,690	0.1264	21	28.46	810.1	13.05
2	257.6	66,370	0.1593	22	25.35	642.4	16.46
3	229.4	52,640	0.2009	23	22.57	509.5	20.76
4	204.3	41,740	0.2533	24	20.10	404.0	26.17
5	181.9	33,100	0.3195	25	17.90	320.4	33.00
6	162.0	26,250	0.4028	26	15.94	254.1	41.62
7	144.3	20,820	0.5080	27	14.20	201.5	52.48
8	128.5	16,510	0.6405	28	12.64	159.8	66.17
9	114.4	13,090	0.8077	29	11.26	126.7	83.44
10	101.9	10,380	1.018	30	10.03	100.5	105.2
11	90.74	8,234	1.284	31	8.928	79.70	132.7
12	80.81	6,530	1.619	32	7.950	63.21	167.3
13	71.96	5,178	2.042	33	7.080	50.13	211.0
14	64.08	4,107	2.575	34	6.305	39.75	266.0
15	57.07	3,257	3.247	35	5.615	31.52	335.0
16	50.82	2,583	4.094	36	5.000	25.00	423.0
17	45.26	2,048	5.163	37	4.453	19.83	533.4
18	40.30	1,624	6.510	38	3.956	15.72	672.6
19	35.89	1,288	8.210	39	3.531	12.47	848.1
20	31.96	1,022	10.35	40	3.145	9.88	1,069

*20–25°C or 68–77°F is considered average room temperature.

Figure 3–7 Copper wire table.

is measured in circular mils (abbreviation is cmil). A mil is one thousandth of an inch, or 0.001 inch. One cmil is the cross-sectional area of a wire with a diameter of one mil. The number of cmil in any circular area is equal to the square of the diameter in mils.

The total resistance of a segment of conductor (or wire, or cable) is given by Equation 3–2:

$$R = \frac{\rho l}{A} \tag{3–2}$$

where R = resistance in ohms (Ω)

ρ = specific resistance in circular-mil ohms per foot

l = length of the conductor in feet

A = cross-sectional area in circular-mil (cmil)

Example 3-1

Problem

Let us calculate the resistance of 100 ft of No. 20 copper wire. Note that from Figure 3–1, the ρ for copper is 10.4; from Figure 3–2, the cross-sectional area for AWG 20 wire is 1022 cmil.

Solution

$$R = \frac{\rho l}{A}$$

$$= 10.4 \text{ (cmil } \Omega\text{/ft)} \times 100 \text{ ft/1022 cmil}$$

$$= 1.02\ \Omega$$

We see that resistance increases with length but decreases with thickness. A higher gauge number implies a smaller diameter, higher resistance, and lower current-carrying capacity.

To better understand how the resistance of a conductor is related to other factors, compare a coffee stirrer with a regular drinking straw. Imagine drinking soda with a coffee stirrer, as opposed to a regular straw. Obviously, the coffee stirrer, which is thinner (has a higher AWG number), will need greater pressure (because it offers higher resistance) and draw less liquid (or less current). Telephone cable used indoors is typically 24 or 26 AWG, whereas household electrical wiring is typically 12 or 14 AWG. Most networking cable, such as Category 5 Unshielded Twisted Pair, is 22 or 24 AWG wire. As the resistance increases, signal attenuation increases, or the strength of the signal decreases. Therefore, maximum segment length and AWG specifications for cables must be strictly observed.

Impedance (Z)

Impedance (Z), expressed in Ω, can be defined as opposition to alternating current as a result of resistance, capacitance, and inductance in a component. Characteristic impedance, Z_0, is determined by the square-root of the ratio for inductance in the line to the capacitance between the conductors. For most transmission cables, the size of the conductors, and the spacing and insulation between the conductors remains constant. Therefore, its characteristic impedance, Z_0, is a constant, irrespective of the cable length, as shown in Equation 3–3:

$$Z_0 = \sqrt{\frac{L}{C}} \tag{3–3}$$

where Z_0 = characteristic impedance in Ω
L = inductance in Henry
C = capacitance in Farad

Example 3-2

Problem

If the inductance of a 500 ft cable is 100 mH and its capacitance is 35 μF, find its characteristic impedance.

Solution

$$Z_0 = \sqrt{\frac{L}{C}}$$

$$= \sqrt{\frac{100}{.035}}$$

$$= 53.45\ \Omega$$

Z_0 is an important variable when terminating cables. There is maximum transfer of power from an input to an output when the impedance of the input equals that of the output, or in other words, there is **impedance matching.** To use a transmission line properly, it must be terminated in load impedance equal in value to its characteristic impedance. If different, power is either absorbed by the load or reflected back to the source, or both. In any case, it results in a power loss or a loss in signal strength, which is certainly undesirable. For example, since the characteristic impedance of a 10Base2 coaxial cable is 50 Ω, it must be connected to a 50 Ω cable-terminator so that all of the transmitted power can be absorbed by the load. Also, ¼-inch coaxial cable used in cable distribution systems for television has a characteristic impedance of approximately 75 Ω and must be terminated with a 75 Ω connector.

Echo

Echo or *return loss* is a reflection, as shown in Figure 3–8, that occurs when an electrical signal encounters an impedance irregularity. The greater the distance from a source to an irregularity, the greater the time-delay in the reflected signal. Echo is detrimental to transmission in proportion to the amount of delay suffered by the signal and the amplitude of the echoed signal.

In voice communications, the most serious form of echo arises from imperfect *hybrid* balance in telephones. **Four-wire terminating sets** or hybrids are devices that convert the transmission circuit from four-wire to two-wire, as shown in Figure 3–9. Economics impels the designer to reduce the number of wires as much as possible to minimize costs. The four-wire circuit with two directions of transmission at the local switching office must be combined into a single two-way two-wire circuit for extension through two-wire switching systems and two-wire local loops. When the balancing network fails to perfectly match the actual two-wire loop impedance and the termination impedance of the line card, a signal feeds back to the talker at the distant end as an echo. The impedance of the two-wire loop depends on the length of the loop, the type of cable used, whether or not loading coils are used, the impedance of the customer premises equipment such as telephone sets and modems, and the number of telephones in use (off hook). All these variables make it impossible to know the impedance characteristic exactly, which can result in poor hybrid balance and echo for some loops.

Figure 3–8
Effect of echo or return loss.

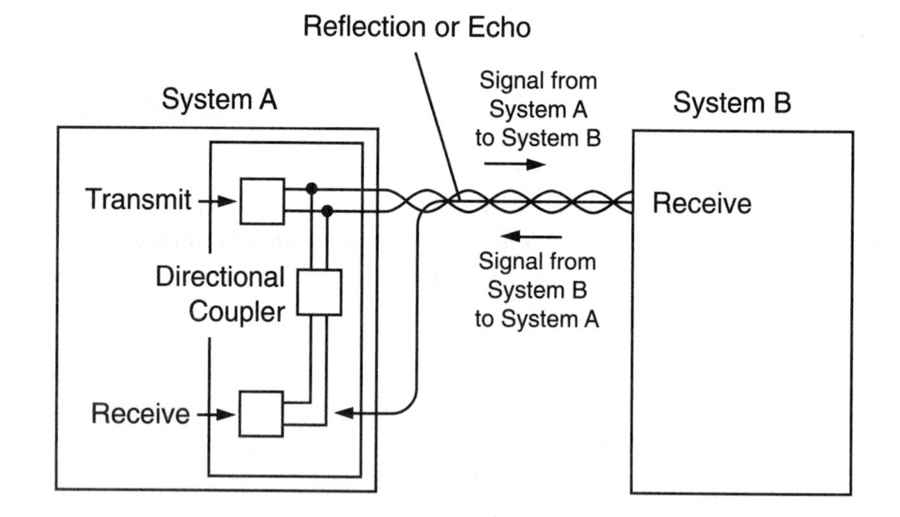

Another source of echo is acoustic echo, which arises because of the coupling from the speaker to the microphone of a telephone set. In a conventional telephone set, this coupling path has a large amount of loss (over 20 dB) and a small delay (less than 1 ms), but its effect is accounted for by lumping it with hybrid balance. Speakerphones have the

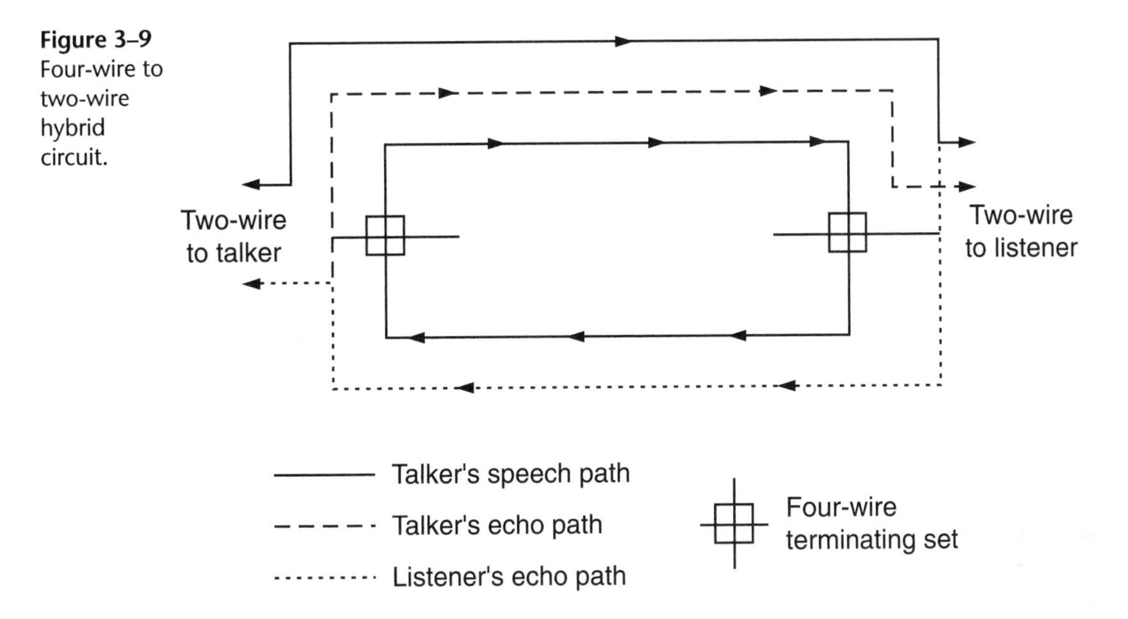

Figure 3–9 Four-wire to two-wire hybrid circuit.

potential of much longer delay paths and loss, which can create echo. Echo control for these types of devices is normally done at the device itself by turning off the transmission path when receiver is active or vice versa.

The degree to which echo is objectionable depends on echo loudness and total delay. The total delay is associated with the time required for analog-to-digital conversion and encoding at both ends, and the transmission time. Delay from transmission is in the range of 30 ms for transcontinental domestic calls, from 50 to 100 ms for international calls, and 500 ms for satellite calls. This delay affects the customer's perception of echo. If the delay is small (less than 10 or 20 ms), the customer hears almost nothing. Larger delays lead to a subjective annoyance perceived as echo. The larger the delay, the less masking there is by the direct speech and the more annoying the echo.

Echo suppressors in analog circuits and *echo cancellers* in digital circuits control echo. Echo suppressors attenuate the reflected signal by approximately 15 dB. Long circuits, such as satellite circuits with round-trip delays of about 0.5 second, require a more effective method of eliminating echo. Such circuits use echo cancellers that perform the same function as echo suppressors but operate by creating a replica of the near-end signal and subtracting it from the echo to cancel the effect.

Crosstalk

Crosstalk refers to the amount of coupling between adjacent wire pairs, which occurs when a wire absorbs signals from adjacent wires. Crosstalk is measured by injecting a signal into one pair and then measuring the strength of that signal on each of the other pairs in the cable. It is classified as either near-end crosstalk (NEXT), depicted in

Figure 3–10, or as far-end crosstalk (FEXT), depicted in Figure 3–11. In wire installations, NEXT is the most important because at the near end the signal source is at its highest level, while the received signal is lowest, having been attenuated by a loss in the wire. Thus, crosstalk is highest at the near end.

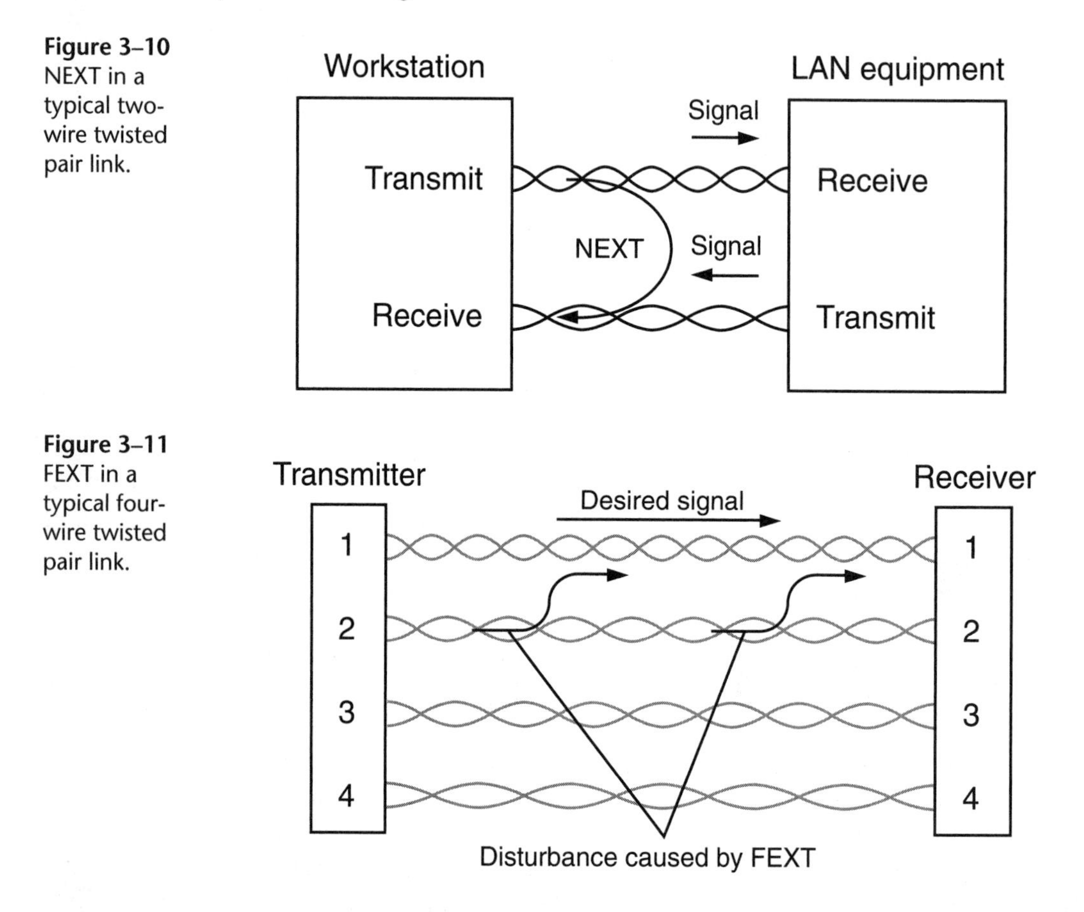

Figure 3–10 NEXT in a typical two-wire twisted pair link.

Figure 3–11 FEXT in a typical four-wire twisted pair link.

The EIA/TIA standard specifies strict crosstalk specifications between pairs. The amount of crosstalk coupling is a function of both the wire itself and the telecommunications outlets in which it is terminated. One cause of NEXT is cabling that has been installed with an insufficient **bend radius**, which can press wire pairs flat inside the cabling or untwist them. The bend radius refers to the radius of the loop when there are bends or angles in the cable route, such as at manholes or pullboxes, as shown in Figure 3–12.

You will find that the twists per foot vary for different pairs in a CAT 3 or CAT 5 UTP cable. Careful control of twists per foot and spacing between adjacent pairs reduces radiation, noise pickup, and crosstalk. In the case of Enhanced CAT 5 and CAT 6, where crosstalk levels are kept to a minimum, the keys are manufacturing techniques. Primarily, the twist ratio between the four pairs is refined: the twists are tighter, and the pairs are

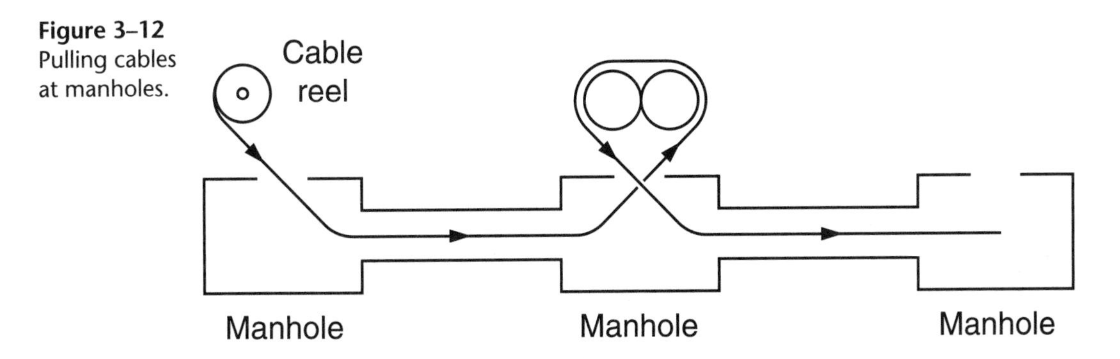

Figure 3–12 Pulling cables at manholes.

balanced out in relation to each other for optimum performance. The quality and consistency of the copper wire is also very important. Basically, a higher grade of cable is less susceptible to the data loss that imperfect installation might lead to.

Unlike Ethernet and Fast Ethernet, Gigabit Ethernet uses all four pairs of CAT 6 for data transmission. Gigabit Ethernet over copper is susceptible to transmission problems such as echo, FEXT, and delay skew. FEXT is not difficult to quantify, but it can be a bit tricky since this measurement varies depending on the length of the cable being tested. Therefore, TSB-95 defined Equal-Level Far-End Crosstalk (ELFEXT) and power sum ELFEXT to make certification in the field much more practical. ELFEXT makes up for the natural discrepancies in cabling lengths by providing more standardized parameters regardless of cable length. Delay skew usually occurs as a result of different insulation materials being used within a cabling plant.

Electromagnetic Interference (EMI)

Electromagnetic Interference (EMI) is a result of electromagnetic (E/M) emissions: every piece of electrically powered equipment transmits and receives E/M energy because all conductors have the potential to act as radio antennas, whether they are tiny filaments on a circuit board or lengths of cable. Also, a power line, which is a source of AC voltage at 60 Hz, is a conductor of interfering RF currents. When a receiver is connected to the power line, the RF interference can produce noise and whistles in the receiver output. This can be minimized through the use of a filter, which is plugged into a wall outlet.

In general, conductors become better antennas as the frequency increases, which is why EMI becomes more of a problem in LANs operating at higher speeds. As E/M emissions increase, they can cause a range of problems, including performance degradation, software crashes, and data corruption. In case of STP, the shield itself can radiate E/M energy if it is not properly grounded. This is a common installation problem in STP.

UTP relies solely on twists or balance to minimize its susceptibility to EMI. During testing, signals of opposite polarity are sent along a twisted pair of the cable. The two wires of a pair receive an equal voltage since they are twisted. If the receiving device does not detect a voltage difference across the pair, it can be concluded that the interference will have no effect. In other words, if the interfering signals are perfectly balanced, they cancel each other out, eliminating the tendency of the cable to act as a radio antenna. Although the balance is rarely perfect in real life, the circuit has sufficient margin that it does not have to be so.

International Cabling Specifications

Although there are many cabling specifications, as a case in point, let us examine the electromagnetic compatibility (EMC) issues. Anyone who has firsthand experience with a cordless phone that picks up strange transmissions or an electric garage door opener that seemingly operates on its own is well aware that many new products still reach production without consideration of EMC. Meeting EMC requirements in the United States and Europe poses real challenges for manufacturers trying to sell into both markets.

Tackling these EMC problems has become all the more difficult as standards keep evolving and markets become increasingly global. For instance, many manufacturers build products for sale in both the United States and Europe and have to address the EMC regulatory requirements of both regions. While there are many similarities between the EMC requirements for these two continents, there remain many differences as well. In an effort to remove international barriers to trade, U.S. and European officials signed a mutual recognition agreement (MRA) in May 1998. One of the areas covered in the MRA is EMC.

Instead of a direct equivalence between European and U.S. regulations, the MRA supports the mutual recognition of test results and other conformity assessment documentation. Once implemented, this would mean that U.S. manufacturers could go to a local test facility to be certified by U.S. agencies for compliance with European regulations. Similarly, European laboratories would be certified to carry out EMC testing as defined by the FCC for manufacture of products sold in the United States. One of the major points of departure today between U.S. and European regulations lies in testing limits. Traditionally, the upper frequency limit for commercial EMC testing of emissions has been one GHz. But the FCC recently raised the requirements for radiated immunity testing to 40 GHz. This decision was driven by the proliferation of new communications equipment operating above one GHz, such as wireless devices, as well as the rising clock rates of desktop computers. European regulations, on the other hand, presently stop at 1 GHz. But new limits, envisioned in EB55022, will gradually raise the limit to 2.5 GHz, 5 GHz and eventually 18 GHz.

Devices that intentionally radiate, such as cell phones or other radio transmitters, typically fall under the third compliance category called Type Acceptance. In this case, the manufacturer or its agent must obtain certification from a notified body or government agency. In fact, the emissions characteristics of intentional transmitters and receivers are covered by a variety of documents in both the United States and Europe depending upon power output, operating frequency, type, location of use, and antenna type. One of the major forces driving the introduction of new test equipment for EMC compliance are the stringent regulations for intentional radiators. The high risks associated with failure to comply with EMC regulations are driving more design teams to employ pre-compliance test equipment early in the product development cycle. Whether design teams decide to use a pre-compliance system, perform a full EMI test themselves, or resort to a third-party laboratory, one issue to keep in mind is that regulations are in a state of transition. One must keep abreast of changing EMI requirements both in the United States and Europe to help eliminate any unpleasant surprises.

COPPER VERSUS FIBER

Optical fibers have several advantages over copper cables: immunity to EMI, lightning, electrical discharges, and crosstalk, no electrical ground loop or short circuit problems, and resistance to nuclear radiation and high temperatures. Also, there is no electrical hazard when a fiber-optic cable is cut or damaged. More importantly, a fiber can carry thousands of times more information than can a copper wire of the same size. Optical fiber cables are lighter and take less space than copper cables for the same information capacity. Fibers also have longer cable runs between repeaters because a signal loses very little strength as it travels down a fiber, as compared with copper. Optical fiber losses are independent of the transmission frequency on a network; there is no crosstalk that can degrade or limit the performance of fiber as network speeds increase. As a result of the crucial advantages that fiber offers over copper, it is often used in backbone wiring, noisy environments such as factory floors, and security-conscious installations like military and banking.

Yet, the reason copper remains at the cabling forefront is price. Fiber-optic cable is relatively expensive and has higher installation cost because of the need for specialized personnel and test equipment. Although fiber is only about 30 percent more expensive than copper, the cost of fiber networking hardware like Network Interface Cards (NICs) and fiber hubs is anywhere from two to five times as much as their copper equivalents. However, as these prices continue to decrease and our bandwidth requirements continue to increase, we see more and more fiber being installed in networks.

FIBER-OPTIC CABLES

The rapid implementation of optical telecommunications has significantly aided the growth of information technology in the 21st century. Ready access to the Internet and the decreasing cost of long-distance telephone calls are in part a result of the high capacity of optical fiber telecommunications links. Optical carriers are designated according to their transmission capacity. **Fiber-optic cable** is a transmission media designed to transmit digital signals in the form of pulses of light. It was not until the 1950s that the first optical fiber was made. Although this optical fiber could transmit light, it did not carry information very far, as most of the signal was attenuated or lost in transmission. In fiber, the loss or attenuation is measured in decibels per kilometer (dB/km). In 1970, three

Corning scientists, Robert Maurer, Donald Keck, and Peter Schultz developed the first optical fiber with losses less than 20 dB/km. Today, losses typically range from 0.2 to 2.0 dB/km, which vary with the wavelength of light. Fiber optic communications use the wavelengths in the near-infrared region: 850, 1300, and 1550, nanometers (nm).

Fiber Construction and Types

An optical fiber, illustrated in Figure 3–13, is made of either glass (extremely pure silica) or plastic. It consists of an inner layer called **core** through which light travels. The core is surrounded by an outer layer called **cladding**, which contains the light within the core. These layers are protected by a jacket or coating. A fiber is thinner than a human hair but stronger than a steel fiber of similar thickness. The sizes of the fiber have been standardized nationally and internationally. For example, when expressed as 62.5/125, the first number is the core diameter and the second number is the cladding diameter in microns or μm. Basically, there are two types of fibers: singlemode and multimode.

Figure 3–13
Typical fiber cross-section.

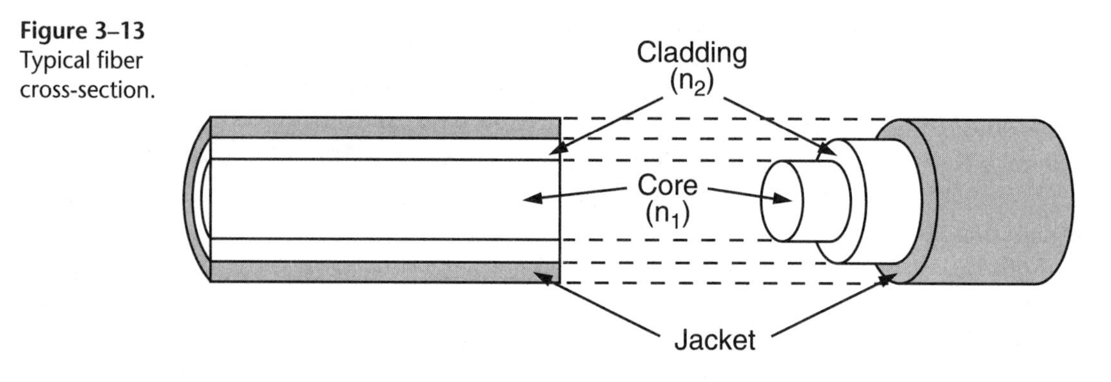

Singlemode Fiber

Singlemode fiber has a very small core and is designed to carry only a single light ray, as shown in Figure 3–14. Typical core diameters are 2 to 9 μm. Singlemode fiber has a much higher capacity (GHz to THz), is most efficient, and allows longer distances than does multimode fiber. However, it is difficult to work with because of its small core diameter, especially when it comes to splicing (permanent joining of two fibers) or terminating the fiber. It is typically used for applications such as LAN backbones, WANs, telephone company switch-to-switch connections, and CATV.

Multimode Fiber

Multimode fiber can be classified into either step-index or graded-index fiber. It is designed to carry multiple light rays or modes concurrently, each at a slightly different reflection angle within the optical fiber core, as shown in Figure 3–15 (a) and (b). The step-index fiber has a sharply defined boundary between the core and the cladding when

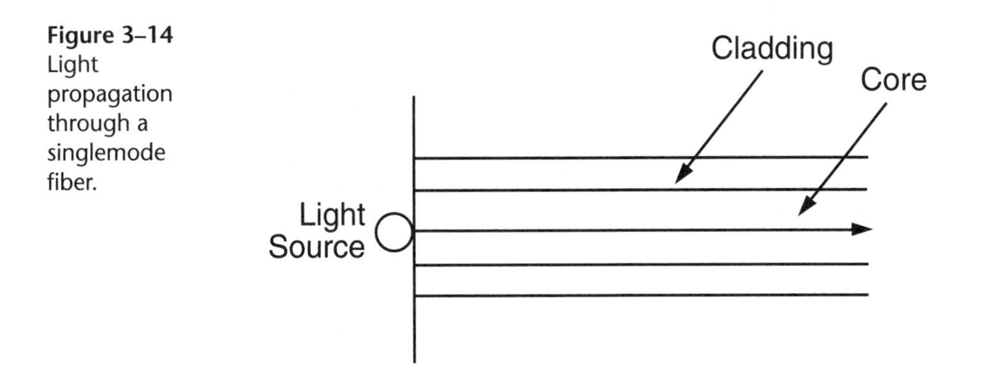

Figure 3–14 Light propagation through a singlemode fiber.

compared with the graded-index. In a multimode fiber, the glass core diameter varies from 50 to 200 μm. In North America, the most common size is 62.5/125; in Europe, 50/125 is often used. When compared to singlemode, multimode is less expensive, easy to terminate, lends itself to addition of end connectors, and can result in more modes of light than can be accomplished with small-core diameters. These fibers are typically used in LANs for short runs less than few km, where the required signal bandwidths are smaller (a few hundred MHz).

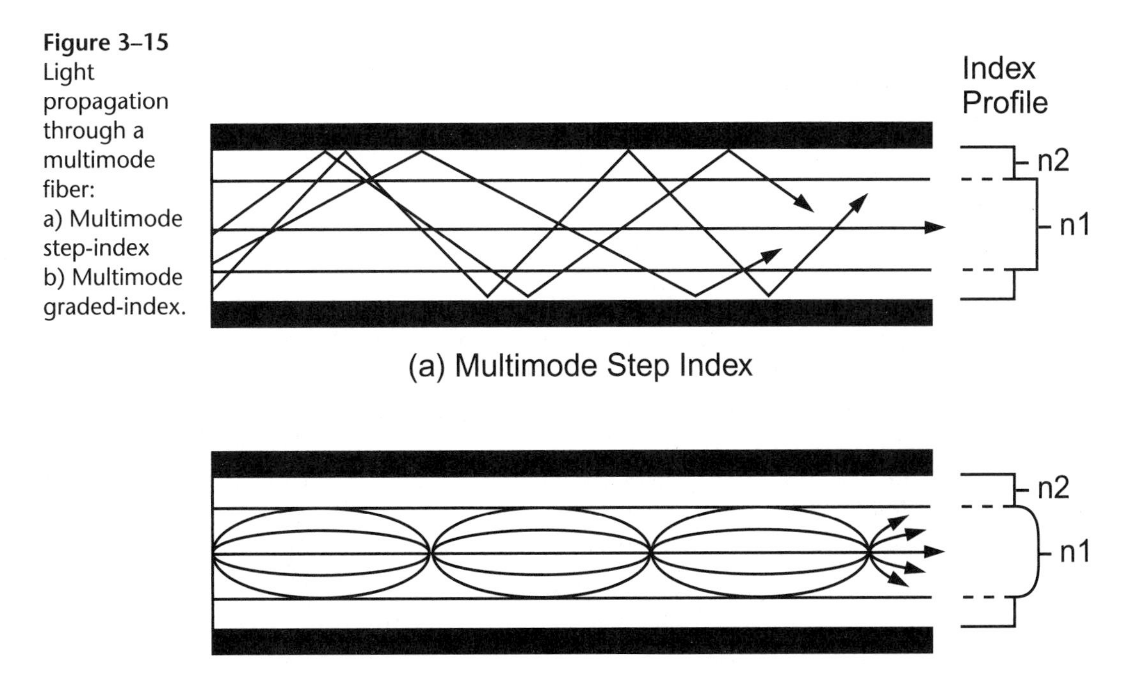

Figure 3–15 Light propagation through a multimode fiber: a) Multimode step-index b) Multimode graded-index.

Light Propagation through Fiber

When a light ray travels from one medium to another, both *reflection* and *refraction* can take place, as shown in Figure 3–16. Reflection occurs when light bounces back in the same medium; refraction occurs when light changes speed as it travels in the second medium and is bent or refracted. The factor by which light changes speed is the **refractive index** or *index of refraction*, n, which is a ratio between the speed of light in free space, *c*, and the speed of light in the medium. A good example of refraction is the case of a fisherman looking at a fish in a pond from his boat. The fish is not exactly where the fisherman sees it, so the fisherman must compensate by putting the rod in at a different angle.

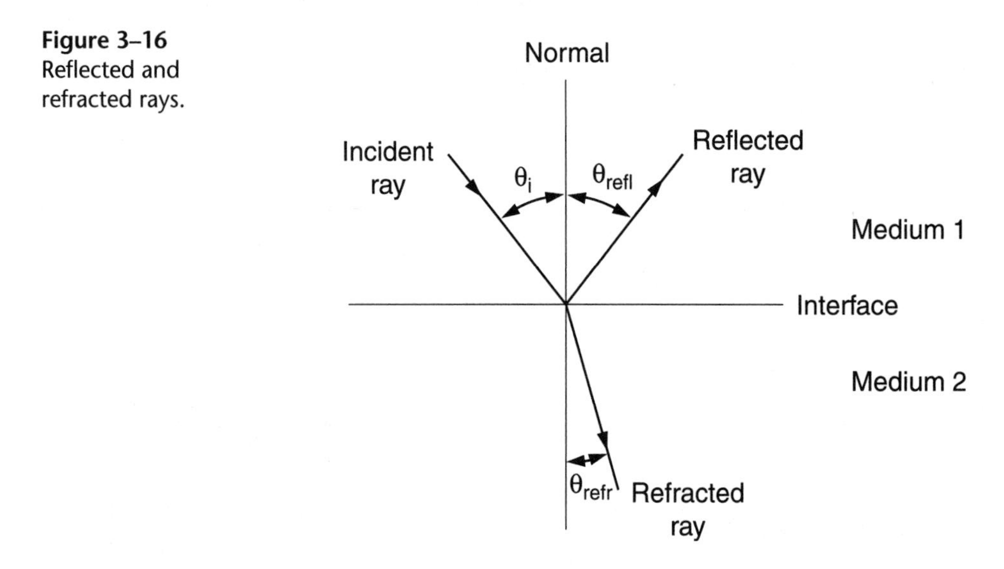

Figure 3–16 Reflected and refracted rays.

There is a correlation between the path the light will follow as it travels from one medium to another and the refractive indices of two media. There are three important cases that define the type of reflection/refraction that can be obtained when light goes from one type of medium to another. As shown in Figure 3–17(a), when $n_1 < n_2$, light bends toward the normal, so that the angle of refraction (θ_{refr}) is less than the angle of incidence (θ_i). The angle of incidence is the angle between the light in the first medium and the normal, which is an imaginary line perpendicular to the interface between the two media. The angle of refraction is the angle between the light in the second medium and the normal.

When $n_1 > n_2$, as illustrated in Figure 3–17(b), light bends away from the normal so that the angle of refraction (θ_{refr}) is greater than the angle of incidence (θ_i). As the angle of incidence increases, the angle of refraction approaches 90°. When the angle of refraction is exactly 90°, the light does not enter the second medium but is reflected along the interface, as depicted in Figure 3–17(c). The angle of incidence when this occurs is known as the *critical angle* (θ_c).

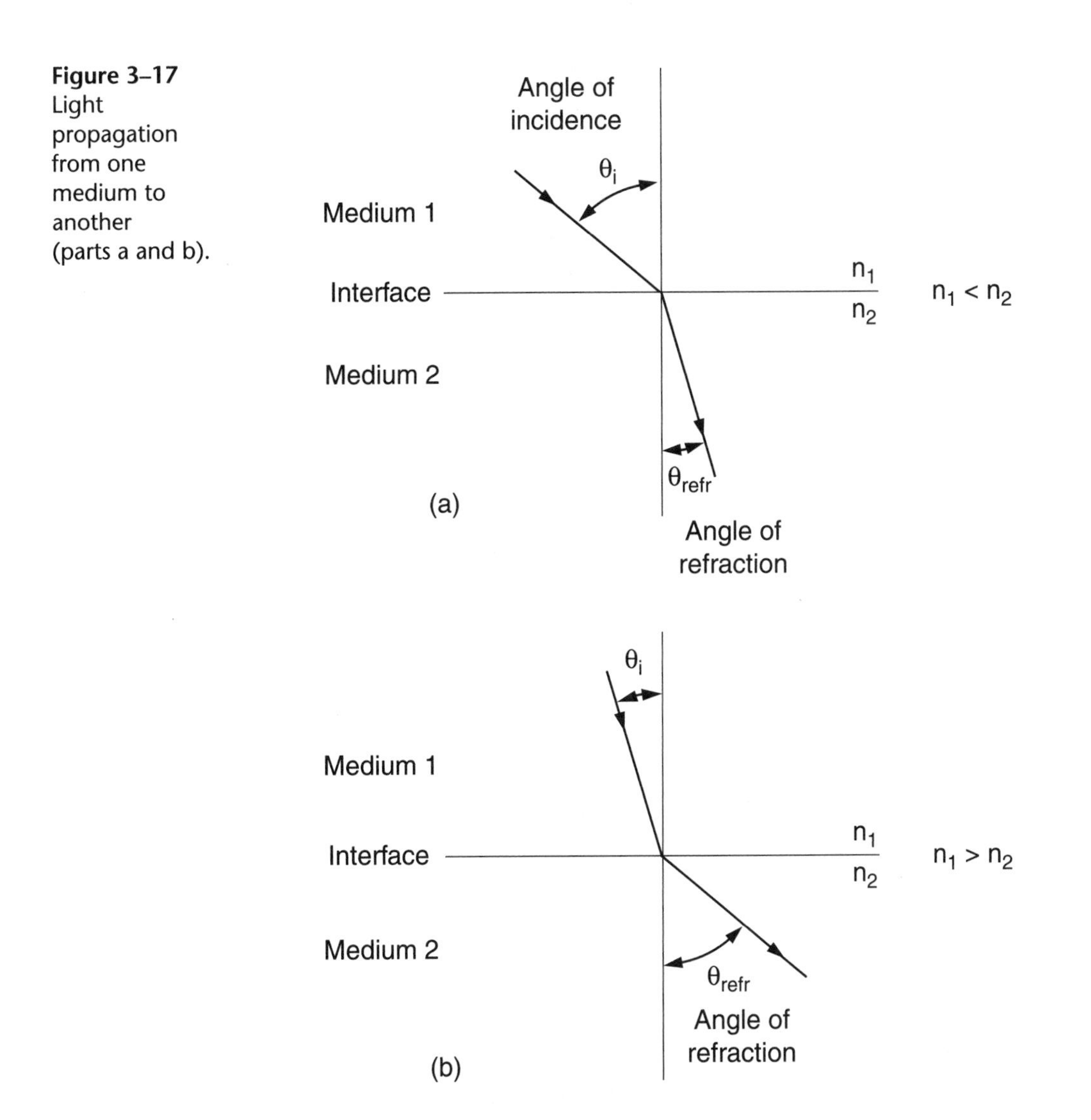

Figure 3–17 Light propagation from one medium to another (parts a and b).

As the angle of incidence increases past the critical angle, light is reflected at the interface and does not enter the second medium, as shown in Figure 3–17(d). This is *total internal reflection.* The angle between the reflected light and the normal is the angle of reflection, which is always equal to the angle of incidence as long as the angle of incidence is greater than the critical angle. In case of fiber optics, light is refracted from a light source into the cable end and then propagates down the cable by total internal reflection.

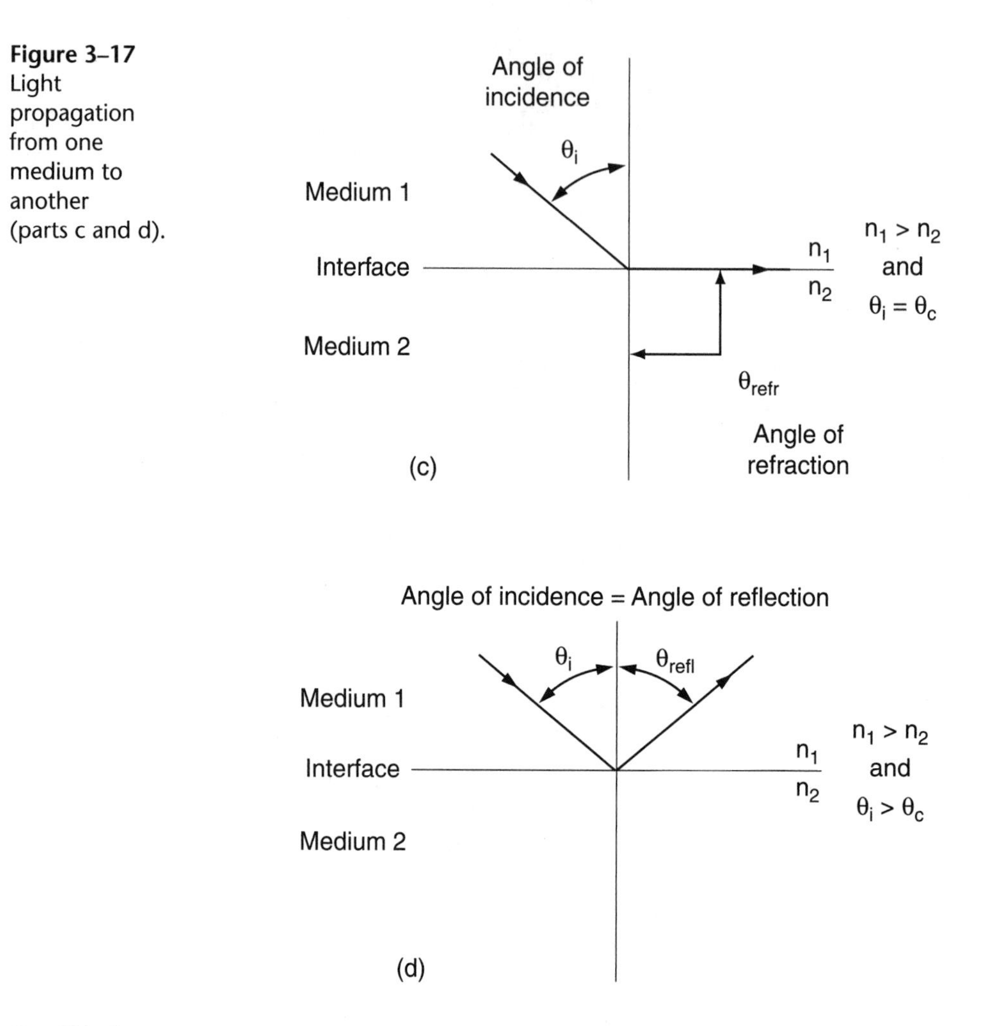

Figure 3–17 Light propagation from one medium to another (parts c and d).

Snell's Law

Snell's Law states that a relationship exists between the refractive indices of the two media, n_1 and n_2, and the angle of incidence and refraction, θ_i and θ_{refr}. This relationship is algebraically expressed as shown in Equations 3–4, 3–5, and 3–6:

$$\frac{\sin\theta_i}{\sin\theta_{refr}} = \frac{n_2}{n_1} \tag{3–4}$$

As shown in Figure 3–17(c), when $\theta_i = \theta_c$ $\theta_{refr} = 90^\circ$, and $\sin\theta_{refr} = 1$. Therefore,

$$\sin\theta_c = \frac{n_2}{n_1} \tag{3–5}$$

or

$$\theta_c = \sin^{-1}\left(\frac{n_2}{n_1}\right) \tag{3–6}$$

Snell's law is one of the key theories behind the propagation of light along a fiber. To make light travel down a fiber, the angle of incidence has to be greater than the critical angle. If the critical angle is known, the ratio of refractive indices is also known. This provides a value needed to decide what types of materials will become the core or the cladding.

The core material has a refractive index of n_1, and the cladding has a refractive index of n_2. The core has a higher refractive index than the cladding, which results in **total internal reflection** only when light strikes the core-cladding interface at an angle greater than the critical angle. Since the angle of incidence is equal to the angle of reflection, the light will continue to travel down the fiber cable by total internal reflection. Any light striking the interface at less than the critical, that is, not within a region called the *acceptance cone*, will be absorbed or lost into the cladding.

For light to be guided in the core, it must be launched in the fiber from the outside. The *acceptance angle* (Θ) is the greatest possible angle at which light can be launched into the core and still be guided through total internal reflection. It can be derived by using the law of refraction, which is represented in Equation 3–7:

$$\frac{\sin\theta_i}{\sin(90^\circ - \theta_c)} = \frac{n_2}{n_1} \tag{3–7}$$

But, $\sin(90^\circ - \theta_c) = \cos\theta_c$, and $\sin^2\theta + \cos^2\theta = 1$. Therefore,

$$\sin(90^\circ - \theta_c) = \sqrt{1 - \sin^2\theta_c}$$

Also, since the refractive index of air is 1, $n_1 = 1$, and the equation becomes:

$$\sin\Theta = n_2\sqrt{1 - \sin^2\theta_c} \tag{3–8}$$

Using the critical angle formula in Equation 3–5, we have:

$$\sin\Theta = \sqrt{n_1{}^2 - n_2{}^2} \tag{3–9}$$

where n_1 is the refractive index of the core

n_2 is the refractive index of the cladding

Θ is the acceptance angle

Numerical Aperture (NA)

Numerical Aperture (NA) is the sine of the acceptance angle, which can be described as the light-gathering ability of an optical fiber. The larger the NA, the greater the amount of light that can be accepted into the fiber, hence, the greater the transmission distance that can be achieved. But if the NA is too great, the bandwidth of the system degrades. The NA value is always less than one, 0.21 for graded-index fibers, and 0.5 for plastic. In a single-mode fiber, since light is not reflected or refracted, there is no acceptance angle, and the NA is rarely specified.

Example 3–3

Problem

A light ray is incident from air into a fiber. The index of refraction of air is equal to 1. The index of refraction of the fiber core is 1.5, while that of the cladding is 1.48. Find the critical angle, the acceptance angle, and the light-gathering ability of the fiber.

Solution

Critical angle:

$$\sin\theta_c = \frac{n_2}{n_1} = \frac{1.48}{1.5}$$

$$\theta_c = \sin^{-1}\left(\frac{n_2}{n_1}\right) = \sin^{-1}\left(\frac{1.48}{1.5}\right)$$

$$= 80.63°$$

Acceptance angle:

$$\sin\Theta = \sqrt{n_1^2 - n_2^2}$$

$$= (1.5)^2 - (1.48)^2$$

$$= 0.244$$

$$\Theta = 14.13°$$

Light-gathering ability is same as Numerical Aperture:

$$NA = \sin 14.13° = 0.244$$

Optical Sources and Detectors

In fiber-optic transmission, attenuation varies with the wavelength of light; there are three low-loss *windows* of interest: 850 nm, 1300 nm, and 1550 nm, as shown in Figure 3–18. The 850 nm window is perhaps the most widely used because 850 nm devices are inexpensive. The 1300 nm window offers lower loss, but at a modest increase in cost for LEDs. The 1550 nm window is mainly of interest to long-distance telecommunications applications and requires the use of laser diodes.

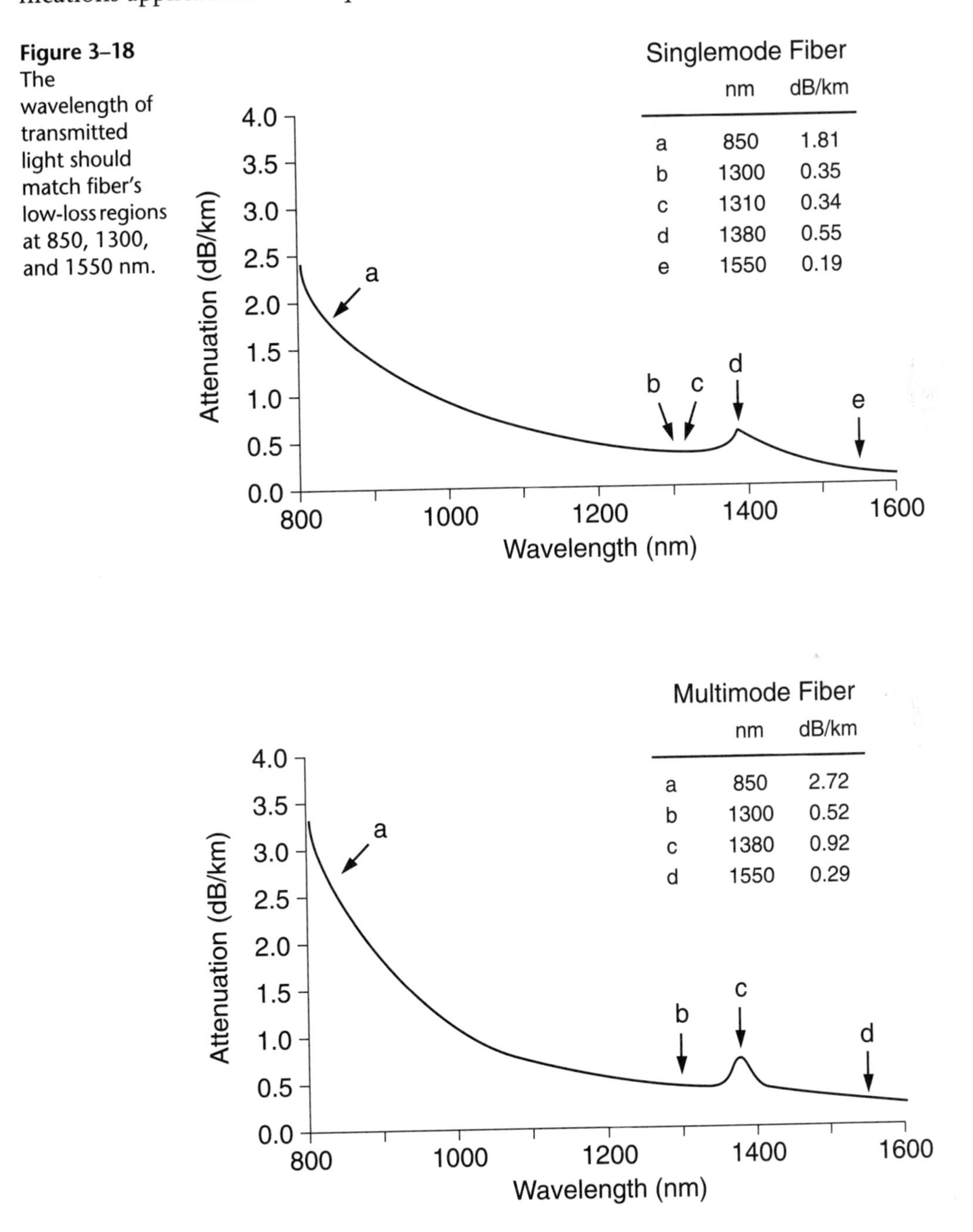

Figure 3–18 The wavelength of transmitted light should match fiber's low-loss regions at 850, 1300, and 1550 nm.

For bit rates less than 50 Mbps, a LED is an excellent choice for an optical source for use in a fiber optic link because of its long life span (10^6 hours), operational stability, wide temperature range, and low cost. A laser diode is used in fiber-optic links primarily because they are capable of producing 10 dB more power than an LED, in addition to emitting *coherent* or monochromatic light, as illustrated in Figure 3–19. Coherence means that all the light emitted is of the same wavelength. The choice of the source type (LED or laser diode) is based primarily on distance and bandwidth. Figure 3–20 shows the general principles of launching laser light into fiber. The laser diode chip emits light in two directions: to be focused by the lens onto the fiber in one direction, and in the other direction onto a photodiode. The photodiode, which is angled to reduce back reflections into the laser cavity, provides a way of monitoring the output of the lasers and providing feedback so that adjustments can be made.

Optical detectors, typically photodiodes, are used as light receivers to convert the optical energy into electrical energy. Since the optical signal is weak, the sensitivity of the detector is critical for the overall fiber optic link performance. Two types of optical detectors are widely deployed, the positive-intrinsic-negative (PIN) photodiode and the avalanche photodiode (APD). In a PIN photodiode, light is absorbed and photons are converted to electrons in a 1:1 relationship. APDs are similar devices, but provide gain through an amplification process—one photon releases many electrons. PIN photodiodes are low cost but less efficient, while APDs, although more expensive, have higher sensitivity and accuracy.

In some applications, it is desirable to have a transmitter and receiver in a single package called a *transceiver.* A transceiver sends and receives a signal usually over two separate fiber cables, and the dual circuits are isolated from one another. A *repeater* contains a receiver and a transmitter that are connected in series. The receiver detects the signal, amplifies and regenerates it, and produces an electrical signal that drives the transmitter in the repeater. Repeaters are used in long-span links to overcome distance limitations.

Construction of a Fiber-Optic Cable

Choosing the right fiber-optic cable has become more challenging than ever because of the advent of new cable designs, the number of suppliers, and changes in fiber specifications. From a technical standpoint, more than one type of cable may be appropriate for many applications. In that case, other factors such as ease of use, size, and cost should be considered in the evaluation and selection process. A typical fiber-optic cable, shown in Figure 3–21, contains fibers, coating, buffer tube, strength member, and an outer jacket. The innermost member of the cable is a support element made of steel or fiberglass/epoxy material. The individual fiber cables are stranded around the central member and consist of just the optical fiber, coating, and buffer tube. The buffer is used as a cushion to provide radial protection and enhance the tensile strength of the fiber.

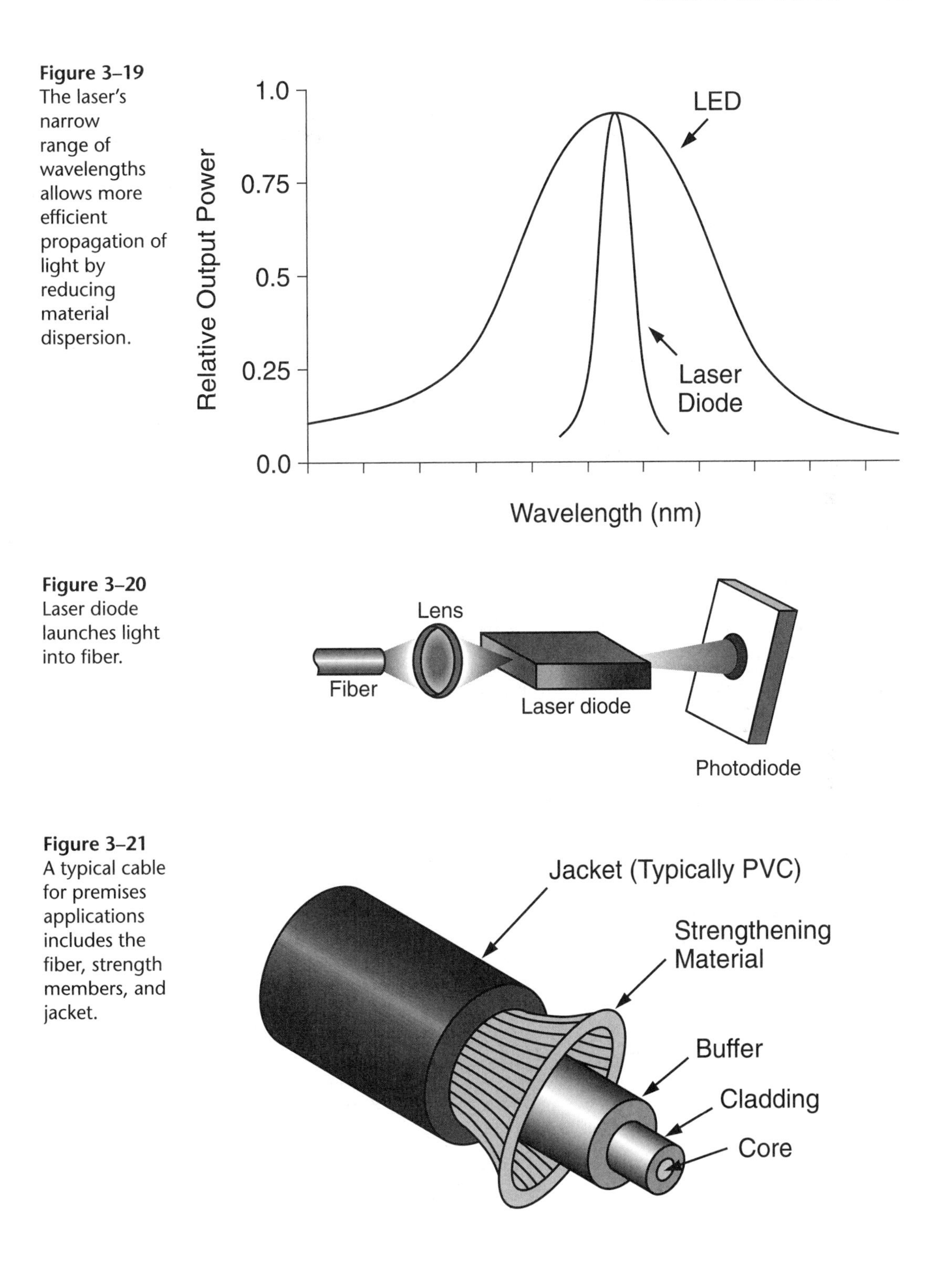

Figure 3–19 The laser's narrow range of wavelengths allows more efficient propagation of light by reducing material dispersion.

Figure 3–20 Laser diode launches light into fiber.

Figure 3–21 A typical cable for premises applications includes the fiber, strength members, and jacket.

Loose Buffer

Loose buffer allows the fiber to move inside, which relieves the cable from stresses occurring during installation and frequent handling. Typical applications of loose buffer cable are outside installations, so the space between the tubes is sometimes gel-filled to give better waterproofing protection to the fiber. The outdoor environment subjects a cable to the most extreme range of environmental conditions—a wide operation-temperature range, thermal shock, wind loading, ice loading, moisture, and lightning. Therefore, protecting and preserving the optical properties of the fiber is a design priority.

Loose-tube cables, whether flooded under the jacket or water-blocked with dry, swellable materials, protect the fibers from moisture and the long-term degradation moisture can cause. The gel within the loose-tube construction stops the penetration of water and keeps it away from the fiber. In cold temperatures, the protection keeps water from freezing near the fiber and eliminates possible stress fractures. The loose buffer construction also results in reduced *macrobending* stresses, which occur when optical fibers are wound on reels for transportation and during the installation process. However, gel-filled cable requires the installer to spend time cleaning and drying the individual cables, and cleaning up the site afterward.

Tight Buffer

Tight buffer refers to layers of plastic and yarn material applied over the fiber. This results in a smaller cable diameter, a smaller bend radius, and greater flexibility. Tight-buffered cable is generally easier to prepare for connection or termination, but it does not provide protection from water migration, nor does it isolate fibers well from the expansion and contraction of other materials as a result of temperature extremes. Tight-buffered cables, often called premises or distribution cables, are ideally suited for indoor-cable runs such as patch cords and LAN connections because the indoor environment is less hostile and not subject to the extremes seen outdoors. These cables must conform to the NEC requirements.

Joining Fibers

Connecting fibers is a critical part of fiber optic cabling. No matter what type of joining technique is used, the ultimate goal is to let the light go from one point to another with as little loss as possible. A *splice* welds, glues, or fuses together two ends of a fiber and unites two fibers into one continuous length. A fusion splicer is depicted in Figure 3–22. *Connectors* are nonpermanent joints used to connect optical fibers to transmitters and receivers or panels and mounts. A splice is considered a more permanent joint than that created by a connector; splices are used for long-haul, high-capacity systems, while connectors are used for short-distance and end terminal equipment. Connectors are becoming increasingly easier to handle, mount, and install. However, one must follow specific directions to prepare the fiber for a particular type of connector—the type of epoxy or

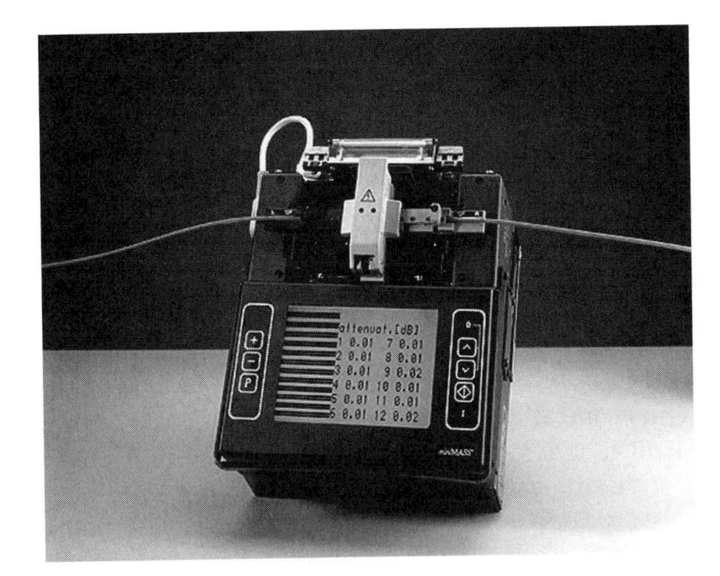

Figure 3–22
Fusion splicer.

(photo courtesy of Corning Cable Systems)

cementing agent, the length of the jacket, the strength member, and the fiber that must be stripped back. A single-mode fiber, because of its small core diameter, is more difficult to connect or splice than a multimode fiber.

Couplers are used to split information in many directions. When using WDM, one fiber can carry more than one signal simultaneously using different wavelengths. A single fiber, using a bi-directional coupler, may be used to both send and receive optical signals. Couplers may also be used to divide an optical signal from a single fiber across multiple fibers. For instance, a three-port coupler splits the incoming signal into two outgoing channels, which has applications in LANs.

Transmission Impairments in Fiber-Optic Cables

Although fiber-optic cables are immune to EMI and crosstalk effects, they are susceptible to other factors such as dispersion—which limits the bandwidth of a fiber—and absorption, scattering, and bending losses that contribute to a loss of signal.

Dispersion

Dispersion refers to pulse broadening or spreading of the light as it travels down the optical fiber. The fiber acts like a low-pass filter, letting low frequencies pass and attenuating the rest. A light ray tends to disperse more over longer lengths. Dispersion influences the bandwidth, bit rate, and pulse shape of the fiber. It is most often measured in picoseconds per nanometer-kilometer (ps/nm-km), where ps describe the increase in pulse width, nm measures the pulse width of a typical light source, and km represents the length of the cable. There are different types of dispersion that occur in different types of fiber.

Material dispersion, found in both singlemode and multimode fibers, is dependent on the dopants of the core glass. In multimode fiber, different modes propagating at different speeds result in *modal dispersion* or *Differential Mode Delay (DMD).* In DMD, a single wavelength is split into multiple beams, typically because of the structure of the fiber core. These beams travel on two or more paths, which may vary in length and may have different transmission delays, as shown in Figure 3–23 (a) and (b). In effect, a signal injected into the fiber will travel over several different paths and be received at the end at slightly different times. This variation can cause jitter, a condition where data transmission is impaired or even prevented altogether.

DMD is typically 15 to 30 ns/km, and if the distance is doubled, the dispersion time doubles. DMD can also be expressed in frequency, such as 100 MHz-km, which indicates that the highest operating bandwidth is 100 MHz for a 1 km fiber. Although its effect may be insignificant at short distances, DMD could limit the bandwidth of a fiber-optic system that transmits data over longer distances. In some cases, it can be addressed by using a special type of patch cord that conditions the laser signal. The graded-index fiber can be made to have lower DMD, and, therefore, higher bandwidth than step-index fiber. As single-mode fiber has only one transmission path there is no DMD, resulting in highest bandwidth (GHz to THz).

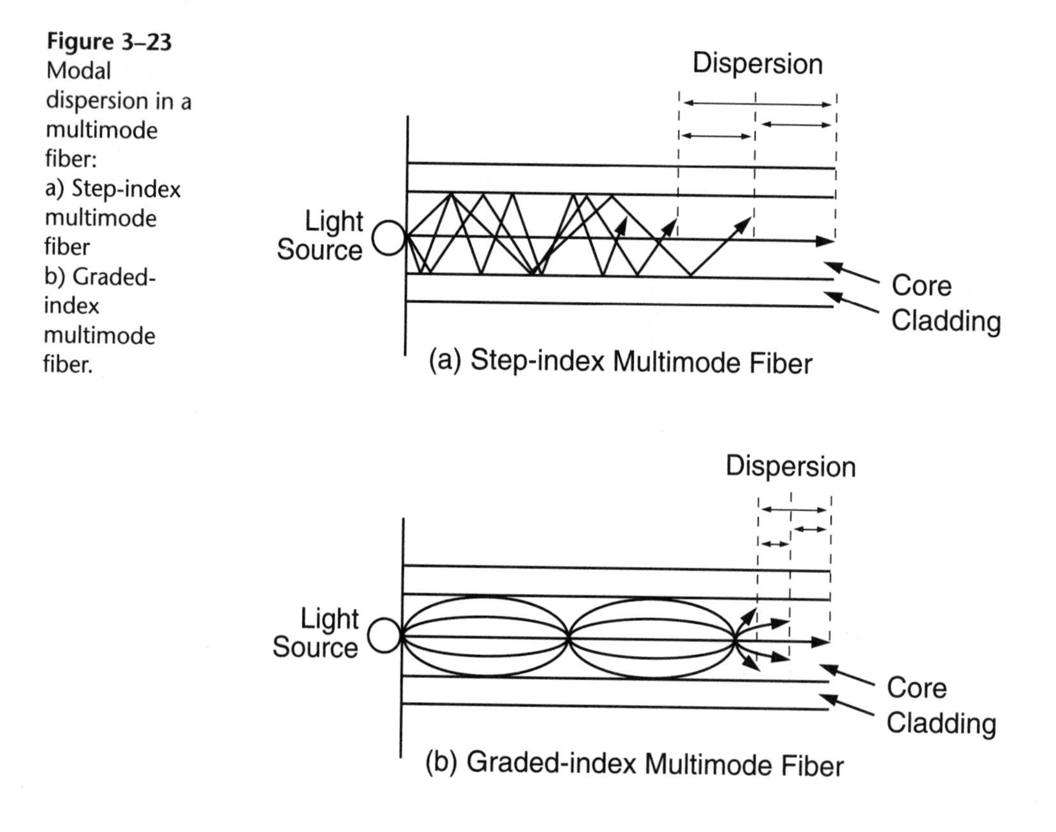

Figure 3–23 Modal dispersion in a multimode fiber: a) Step-index multimode fiber b) Graded-index multimode fiber.

Attenuation in fiber-optic cables is a result of scattering, absorption, and bending losses. *Scattering* is a result of imperfections in the glass fiber as it is heated in the forming process. The microscopic variations become fixed in the glass causing mirror-like reflections in the fiber. This loss can be reduced by controlling the cooling of the fiber. *Absorption* is basically a material property, primarily a result of the atomic resonance in the glass structure. *Bending losses* occur because of improper installation and can be reduced by refining the manufacturing and installation techniques.

CABLING ARCHITECTURE

Network cabling infrastructure is the most important part of a network—an essential building block without which a system may fail to function properly or at optimum efficiency. All too often, organizations fail to identify the importance of structured cabling systems for future expansion. Although the sluggish movement of data is usually blamed on hardware or software, over half of all network problems are related to cabling systems.

The ANSI/TIA/EIA 568-B, published in March 2001, replaces the current standards document ANSI/TIA/EIA 568-A, dated October 1995. TIA is chartered to review the content of industry standards every five years and re-publish the documents. Although there is overlap between 568-A and 568-B, a significant technical update in 568-B is that CAT 5 is no longer recognized, and it has been replaced by CAT 5E. Also, the term *Telecommunications Closet* has been replaced with *Telecommunications Room (TR)*. The Telecommunications Room is generally the connection point between the building backbone cable and the horizontal cable.

When it comes to cabling architectures, network managers basically have three choices: a conventional distributed copper setup, a fiber distributed scheme, and a centralized setup in which all LAN equipment is in one place. Conventional distributed cabling schemes are based on a fiber backbone and star-wired copper to the desk. Fiber distributed schemes are like conventional distributed schemes except that fiber is used everywhere. In centralized cabling, all switches, hubs, and cable connecting equipment are in a central location like the building's basement. Centralized architecture provides greater security and is easier to maintain and troubleshoot since all the telecommunications devices are in one spot.

Structured Wiring

Although **structured wiring** has been used to connect telephones for decades, for many years the practice of wiring between data equipment was largely unstructured and improvised to satisfy short-term needs. Just a few years ago, prior to structured wiring, it was very simple to install new telecommunications cabling. There were no stringent distance limitations, no pathway constraints, and no closet requirements. However, with the increase in desktop equipment throughout the workplace, structured wiring has become a critical focal point of effective site planning. With the introduction of cabling standards, specifically TIA/EIA 568, an installer is required to meet more stringent installation standards to protect the integrity of the cabling system and to eliminate the need for constant recabling with the addition of each new application.

As a result of the standards, many companies now have well-defined, structured cabling systems as an integral part of their building structure. In a structured environment, active equipment like routers, switches, bridges, repeaters, and servers are located in TCs for security reasons. In a distributed network with a 10,000 sq. ft. serving area, a 10×11 ft TC is recommended. Using collapsed backbone architecture, it is possible to decrease the size of these closets by referring to sizes in an annex of the TIA/EIA-569-A standard. Refer to Figure 3–24 for a typical structured wiring layout.

In addition to being easier and cheaper to maintain and upgrade, structured wiring offers significant advantages over unstructured wiring:

- Promotes an efficient and economical wiring layout that technicians can easily follow
- Enhances problem detection and isolation with standardized layout and documentation
- Ensures compatibility with future equipment and applications

The TIA/EIA 568 standard addresses voice, data, and video distribution. Its goal is to define a wiring system that supports a multivendor, multiproduct environment. There is a consensus within the ISO wiring committee to conform the TIA/EIA standard to its international equivalent (IEC) specifications to form a unified international wiring standard. The recommended wiring system topology is a hierarchical star, which supports both centralized and distributed systems and provides central points for management and maintenance. Using cross-connects, the star topology can be configured as a bus, ring, or tree. The wiring system is classified into three main elements:

1. Backbone wiring,
2. Horizontal wiring, and
3. Work Area wiring

Backbone wiring is the connection between the Telecommunications Room (TR) and the equipment room within a building, and the connection between buildings. A maximum of two levels of cross-connecting is recommended for the backbone; the intermediate cross-connect and the main cross-connect. This is exclusive of any cross-connect in the TR where the horizontal connects to the backbone. The maximum distance from the TR to the intermediate cross-connect is 500 m for all media types. Distances to the main cross-connect are media dependent.

Horizontal wiring refers to the connection between the work area and the termination in the TR. It is limited to a maximum of 90 m. This is independent of the media type so that the TR is common to all media and all applications operating over the media. In addition, there is an allowance for 3 m in the work area and 6 m for cross-connecting in

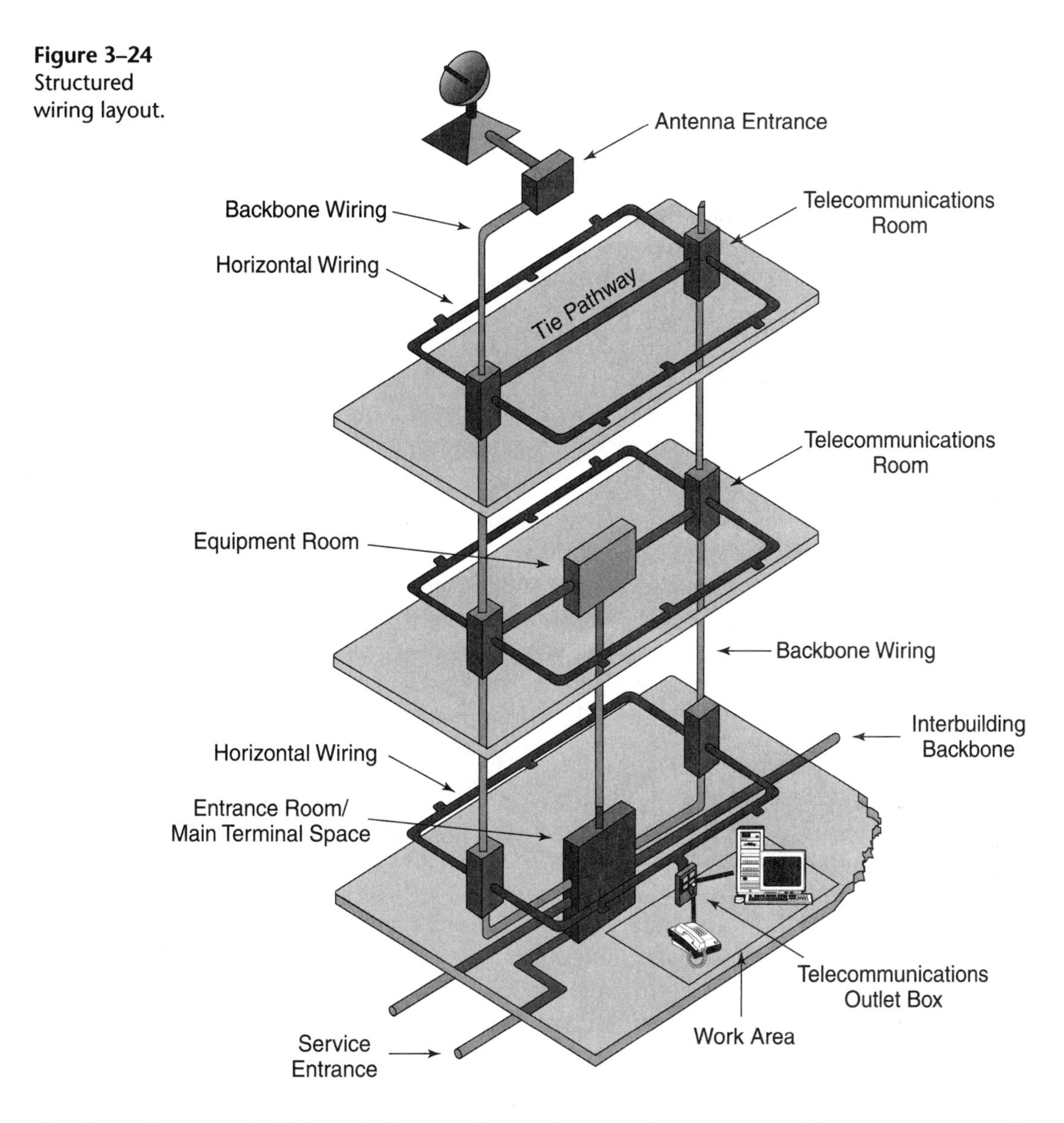

Figure 3–24 Structured wiring layout.

the closet for a total of 99 m. When applying this specification, it yields a maximum end-to-end length of 100 m including patch cords. As an example, let us consider a typical LAN grade multimode fiber with a bandwidth of 200 MHz per km. Since the current structured cabling standard allows 100 m (or 0.1 km) lengths of horizontal fiber cabling, each length can support 2 GHz (or 2000 MHz) of bandwidth. Since users do not yet feel the need for 2 GHz bandwidth to the desktop, there are very few fiber-to-the-desk cabling systems today. Most of the horizontal wiring is CAT 5 UTP cable.

Work Area wiring refers to the connection between a user station and the outlet. In most commercial installations one CAT 3 for voice and one CAT 5 for data should be the minimum to be installed. Work Area wiring is not permanent wiring, and the standard provides a means for the specific application (communication system) to adapt to the building wiring. The telecommunications outlets in the work area must also meet the specified physical jack arrangement. Normally, jacks and crossconnects are designed so that the installer always punches down the cable pairs in a standard order, from left to right: pair 1 (Blue), pair 2 (Orange), pair 3 (Green), and pair 4 (Brown). RJ-45 connectors are pinned in either of two specific ways (T568A or T568B), as illustrated in Figure 3–25. T568A and T568B are the two wiring standards for an 8-position modular connector, permitted under the TIA/EIA 568-A wiring standards document. The only difference between T568A and T568B is that the orange and green wire pairs (pairs two and three) are interchanged. T568B is commonly used in commercial installations, while T568A is prevalent in residential installations.

Centralized Cabling

In a *centralized cabling* system, the highest functionality networking components reside in the main distribution center interconnected to intermediate distribution centers or to TR. The idea is to connect the user directly from the desktop or workgroup to the centralized network electronics. There are no active components at floor level. Connections are made between horizontal and riser cables through splice points or interconnect centers located in a TR.

Fiber Zone

Fiber zone is a combination of collapsed backbone and a centralized cabling scheme. Fiber zone cabling is a very effective way to bring fiber to a work area. It utilizes low-cost, copper-based electronics for Ethernet data communications while providing a clear migration path to higher speed technologies. Like centralized cabling, a fiber zone cabling scheme has one central Main Distribution Center (MDC). Multifiber cables are deployed from the MDC through a TR to the user group. A typical cable might contain 12 or 24 fibers. At the workgroup the fiber cable is terminated in a multi-user outlet (MUO) and two of the fibers are connected to a workgroup hub. This local hub, supporting six to twelve users, has a fiber backbone connection and UTP user ports.

Connections are made between the hub and workstation with UTP cables. The station NIC is also UTP-based. The remaining optical fibers are unused or left *dark* in the MUO. Dark fibers provide a simple mechanism for adding user channels to the workgroup or for upgrading the workgroup to more advanced high-speed network architectures like video teleconferencing. Upgrades can be accomplished by removing the hub and installing fiber jumper cables from multi-user outlets to workstations.

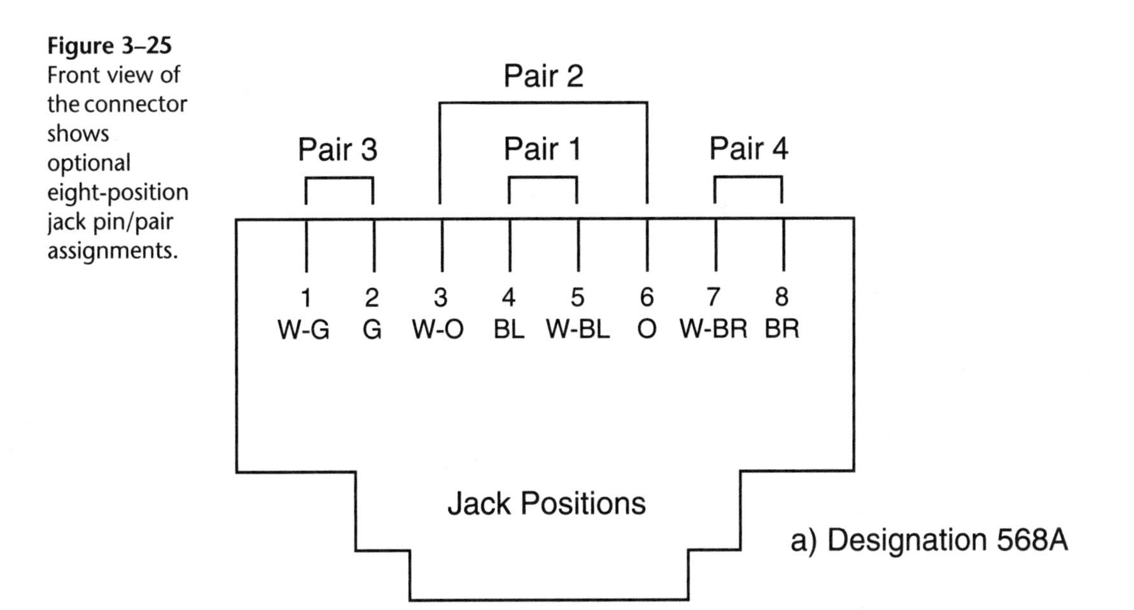

Figure 3–25 Front view of the connector shows optional eight-position jack pin/pair assignments.

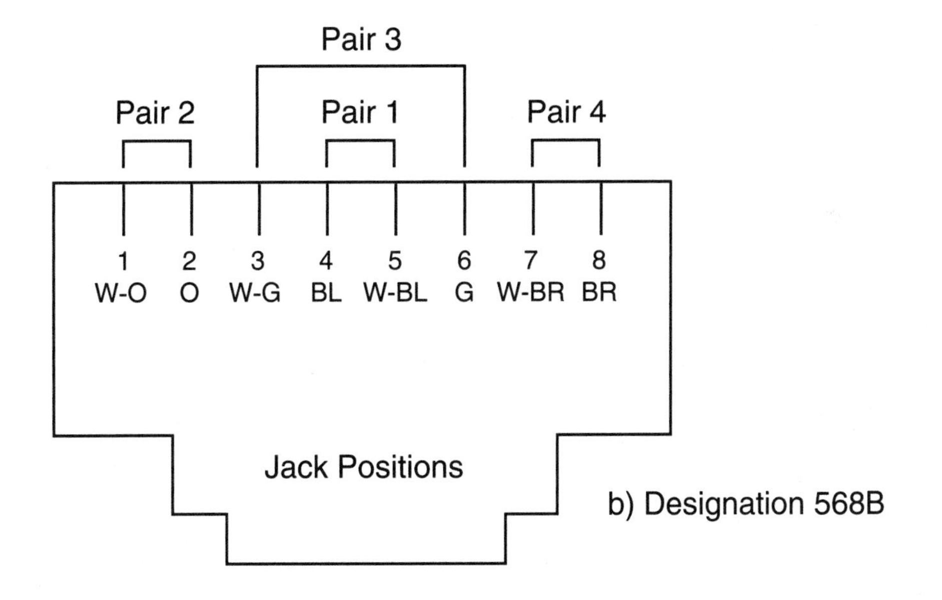

Cable Facilities Hardware

The cable installation hardware is used to organize and control the placement of cable in a facility. It is an important part of cable installation and maintenance, as it facilitates troubleshooting and network expansion.

Conduit is a pre-installed plastic or metal pipe that runs between or through buildings to ease cable installation. It comes in many diameters ranging from 0.5 to 6 inches depending upon the application. Within buildings, it is commonly used to provide readily accessible paths for cable between floors, through firewalls, and around structural supports. All conduits should contain pull-strings for cable installation.

Relay Rack is a metal frame that is used to secure and support networking equipment. Most telecommunications devices are designed so that they can be mounted directly in the relay rack or placed on shelves set up in the rack. The racks are sometimes enclosed in cabinets or TRs. This ensures security and prevents unauthorized access to the equipment.

Patch Panel is a piece of cable termination equipment that connects raw cables to standard ports or connectors. This allows a single, manageable point of access for several cables. Patch panels are usually mounted in relay racks or in enclosed equipment cabinets. The front surface, or faceplate, of the patch panel provides a series of modular ports or connectors, depending upon the media being connected. The back of the patch panel is made up of a number of connection points for facility cable.

Cable Installation

Many of the transmission problems occur as a result of poor installation practices. As a baseline, it is crucial to follow the TIA/EIA 568A guidelines related to factors such as degree of twist, bend radius, and termination. A typical 10BaseT network has a huge safety margin. Components, connections, cabling, and installation can each be off spec, and the network will still work. This convenient fact has changed as network speeds have increased. Many 100 Mbps Fast Ethernet networks see that about 10% of their CAT 5 nodes fail to operate at the anticipated higher speed, although both Fast and Gigabit Ethernet are supposed to run on the installed base of Category 5 networks. The reason is that at the higher speeds, performance margins begin to shrink dramatically. Stated simply, bidirectional signaling with four pairs adds new network complexity, and the higher-speed signals are weaker while the noise accompanying them is relatively strong.

Installing cabling and hardware for high-speed networks is a critical skill. Pulling tensions, bend diameters, fill ratios, separation from power circuits, grounding, termination techniques, and many other skills must be studied, practiced, and mastered. In addition, each installation will have a greater margin if the very best hardware, connectors, and cabling are specified and installed.

Miswired patch cables, jacks, and crossconnects are common in UTP. A patch cable is a twisted-pair or fiber optic jumper cable used to connect a computer to a network or a hub to a distribution panel. In general, the patch cords are *straight-through*, which means that pin one of the plug on one end is connected to pin one of the plug on the other end, as shown in Figure 3–26. *Crossover* cable crosses the transmit and receive pairs which are the orange and green pairs in standard cabling, as depicted in Figure 3–27. The only time you cross-connect is when you connect two Ethernet devices directly together without a hub. This can be two computers connected without a hub, or two hubs via standard Ethernet ports in the hubs. The jack's internal wiring connects each pair to the correct pins, according to the assignment scheme for which the jack is designed. The white striped lead is usually punched down first, followed by the solid color. The minimum bend radius for UTP is 4× cable outside diameter. For standard four-pair CAT 5 cabling, the bend radius should exceed 1 inch. If the bend radius is too tight, the wiring inside the jacket could be pressed flat or begin to untwist, resulting in the potential for attenuation and crosstalk.

Figure 3–26
Configuration of a patch cable.

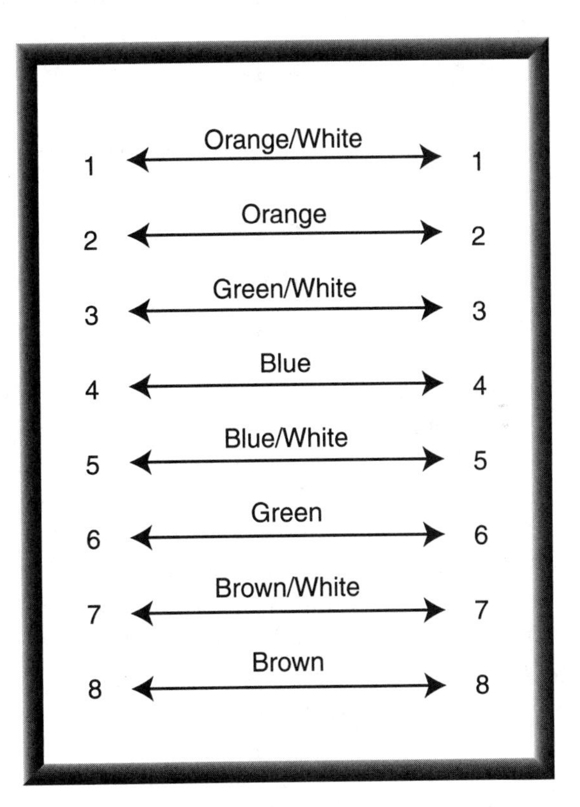

Pulling the cabling too tightly during installation can also cause the wiring to untwist. TIA/EIA 568 specifies that the wire pairs within CAT 5 cabling should not be untwisted more than a half-inch from the point of termination. Exceeding this limit could increase the potential for crosstalk and susceptibility to RFI and EMI. Untwisting of the wire pairs can also cause impedance mismatch. Also, jacket removal at the termination point should

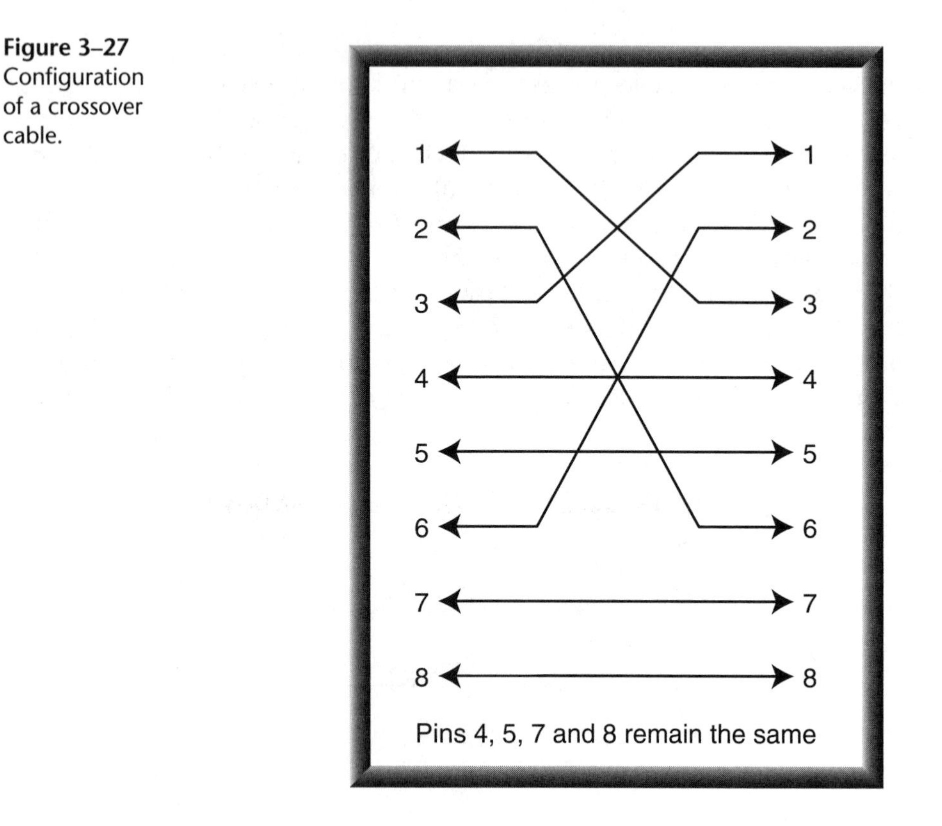

Figure 3–27 Configuration of a crossover cable.

be kept to a minimum. When using cable ties to join a bundle of cables, avoid cinching the ties too tightly. Over-cinching the ties can have the same effect as an insufficient bend radius, particularly with the cables on the outside of the bundle. When installing cabling to patch panels, make sure to provide adequate strain relief. Reinforcing support becomes increasingly important as you add more cables to a patch panel over time.

For fiber-optic cables not in tension, the minimum bend radius is 10× cable outside diameter. An insufficient bend radius can cause broken fibers. Corners and sags between poles put a lot of strain on the fiber. When the tension on the fiber exceeds the allowable limit specified by the manufacturer, some modes cannot propagate because of cracks in the fiber, resulting in signal attenuation.

Cable Tests

Higher-speed, higher-bandwidth technologies require higher-powered testing. The integrity of links between infrastructure elements such as connectors, cabling, patch cords, patch panels, and cross-connects is becoming increasingly critical. Therefore, thorough end-to-end testing customized for the requirements of the network is a must. A network must provide easy access to test points in the wiring closet so that it can be tested after all the components have been installed. Using sophisticated, properly-calibrated test equipment that produces detailed reports is important because these reports can be used for future troubleshooting. According to the TIA/EIA standard, every cable tester is required to run a suite of four tests:

- Length
- NEXT
- Wiremap
- Attenuation

The cable length is checked using a Time Domain Reflectometer (TDR), which transmits a pulse down the cable and measures the elapsed time until it receives a reflection from the far end of the cable. The standard requires that NEXT be measured from both ends of the link. Wiremap checks for open, short, crossed-pair, reversed-pair, and split-pair, and verifies a match between the pin-and-connector pairs on either end of the link. All testers verify that the maximum attenuation value, as defined in the specification, is not exceeded. A failure probably indicates a kink or bend in the cable, poor termination, or a cable grade that is unsuitable for the data rate.

Lastly, it is vital to measure the *Return Loss (RL)*. RL limits have recently been defined for both CAT 5 and 5E cables. It was not previously specified because it has no effect on 10BaseT signaling. For high-speed protocols, it is a critical measurement. It is a strong indication of an installation's performance margin. Many testers offer additional features such as customization of autotests, measure of traffic, built-in talk set, and a tone-generator tool. The minimum performance requirements depend on the type of cable such as fiber-optic, UTP, STP, and Coax. A cable is a passive component, and transmission impairments can only be measured when signals are transmitted by equipment attached to either end of the wire. For this reason, cabling cannot be tested and certified in isolation.

SUMMARY

The process of transporting information in any form including voice, video, and data between users is called *transmission* in the telecommunications industry. Cabling systems are the backbone of a communications network. The type of communications wiring should always be dictated by the application. Any transmission medium offers a trade-off between bandwidth and distance. The greater the bandwidth requirement, the shorter the distance it can support with other factors being equal. High-bandwidth applications are fueling the migration to fiber-optic cabling. Though fiber is used increasingly in backbone networks, copper remains at the cabling forefront because of lower cost and ease of installation. However, with network requirements changing constantly, it is important to employ a cabling system that can keep up with the demand. One must remember that labor is usually more than half the cost of an installation. During installation, cabling-standards compliance saves an end user from expensive recabling each time a new application is added.

As a result of the variety of transmission media and network design methods, selecting the most appropriate medium can be confusing. When choosing the transmission medium, we must consider several factors such as transmission rate, distance, cost and ease of installation and maintenance, and resistance to environmental conditions. Physical cable is not always the most effective way to accomplish long distance distribution of information. Installing cable in uninhabited or inclement terrain is inefficient in terms of initial installation or maintenance. In these circumstances, the most common method for transmission is a wireless link. However, most existing wireless services are more expensive, less functional, and offer limited coverage when compared with their wireline counterpart.

The cabling industry has been experiencing a quiet revolution in the past few years. More and more corporate clients are demanding real-time intelligence in their cabling systems, and cabling vendors are hurrying to meet these demands. A real-time cabling-management system provides real-time information on the status of connections at the wiring closet, reports all connectivity changes to the network-management station in real time, and guides the system administrator in planning and implementing wiring changes. With high-speed technologies, a cabling infrastructure must maintain consistent performance levels throughout the entire system—including the cabling itself, as well as patch panels, cross-connects, connectors, and connector interfaces.

Case Study

At a doctoral/research-extensive institution, the cabling system provides a framework that supports Gigabit Ethernet to the desktop to enable a host of high-tech, data-intensive applications, such as distance learning, teleconferencing (video over IP), telemedicine, digital radiology, and medical imaging research. In selecting a cabling system and connectivity components, high performance and maximum uptime were considered to be most important. All cables and components are installed to meet TIA/EIA 568-B.2-1 standards. To maintain maximum flexibility for current and future needs, there are a minimum of three cabling runs per workstation: for voice, for data, and a multipurpose spare. The institution standardized on the CAT 6 system because it was less expensive to buy, install, and maintain than a fiber-optic system, while offering a high-value proposition in terms of capacity and performance. According to the institution, this project ensures a cabling framework with a lifespan of ten years.

Question

1. Compare the cost, benefits and lifespan of installing CAT 6 versus fiber to the desktop at your institution.

REVIEW QUESTIONS

1. Explain the construction of each of the following cables and discuss its applications:
   - A. Thin Coax
   - B. STP
   - C. Category 3 UTP
   - D. Category 5 UTP
   - E. Singlemode Fiber-optic
   - F. Multimode Fiber-optic
2. Define the following terms and discuss their applications:
   - A. Echo
   - B. Crosstalk
   - C. Bend Radius
   - D. Electromagnetic Interference
   - E. Coherence
   - F. Modal Dispersion
3. Discuss the current status of Enhanced CAT 5 and higher grade UTP cable.
4. Calculate the resistance of 500 ft length of AWG 24 copper wire.
5. Analyze the implications of *impedance matching* for telecommunications cables.
6. Construct an argument for international cabling specifications. You may use electromagnetic compatibility as a case in point.
7. Assess the advantages and disadvantages of fiber versus copper.
8. Distinguish between reflection and refraction using schematics.
9. Is the velocity of light higher in water or in air?
10. What is the speed of light in a glass fiber-optic cable with a refractive index of 1.52?
11. Describe the propagation of light through fiber.
12. A fiber-optic cable core has a refractive index of 1.45 and its cladding has a refractive index of 1.43. Determine the following:
    - A. Critical Angle
    - B. Numerical Aperture
    - C. Acceptance Angle
13. Develop a rationale for implementing *Structured Wiring*.
14. Analyze the various components of *Structured Wiring*.
15. Identify some of the cable installation hardware in cabling facilities.
16. Which are some of the critical components of cable tests?

4 VOICE COMMUNICATIONS

KEY TERMS

- Public Switched Telephone Network (PSTN)
- Switching
- Blocking and Non-blocking Networks
- Local Exchange
- Local Area
- Toll Center
- Trunks
- Tip and Ring
- Pulse Dial
- Dual Tone Multiple Frequency (DTMF)
- Loop Start
- Ground Start
- E&M Signaling
- Common Channel Signaling
- Signaling System Seven (SS7)
- Voice Processing Systems
- Tie Lines/Trunks
- Direct Inward Dial (DID) Lines/Trunks
- Direct Outward Dial (DOD) Lines/Trunks
- Private Branch Exchange (PBX)
- Centrex
- Grade of Service (GoS)

OBJECTIVES

Upon completion of this chapter, you should be able to:

- ✦ Discuss the public telephone network infrastructure
- ✦ Describe the operation of the telephone set with address signaling
- ✦ Analyze analog signaling schemes for telephone lines and trunks
- ✦ Identify the protocol architecture and major components of Signaling System Seven
- ✦ Discuss the applications of intelligent network services
- ✦ Explain different business telephone systems including Private Branch Exchange (PBX) and Centrex
- ✦ Evaluate the telephone network design parameters

INTRODUCTION

A common activity that relies on telecommunications technology is a telephone call. To a telephone user, the **Public Switched Telephone Network (PSTN)** looks like a black box with telephone lines that mysteriously get connected to one another. The earliest telephone network started with dedicated lines being installed from one telephone to another. It quickly became evident that such a one-on-one arrangement was inefficient to meet growing economic and social needs. A network began to evolve based on the basic assumption that at a given instant in time, there are only a limited number of subscribers using the service. Therefore, it is more efficient to connect the customer line to the telephone service only when there is an incoming call or the user wants to make an outgoing call, signaled by lifting the phone off the hook. Once a line is connected it stays up for the entire duration of the call. This concept, referred to as *circuit-switching* technology, is the basis for telephone networks.

The evolving layout of the network has been primarily a function of economics. Subscribers share common transmission facilities with switches or nodes that permit sharing. Within the United States, the network is known as the PSTN. The telephone, which serves as a user interface, is an integral part of the network. Yet, the telephone instrument is simple, rugged, and economical. It is a tribute to the inventive genius of the millions of scientists, engineers, and technologists that the complexity of the PSTN is hidden so well that users do not have to know any more about the network than how to dial and answer telephone calls. In reality, however, the network is composed of countless switching circuits, line and trunk signaling devices, and voice processing systems that enable telephone companies to offer intelligent network services and voice communications options to consumers and businesses.

PUBLIC TELEPHONE NETWORK

The public telephone network, also called PSTN, is a systematic set of interconnecting transmission lines arranged so that one telephone user can talk to any other within that network. It is interconnected to hundreds of other national networks to form a gigantic international network. As depicted in Figure 4–1, the PSTN is based on star, ring, and mesh topologies. The circuit switched network consists of *transmission paths* and *nodes*. Nodes are *exchange* or *switching* points where two or more paths meet, enabling the users to share transmission paths. The word *exchange* refers to the process of changing or exchanging wire pairs, which, in the early days, was done manually by an operator. Automatic electronic switches soon took over.

A *switch* sets up a communication path on demand and takes it down when the path is no longer needed. It performs logical operations to establish the path and automatically charges the subscriber for usage. The term **switching** refers to the process of routing communications to different parties. If the preferred route is busy, the switch must be

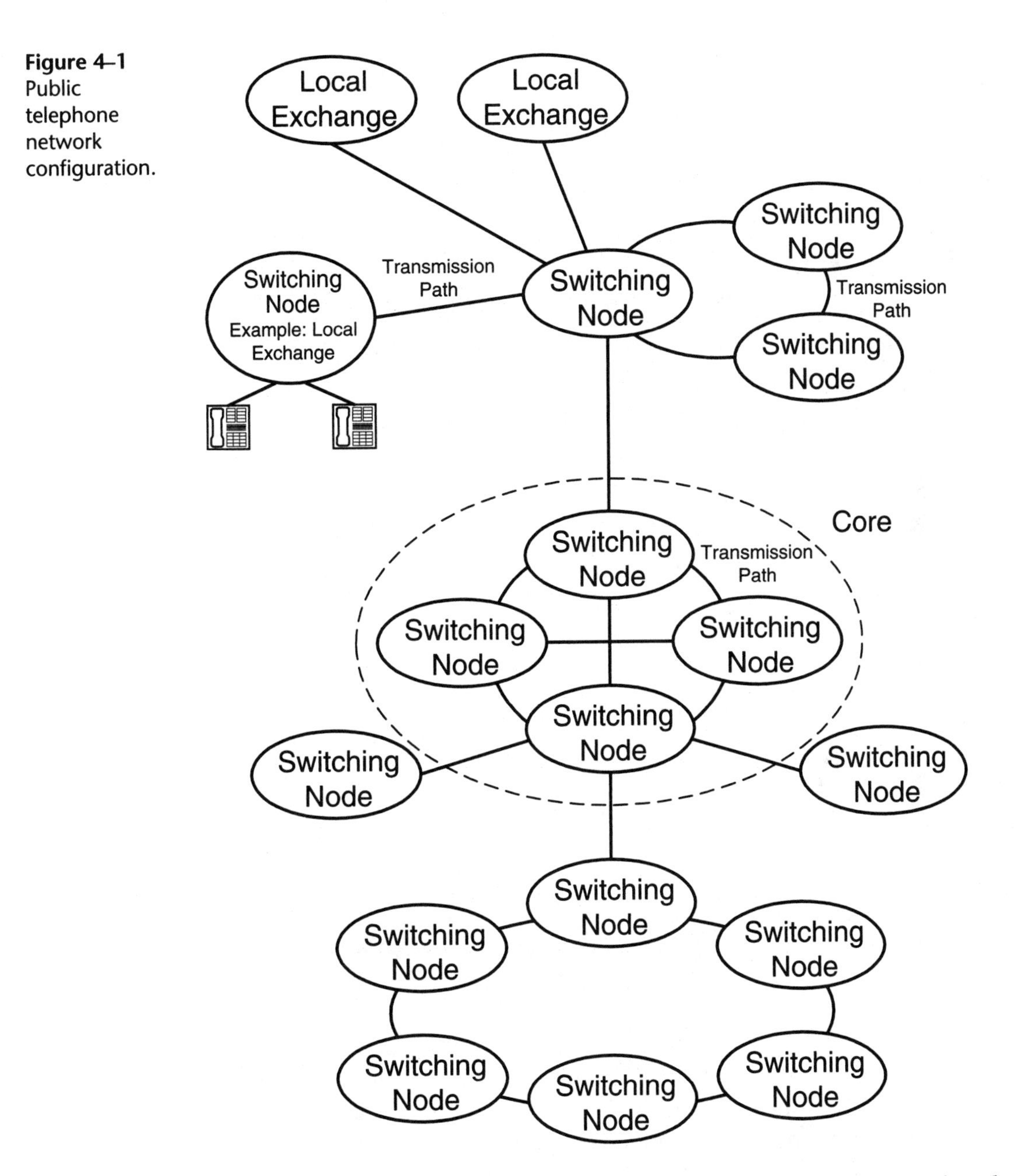

Figure 4–1 Public telephone network configuration.

capable of choosing an alternate route. Today, most switching systems are digital and contain the intelligence to route calls over the most economical path.

Although originally designed to carry voice, PSTN is used more and more to carry data. The FCC, a regulating body, has drawn a distinction between plain old telephone services (POTS) and *enhanced services* (ES). The FCC has defined ES as the use of the existing telephone network to deliver services other than basic transmission, such as voice mail, email, voice or fax store-and-forward, data processing, and gateways to on-line databases. The increasing levels of Internet use are beginning to affect the telephone network. Internet usage is placing unexpected demands on local exchange carriers' switches. Switch congestion can threaten service quality for all users, including PSTN telephony.

Switching Systems

All switching systems include the following elements, as shown in Figure 4–2:

- *Switching matrix* connects paths between input and output nodes
- *Controller* directs the connection of paths through the switching network
- *Database* stores the system configuration and addresses, and features of lines and trunks
- *Line circuits* interface with outside plant for connection to users. All local and PBX switching systems include line ports
- *Trunk circuits* interface interoffice trunks and service circuits
- *Common equipment* includes power supplies, testing equipment, and distributing frames

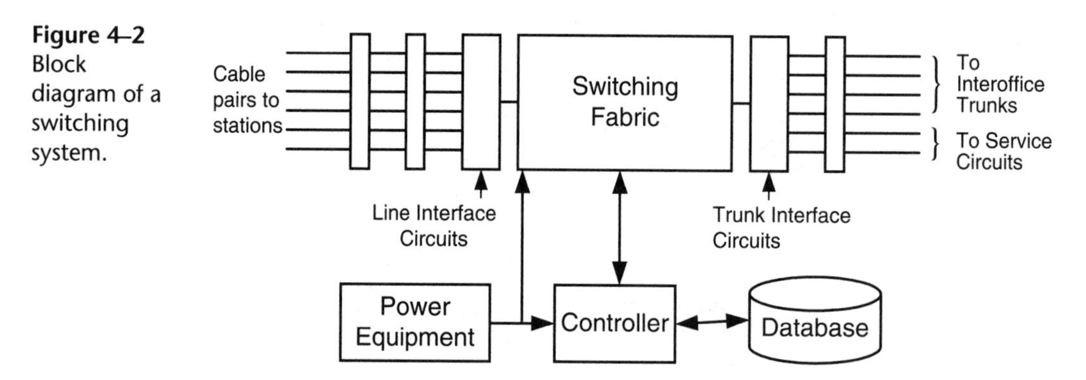

Figure 4–2 Block diagram of a switching system.

When a user signals a switching system with a service request, the switch determines the terminating station's address from the telephone number dialed and translates the number to determine call routing. Translation tables specify the trunk group that serves the destination, the most economical route, an alternate route if the first choice is blocked, the number of digits to dial, any digit conversions needed, and the type of signaling to use on the trunk. These switches are called *common control* systems. Some older switches called *direct control* lack alternate routing and digit translation capabilities.

Concentration, a line-to-trunk ratio, is one factor that determines the probability that a call will be completed. Older electromechanical networks that contain fewer paths than terminations are called **blocking networks** because all users cannot be served simultaneously. Digital networks enable a connection to be made between any two ports independently of the amount of traffic and are therefore known as **non-blocking networks**. However, this does not ensure that users will not encounter blockage. Trunks and lines are always designed for some level of blockage. The term *virtually non-blocking* has evolved to describe a network that is not designed to be totally non-blocking but provides enough paths that users rarely find themselves blocked by the network.

If the switching system is configured without enough common equipment, users will encounter delays that they interpret as blockage. Switches are also subject to processor overloads, which can result in a variety of call-processing delays. Switches are rated by the number of *Busy Hour Call Attempts (BHCA)*, which describes the number of calls the system can handle during the peak hour of the day. The term *call attempts* refers not only to call originations but accesses to features such as call pickup, call transfer, and call waiting, which require attention by the central processor.

Telecom Infrastructure

The fundamental network design problem is determining how to assemble the most economical configuration of circuits and equipment based on peak and average traffic load, grade of service required, and circuit and administrative costs. It is practical to connect a few nodes with direct trunks between nodes, but this is feasible only up to a point. As the number of nodes increases, the number of direct connections increases in direct proportion to the square of the number of nodes, which soon becomes difficult to manage.

To control costs, a hierarchical network can be formed using tandem switches to interconnect the nodes. The number of levels in a network hierarchy is determined by the network's owner and is based on a cost/service balance. Prior to 1984 the public telephone network operated by AT&T and connecting RBOCs was a network hierarchy with five different classes of switching systems, as illustrated in Figure 4–3. With the increasing power and intelligence of switching systems, the traditional hierarchical arrangement is giving way to a more flat network structure built by the long distance carriers.

Before 1996, the telephone line and the dial tone were provided to the customer only by the RBOCs, which are also called Incumbent Local Exchange Carriers (ILECs). The 1996 Telecommunications Act removed this restriction. The non-incumbent LECs are called Competitive Local Exchange Carriers (CLECs). Although most CLECs resell ILEC services, some are starting to build their own facilities and infrastructure. In spite of the fact that there is no longer a true distinction between local and long distance carriers, the traditional infrastructure is still prevalent.

A long distance telephone call uses the circuits of the local telephone companies at both ends of the call. The local telephone companies at both ends provide the local leg or local loop of the circuit. Long distance companies, commonly referred to as Inter-Exchange Carriers (IECs or IXCs), pay the local companies for the use of these circuits. In the United States, AT&T, Sprint, and MCI WorldCom are the most prominent IXCs. Also,

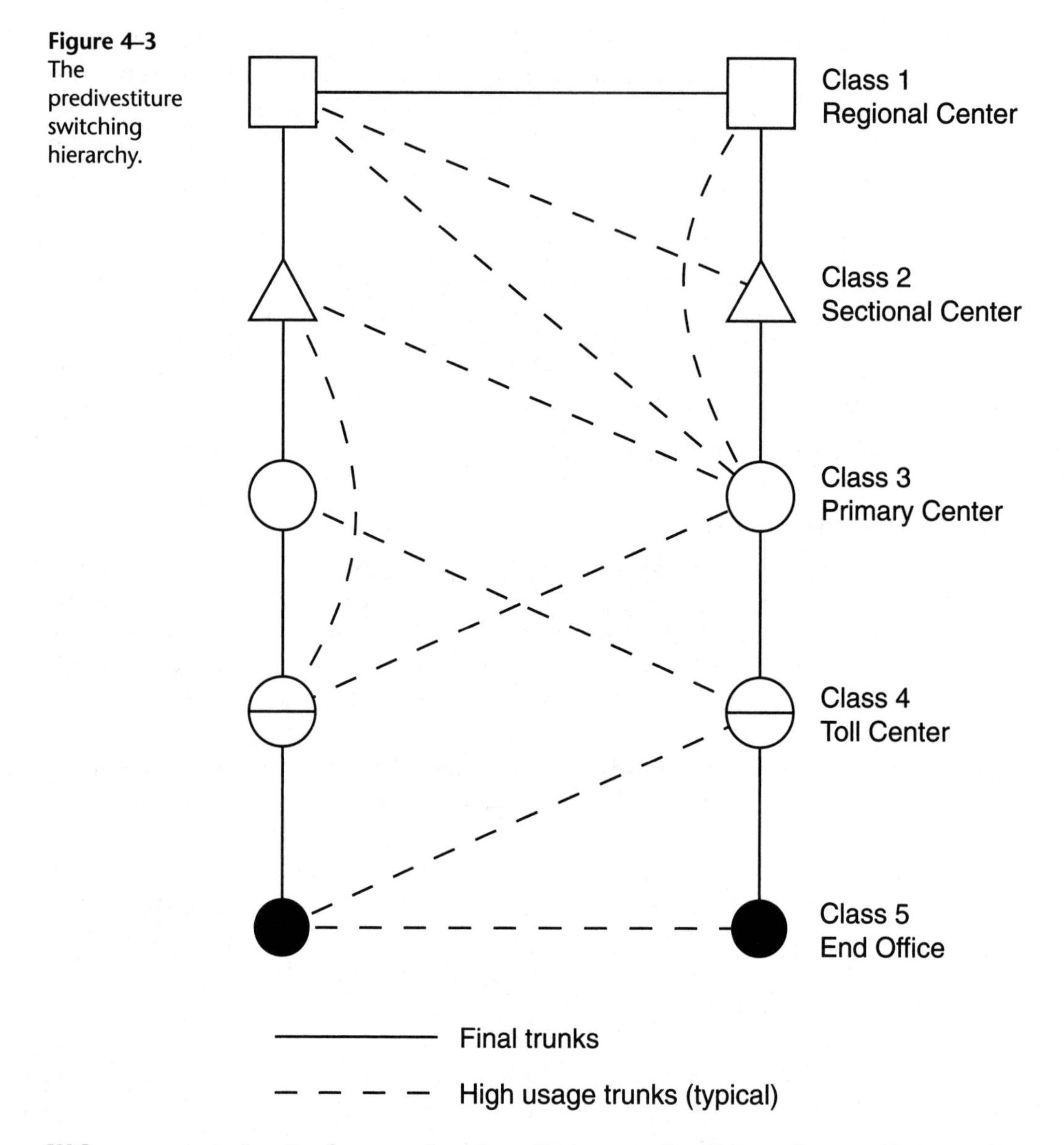

Figure 4–3 The predivestiture switching hierarchy.

IXCs may rent circuits from each other. For example, although a user's long distance company is MCI, the person may be using circuits owned by AT&T.

Every telephone subscriber is first served by a class-5 office called a **local exchange** (or *central office* or *end office*). From this point on, it will be referred to as a local exchange. This local exchange is a switching center that typically supports thousands of subscribers in a localized area and also provides dial tone services to the user. **Local area** is that geographical area within which subscribers can call each other without incurring tolls (extra charges for a call). Toll calls and long distance calls are calls that go outside the local exchange. These can be further classified into IntraLATA and InterLATA. Every local exchange is connected to a class-4 office called **toll center** or *tandem office* in order to enable calls to go over the long-distance network. All calls that pass through (or tandem through) the toll center incur added costs.

Toll centers are connected to a POP or POI where LECs and IXCs meet, as shown in Figure 4–4. In most U.S. locations, smaller companies and residences reach their long-distance carrier or IXC networks through equal access. Each IXC doing business in the local area rents space in the local exchange switch, and the cost for using this circuit is called an *access charge*. After this point, the call then travels over the long distance network of the IXC selected by the caller. This switching process is transparent to the user. The actual cost of a long distance call is based upon many variables such as distance, time of day, duration, total number of calls made (higher volume = lower costs), and special promotions in effect at the time one selects a long distance carrier.

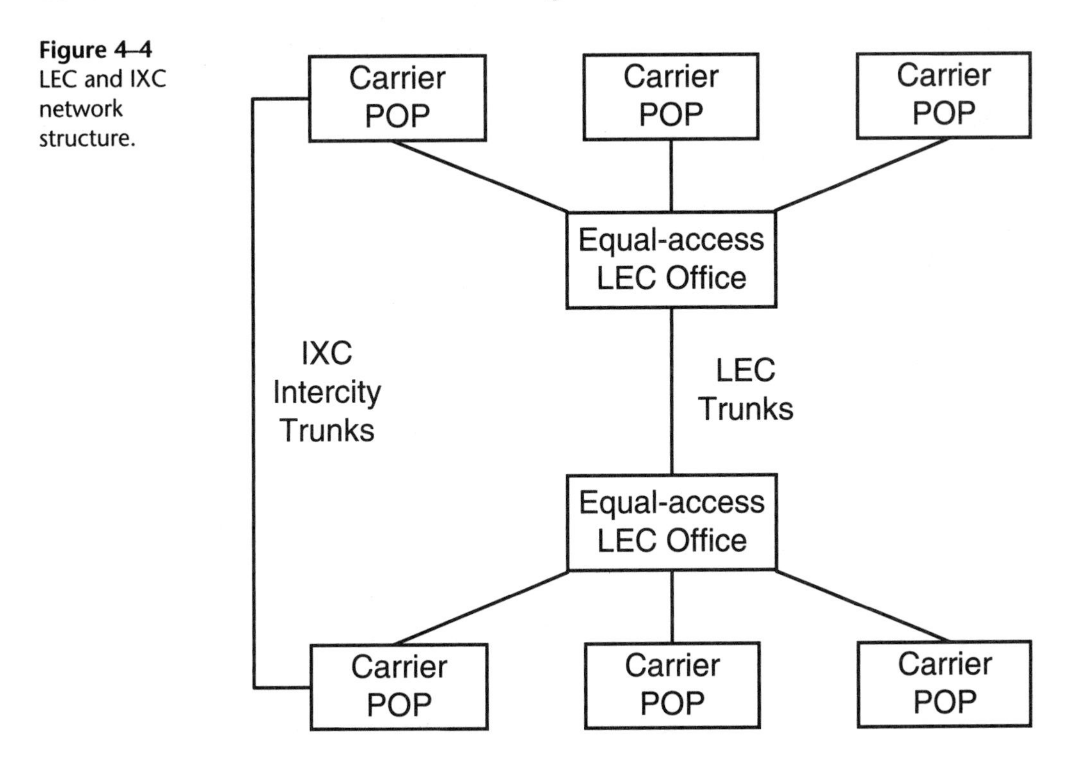

Figure 4–4 LEC and IXC network structure.

Telephone Cable Architecture

Most commonly, cables are routed over poles. These telephone pole structures vary depending upon the load, weather factors, and terrain factors. Recently, the trend has been to bury cables underground. The depth of these cables is usually between two and four feet, depending upon location, regulatory requirements, and future requirements.

Cable splicing takes place in the ground or in above-ground pedestals. In more congested areas, cables are run in conduit, with manholes used for splicing. The advantage of a conduit is the ability to rerun cables—for example, to replace copper wiring with fiber optic cables. Typically, plastic conduit is used for longer cable runs since it has a lower coefficient of friction than does concrete or fiberglass conduit. Submarine cable is used where land routing is not practical. Protective materials are placed around the cable, especially in shallow water. Clamps anchor the cable from the action of currents.

The telephone lines that connect users to the local exchange that serves them are called *subscriber lines* or *local loops*. These voice-grade distribution cables are twisted-pair copper wire (19 to 26 gauge) that run overhead (aerial) or are buried underground. For rural subscriber loops, most LECs used analog amplification devices called line extenders at the local exchange to boost signal levels. *Lines* are circuits or paths that connect stations to the nodes. *Stations* are terminal points in a network. Telephone sets, PBX, data terminals, facsimile machines, and computers all fall under the station definition.

All exchanges and switching centers are connected to each other with multiple lines called *trunks* in North America and *junctions* in Europe. The trunks can be twisted pairs of wires, coaxial cables, fiber-optic cables, or radio links. The trunks carry the signals from one point to another without directly serving subscribers. *Feeder* cable connects the main trunk line to the drop cable, which feeds into the subscriber's home and the terminal equipment. Feeders are large bundles of twisted pair ranging anywhere from 50 to 3000 pairs of wires. The feeders are run to breakout points called manholes. The distribution cable or branch feeder is then strung to various customer locations. The end of the pair to the final customer location is called *station drop*. The telephone cable architecture described above is depicted in Figure 4–5.

Trunks

Trunks are circuits or links that interconnect nodes. These high-speed digital transmission lines or digital carriers have been evolving over the past 40 years. The T-Carriers system in North America, as well as the International CCITT (now ITU) specifications begin with a transmission rate of 64 kbps. The ITU classification is mostly prevalent in Europe; these lines are designated according to their transmission capacity, as indicated in Figure 4–6. T-carriers are designated according to their transmission capacity, as shown in Figure 4–7. The T-2 and T-4 lines are used primarily by the carriers, while the T-1 and T-3 lines are used by both the carriers and their customers. A T-3 or DS-3 line consists of 672 channels, while the T-1 or DS-1 line consists of 24 channels. Each channel supports 64 kbps. Optical carriers support much higher bit rates. Synchronous Optical Network (SONET) is a hierarchy of optical standards over single-mode fiber-optic cable. Internationally, SONET is known as Synchronous Digital Hierarchy (SDH). The optical carriers are also designated according to their transmission capacity, as shown in Figure 4–8.

Figure 4–5 Telephone cable architecture.

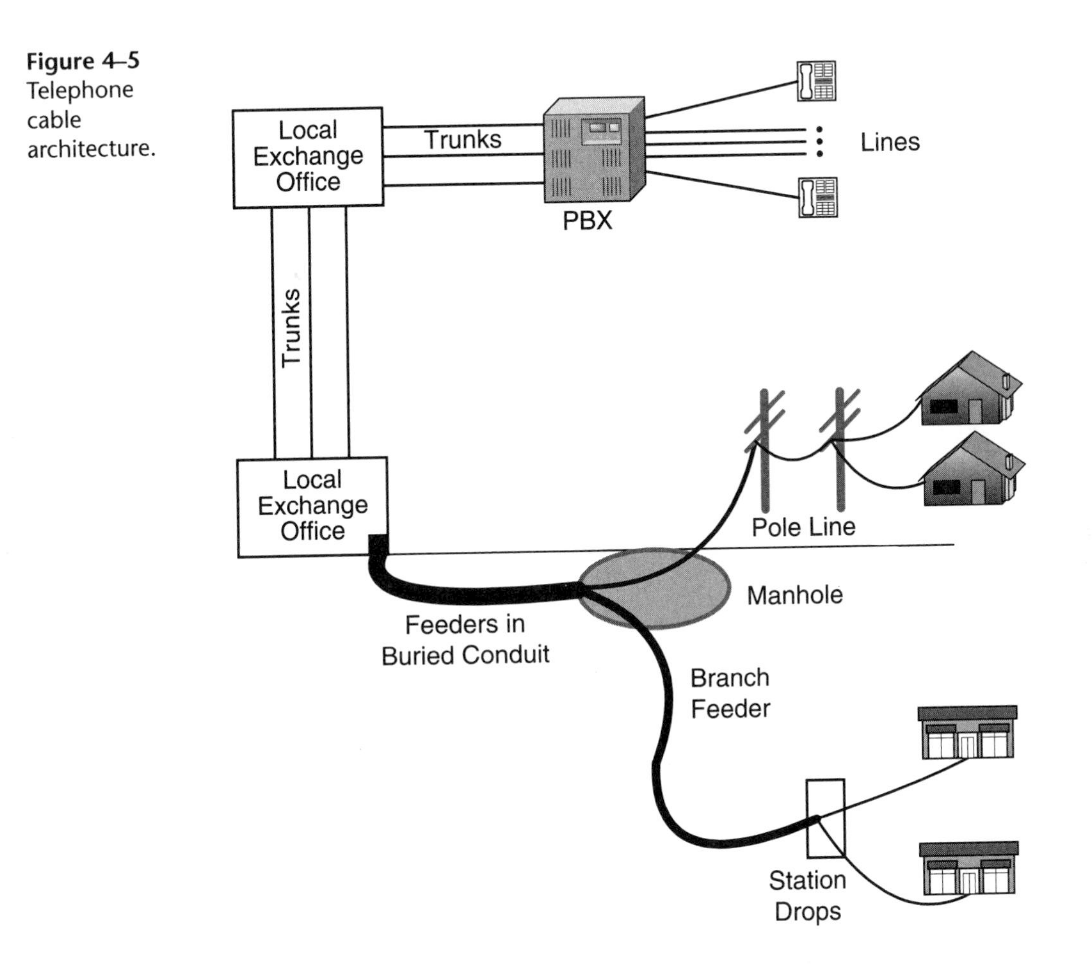

Digital Carrier	Transmission Capacity
DSO	64 kbps
E1	2.048 Mbps
E2	8.448 Mbps
E3	34.368 Mbps
E4	139.264 Mbps

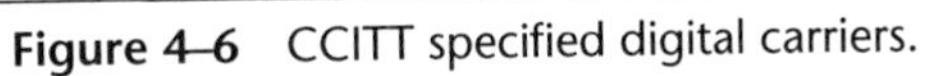

Figure 4–6 CCITT specified digital carriers.

T-Carrier	Transmission Capacity
DSO	64 kbps
T-1	1.544 Mbps
T-2	6.312 Mbps
T-3	44.736 Mbps
T-4	274.176 Mbps

Figure 4–7 T-carriers and their transmission capacity.

Optical Carriers	Transmission Capacity
OC-1	51.84 Mbps
OC-3	155.52 Mbps
OC-9	466.56 Mbps
OC-12	622.08 Mbps
OC-24	1.244 Gbps
OC-36	1.866 Gbps
OC-48	2.488 Gbps
OC-96	4.976 Gbps
OC-192	9.953 Gbps
OC-255	13.92 Gbps

Figure 4–8 Optical carriers and their transmission capacity.

Subscriber Line or Local Loop

Subscriber Line or *Local Loop* is usually one pair of UTP wire and is mostly analog. More recently, local providers have been installing two-pair (four wires) or four-pair (eight wires) connections to residences. The reason for this is that customers often request separate fax lines and data communications hook-ups that require additional phone lines. Since each of these applications require a two-wire interface, it is far more cost effective to install multiple pairs the first time than to install a single pair of wires every time a customer asks for a new service.

Line Conditioning

Dedicated or leased lines, because of their permanent status, are conditioned to tighten telephone company parameters so that they can transfer data at higher transmission rates with reduced errors. As depicted in Figure 4–9, different frequencies encounter different attenuation and propagation delay times through the telephone network. Propagation delay is the time it takes for a signal to travel from source to destination. Although this does not affect voice communication, it can be detrimental to data transmission, particularly at speeds above 2400 bps. Conditioning levels C1 through C5, shown in Figure 4–10, specify stringent guidelines for attenuation distortion and envelope delay distortion. Attenuation distortion refers to gain fluctuations with frequency, and envelope delay distortion is a method used to test and measure the variance in propagation delay of different signals within the voice band. For high-speed data transmission, equalizer circuits in the modem compensate for these distortions.

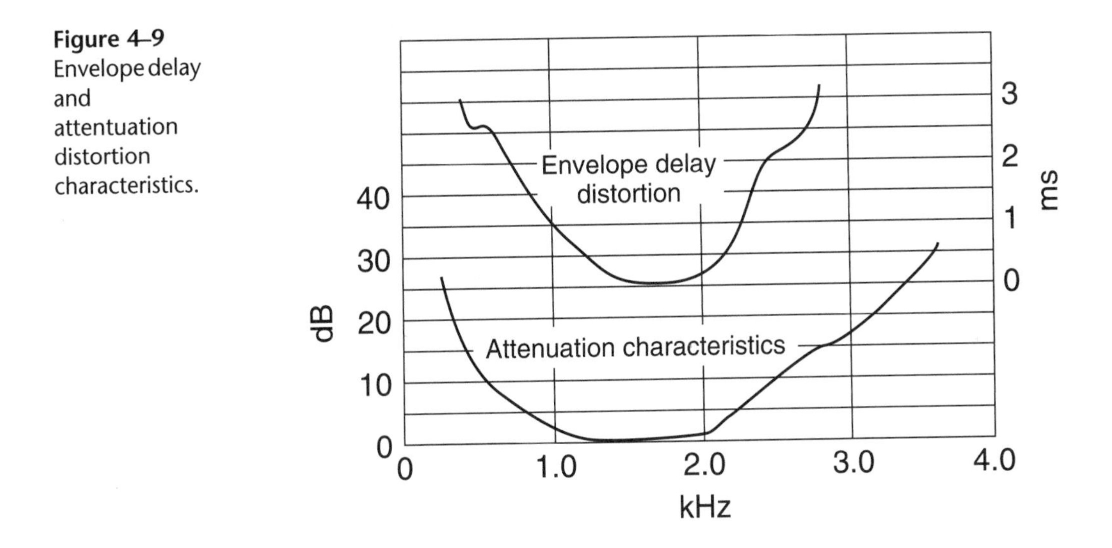

Figure 4–9 Envelope delay and attentuation distortion characteristics.

THE TELEPHONE

The telephone is also referred to as a station or set. A typical telephone consists of a handset and its cradle or base integrated with a signaling device consisting of either a dial or push buttons. The handset is designed to comfortably fit the distance between the mouth to the ear. It consists of two electroacoustic transducers: the mouthpiece or transmitter and the earpiece or receiver. Functional diagrams of telephone sets are illustrated in Figure 4–11.

Conditioning Type	Envelope Delay Distortion		Attenuation Distortion	
	Frequency Range (Hz)	Specification Limits (μs)	Frequency Range (Hz)	Specification Limits (dB)
C1	1000–2400	1000	300–2700	–2 to +6
			1000–2400	–1 to +3
C2	500–2800	3000	300–3000	–2 to +6
	600–2600	1500	500–2800	–1 to +3
	1000–2600	500		
C3 (access lines)	500–2800	650	300–3000	–0.8 to +3
	600–2600	300	500–2800	–0.5 to +1.5
	1000–2600	110		
C3 (trunks)	500–2800	500	300–3000	–0.8 to +2
	600–3000	260	500–2800	–0.5 to +1
	1000–2600	260		
C4	500–3000	3000	300–3200	–2 to +6
	600–3000	1500	500–3000	–2 to +3
	800–2800	500		
	1000–2600	300		
C5	500–2800	600	300–3000	–1 to +3
	600–2600	300	500–2800	–0.5 to +1.5
	1000–2600	100		

Figure 4–10 C1 through C5 line conditioning for private leased lines.

Analog versus Digital Telephones

The distinction between an analog and digital telephone is how voice is processed in the telephone, as shown in Figure 4–12. The distinguishing feature between an analog and digital telephone is the location of the Codec (an analog-to-digital converter). If the Codec is located in the telephone company's switching equipment on a line card, the phone is analog. If the Codec is in the telephone, the telephone is digital. Voice is an analog quantity and is processed as such in an analog telephone. In a digital telephone, voice is converted from an analog signal to a digital signal by sampling it 8,000 times per second with each sample being represented by a binary number. The setting 'Station Type' is used to specify the type of telephone attached to the port on the back of the phone system. Most systems give a choice between analog and digital phones.

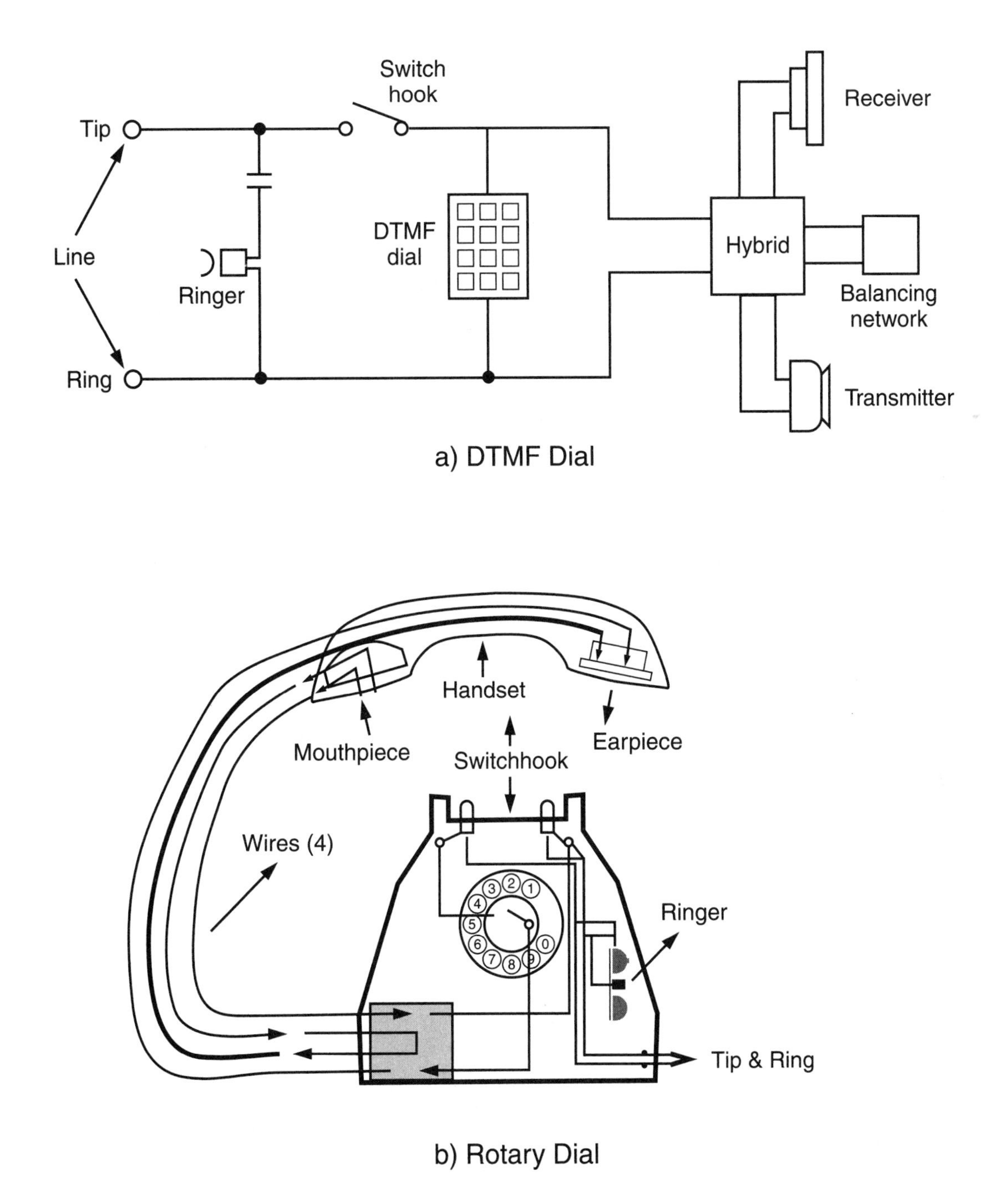

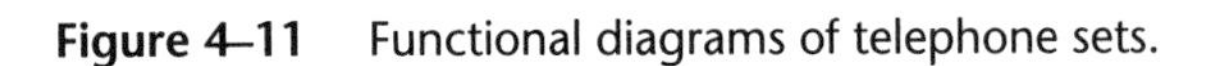
Figure 4–11 Functional diagrams of telephone sets.

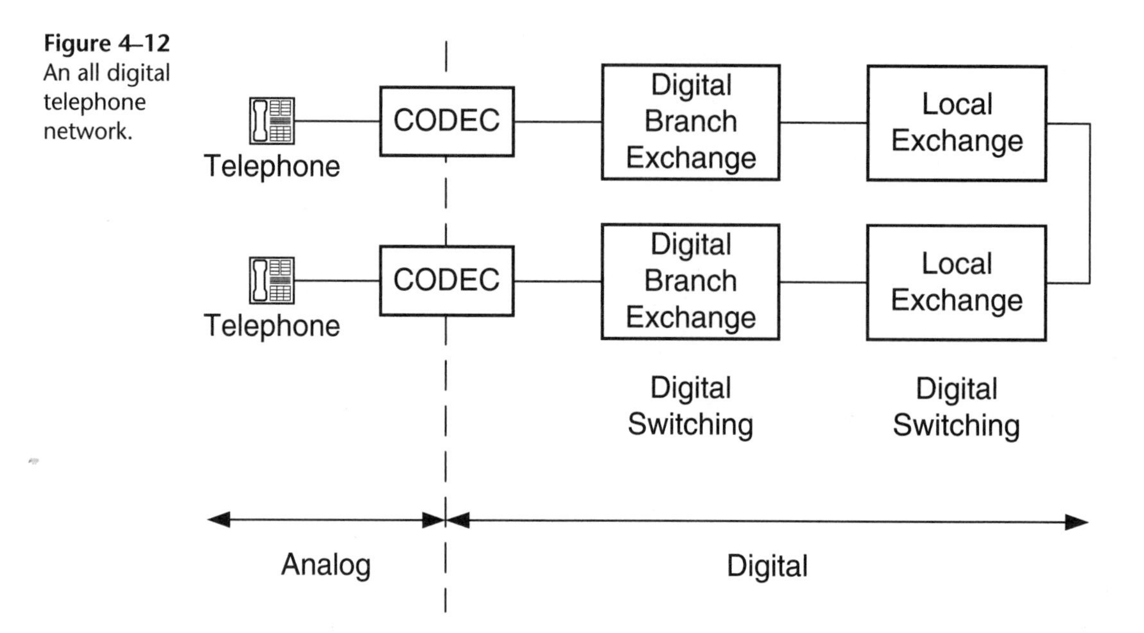

Figure 4–12 An all digital telephone network.

Digital telephones have a wide range of capabilities in contrast to the analog telephones. Many of these features apply to sets connected to a business telephone system, while some are available on single-line telephones. The business telephone is designed with function keys, also called feature buttons, which serve different purposes. These include call hold, call transfer, call conferencing, call forwarding, speed dial, and last number redial. Some buttons are programmable, while others have soft keys, which means that the same button performs different functions at different times. Most sets are equipped with visual indicators such as LEDs, while some may also be equipped with a Liquid Crystal Display (LCD).

In the idle state, the LCD simply displays the date and time. If the Caller Identification (Caller ID) feature is activated on an incoming call, then it may show the name and/or telephone number of the caller. Some business telephone systems enable the user to leave a pre-selected message such as "In a Meeting" or "On Vacation"; when someone calls that number from within the office, the caller's display shows the message the user has pre-selected. The Custom Local Area Signaling Services (CLASS), such as selective call acceptance, forwarding, or rejection, and automatic callback can only be implemented on digital sets. As compared with the analog telephones, the features available on digital phones provide a more effective means of communication within a business. Some of the features discussed next are available on both analog and digital telephones.

- *Visual Indicators* such as LEDs are used to convey a message to the user. For example, a blinking light indicates a new incoming call while the user is on another call. If the set is connected to a voicemail system, a blinking light indicates a message waiting in the user's voice mailbox.

- A *Speakerphone* enables the user to have hands-free conversation. The one-way speakerphone can hear the caller but cannot speak back to the caller without picking up the handset.
- A *Last Number Redial* feature enables the caller to automatically dial the last number called even if it is not a preprogrammed speed number. This feature is very useful when one encounters a busy tone on the network and is certainly a time saver.
- *Speed Dialing* allows a user to preprogram telephone numbers into the memory of the telephone set. By using a feature button or a preprogrammed button on the set, the telephone will dial the string of digits in a much shorter time.
- A *Call Hold* feature, when activated, will disengage both the transmitter and the receiver for a period of time. A visual indicator such as a flashing light will let the user know that the call is on hold. This mechanical function keeps the line active until the line button is pressed again to reconnect the parties.
- *Call Transfer* is used to transfer the call to another line. A feature button or a preprogrammed button on the set will perform this function.
- A *Conference Call* is set up among three or more people. There is a separate button typically labeled *"conference"* that is used to activate this feature. Usually, the more people you add to the conference, the harder it is for everyone to hear. Therefore, it is advisable to use special conferencing equipment for calls among more than four people. It is important to know how to drop off one of the conferees from your conference call without ending the call. If you have set up the conference call from your telephone and you hang up, the other call participants are automatically disconnected.
- *Call Forwarding* is used to specify the extension to which a call will forward if the extension rings with no answer for longer than a specified period of time. For example, this feature can be used to forward a call to voice mail if a phone is unanswered.
- *Busy Forwarding* is used to forward a call to another extension if the station is busy. This feature can be used to set up a hunt group; for example, if Jim is busy when a call arrives, ring Mary's phone, if she is busy go to Steve, and so on.

Telephone Circuit and Speech Characteristics

Telephony is the science of translating sound into electrical signals, transmitting them, and then converting them back to sound. The electrical signals are transmitted over the *Tip* or transmit wire. The *Ring* or receive wire connects to the receiver of the telephone. A standard RJ-11 jack wired with a pair of conductors, traditionally called **Tip and Ring**, connects the telephone to the plug in the wall. Figure 4–13 shows how the Tip and Ring are connected to the components within the telephone set. The Tip is connected to the dial pad and in turn to the mouthpiece, while the Ring is connected to the Ringer and, in turn, to the earpiece. The Tip and Ring wires are also cross-connected to the switch-hook

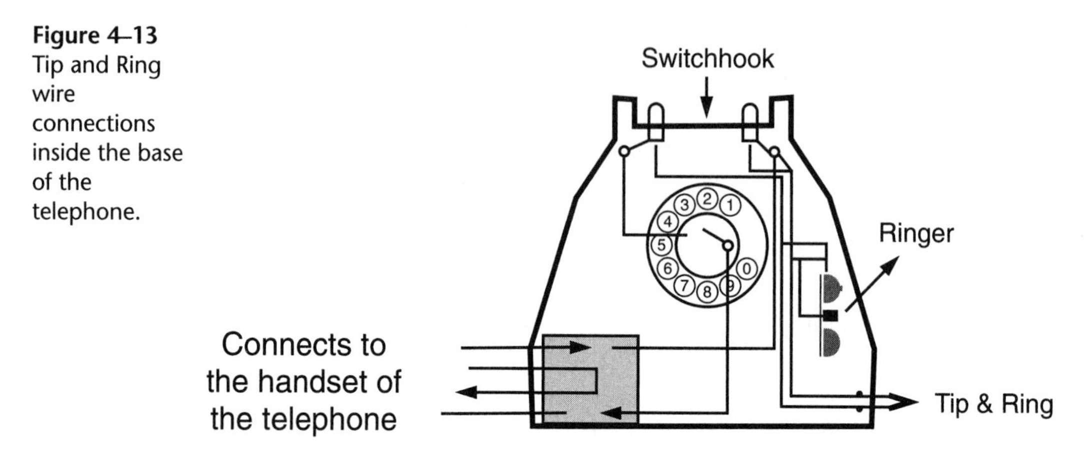

Figure 4–13 Tip and Ring wire connections inside the base of the telephone.

with the spring-loaded buttons. When the spring-loaded buttons are up, the circuit is complete; when they are down they break the circuit. Thus, the switch-hook accomplishes a very important function of connecting/disconnecting the customer line with the local exchange office.

As specified in EIA RS-470, the telephone has a Tip-to-Ring impedance of 600 ohms. It is necessary to achieve impedance matching between the input and the output to enable maximum transfer of power. However, the impedance of the line is complex and will rarely match that of the set, due to cable characteristics. This results in losses in the objective range of 4 to 6 dB. Usually the total impedance of a typical local exchange POTS line exceeds 900 ohms. Therefore, many central office hybrid alignments are performed using 900 ohms terminations with capacitance.

In the handset, the two-wire circuit is connected to a four-wire hybrid with one transmit wire connected to one receive wire in a loopback arrangement. This is shown in Figure 4–14. With the transmitted sound looped back to our ear, we can hear everything we say. This process produces side tone, which is a means of letting us know that a connection across the telephone network still exists. If there is no side tone, we hear nothing and can conclude that the connection is cut off. This basic learning in telephone usage is acquired through experience.

The mouthpiece converts acoustic energy into electric energy by means of a microphone or carbon granule transmitter. A DC current provided by the telephone company is passed through two electrodes: one attached to a diaphragm, and the other supported by the handset. The two electrodes are separated by thousands of carbon granules, as illustrated in Figure 4–15. When sound impinges on the diaphragm, it produces variations of air pressure that are transferred to the carbon; the resistance of the carbon varies inversely with pressure. The current varies inversely with resistance, resulting in an analog sine wave proportional to the frequency and amplitude of the voice. The receiver consists of a diaphragm of magnetic material, which converts electrical energy to acoustic energy, as illustrated in Figure 4–16.

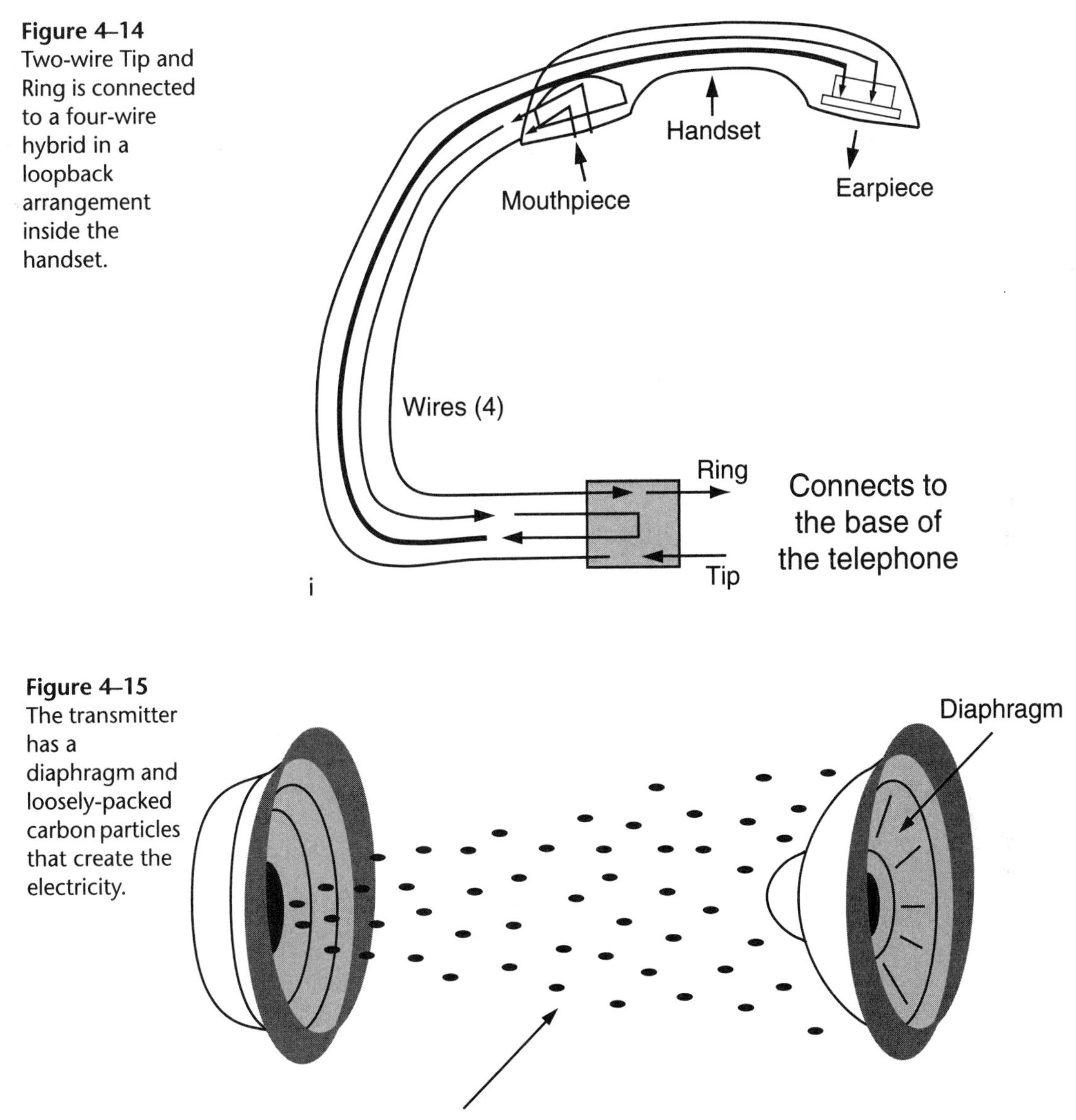

Figure 4–14 Two-wire Tip and Ring is connected to a four-wire hybrid in a loopback arrangement inside the handset.

Figure 4–15 The transmitter has a diaphragm and loosely-packed carbon particles that create the electricity.

The efficiency of converting speech into electrical signals is determined by the Transmit Loudness Objective Rating (TLOR), and the efficiency of converting electrical signals into speech signals is defined as the Receive Loudness Objective Rating (RLOR). Since the TLOR and RLOR represent losses, the more positive the number, the more the loss presented. TLOR and RLOR rating methods are specified in IEEE Standard 661-1979.

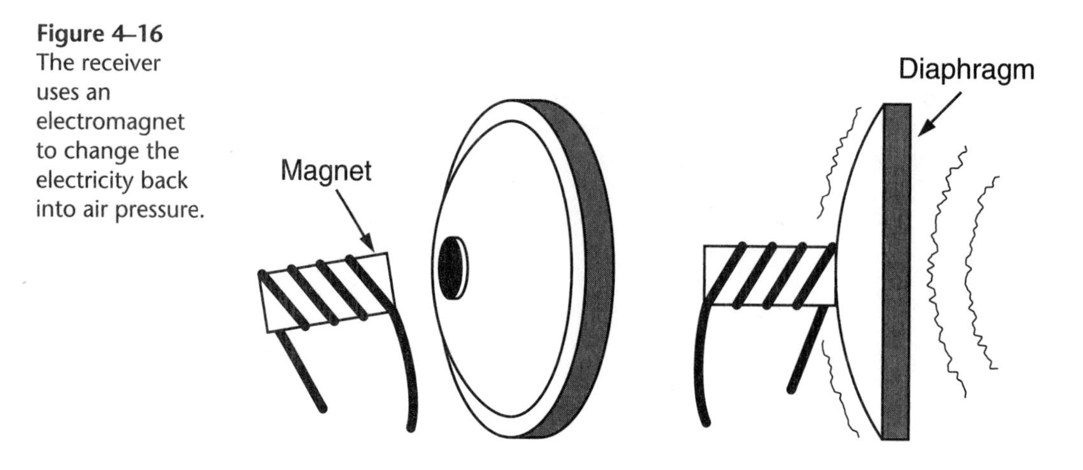

Figure 4–16 The receiver uses an electromagnet to change the electricity back into air pressure.

Outgoing Call

The dial pad, also called keypad or touch-tone pad, is used to dial telephone numbers as well as to interact with voice processing systems such as voice mail and Interactive Voice Response (IVR). The address signaling for an outgoing call can be accomplished by three different methods: Pulse Dial (Rotary), Dual Tone Multi-Frequency (DTMF), or Multi-Frequency (MF). In the United States, most telephones use the DTMF method referred to as touch-tone signals. But internationally, pulse dialing is still dominant since most of the telephone systems are older and have not adopted DTMF dialing. Some telephones with a touch-tone pad have a slide switch on the side that lets the caller select DTMF or pulse dial.

Pulse Dial

Pulse Dial specifications are recommended by the EIA Standard RS-470 for use within the United States. In general, the pulse repetition rate is between 8 and 11 pulses per second (pps), although the exact allowable range depends upon the type of capacitance used in the telephone's protective network. As an example, at 10 pps, the 0 key may output 10 pulses taking approximately 1 second, while the 5 key may output 5 pulses and take approximately 1/2 second to transmit. The interdigit interval is specified between 600 ms and 3 s.

Dual Tone Multiple Frequency (DTMF)

Dual Tone Multiple Frequency (DTMF) consists of a frequency matrix, where every key transmits a unique combination of two frequencies or tones. Figure 4–17 describes the tone pairs used. For most systems, only columns 1–3 are used, providing 12 possible signaling states. Except where denoted with parenthesis ( ) or brackets [ ], the lettering above the digit number appears on standard telephone sets. Within the North American phone system, telephone companies often use the octothorpe (#) to support special fea-

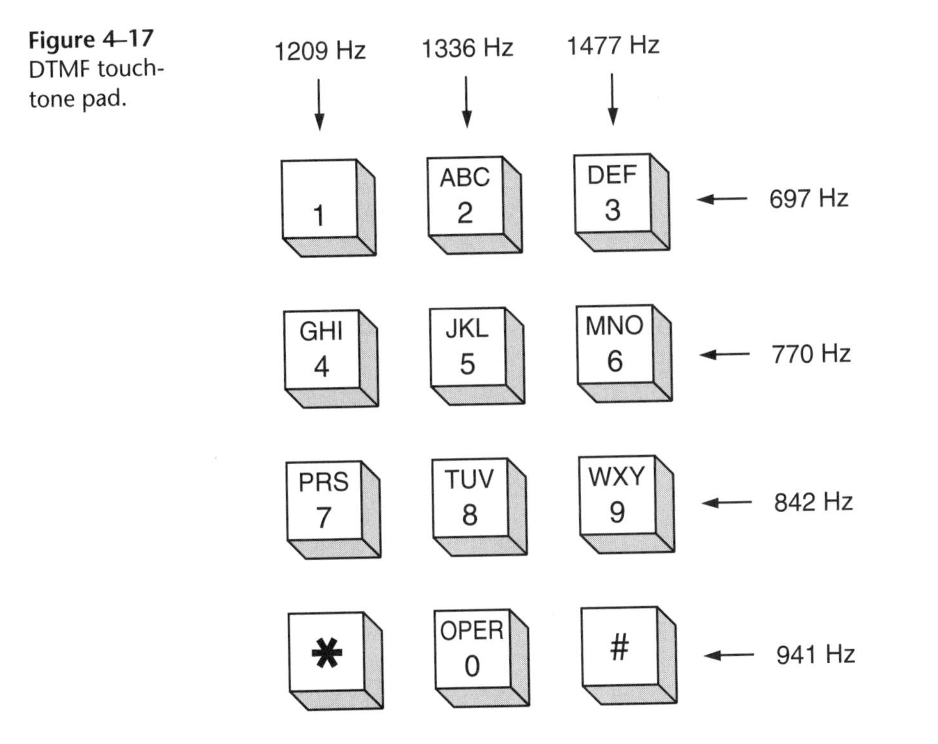

Figure 4–17 DTMF touch-tone pad.

tures. For example, when the # is the first digit in a dialing sequence, it notifies the switch of a call requiring special treatment, such as a data call.

The digits in brackets (A–D) indicate the additional fourth column of DTMF signaling, allowing a total of 16 possible digits, as shown in Figure 4–18. The U.S. government's AUTOVON network uses these digits (A–D) to represent call priorities of Flash Override (A), Flash (B), Immediate (C), and Priority (D), respectively. AUTOVON calls placed without using these digits are treated as normal priority. For automatic dialing, the minimum digit cycle time is 100 ms. The duration of the DTMF signal must be at least 50 ms, and the interdigit interval is specified between 45 ms and 3 s.

Multi-Frequency (MF)

Multi-Frequency (MF) signaling is similar to DTMF and is used on trunk circuits, pay telephones, and some CCITT (ITU) signaling schemes (CCITT #5). Combinations of two tones are used to transmit signaling information, as indicated in Figure 4–19. MF and DTMF signals are more reliable and considerably faster than pulse dial. In both methods, digits are transmitted at the rate of about 7 digits per second.

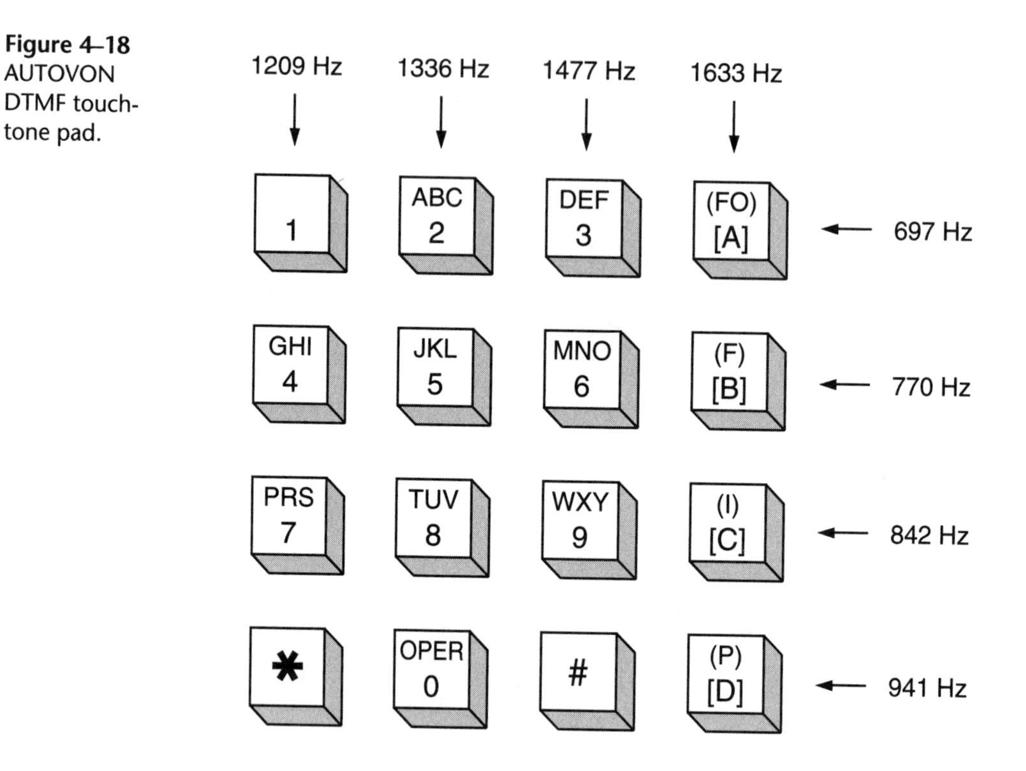

Figure 4–18 AUTOVON DTMF touch-tone pad.

Incoming Call

When a telephone set is called, the local exchange generates a high Ring voltage of about 90 to 105 volts AC with a frequency of 20 Hz superimposed over the –48 volts DC. A capacitor in the phone passes the AC but blocks DC. Upon detecting the AC voltage, the phone provides an audible signal that alerts the user of an incoming call. Each telephone set provides a Ringer Equivalence Number (REN), as defined within FCC Part 68 and EIA RS-470. Some examples of RENs are 1A, 1.5A, 0.5B, and 2L. The REN is used to ensure that the local exchange can provide the correct amount of power required to ring the telephone. It describes the frequency range, power range, and bias voltage range of the telephone set. When the phone is answered, DC current begins to flow in the loop. The local exchange senses the current flow and removes the superimposed Ring voltage. The –48 volts DC that is always on the line operates the telephone when it is being used.

LINE SIGNALING

The telephone line is a connection from the local exchange office to a single-line telephone or a business telephone system such as a PBX. The most common line signaling method provided by the local exchange to communicate with a single-line telephone set is **Loop Start**. **Ground Start** is more common on PBXs and trunks. Both methods use DC

Tone Combination	Digit
700 + 900	1
700 + 1100	2
900 + 1100	3
700 + 1300	4
900 + 1300	5
1100 + 1300	6
700 + 1500	7
900 + 1500	8
1100 + 1500	9
1300 + 1500	0

Figure 4–19 Multifrequency signaling.

currents to represent specific states. These states include on-hook and off-hook supervision, disconnect supervision, and dialing supervision.

Loop Start

A *loop* implies a closed circuit. If the phone is on-hook, the circuit is broken and no current flows, which is referred to as an idle state. When the phone is off-hook, the circuit is complete and current flows through the wires. The local exchange switch senses the current and provides an audible ready signal called *dial tone*. A schematic for Loop Start signaling is represented in Figure 4–20. The advantage of Loop Start is that there is no need for accurate ground references between the local exchange and the telephone. Also, the Tip and Ring wires may be reversed with no adverse impact on operations. However, a drawback of Loop Start is the poor *glare* resolution. Glare occurs when both the local end and the remote end attempt to access the circuit at the same time. With Loop Start, the telephone is not informed of an inbound call until ringing is detected. As a result, on some occasions when we pick up the telephone to place a call we may find that the line is already connected to an incoming call although the bell has not rung.

Ground Start

When implemented by a local exchange, the Ground Start interface is usually incorporated only on trunks and PBXs, as shown in Figure 4–21. This DC line signaling method operates as follows. When a call is placed from the local switching exchange to the PBX, the local switch immediately grounds the conductor tip to seize the line. With some seconds delay,

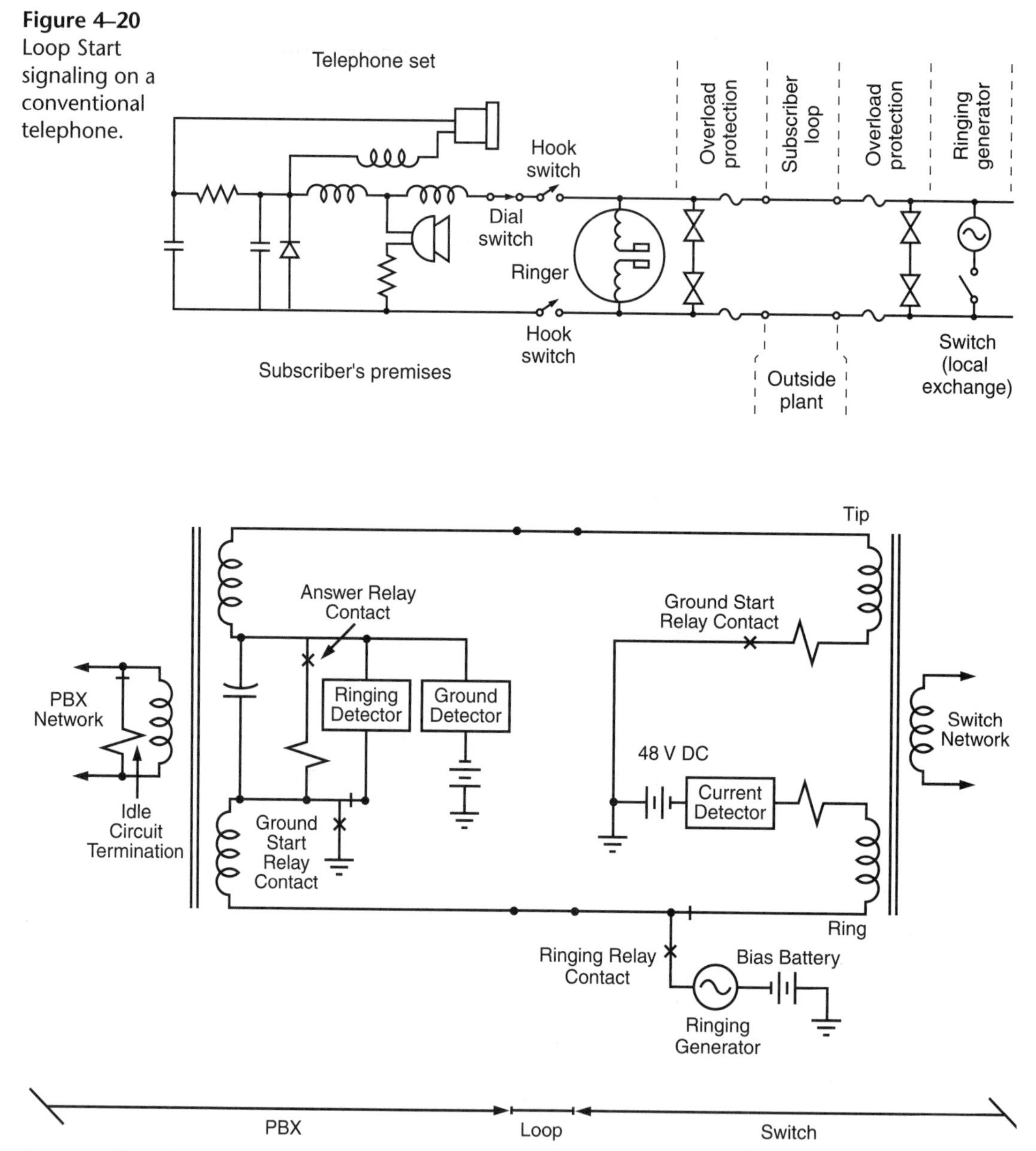

Figure 4–20 Loop Start signaling on a conventional telephone.

Figure 4–21 Ground Start interface block diagram.

ringing voltage is applied to the line, after which the PBX immediately detects the grounded-tip conductor and does not allow an outgoing call from the PBX to use this circuit, thus avoiding glare. In a similar fashion, if a call originates at the PBX and is outgoing, the PBX grounds the ring conductor to seize the line; the local switching exchange recognizes this condition and prevents other calls from attempting to terminate on the circuit. Ground Start minimizes the possibility of glare by providing remote-end disconnect supervision. The disadvantage of Ground Start is that the Tip and Ring leads cannot be reversed. Also, the local-end and remote-end grounds must be at the same potential: earth ground.

TRUNK SIGNALING

Trunk signaling can be separated into two categories: *out-of-band* and *in-band*. An out-of-band signaling system uses a separate network to pass call setup, charging, and supervision information. In-band signaling carries these signals over the same circuit that carries information. In-band signaling has several drawbacks that are overcome by out-of-band signaling. First is its susceptibility to fraud. Toll thieves are able to defeat automatic message accounting systems by using devices that emulate signaling tones. The second drawback is setup time, which is reduced by using a separate signaling network. Reducing call setup time, in turn, reduces the access charges that IXCs pay to the LECs. Lastly, in-band signaling is incapable of supporting virtual networks. Virtual network applications include credit-card authorization, 800 number portability, cellular phone roaming, Intelligent Network (IN) services, and CLASS.

The trunk signals can be divided into different groups according to the type of information being carried. *Supervising signals* monitor the status of a line or circuit to determine if it is busy, idle, or requesting service. *Call Alert* and *Call Progress signals* establish the process for incoming and outgoing calls. They include dial tone, ringer tone, and special identification tones. *Addressing signals* refer to the destination information such as the actual telephone number being called. For in-band systems, addressing signals are transmitted over trunks as DTMF or MF signals, while out-of-band systems transmit them as a data signal.

Single Frequency (SF)

The most common in-band, analog trunk signaling system is called Single Frequency (SF). Here, all information is transmitted in the voice band. In the United States, a frequency of 2600 Hz is used, while in the U.K., it is 2280 Hz. An idle status is indicated by a 2600 Hz tone, while trunk seizure drops the 2600 Hz tone from the channel to indicate a busy status. The SF system contains the circuitry to change the state of the E&M leads in response to the presence or absence of the SF tone and vice versa.

E&M Signaling

By long-standing convention, trunks use an E&M *(recEive and transMit)* signaling scheme to communicate its status to the attached local exchange office equipment. **E&M Signaling** is a two-state signaling scheme (On-Hook/Off-Hook) commonly used on digital four-wire local exchange and tie trunks. Although there are five different types of E&M signaling interfaces, some are more popular. They are illustrated in Figure 4–22.

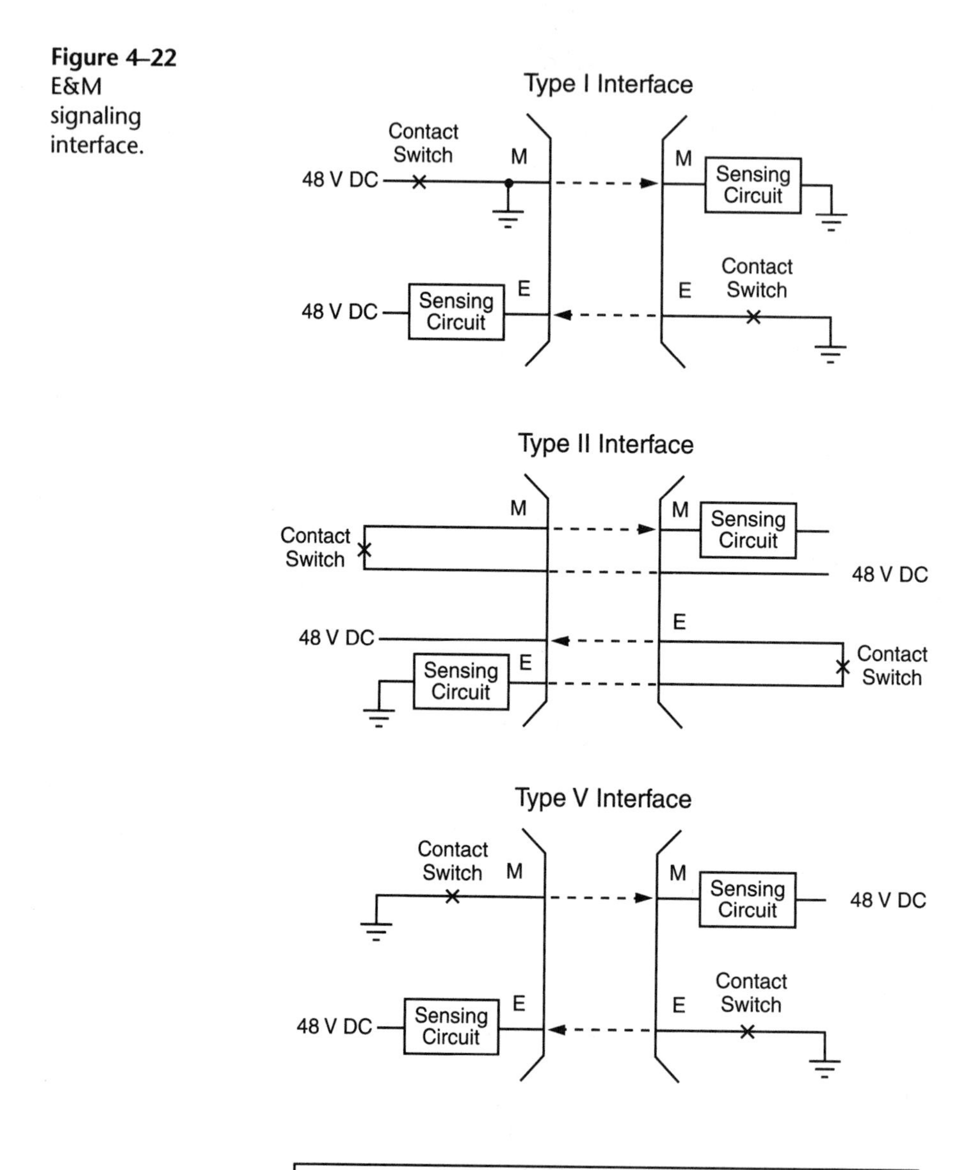

Figure 4–22
E&M signaling interface.

E&M Signaling	Condition	M-lead	E-lead
Type I	on-hook	Ground	Open
	off-hook	48 V DC	Ground
Type II	on-hook	Open	Open
	off-hook	Battery	Ground
Type IV	on-hook	Open	Open
	off-hook	Ground	Ground

Type I Signaling is the most commonly used four-wire trunk signaling interface in North America. When idle, the M lead is grounded while the E lead is open. When busy, a –48 volt DC signal is applied to the M lead and the E lead is grounded. This causes high return current through the grounding system. The asymmetrical nature of this signaling scheme is thought to be a potential source of interference.

Type II Signaling is least likely to cause interference problems in sensitive environments as it provides almost complete isolation of signaling power systems. It is seen occasionally in North America, usually on Centrex trunk circuits.

Type V Signaling is the most popular interface outside North America. While it does not provide isolation between power systems, there is minimal return current in this symmetrical signaling scheme.

Common Channel Signaling

Common channel is the most prominent out-of-band signaling system. Within the United States, the original implementation of **Common Channel Signaling** started in 1976, and was known as CCIS (Common Channel Interoffice Signaling). This is similar to CCITT's Signaling System #6 (SS6). The CCIS protocol operated at relatively low bit rates (2.4K, 4.8K, 9.6K) and transported messages that were only 28 bits long. However, CCIS could not adequately support an integrated voice and data environment. Therefore, a new High-level Data Link Control (HDLC) protocol based signaling standard was developed by CCITT: Signaling System #7 (SS7). First defined by the CCITT in 1980, the Swedish PTT started SS7 trials in 1983, and some European countries are now entirely SS7-based. Within the United States, telephone companies began implementing SS7 in 1988. Currently, a large majority of the IXCs and LECs have migrated to SS7. However, some LECs are still in the process of upgrading their networks to SS7 because the number of switch upgrades required for SS7 support impacts the LECs much more heavily than the IXCs. The slow deployment of SS7 within the LECs is also, in part, responsible for delays in incorporating Integrated Services Digital Network (ISDN) within the United States.

Signaling System #7 (SS7)

Signaling System 7 (SS7) is a control system independent of the network hardware. It is a data communications protocol that underlies the Intelligent Network. It handles end-to-end call supervision, and call timing and billing. It is also responsible for call routing and congestion control. In addition, it provides database information to form virtual networks and enhanced services such as identifying and verifying callers. SS7 uses a layered protocol that resembles the OSI model but has four layers instead of seven.

Figure 4–23 shows the SS7 protocol architecture compared with the OSI model. The first three layers are called the *Message Transfer Part (MTP)*. In the MTP, the bottom *physical layer* provides a full-duplex connection between network nodes. The *link layer* is responsible for flow control and error correction. The *network layer* routes messages from source to destina-

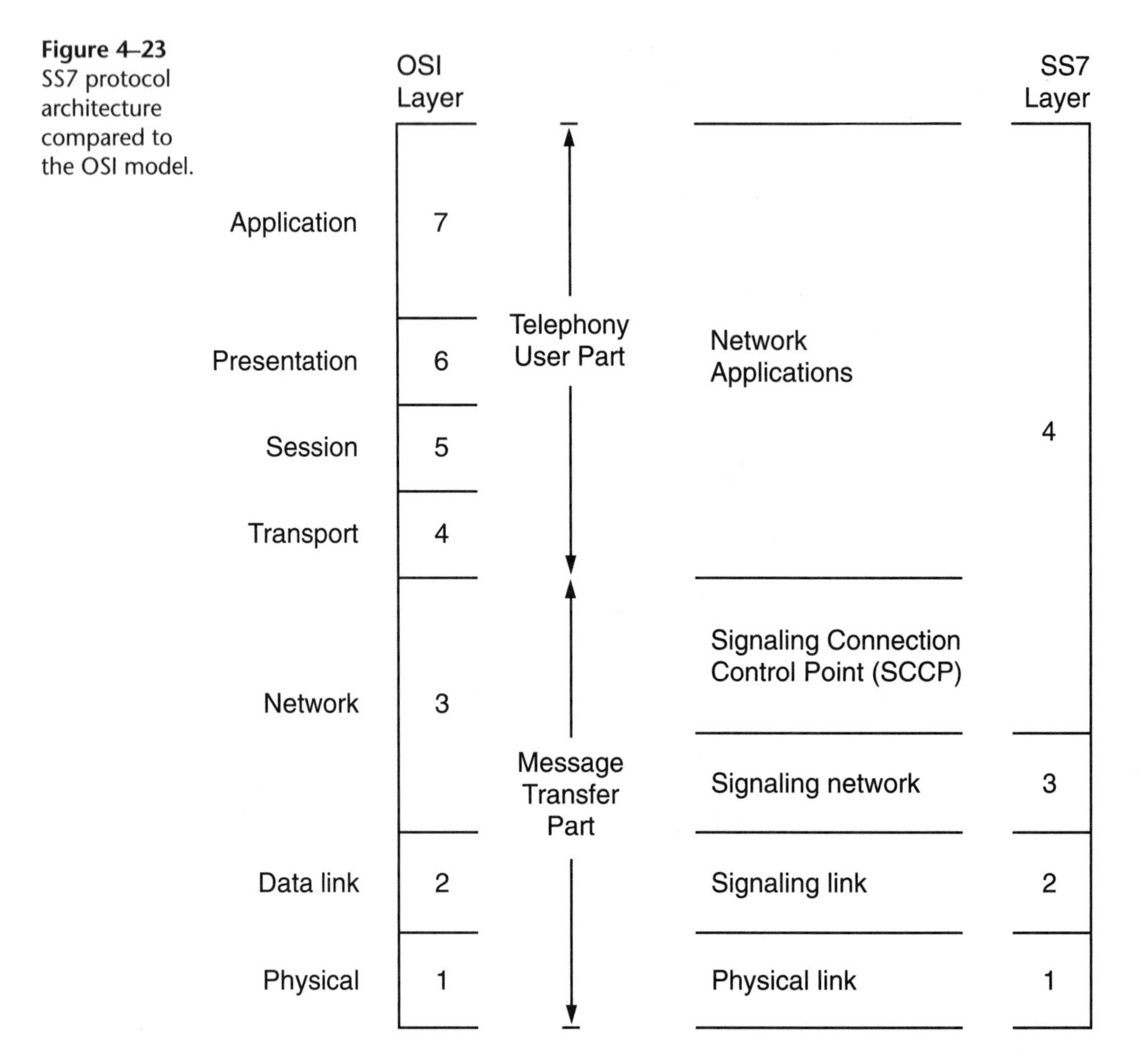

Figure 4–23 SS7 protocol architecture compared to the OSI model.

tion, and provides an interface between lower layers and the fourth layer. The top layer called the *Signaling Connection Control Part (SCCP)* is responsible for addressing requests to the appropriate application and determining its status. Beyond this layer are the telephony, ISDN, and data user parts, which comprise of network applications. Note that while many of the elements of SS7 are common, there can be some significant regional variations in its deployment.

SS7 system has three major components: *Service Switching Point (SSP)* also called *Action Control Point (ACP)*, *Signal Transfer Point (STP)*, and *Service Control Point (SCP)* or *Network Control Point (NCP)*. These are interconnected, as shown in Figure 4–24. The SSP or ACP is a tandem switch in an IXC network or a local exchange switch in a LEC network that can provide switching and routing functions. STPs are packet-switching nodes that facilitate

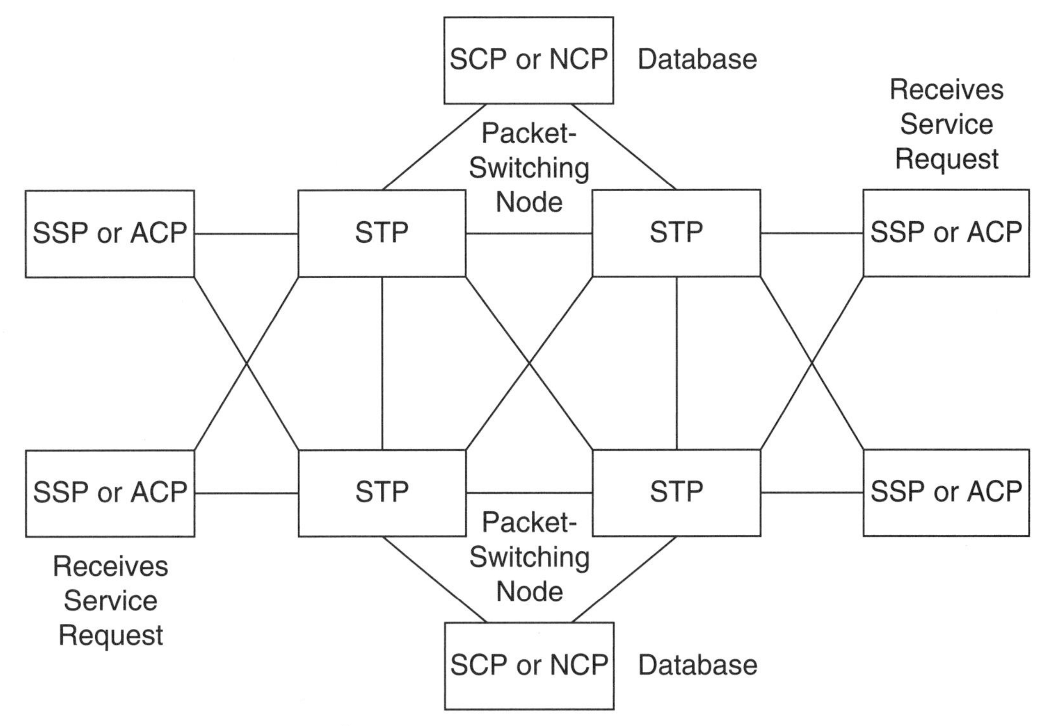

Figure 4–24 Interconnection of SS7 system components.

highly reliable and cost-effective network architectures. SCPs or NCPs are typically data processing CPUs connected to databases of circuit parameters, routing, and customer profiles. They are responsible for handling different types of calls, for example, virtual network, calling card verifications, and others. Communications between an ACP and a NCP may traverse several STPs. Although carriers have their own signaling networks, they are interconnected to enable carriers to interoperate. As a result, there is a 100 percent redundancy built into the system to protect it from failure.

INTELLIGENT NETWORK (IN) SERVICES

In the late 1970s, telcos became fixated on their expensive investments in computer-controlled switching and were intrigued by the prospect of providing intelligent services using the CCIS protocol. The concept of network control was extended to let digital switches communicate with databases (known as SCPs) and signal processing systems (Intelligent Peripherals). Services based on IN, also referred to as Advanced Intelligent Network (AIN) in the United States, have been almost universally adopted by wireline and wireless telephone companies across the globe. These services control the carrier's networks, interconnect with other networks, and most importantly, provide business and residential subscribers with a broad range of sophisticated calling services. Operators adopt IN services either because they generate substantial revenue or because of regulatory requirements.

The most common IN service to the residential subscriber is caller ID, which provides the called party with the name and telephone number of the caller, or other information pertaining to the call. Some common IN services for businesses include Automatic Call Distribution (ACD), IVR, Automated Attendant (AA), and voice mail. The amount of SS7 traffic being generated by telecommunications networks is growing fast. As ILECs and CLECs are now unable to compete effectively on price, they are distinguishing themselves based on the services they offer. The sophisticated services are made available using the AIN functionality, which increases the amount of SS7 traffic on the network. Let us study how some of these services are implemented.

Caller Identification

Caller ID is also called Incoming Caller Line ID (ICLID) service and can be provided over standard analog two-wire loops. The local exchange office switch transmits the ICLID information through modem bursts between the first and second ringing cycles. The user's telephone must have a special phone or adapter box that provides the decode/display functions. If the user picks up the handset during the first ring cycle, there is a possibility that the ICLID information may not be received properly. When calls are placed from PBX users, the calling party information transmitted through the PSTN usually includes only the main PBX number. Thus, the PSTN does not provide information regarding the PBX extension.

Automatic Call Distribution (ACD) for Businesses

Automatic Call Distribution (ACD) or Automated Call Queuing is the process of distributing incoming calls to a specific group of terminals and to queue calls (put them on hold) if all terminals are busy. This is an extremely useful feature for companies that receive many incoming calls for a group of people. Most of us have experienced an ACD queue. If you have ever called an airline company, a hotel company, or the IRS, then you've encountered an ACD system. An ACD system usually greets you with a friendly prompt such as "Thank you for calling XYZ Corporation. Your call is important to us. Please hold for the next available agent." However, ACD is not only a tool to keep callers on hold while the call center tries to catch up with the call volume, but also can be a handy tool to answer calls promptly, efficiently, and professionally. Businesses use an ACD system to achieve various objectives: To distribute calls evenly throughout the workgroup; to ensure that all callers are greeted by a human being right from the start; and to route calls to agents that the company wants handling most of its incoming calls. A scripted flowchart for an ACD system is illustrated in Figure 4–25.

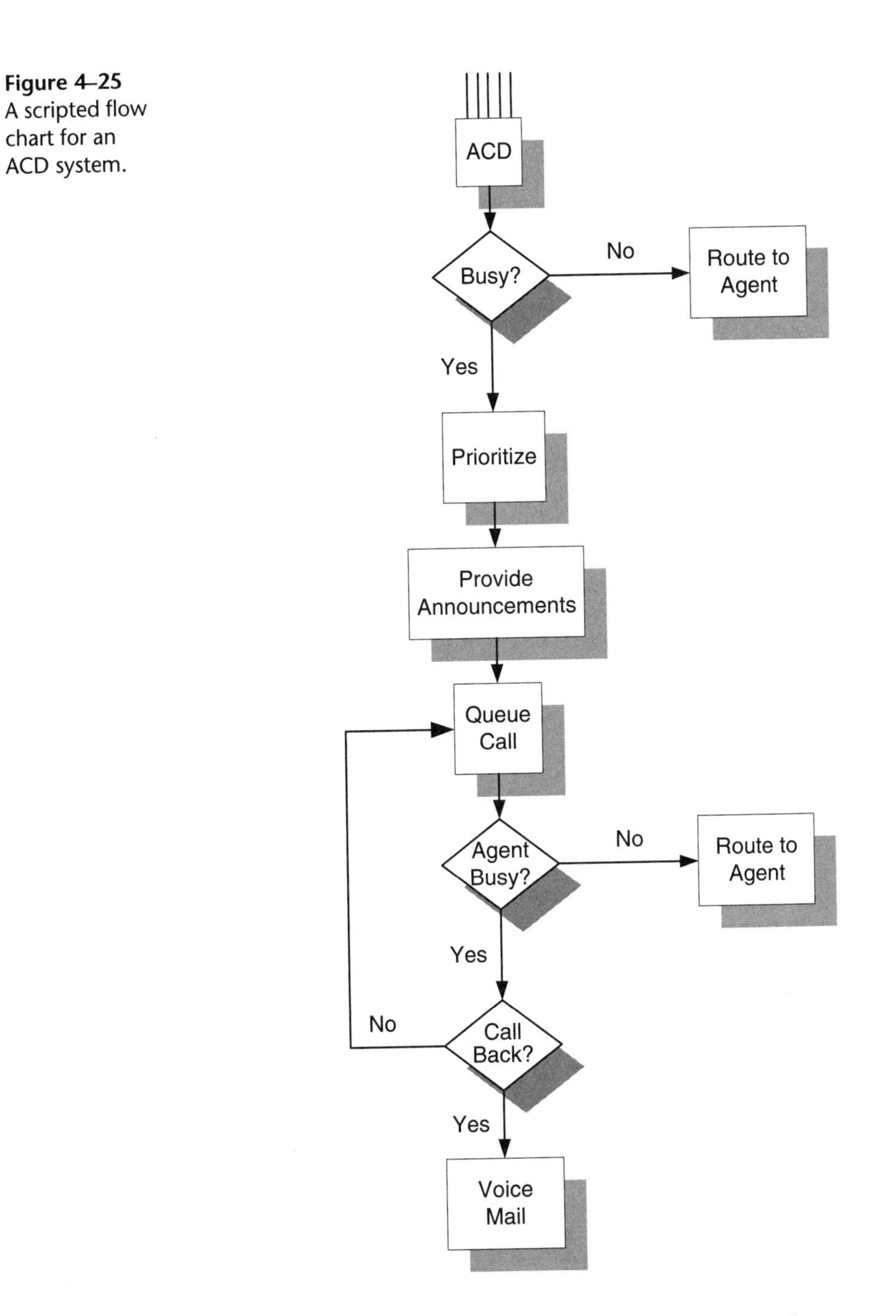

Figure 4–25 A scripted flow chart for an ACD system.

The most common use of ACD is for sales or technical support departments. A group of employees called a "workgroup" share all of the calls that are placed into the department. The ACD system decides which agent's phone will ring next and puts overflow calls into a holding pattern until there is a free agent. The queue keeps track of the order of the callers, plays music-on-hold, and can even play update greetings for the callers. This assures the caller that he/she has not been forgotten. Some ACD systems may inform the callers of the average expected wait time and update them whenever their position in the queue changes. These features enable callers to judge whether or not they want to continue holding. Although most large call centers have workgroup with 50 to 100 agents or more, even small groups can benefit from ACD technology.

In case of telecommuters, there is some variation of this technology. Whenever the PBX applies the ring voltage to the circuit, it routes the call to an offsite phone number. The PBX is tricked into seeing an offsite telephone as a local extension. Using this approach, telecommuters appear to the system to be local, and thus can participate in ACD groups and use the PBX's voice messaging services. The downside is that the business must provision one of these circuits for each telecommuter, which can get expensive.

Voice Processing Systems for Businesses

Voice Processing Systems include IVR, AA, and voice mail. Many companies install these systems to reduce the number of agents required "on the phones." It is important to have sufficient number of ports or connections between any voice-processing system and the PBX or Centrex.

Interactive Voice Response (IVR)

Interactive Voice Response (IVR) refers to the technology supporting the interaction of users with a system that provides information to the caller. A functional diagram of an IVR system integrated with an ACD and computer database, all networked with a PBX, is shown in Figure 4–26. For example, many banks have implemented IVR to allow clients to access their own records. These systems require entry of a Personal Identification Number (PIN) with a minimum of four digits. An IVR system's main function is to directly provide the caller with automated information. It interfaces with computer databases to retrieve records and manipulate data if necessary.

Automated Attendant or Auto Answer (AA)

Automated Attendant or Auto Answer (AA) is similar to a switchboard operator who directs calls to the appropriate extension, as illustrated in Figure 4–27. The main difference is that it can direct multiple calls at one time depending upon the number of ports. Most AA systems respond to touch-tone telephones although some respond to rotary dial. As an example, airline companies use ACD and AA in combination. The caller is initially greeted by the AA system: "Thank you for calling XYZ Corporation. This call may be monitored or recorded to ensure quality service. For today's flight arrivals, departures, and gate information, press 1. For domestic reservations and fares, press 2. For international reservations and fares, press 3. For all other inquiries, press 4. If you do not have a touch-tone telephone

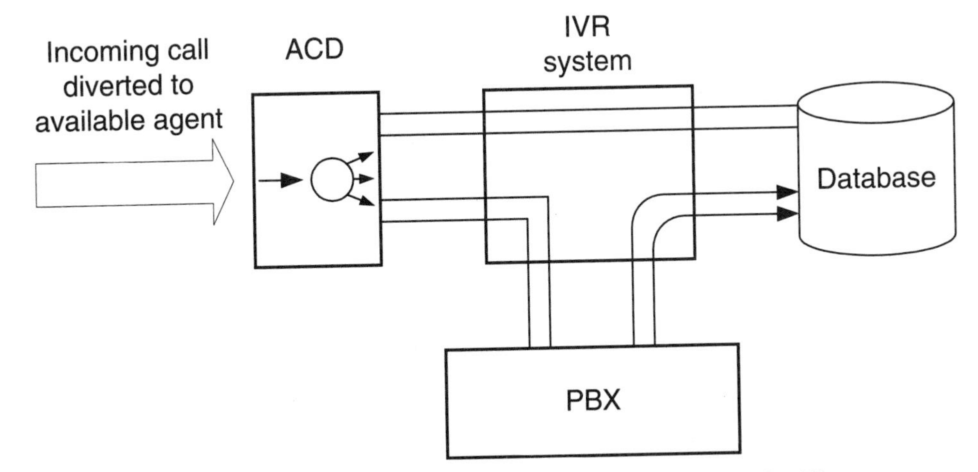

Figure 4–26 Functional diagram of IVR, ACD, and database integration with PBX.

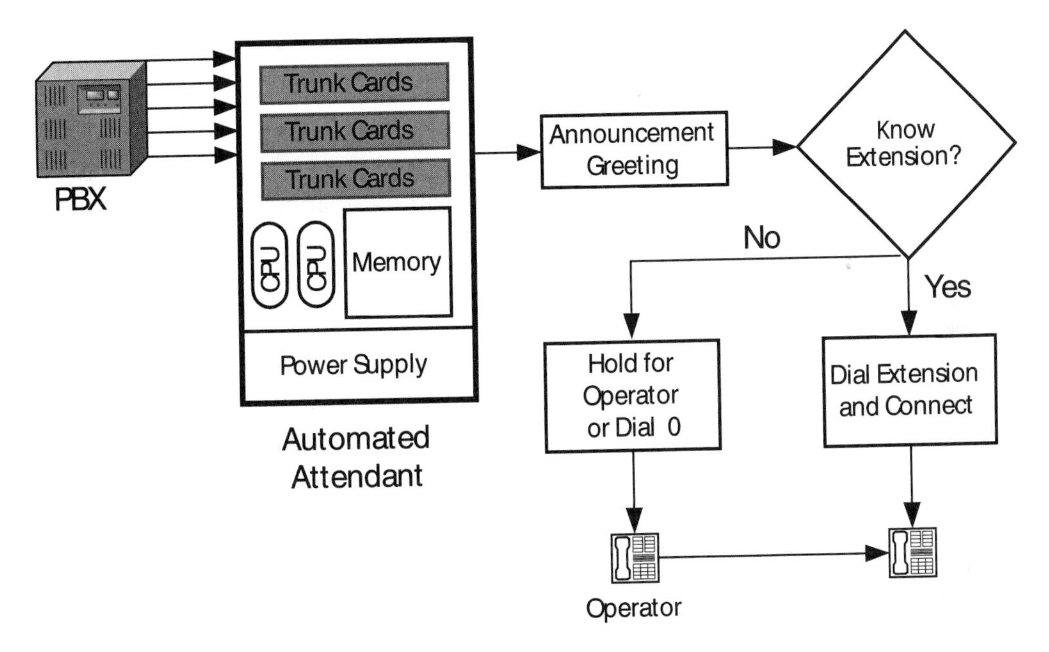

Figure 4–27 An automated attendant configuration.

please stay on the line and an agent will be with you shortly." Then, upon selecting the appropriate option, the call may be directed to another AA system or ACD system. The AA system is efficient, as the company may use it to provide callers with routine information such as flight arrival and departure times without involving an agent.

Voice Mail

Voice mail systems differ in memory and control functions. The voice mailbox is an electronic location in the system on which messages are held for each system user. The mailbox owner calls in and retrieves messages using a password entered from a touch-tone telephone. The user can listen to the messages, save them, erase them, send a reply, or forward them to another user's mailbox in that system.

BUSINESS TELEPHONE SYSTEMS

The decision-making process for a business telephone system involves evaluating different options offered by telephone companies: type of leased lines, voice communications equipment, and services. These systems are simple overlays on the PSTN, where a corporation just leases transmission capacity and provides their own switching. In some cases, for example, international banks, an enterprise may decide to build its own private network providing many of the services of the PSTN but dedicated to that one corporation. Such single-corporation networks may lease their excess capacity to other corporations, possibly even to the PSTN. Private networks are set up mostly to save money but sometimes to provide acceptable service in situations where the PSTN provides substandard service.

Telephone Lines

Telephone companies may rent their circuits that permanently connect the local exchange to an organizational voice facility or two or more offices of a large organization. These circuits may carry voice, data, or video signals. The circuits are called *leased lines* or *dedicated lines* and are rented from the LEC on a flat-rate basis. There are different types of leased lines: analog, ISDN, T-1 and **Tie lines** with additional options such as **Direct Inward Dial (DID)**, **Direct Outward Dial (DOD)**, *Foreign Exchange (FX)*, and *toll-free*. Both DID and DOD lines are usually referred to in the context of a PBX. The leased lines establish physical and logical connections between two switches across which network traffic travels.

Two-wire or half-duplex transmission is where both transmit and receive paths are carried on the same wire pair or other single medium. Four-wire or full-duplex transmission is where transmit and receive paths are separate and a wire pair is assigned to each path. Older connections required two sets of wires for full-duplex transmission, one set for sending and one set for receiving. Now, however, modem technology allows for full-duplex transmission to occur over only two wires (one set) through a process known as *echo cancellation*. There are also two-wire and four-wire switches. Although uncommon today, two-wire switches can be found in the LEC circuit. Four-wire and six-wire switches used in conjunction with multiplexing equipment are found in the tandem and IXC circuits. All digital switches are at least four-wire. In six-wire switches, there are two wires for transmit, two wires for receive, and two wires for supervisory signaling.

ISDN Line

Based on all-digital transmission that allows the transport of voice and non-voice services on a call-by-call basis, ISDN line provides the user with easy access to multiple services over a single connection to the network.

T-1 Line

A special type of high-capacity phone line for larger offices, T-1 line delivers 24 phone lines on two pairs of wires. T-1 simplifies the cabling coming into the office. It is digital while ordinary phone lines are analog, so it provides slightly better sound quality.

Tie Trunk

Point-to-point line connecting two voice facilities, typically PBXs or Centrex, so that the users can communicate without dialing an outside call, as shown in Figure 4–28, is called tie trunk. It connects both trunks and station lines within a customer's network. In most cases, tie lines are considered the same as tie trunks. However, some make a distinction between tie trunks and tie lines, referring to tie lines as only those that connect lines throughout a network connection. Tie circuits normally support both incoming and outgoing calls. Tie trunks are typically four-wire E&M type circuits, but may occasionally be two-wire lines using Ground Start signaling. Satellite Tie Trunks (STTs) are ideal for use within campus environments and allow for functional transparency between remote and host PBXs. In order to provide transparency, HDLC protocol-based signaling methods usually are employed.

DID Line

A special type of line used for incoming calls only, DID line enables the direct dialing of each individual extension in the PBX, as illustrated in Figure 4–29. The phone company delivers a three or four digit code along with the call. For example, if the main phone number is 555-1100, extension 101 can be dialed directly by dialing 555-1101. A DID line simply sends a code, in this example, 101, which tells the phone system that there is a direct call for extension 101. The call can bypass the receptionist or auto-attendant at 555-1100 and ring directly at the 101 station. If the directly dialed extension is not answered or is busy, the call may then go to the switchboard operator or it may be answered by voice mail. An organization could also do this by getting one physical phone line for each employee, but this is a waste of money. For example, DID enables sharing 50 phone numbers on 4 physical telephone circuits. However, caller ID is not available on DID lines.

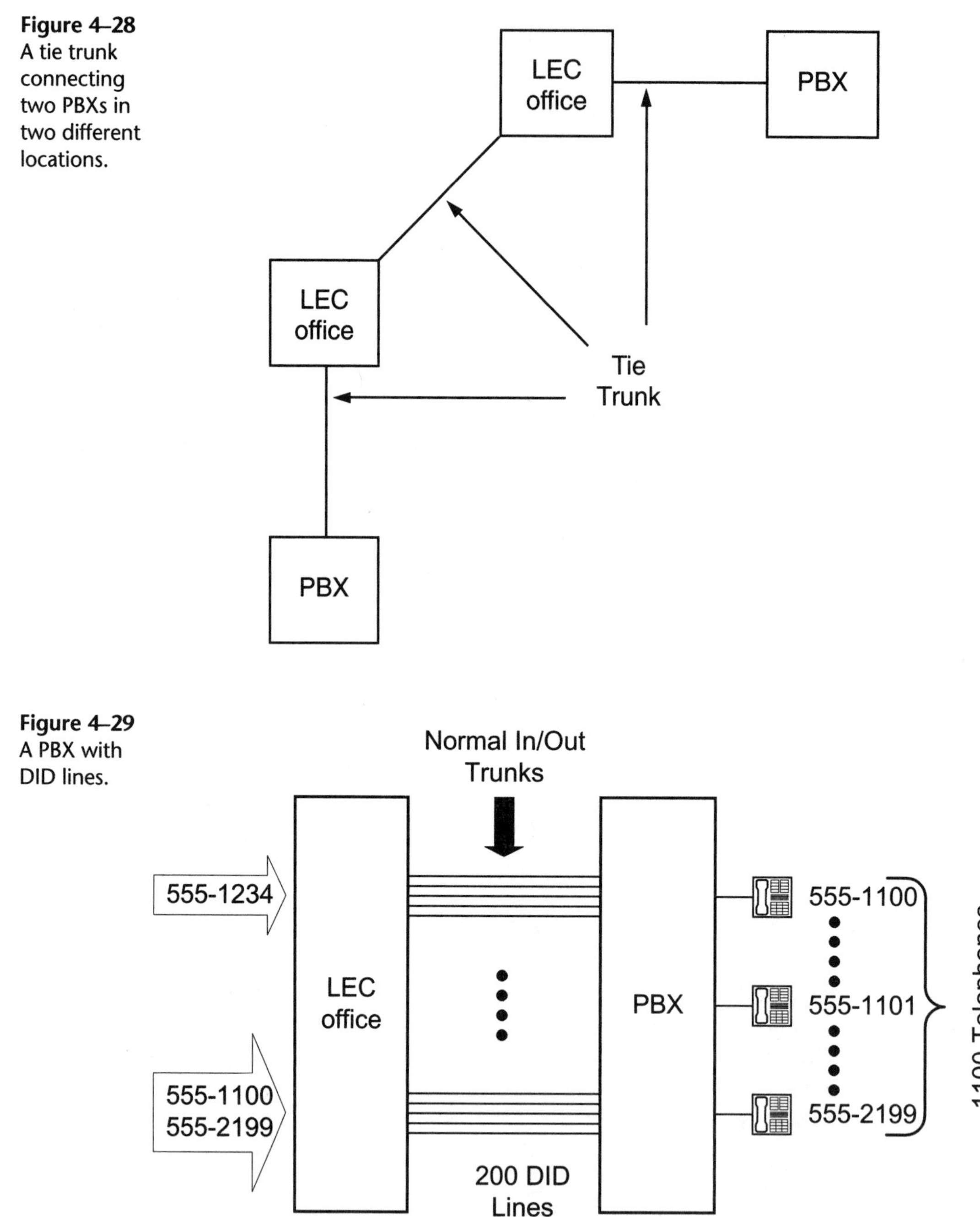

Figure 4–28
A tie trunk connecting two PBXs in two different locations.

Figure 4–29
A PBX with DID lines.

DOD Line

Extension users can dial directly outside the system with DOD line. The users dial an access code, for example, number 9, that signals the telephone system to find a line to connect to the outside world. The station user then dials a local seven-digit or a ten-digit telephone number. Some systems require that the user enter a pin number for toll calls. No operator or manual assistance is required via this access method. In the past, when less-sophisticated telephone systems were available, it was not uncommon for a company to route all of its outbound calls through an internal operator. The operator's responsibility was both to screen calls and to route calls over appropriate facilities. With the advent of intelligent PBXs and Centrex, such limitations can be programmed on a telephone-by-telephone basis, or even user-by-user basis, thereby eliminating the operator's involvement.

FX Circuit

A leased-line service that allows a customer to draw dial tone from a remote local exchange in the LEC's service area is an FX circuit, as represented in Figure 4–30. The primary use of FX is to provide customers with a local telephone number to reach the company that has an office in a different city—for example, a customer located in Peoria, Illinois wants to call a supplier in Chicago, Illinois. Normally, this is a long-distance IntraLATA call between the two cities served by the LEC. But if the supplier has a FX service the local telephone number is connected to a leased line. The local exchange office transfers the call to the supplier's office in Chicago. The supplier pays for a local dial-tone line at the Peoria local exchange and the mileage charges for the dedicated private line to the supplier's location in Chicago.

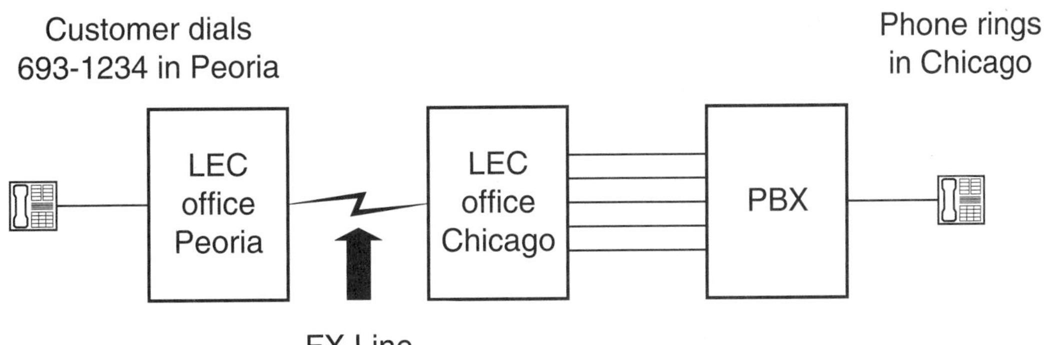

Figure 4–30 An FX connection between two cities.

Toll Free Line

The most widely used service today, toll free line is a reverse-billing service where the callers can call at any time without worrying about the cost of the call. The subscriber of the toll-free calling service, either an individual or an organization, pays according to the volume of calls and origin of the calling party.

Private Branch Exchange (PBX)

Private Branch Exchange (PBX) is also referred to as Private Automatic Branch Exchange (PABX), or Electronic Private Automatic Branch Exchange (EPABX). The PBX is an on-premises telephone exchange (a digital device) that supports digital data services as well as analog telephone service and provides dial tone to the telephones. It serves as a point of entry into the PSTN. PBX is a popular choice of most businesses for multiple reasons. First, most calls are placed between users in the same organization. PBX systems pool outside line resources (PSTN lines) but also provide switching of in-house calls. It is easier to call someone within a PBX because all internal calls are routed through the PBX and never go to the local exchange. Therefore, we typically dial just 4 digits for all in-house numbers. Second, all the users do not need simultaneous access to the phone system at the same time. On an average, users will use the phone for about five minutes or less each hour. PBX is much less expensive than connecting an external telephone line to every telephone in the organization. Third, PBX has the ability to handle many users (several hundred to several thousand) on a single system and provide centralized support such as voice mail for digital-feature phones.

Testing a PBX system involves three different configurations: call from station to station, call from a station to an outside line, and call from an outside line into the office. One must make sure that none of the calls go into a black hole, that is, callers hear a ring but the called station does not ring. A PBX is appropriate for mid-size and large businesses that need to provide fairly basic telecommunications services for several hundreds to thousands of users. The PBX and the circuit-switched telephone network it is connected to may be displaced by next-generation technologies that do more, cost less, and are easier to administer. PBXs are to telephony what mainframes are to computing.

The typical components of a PBX are shown in Figure 4–31. A PBX cabinet comprises mainly of shelves that are made up of a group of slots that house cards or circuit boards for T-1s and DID lines. Voice mail, AA, and other voice-processing systems may also reside within the PBX. In addition, there is a power supply, a cooling fan, and supporting electronic components. They are big and expensive but provide highly reliable service to a large number of users. PBX technology although highly reliable, is limited in scope. The problem is that PBXs are an increasingly outmoded technology in an era when data communication is as important, or more so, than voice. The PBX, in order to survive, needs to be reinvented as a multipurpose platform that interoperates with computer networks and with next-generation telephone networks.

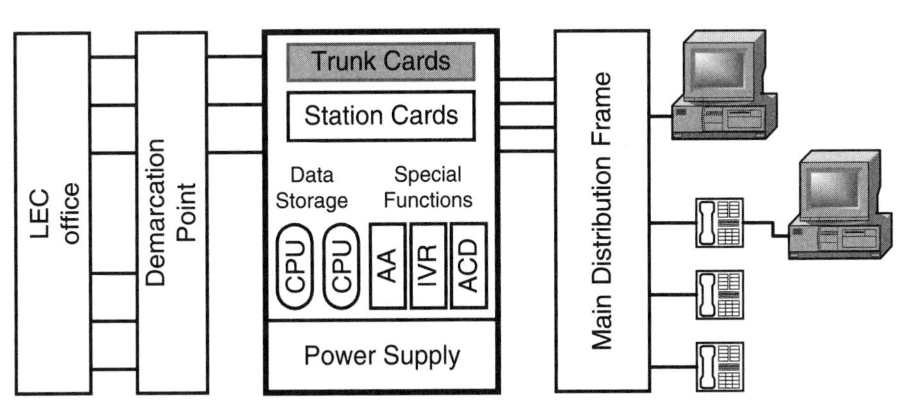

Figure 4–31 PBX system components.

PBXs, the most widely implemented business telephone systems, are not without weaknesses. Enhanced services such as voice mail, ACD, and AA can be expensive and difficult to configure. In addition, PBX management tasks are not always user-friendly. But most importantly, on many PBXs, the extension number assigned to a particular phone (e.g. 101, 102, etc.) is hard-wired, that is, it corresponds to a specific physical phone jack. So if employees move to another desk or office, their extensions change, thereby causing disruption within a business.

Centrex

Many companies, particularly smaller businesses, may not want to purchase and manage their own telephone system because of the capital investment, technical requirements, or time limitations. Most local telephone companies offer PBX-like services to end-users without requiring customers to buy their own telephone system. **Centrex**, which stands for Central Office Exchange Service, is a brand name for a set of services offered by local telephone companies on business telephone lines. The components of Centrex at the LEC office and customer premises are shown in Figure 4–32.

Centrex was originally introduced in the early 1960s to large customers, but it has migrated to targeting low- and mid-range customers. It can basically be described as LEC-provided PBX services. These services are heavily marketed by the LECs and are called by a variety of different names. In effect, Centrex allows a business to use the local telephone company as its telephone system. Each employee is given a private phone line with DID and DOD features. Centrex telephone lines are really just normal telephone lines that provide extended services such as three-way calling for conference calls, call transfer, caller ID, and voice mail. A functional diagram of an ACD system used with Centrex service is illustrated in Figure 4–33.

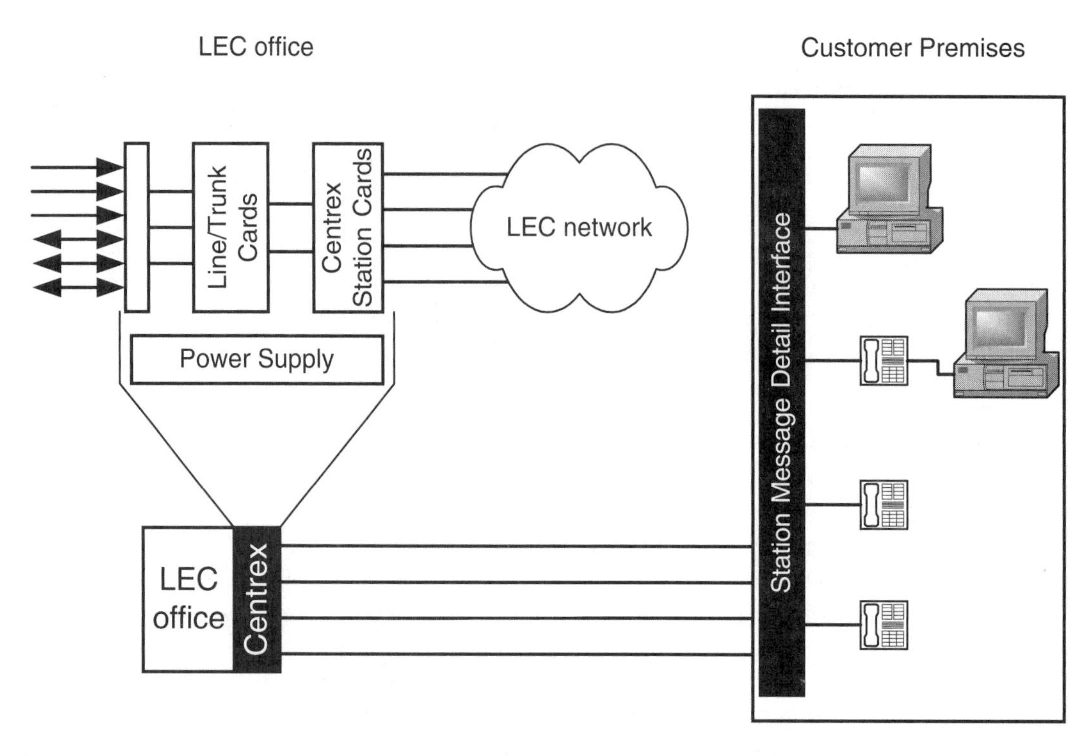

Figure 4–32 Centrex system components.

Centrex lines usually cost about 20% to 50% more per month than plain analog phone lines. It is a cost-effective way to get the features of a PBX without having to buy a PBX. The advantage is that businesses do not need to make any capital investment, as the local telephone company provides support and service. A Centrex is connected to the LEC office with a special circuit called Station Message Detail Interface (SMDI) link. It is appropriate for small businesses or companies, but it can be expensive for larger workgroups.

Virtual PBX Service

Companies that work out of several locations, have a lot of mobile workers, or need to accommodate telecommuters have the option of a virtual PBX service. The service emulates the auto-attendant, voice mail, directory service, and ACD services provided by high-end PBXs without requiring customers to invest in on-site equipment. If one owns a small business and does not want to invest in a new phone system, this service provides all of the features of a PBX and ACD system through a toll-free number, so there is no equipment to purchase. The employees need just a telephone and a phone line. The virtual PBX system creates the appearance of a single office phone system, when in fact, users can be scattered throughout North America. To outside callers, the service sounds like an expensive PBX system with sophisticated call routing, messaging, and ACD fea-

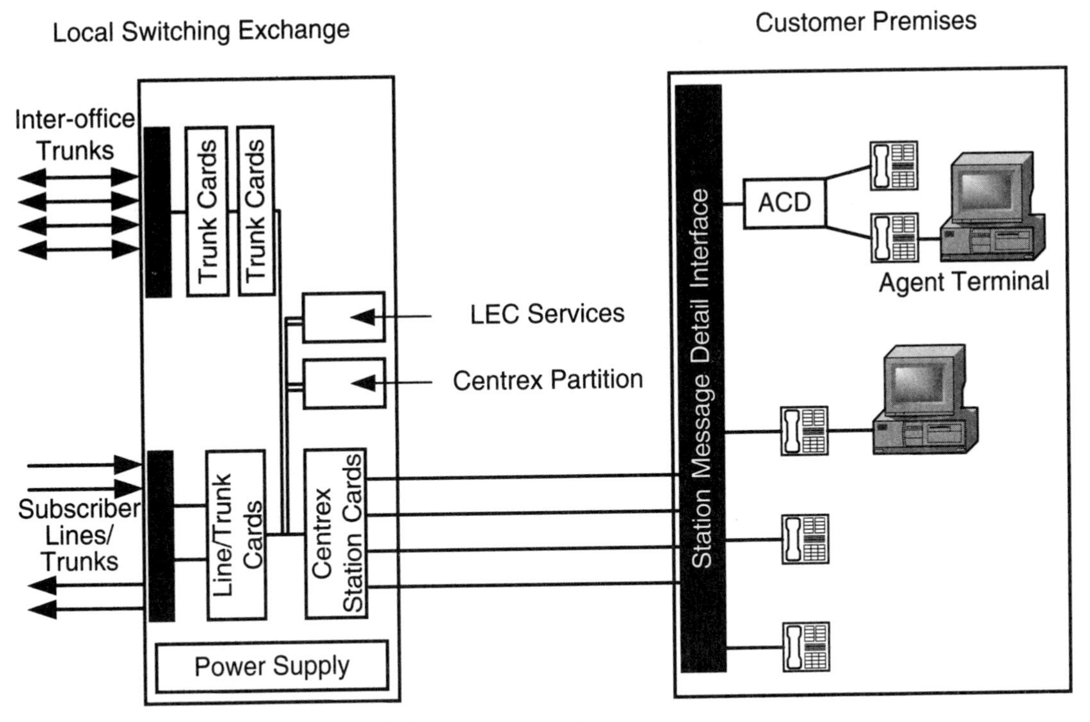

Figure 4–33 An ACD system used with Centrex service.

tures. Every employee has an extension number that can be dynamically mapped out to one or more telephone numbers. Thus, a small business can outfit itself with a complete call processing/ACD solution without buying any equipment using Virtual PBX. Its main drawback is that over a period of time, it can turn out to be quite expensive.

NETWORK DESIGN PARAMETERS

In telecommunications, careful circuit design, high-quality equipment, and systematic maintenance help assure quality. The telecommunications system always has some degree of impairments in the path between users. Transmission design is a compromise between quality and cost. Although it is possible to build a telecommunications system that will reproduce voices with near-perfect fidelity and clarity, such quality is neither necessary nor economical. Transmission quality is entirely different from switching-system quality. Switching systems are go/no-go devices. The connection is either established or it is not. In switching systems, quality is a statistical measure.

We are all aware that service demands are not constant—they vary by time of the day, day of the week, and season of the year. Demand is continually evolving in response to changing calling habits, and business conditions. Competition and technology have a substantial effect on cost and demand. Network design is always a compromise that seeks to balance the use of existing equipment with the provision of satisfactory service. Thus, a typical networking environment is a combination of modern and obsolete pieces. Even a new private network assembled with the latest technology can quickly become partially obsolete due to continuous technical advances. The size of the network is continually expanded or shrunk to match the demand.

Grade of Service

Users compete for a limited number of trunks. If all trunks are in use, the next outgoing call is blocked and the next incoming call receives a busy signal. This can be detrimental for businesses. In telephone systems, it is important to understand that the total number of telephone calls, including incoming and outgoing, cannot exceed the total number of outside lines. If a business has twenty outside lines, and it receives eleven telephone calls at the same time that nine people are placing outside calls, then the next caller will get a busy signal. One of the most important considerations in network design is the trade-off between network size and the probability that the calls will not be completed, referred to as **Grade of Service (GoS)**. Thus, GoS is the ratio of the number of calls that cannot be completed, or lost calls, to the total number of attempted calls during the busiest hours of the day, as shown in Equation 4–1. GoS is more accurately defined as the *probability of blockage*. For example, a GoS probability = 0.01, meaning that one call in 100 will not be completed.

$$\text{Grade of Service} = \frac{\text{Number of lost calls}}{\text{Number of attempted calls}} \qquad (4\text{–}1)$$

Example 4–1

Problem

If there are 354 seizures (lines connected for service, or completed calls), and six blocked calls (lost calls), during the busy hour, what is the Grade of Service?

Solution

Grade of Service = Number of lost calls/Total number of attempted calls

$$= \frac{6}{354 + 6}$$

$$= \frac{6}{360}$$

$$= 0.017$$

Estimated Traffic

The GoS parameter, which is a measurement of telephone service quality, can be estimated from usage or *total traffic intensity. Traffic* is the term that quantifies usage. For a particular organization, the best predictor of traffic load is a historical analysis of past usage. If data on traffic is unavailable, one can use average industry estimates. In communications, a *call* is any demand to set up a connection. It is also a unit of traffic measurement. Usage or total traffic intensity is measured in centi-call seconds (CCS). One CCS equals 100 call seconds of traffic in one hour. Thus, 36 CCS of traffic per hour means 3600 seconds of traffic every hour, which implies 100% utilization of the trunk. Thirty-six CCS is also expressed as one Erlang.

Network Design

Since the behavior of users is random, it is necessary to have sufficient data in order to make calling pattern predictions. Given the GoS and estimated usage, one can calculate required capacity by using mathematical modeling techniques. When designing a network, one should try to achieve a balance between the cost of the network and satisfactory service to the user. A typical large switch is designed for a GoS probability of 0.005, while a small private network switch may be designed for a GoS probability of 0.01. Optimized designs result in significant cost savings with essentially the same GoS.

CONVERGED NETWORKS

Deregulation of the telecommunications industry around the world has created tremendous competition among service providers. Companies have been forced to distinguish themselves based on innovative telecommunications services. Premium services made available through the carrier's networks have become vital marketing and customer retention tools as well as important new sources of revenue and profit. These network-based applications include a broad range of offerings, such as voice and fax messaging, electronic assistant, ACD, number portability, debit card services, international call back, paging services implemented at central office sites, caller ID, 800 number portability, wireless roaming, and large geographically-dispersed IVR applications. SS7 is also critical to the growth of wireless networks, and it is the means for allowing people to roam from one cell area or territory to another.

The IN is tightly specified for voice. All other data types require special leased access lines or devices such as modems. IN specifications were meant to encourage telecom equipment vendors to design their equipment to work in a multi-vendor environment so telcos would not be locked into one supplier. In addition, IN equipment was designed to work with certain customer systems and databases. The telecommunications equipment industry has been built around vertically integrated manufacturers who sell proprietary systems through their own network of dealers. This is still the case with well over 90% of phone systems sold to businesses. But there is a shift away from expensive, proprietary servers to modular, low-cost servers that customers can install and maintain themselves. Recently, we have seen the introduction of IP- and ATM-based telephone networks, some of which use PC telephony servers and some of which fully merge voice and data into the same network. Many of these systems provide IN services such as ACD, remote call forwarding, and message management, which enable users to work more effectively.

SUMMARY

This chapter discussed the telecommunications infrastructure from the perspective of the end user as well as the carrier. In the United States, the majority of communications today occur on PSTN. Voice communications is an important part of the total telecommunications picture. The voice evolution started with switching circuits. They permit the network to be built economically by concentration of transmission facilities. Recent developments have involved translation from the analog to the digital world. One example is the IN technology that overlays the public voice network, so as to enable new services. IN services include caller ID, ACD, AA, and IVR, which are powerful and flexible technologies for managing calls in businesses of all sizes. The SS7 protocol, which has been adopted by the ITU as the world standard, is the most widely implemented hardware-independent intelligent network signaling system. SS7 is also critical to the growth of wireless networks, and it is the means for allowing people to roam from one cell area or territory to another.

In spite of the complex and technologically sophisticated voice communications equipment, the telephone handset and cables are still relatively simple. The telephone handset, telephone lines, and the telephone company switching equipment are the three basic elements of a voice communications system. In business telephone systems, there are additional components, such as PBX and Centrex, designed to provide efficient voice services within the organization. In case of Centrex or Virtual PBX, the system can be rented where it resides at the LEC office and not exist on the company's premises. But PBXs provide more functionality than Centrex, although their full potential has not been realized, largely because they involve proprietary hardware and software platforms with proprietary services and applications written on top of them. Today, the traditional voice PBX vendors are competing with PC PBXs, ATM PBXs, IP-based PBXs, hybrid IP/circuit switch PBXs, and virtual PBX services.

REVIEW QUESTIONS

1. Define the following terms:
   - A. Switching
   - B. Busy hour call attempts
   - C. Glare
   - D. Grade of Service
2. Analyze the role of the PSTN in telecommunications today.
3. Describe the cabling and telecom infrastructure with reference to the PSTN.
4. Distinguish between analog and digital telephones. Identify the features available on digital telephones.
5. Discuss three address signaling methods for an outgoing call.
6. Compare and contrast loop start with ground start.
7. Identify the advantages of out-of-band signaling over in-band signaling.
8. Describe the protocol architecture and major components of Signaling System Seven.
9. Discuss the applications of the following:
   - A. Caller ID
   - B. Automatic Call Distribution
   - C. Automated Attendant
   - D. Interactive Voice Response
   - E. Voice Mail
   - F. Tie trunks
   - G. DID lines
   - H. DOD lines
   - I. FX circuit
   - J. Toll-free lines
10. Explain the similarities and differences between the PBX, Centrex, and Virtual PBX.
11. Find the number of calls that will be blocked if 500 calls are attempted during the busy hour for a network with a Grade of Service of 0.01.
12. Discuss the trend towards converged data-voice networks.

5 WIRELESS COMMUNICATIONS

KEY TERMS

Wireless
Improved Mobile Telephone System (IMTS)
Advanced Mobile Phone Service (AMPS)
Cellular Network
Base Station
Mobile Switching Center (MSC)
Cell
Frequency Division Multiple Access (FDMA)
Time Division Multiple Access (TDMA)
Code Division Multiple Access (CDMA)
Soft Handoff
Spread Spectrum
Frequency Hopping Spread Spectrum (FHSS)
Direct Sequence Spread Spectrum (DSSS)
Cellular Digital Packet Data (CDPD)
Wireless Application Protocol (WAP)
Bluetooth
Wireless LAN (WLAN)
Satellite Earth Station
Figure of Merit
Geostationary Orbit (GEO) Satellite
Global System for Mobile Communications (GSM)

OBJECTIVES

Upon completion of this chapter, you should be able to:

- ✦ Develop an understanding of the evolution of the wireless industry
- ✦ Describe the term *cellular* and discuss the advantage of this technology
- ✦ Explain analog cellular technologies
- ✦ Distinguish between different digital cellular technologies
- ✦ Identify different wireless applications and standards
- ✦ Discuss the characteristics and applications of different wireless LAN technologies
- ✦ Assess the role of satellite communications in the wireless industry
- ✦ Discuss the current status of international wireless communications
- ✦ Determine appropriate wireless technologies for different applications

INTRODUCTION

The **wireless** industry is growing at an exponential rate to keep pace with rapidly increasing demand for "information at your fingertips." Wireless refers to a communications, monitoring, or control system in which electromagnetic waves carry a signal through atmospheric space rather than along a wire. The major factors that affect the design and performance of wireless networks are the characteristics of radio or electromagnetic wave propagation over the geographical area. *Propagation* refers to the various ways by which an electromagnetic wave travels from the transmitter to the receiver. The major regions of the earth's atmosphere that affect radio wave propagation are the troposphere and ionosphere. For most communication links the troposphere, which extends up to about 15 km and includes atmospheric precipitation such as fog, raindrops, snow, and hail, makes the most difference.

Most wireless systems use radio frequency (RF) or infrared (IR) waves. RF includes any of the electromagnetic wave frequencies that lie in the range extending from 3 kHz to about 300 GHz (GHz is same as 10^9 Hz), which include the frequencies used for radio and television transmission. IR includes frequencies from 3 THz to 430 THz (THz is same as 10^{12} Hz). IR wireless products have appeal because they do not require any form of licensing by the FCC; in contrast, the FCC regulates and licenses the use of the radio spectrum in the United States. The ITU plays the same role internationally. The FCC and the ITU have played a large role in the development and deployment of wireless technologies. The wireless industry is evolving quickly and rather unpredictably as the products and services are changing. There are competing standards for communications protocols, interfaces, and networks. However, standardization efforts to assist the convergence of these essentially competing wireless access technologies are underway.

CELLULAR MOBILE TELEPHONE SYSTEMS

Mobile communications was mainly applied to military and public safety services until the end of World War II, but after the war it began to be applied to public telephone services. The first public mobile phone service was the Mobile Telephone System (MTS) introduced in the United Sates in 1946. In this system, operation was simplex and call placement was handled by a manual operation. The MTS was then followed by the development of a full duplex automatic switching system—the **Improved Mobile Telephone System (IMTS)**. The IMTS was introduced in 1969 using a 450 MHz band. Although IMTS was widely introduced in the United States as a standard mobile phone system, it was not able to cope with rapidly increasing demand for two reasons: it was a large-zone system, and its assigned bandwidth was insufficient. Therefore, a more advanced and high-capacity land mobile communication system called **Advanced Mobile Phone Service (AMPS)** was introduced in the United States in 1983 in Chicago. The most important feature of AMPS is that it employs a *cellular* concept to achieve high capacity.

Cellular Networks

A **cellular network** is any mobile communications network with a series of overlapping hexagonal cells in a honeycomb pattern, as shown in Figure 5–1. Cellular technology is a type of short-wave analog or digital transmission in which a user has a wireless connection from a mobile device to a relatively nearby base station. The **base station** consists of a transmitter, receiver, controller and antenna system, and has a wireless link to the **Mobile Switching Center (MSC)**. The base station's span of coverage is called a **cell**. Cells can be as small as an individual building (e.g., an airport or arena) or as big as 20 miles across, or any size in between. The base station controller routes the circuit-switched calls to mobile phones. The MSC contains all of the control and switching elements to connect the caller to the receiver and is connected to the local exchange or central office, which is a switching center for wired telephones.

Figure 5–1 The cellular network is a series of overlapping hexagonal cells in a honeycomb pattern.

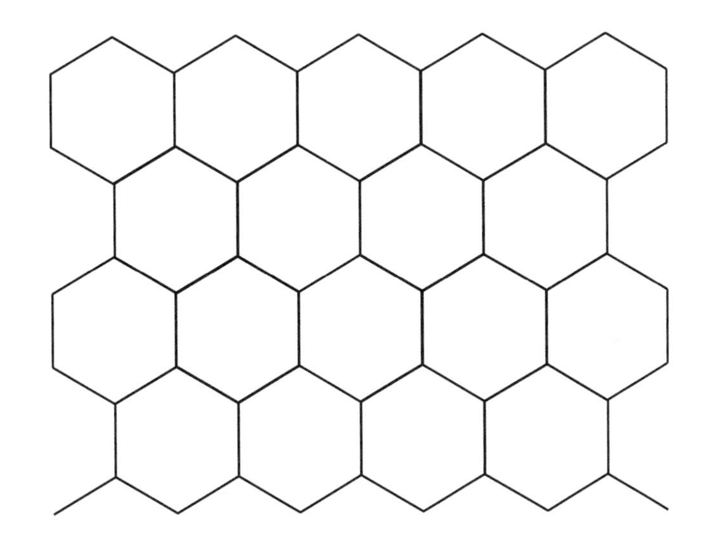

As the cellular device user moves from one cell or area of coverage to another, the MSC senses that the signal is becoming weak and automatically hands off the call to the base station in the next cell into which the user is traveling. This is depicted in Figure 5–2. The purpose of this division of the geographic region into cells is to make the most out of a limited number of transmission frequencies. Cellular systems allocate a set number of frequencies for each cell. Two cells can use the same frequency for different conversations as long as the cells are not adjacent to each other. Typically, within a cellular network, every seventh cell uses the same set of channels or frequencies.

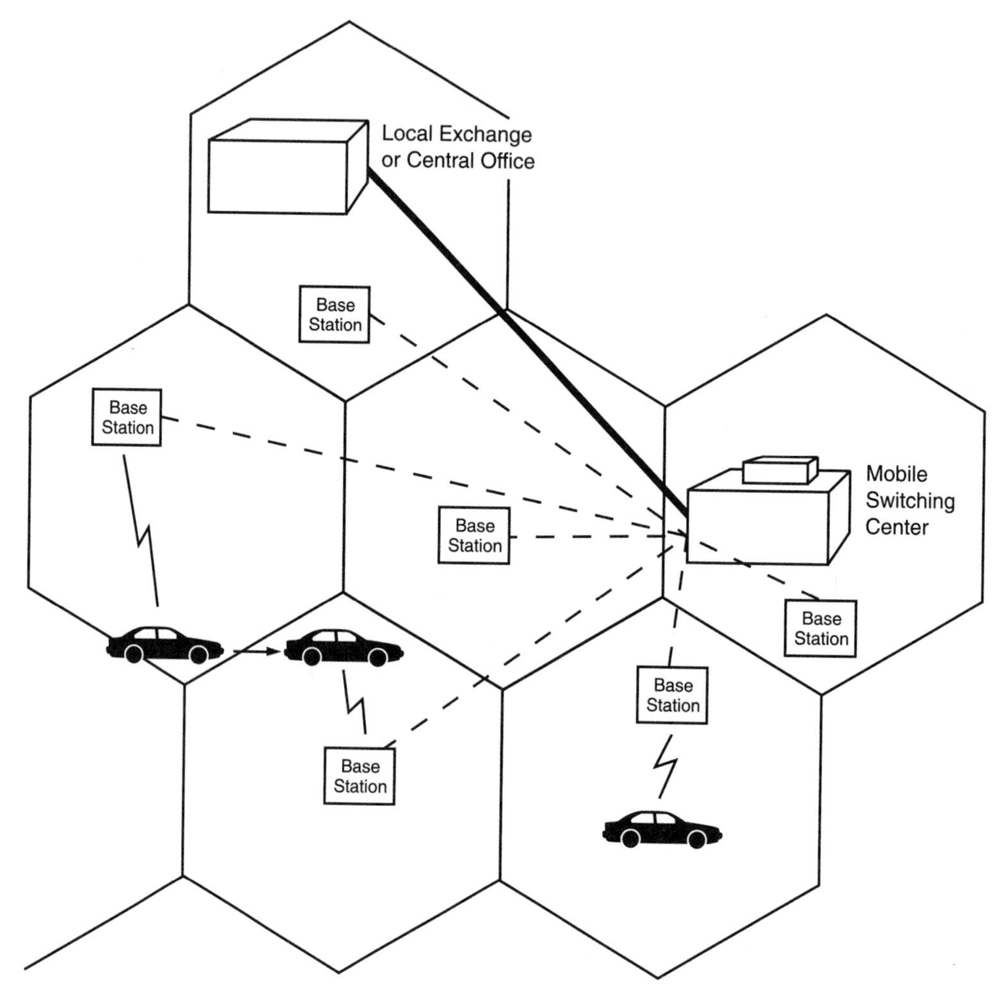

Figure 5–2 The Mobile Switching Center (MSC) hands over the call from one base station to another as the caller moves from one cell to the next. The MSC is also connected to the Local Exchange.

Personal Communications Systems (PCS)

Personal Communications Systems (PCS) has expanded the horizon of wireless communications beyond the limitations of cellular systems. Also called *Personal Communications Networks (PCN)*, its goal is to provide integrated communications (such as voice, data, and video) and near-universal access irrespective of time, location, and mobility patterns. At this time, PCS is not a single standard but a mosaic consisting of several incompatible versions coexisting rather uneasily with one another.

There are three categories of PCS: broadband, narrowband, and unlicensed. Broadband (1900 MHz) addresses both cellular and cordless handset services, while narrowband (900 MHz) focuses on enhanced paging functions. Unlicensed service is allocated from 1910 to 1930 MHz and is designed to allow unlicensed short-distance operation. PCS networks and the existing cellular networks should be regarded as complementary rather than competitive. PCS architecture resembles that of a cellular network with enhancements such as speech quality, flexibility of radio-link architecture, economics of serving high-user-density areas, and lower power consumption of the handsets.

A key feature of PCS is variable cell size and hierarchical cell structure: picocell for low-power indoor applications; microcell for lower-power outdoor pedestrian applications; macrocell for high-power vehicular applications, and supermacro cells with satellites. The combined coverage of different types of cells is called hierarchical cell structure. For instance, the microcells create a second level of coverage under the existing level of macrocells, as shown in Figure 5–3. A PCS microcell has a radius of 1 to 300 meters, and a group of microcells are superimposed by one macrocell, as represented in Figure 5–4.

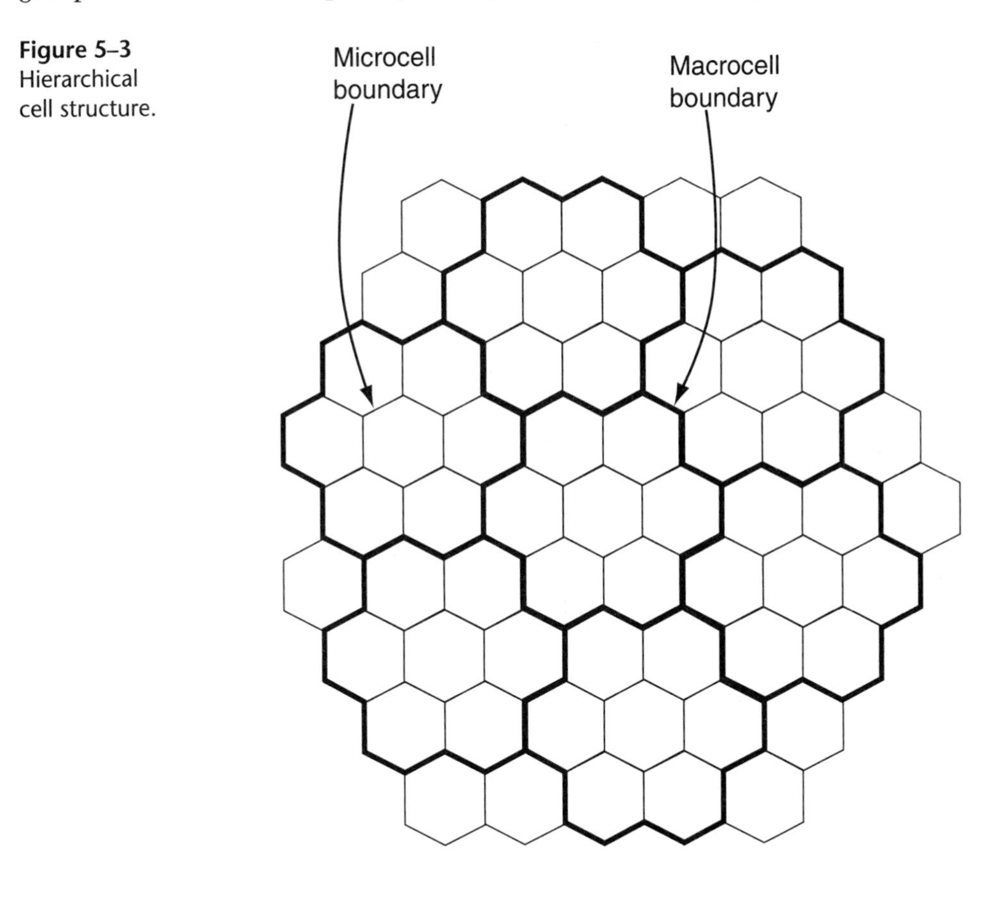

Figure 5–3 Hierarchical cell structure.

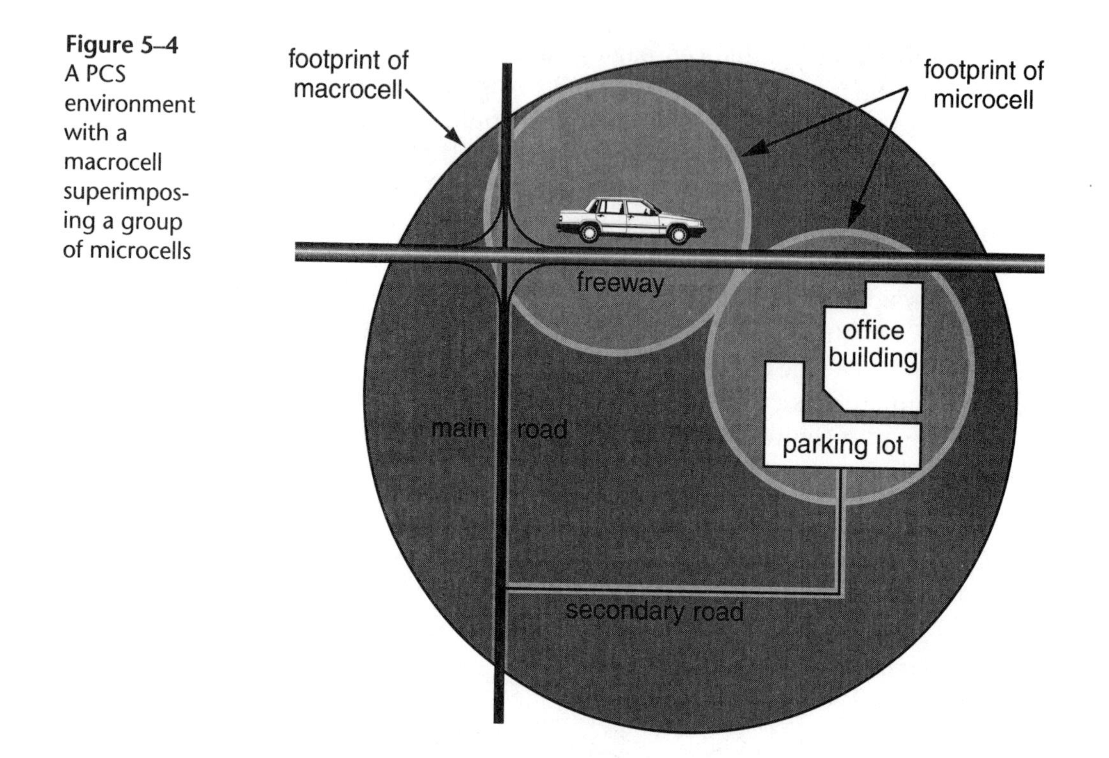

Figure 5–4 A PCS environment with a macrocell superimposing a group of microcells

ANALOG VERSUS DIGITAL ACCESS

Fundamentally, there are three modes of wireless access: 1) Frequency-Division Multiple Access (FDMA); 2) Time-Division Multiple Access (TDMA); and 3) Code-Division Multiple Access (CDMA).

Until recently, most wireless data transmitted through radio communications have been analog. In analog systems, the actual sound of the caller's voice pattern is transmitted over the airwaves by means of a continuous wavelike signal. Next generation wireless systems have turned toward the use of digital signals. Digital systems have several advantages, including better coverage, more calls per channel, broadband communications, less noise interference, and the ability to add new features and functions. Analog and digital technologies and their salient features are identified in Figure 5–5.

Analog Access

The analog cellular systems are referred to as first generation cellular technologies. The North and South American analog cellular systems conform to the AMPS standard, which operates on the 800 MHz or 1800 MHz frequency band. But there are several other types of analog systems standards in the rest of the world. In Europe and Asia, these include Total Access Communications System (TACS), Nordic Mobile Telephone (NMT),

Cellular System Generation	Technology	Operating Frequency	Advantages	Disadvantages
First Generation	AMPS based on FDMA	800 MHz or 1800 MHz	✦ Widest coverage including rural areas	✦ Poor security ✦ Not optimized for data ✦ Limited capacity
Second Generation	TDMA	800 MHz or 1900 MHz	✦ Better security ✦ Higher capacity	✦ May experience an interruption during handoff
Third Generation	CDMA	800 MHz or 1900 MHz	✦ Very high security ✦ Improved capacity ✦ Greater immunity from interference ✦ Soft handoff with no interruption	✦ Limited coverage at this time

Figure 5–5 Cellular technologies and their salient features.

and others. The AMPS and TACS are both based on FDMA. **Frequency Division Multiple Access** is the division of the frequency band allocated for wireless cellular communication into a number of channels separated by guard bands, as shown in Figure 5–6. Each channel can carry only one voice conversation at a time.

By the end of 1995, the total number of subscribers of the AMPS system had reached around 30 million. The analog cellular system was inadequate to satisfy rapidly increasing demand for cellular services. Also, analog cellular architecture has very poor or no security and is not optimized for data. Therefore, United States, Europe, and Japan have independently developed second generation and third generation digital technologies.

Analog access is inherently less optimal than digital for transmitting data. However, it should be considered as a backup solution to digital technologies. The analog cellular system has the widest coverage, with service available in almost any city or town and on most major highways in the United States. About half of the world's cellular subscribers are still using a mobile telephone system based on AMPS. For this reason, analog cellular will remain the only wireless communications option in rural areas for some time to come.

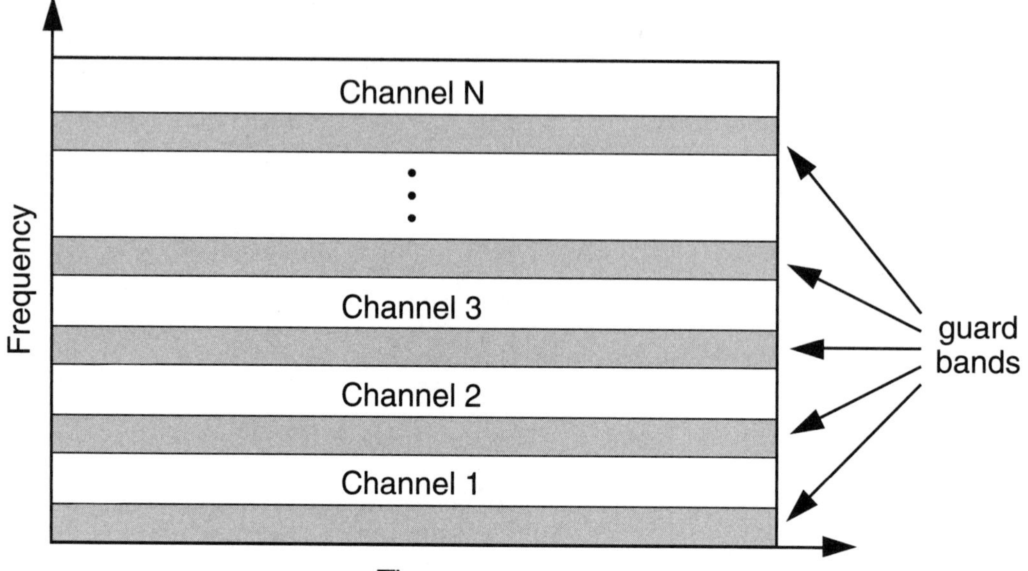

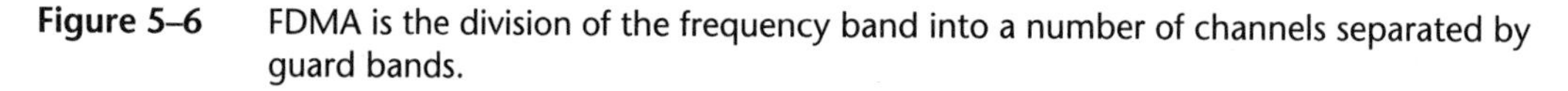
Figure 5–6 FDMA is the division of the frequency band into a number of channels separated by guard bands.

Digital Access

Until now, digital wireless technologies have developed mainly because of the need for increased system capacity for voice transmission. However, this trend is now changing. In recent years, there has been a rapid growth in multimedia communications via the Internet, which has resulted in an increased demand for high-bandwidth transmissions rather than dedicated voice transmissions, even in the wireless world. Another important factor is the increasing need for global coverage.

North America currently has a multitude of digital cellular technologies for wireless radio communication. Collectively referred to as Personal Communications System (PCS), they operate at 1900 MHz, each with different coverage and capacities. By definition, analog cellular technology is not included as a PCS technology because PCS only refers to digital technologies that were designed specifically to provide improvements over analog. Spectrum is a scarce and limited resource, so multiple access schemes are designed to share the resource among a large number of wireless users.

Time Division Multiple Access (TDMA)

Time Division Multiple Access (TDMA)-based second-generation digital cellular technology is designed to coexist with the analog AMPS system. The first implementation of digital AMPS, often known as D-AMPS, used narrowband digital TDMA over the same frequencies as AMPS. The specifications for its operation were provided by the ANSI-136 and the TIA IS-54 standards. D-AMPS uses FDMA but adds TDMA to get three channels for each FDMA channel, tripling the number of calls that can be handled on a channel. In addition to voice conversations, it can carry digital data; Figure 5–7 represents a TDMA system for voice and data transmission. This requires digitizing voice, compressing it, and transmitting it in regular series of bursts that are interspersed with other users' conversations.

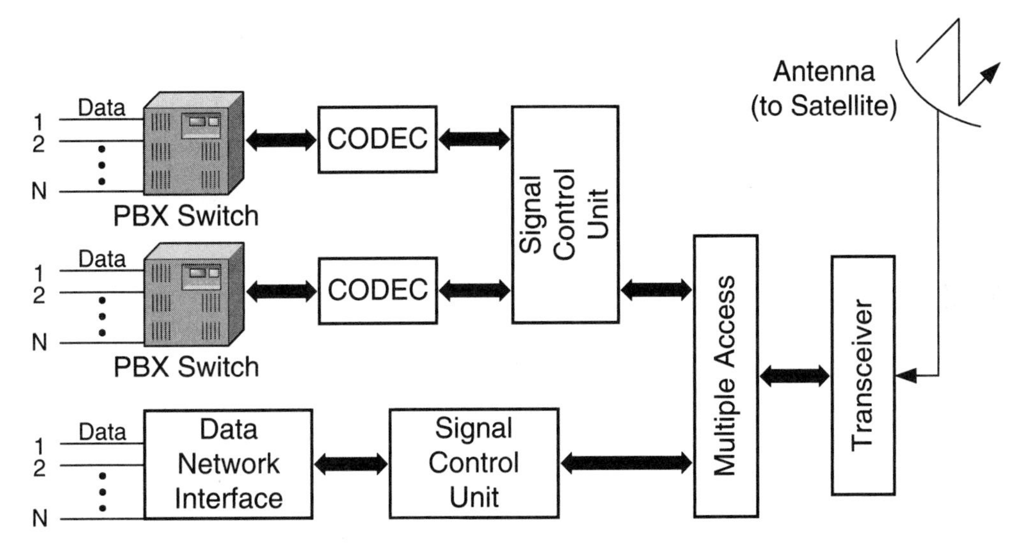

Figure 5–7 TDMA system for voice and data transmission.

TDMA enables users to access the assigned bandwidth on a time basis. D-AMPS divides the available radio spectrum into narrow channels, each of which carries one or more conversations using TDMA. It is designed to allocate three timeslots per 30 kHz channel, yielding a 3-to-1 capacity increase over AMPS. Each user is assigned his/her own time slot on a particular carrier frequency that cannot be accessed by any other subscriber until the original call is finished or handed off. Within each channel, conversations are separated by slight differences in time. Thus, each user occupies the whole channel bandwidth, but only for a fraction of the time, called *slot*, on a periodic basis. TDMA thereby makes more efficient use of available bandwidth than the previous-generation analog technology, in addition to being optimized for both voice and data. The D-AMPS is much more secure than AMPS since encryption is built into the system. It would require sophisticated hackers to modify the scanner for different digitization techniques.

TDMA was first investigated for satellite communication systems in the late 1960s. When the TDMA mode of access is utilized, a single carrier is used by all earth stations in a time-sharing mode of operation, as shown in Figure 5–8. When an earth station transmits a burst of information at a preassigned time to a satellite, the entire bandwidth can be utilized by that earth station. The satellite receives the transmitted burst, and amplifies,

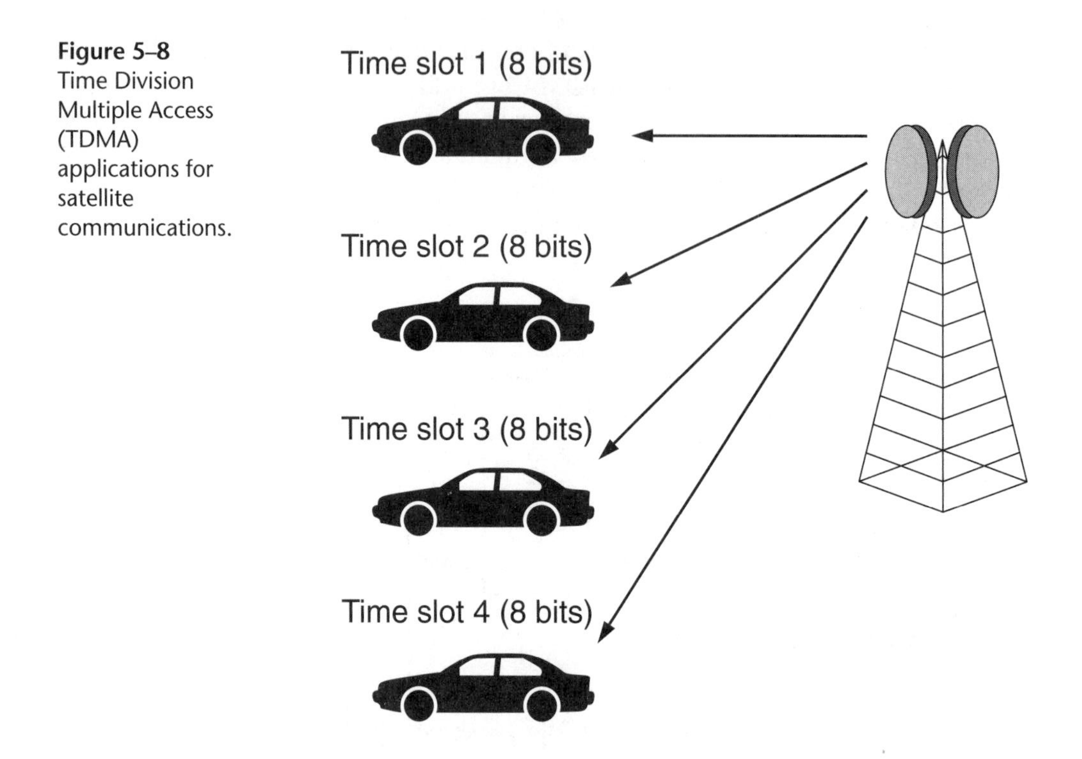

Figure 5–8 Time Division Multiple Access (TDMA) applications for satellite communications.

down-converts, and retransmits it back to earth. The next burst, perhaps originating from another earth station, follows exactly the same pattern. The fact that bursts are transmitted in sequence separated by narrow guard bands necessitates optimum synchronization techniques so that burst overlap can be avoided.

The first commercial TDMA satellite system was operated in Canada in 1976. Investigation into the application of TDMA to land mobile communication systems started around the 1980s. At this time, unfortunately, TDMA was not considered to be suitable for land mobile communications because of its complicated timing controls. However, in 1982, the Conference of European Posts and Telecommunications (CEPT) started to make specifications for the pan-European digital cellular system, in which TDMA was included as an access scheme. Finally, TDMA was selected as the access scheme for the Global System for Mobile Communications (GSM) system in 1988.

As a result of the extensive studies on TDMA, this technology is now a very popular access scheme for land mobile communication systems. It is applied to many second-generation cellular systems and cordless phone systems all over the world. The TIA published the revised TDMA standard IS-136 in 1994. At the core of the IS–136 specification is the DCCH (Digital Control Channel), which provides a platform for PCS, introducing new functionalities and supporting enhanced features that make PCS a powerful digital system.

Like IS-54, IS-136 co-exists with analog channels on the same network. One advantage of this dual-mode technology is that users can benefit from the broad coverage of established analog networks, and at the same time take advantage of the more advanced technology of IS-136 where it exists. TDMA IS-136 is available in North America at both the 800 MHz and 1900 MHz bands.

Code Division Multiple Access (CDMA)

Code Division Multiple Access (CDMA), in layman's terms, is like a room full of people talking in different languages at the same time. However, you can pick out the conversation in the language you understand and ignore all the others. CDMA is a method of sharing a single frequency among users by encrypting each signal with a different code. As a result, it supports many callers along the same carrier. Transmission signals are broken up into coded *packets* of information that hop available frequencies and are reassembled at the receiving end. Each earth station transmits coded information to the satellite, regardless of any overlap with other stations that may be transmitting simultaneously. At the receiver end, the separation of the transmitted information by each station is achieved through the detection of the individual earth station's transmitted identification code.

Like TDMA, CDMA operates in the 1900 MHz band as well as the 800 MHz band. CDMA has three times the capacity of TDMA and ten times that of FDMA. In 1992, the TIA published the first CDMA standard, IS-95. The third generation CDMA systems provide both operators and subscribers with significant advantages over first (analog) and second (TDMA-based) generation systems. The advanced methods utilized in CDMA technology improve capacity, coverage, voice quality, and immunity from interference by other signals introducing a new generation of wireless networks.

Compared with conventional systems, the CDMA system makes frequency assignment easy, as the same frequency can be used for each cell. Moreover, this system allows high-quality communications, since no frequency switching is required when moving from one cell to another. CDMA is the first technology to use a technique called **soft handoff**, which allows a handset to communicate with multiple base stations simultaneously. The system chooses the best signal in order to provide the user with the best audio at all times. In contrast to CDMA users, TDMA users can experience an interruption in the audio when the signal is handed off from one base station to another, resulting in higher interference during handoff and increased dropped calls.

CDMA employs a technique originally created by the military called a **spread spectrum**, which is significantly different from AMPS and TDMA technologies. Rather than dividing RF spectrum into separate user channels by frequency slices as in FDMA or time slots as in TDMA, spread spectrum technology separates users by assigning them digital codes within a broad range of the radio frequency, as shown in Figure 5–9. The system chops up the signal into data packets and spreads it over a wide band of frequencies. The data packets are then reassembled at the receiving end. This makes it very difficult for hackers to grab orderly, meaningful data, and ensures secure communications. In addition, the

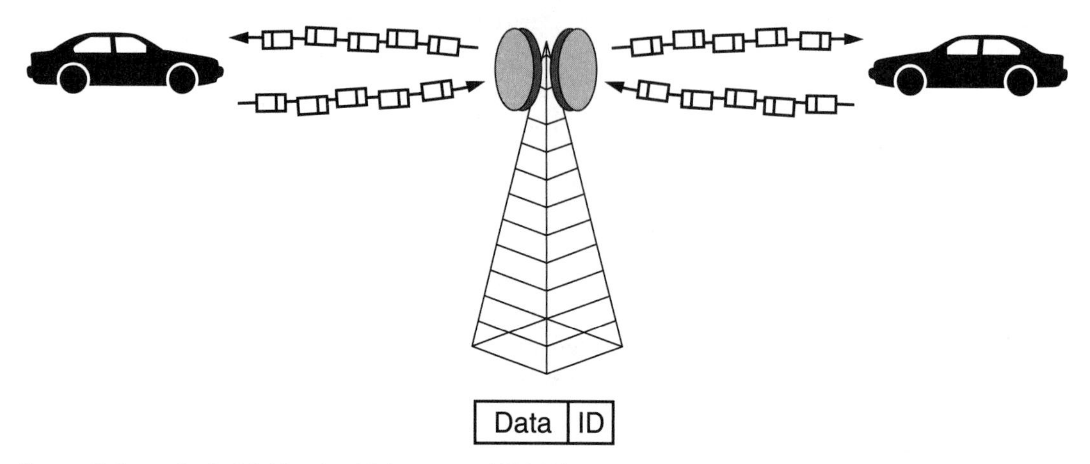

Figure 5–9 Code Division Multiple Access (CDMA).

signal appears like low-level noise to conventional radio receivers. Since the receivers are designed to eliminate noise, the signal goes undetected. Also, this technology is ideal for densely populated areas because it prevents signal interference among mobile units, say, cellular phones and portable computers. The spread spectrum techniques can be divided into two families: Frequency Hopping Spread Spectrum (FHSS) and Direct Sequence Spread Spectrum (DSSS).

Frequency Hopping Spread Spectrum (FHSS)

Frequency Hopping Spread Spectrum (**FHSS**) resists interference by jumping rapidly from frequency to frequency in a pseudo-random way. The receiving system has the same pseudo-random algorithm as the sender and jumps simultaneously. In the frequency-hopping pattern, a very long sequence code is used before the sequence is repeated, over 65,000 hops, making it appear random. Thus, it is very difficult to predict the next frequency at which such a system will transmit/receive data. The system appears to be a noise source to an unauthorized listener, which makes FHSS very secure against interference and interception. Another advantage of FHSS systems is that multiple hopping sequences can typically be assigned within the same physical area, thereby increasing the total amount of available bandwidth. In large facilities, especially those with multiple floors, it is necessary to space antennas, or access points, in an overlapping array. This ensures adequate coverage for wireless devices that are moving from cell to cell. With FHSS, system users can roam between access points on different channels. This makes FHSS technology more flexible than DSSS, in this respect.

Direct Sequence Spread Spectrum (DSSS)

Direct Sequence Spread Spectrum (**DSSS**) resists interference by mixing in a series of pseudo-random bits with the actual data. The receiver, using the same pseudo-random algorithms, strips out the extra bits. A redundant bit pattern (also called a chip) is produced for each bit of data to be transmitted. If one or more bits are damaged in transmission, the original data can be recovered as opposed to having to be retransmitted. This built-in safeguard significantly increases the efficiency of the data transmission process. DSSS can be used as a substitute for leased lines or fiber-optic cables to bridge LAN segments in point-to-point or multipoint connectivity between buildings. In DSSS, roaming can only be done between access points on the same channel, which makes the roaming capabilities of DSSS less robust than those of FHSS.

DSSS systems can transmit at bandwidths of about 11 Mbps for point-to-point connections up to 25 miles. In multipoint applications, where the signal has to be broadcast over a wider arc, the distance the signal can actually travel may be much less than 25 miles, depending on the number of sites and the distances between them. The selection of microwave antennas at the distribution and receiving points therefore becomes crucial. If the network components are not properly balanced, locations close to the antenna could end up monopolizing bandwidth, while locations further away could suffer performance degradation. Another issue of concern with DSSS is signal loss, which can be alleviated by devices such as high-gain antennas and amplifiers.

Cellular Digital Packet Data (CDPD)

Cellular Digital Packet Data (CDPD) has been combined with spread-spectrum radio transmission in several wireless communications products. CDPD allows a packet of information to be transmitted in between voice telephone calls. Even if all the subscribers in a particular area are using their cellular telephones, 30 percent of the frequency is still available for data transfer. For signal transmission, the system starts channel hopping. Anytime it finds an opening, it transmits the information in packets. This approach not only spreads the packets in a predetermined manner, but also causes them to hop frequencies.

CDPD enables data specific technology to be tacked onto the existing cellular telephone infrastructure. It supports wireless access to the Internet and other public packet-switched networks. CDPD is an open specification that adheres to the layered structure of the OSI model and therefore, has the ability to be extended in the future. CDPD supports the Internet's IP protocols for broadcasting and multicasting, including IPv6, and the ISO Connectionless Network Protocol (CLNP).

For the mobile user, CDPD's support for packet-switching means that a persistent link is no longer a requirement. The same broadcast channel can be shared among a number of users at the same time. The user's modem recognizes the packets intended for its user. As data such as e-mail arrives, it is forwarded immediately to the user without a circuit connection having to be established. There is a circuit-switched version, called CS CDPD, which can be used where traffic is expected to be heavy enough to warrant a dedicated connection. Cellular telephone and modem providers that offer CDPD support make it possible for mobile users to get access to the Internet.

WIRELESS APPLICATIONS AND PRODUCTS

There are two distinct, yet complementary, system approaches for addressing the demand for mobility: an enhanced cellular network, and a noncentralized wireless LAN. The two system approaches are complementary because the wide-area cellular network permits high mobility and globalizes communications through handoff and roaming, while wireless LANs can offer orders of magnitude higher data rates through coverage area restriction, which results in substantially reduced signal attenuation and multipath delay spread. Both system approaches have been the focus of extensive research and standards activities.

The features that are the most different in regard to the cellular products, cordless devices, and wireless LANs are terminal mobility, coverage, and bandwidth. Basically, mobility of the cordless phone and wireless LAN is very low compared with the cellular phone. Therefore, these systems do not require any location restriction or handoff functions, which means their system cost is much lower than that for cellular systems. Another feature difference among these three systems is the signal bandwidth. In the case of cellular phone or cordless phone systems, the signal bandwidth is very narrow—approximately 20 to 30 kHz—because they are used only for voice transmission. On the other hand, a desirable signal bandwidth for a wireless LAN system may be 10 MHz. A minimum bit rate of 1 Mbps is required to satisfy intersystem matching between wireless and wired LANs. Regarding Internet access, most vendors are moving toward unifying voice mail and data communications. There are several companies that provide a truly wireless solution by escaping the phone infrastructure.

Wireless Application Protocol (WAP) is a standard for wireless data delivery, loading, and navigation. Before WAP was developed, loading standard HTML pages on mobile phones and other handheld devices could be a problem because HTML pages are designed to be viewed on a computer monitor, not on a 1.5-inch LCD. WAP applications are designed to be viewed on tiny screens found on average mobile phones, but the micro-browser also allows for expanding the viewable area of an application if the application is being accessed from a machine with a larger screen, such as a Personal Digital Assistant (PDA). More important, WAP micro-browsers are configured to use less memory and CPU power, thus extending the battery life of a mobile device. They also allow for easier resumption of an Internet session if a mobile device loses its signal. According to the WAP Forum, a standards-setting body, the WAP protocol eats up a lot less bandwidth than do standard HTTP/TCP/IP pages. An example of a mobile device accessing the Internet through WAP is shown in Figure 5–10.

WAP removes much of the overhead associated with IP, including large message headers and long session startup times. WAP solutions contain a client-side browser for the mobile device and a transmission gateway that uses WAP-over-IP to send and receive data and to gain access to the Internet or corporate Intranets. The browser reaches Web sites using the Wireless Markup Language (WML), which excludes large files such as Java

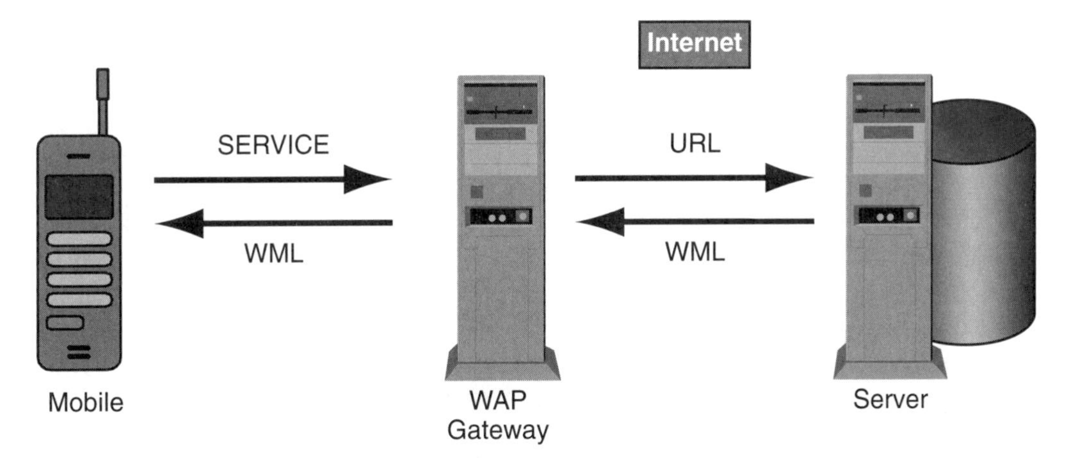

Figure 5–10 A mobile device accessing the Internet through a WAP gateway.

scripts and picture files and sends only plain text information in response to a query. Thus, the information is suited for small display, keyboardless, low-restoration devices used to access information while on the road.

Bluetooth

Bluetooth is a uniting technology that allows any sort of electronic equipment from computers and cell phones to keyboards and headphones to make its own connections, without any wires, cables, or user intervention. A Bluetooth module block diagram is shown in Figure 5–11; the module is built on a 9 × 9 mm microchip incorporated into the mobile device or other electronic devices.

The idea originated in 1994 when mobile phone manufacturer Ericsson began working on a short-distance radio system. With the intent of achieving an open solution, Ericsson formed the Bluetooth Consortium with technology providers IBM, Intel, Nokia, and Toshiba. Each member of the consortium brought a different expertise to the solution, including radio technology and/or software. Bluetooth operates in the unlicensed 2.4 GHz ISM bands, an open frequency band in most countries, ensuring communication compatibility worldwide, although local regulations in Japan, France, and Spain reduce the bandwidth. Bluetooth is intended to be an open standard that works at the two lower layers of the OSI Model, the physical and data link layers. The standard provides agreement on when bits are sent, how many will be sent at a time, and how the devices in a conversation can be sure that the message received is the same as the message sent.

Each Bluetooth device has a unique 48-bit address and uses built-in authentication and encryption at the device level, and not the user level. When a manufacturer wants to make a product based on Bluetooth wireless technology, it contacts the IEEE registration authority and requests a single (or a number of) 24-bit organizationally unique identifiers

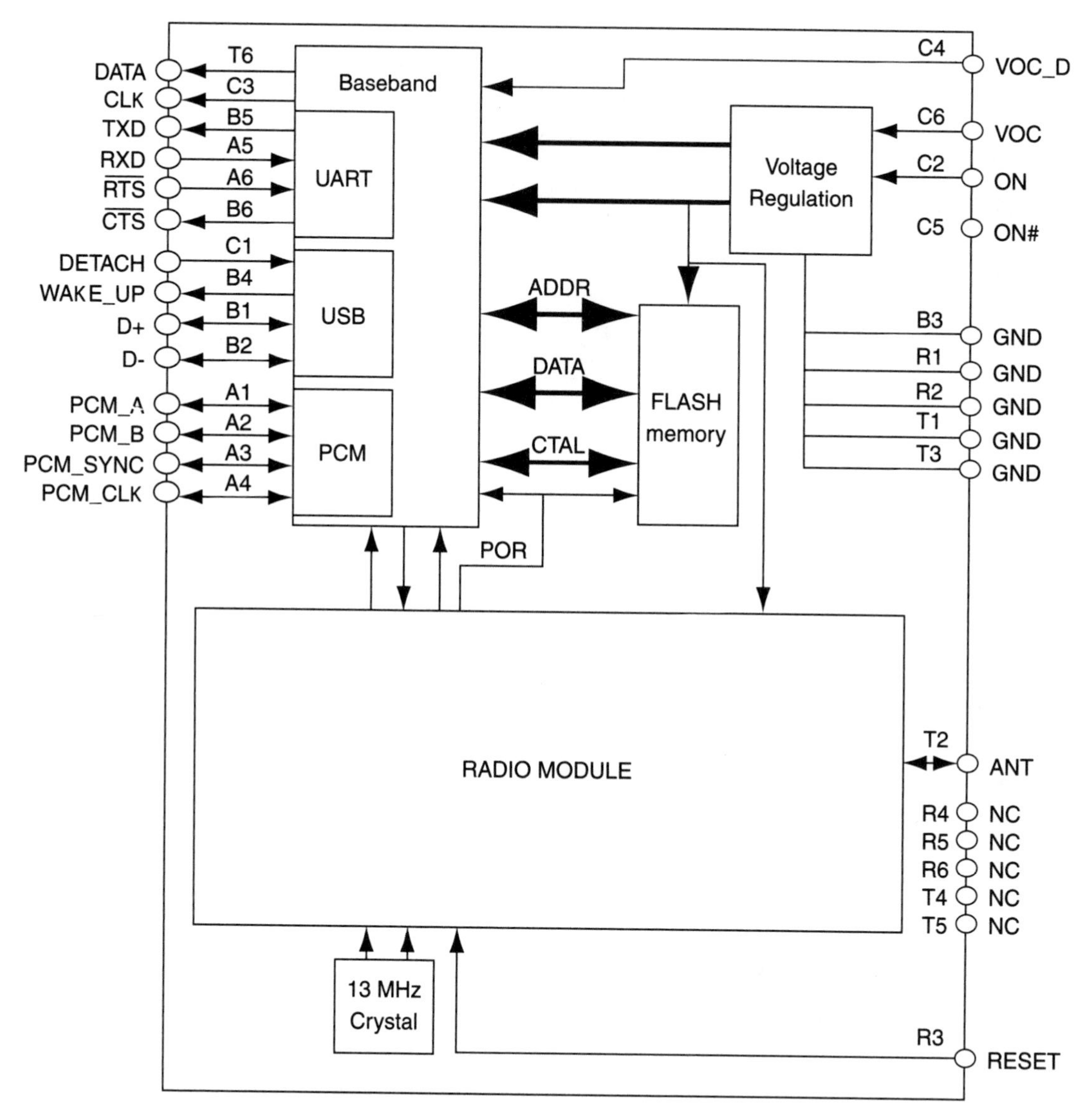

Figure 5–11 Bluetooth module block diagram.

(OUIs). The manufacturer then uniquely assigns the other 24 bits to make up the 48-bit Bluetooth device address. When Bluetooth connects devices to each other, they become paired; one unit acts as a master for synchronization purposes, and other units will be slaves for the duration of the connection, forming a piconet, represented in Figure 5–12. Unlike many other wireless standards, Bluetooth includes application layer definitions for product developers to support data-, voice-, and content-centric applications.

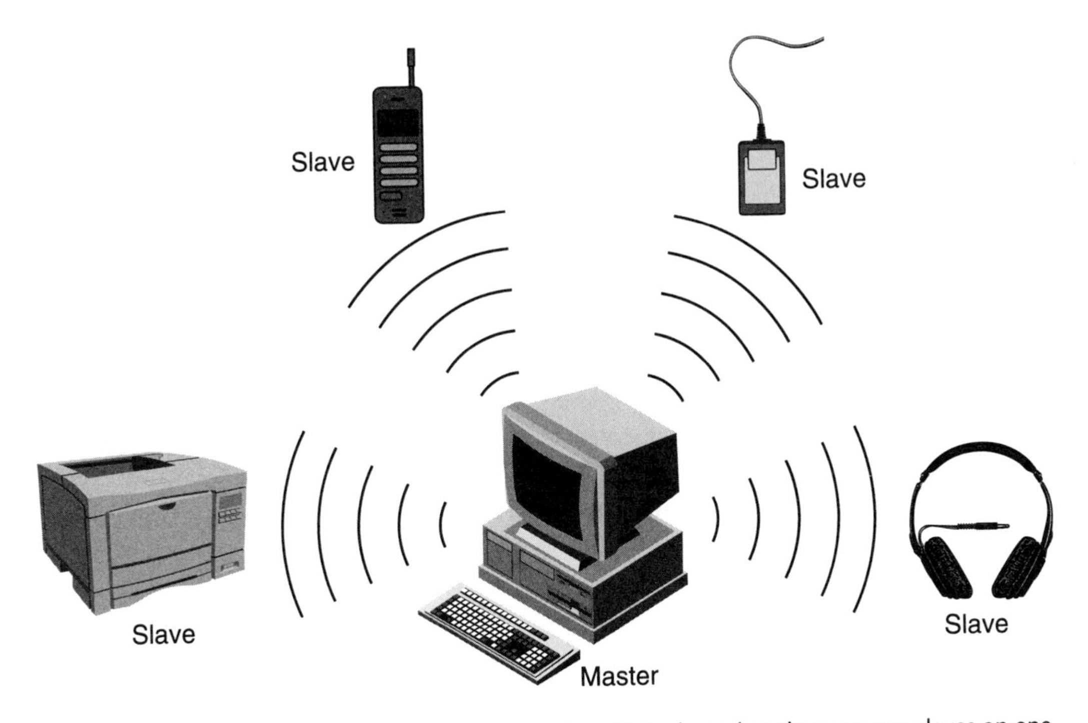

Figure 5–12 Bluetooth piconet with one master and multiple slaves (maximum seven slaves on one master).

A number of devices such as baby monitors, garagedoor openers, microwave ovens, and cordless phones make use of frequencies in the ISM band. Making sure that Bluetooth and these other devices do not interfere with one another has been a crucial part of the design process. Bluetooth radios use a spread spectrum, frequency hopping, full-duplex signal that changes frequencies 1,600 times every second. The signal hops among 79 individual, randomly chosen frequencies at 1 MHz intervals to give a high degree of interference immunity. Bluetooth radios consume very little power: between 30 to 100 mW while transmitting, and between 0.01 to 0.1 mW when not transmitting. The low power limits the range of a Bluetooth device to about 10 meters. But even with the low power, ceilings and walls do not stop the signal, making the standard useful for controlling several devices in different rooms.

Bluetooth wireless technology is supported by product and application development in a wide range of market segments. The specification allows up to seven simultaneous connections to be established and maintained by a single radio. Bluetooth has three generic applications:

- Personal area networks, where two or more Bluetooth products can communicate directly

- Local area networks (LANs), where products communicate with a company's broader network via a Bluetooth LAN access point
- Wide area network (WAN), where a product with Bluetooth-enabled technology can communicate with a wireless WAN device to allow global connectivity, including Internet access

The Bluetooth specification, a continuing process, allows member companies to enhance and extend the technology into new usage models and markets. However, Bluetooth on its own is not a complete solution. A mobile worker wishes to have all the information synchronized between the mobile devices and his networked applications. A common synchronization protocol is also needed to ensure that meaningful connectivity is not limited to devices from only one manufacturer. SyncML is an initiative to develop and promote an open data synchronization protocol. SyncML is an extensible and transport independent technology that allows a device to support a single synchronization standard for both local synchronization over Infrared, Bluetooth, and USB, as well as remote synchronization over Internet and WAP.

The rapidly evolving wireless technology is resulting in two major benefits for the users: lower prices and better functionality. However, there are two major shortcomings of wireless products: speed and limited coverage for certain services. Vendors cite two measures of speed. *Data Rate* is raw radio transmission speed, and it is heavily affected by *overheads,* which consist of the transmission rules and protocols that govern a network. As shown in Equation 5–1, by subtracting overhead from data rate, one arrives at the user rate or *throughput,* which is the capacity available to the user. Bandwidth is critical in maintaining high throughput for each user especially when there are a large number of users. Mostly, wireless data rates are painfully slow when compared with wired options.

$$\text{Overhead} = \text{Data Rate} - \text{Throughput} \tag{5–1}$$

Example 5–1

Problem

A user realizes 8 Mbps throughput on a 10 Mbps network. Find the overhead.

Solution

$$\begin{aligned}\text{Overhead} &= \text{Data Rate} - \text{Throughput}\\ &= (10 - 8)\ \text{Mbps}\\ &= 2\ \text{Mbps}\end{aligned}$$

WIRELESS LANs (WLANs)

Wireless LANs (WLANs), which are increasingly being used by companies and organizations, are flexible communication systems implemented as extensions to or as alternatives for the wired LANs in buildings or campuses. In the past, established vertical applications such as those in warehousing and retail enterprises have dominated the demand for WLANs. The real-time aspects of wireless networks improved the efficiency and productivity of factories, distribution centers, warehouses, and transportation hubs. More and more companies implemented radio frequency data communications technology to provide real-time, on-line inventory and process control.

In recent years, however, the most dramatic demand for wireless applications has been driven by mobile professionals. Through the use of WLANs, company owners, managers, and other executives who spend at least 20 percent of their time away from their desk while at the office are able to maintain constant network connection to key business applications irrespective of where they roam. WLANs have evolved into a secure, high-performance, decision-support tool.

A WLAN transmits and receives data over the air, minimizing the need for wired connections, as depicted in Figure 5–13. It is recommended highly for hard-to-wire sites. A wireless bridge works well for multi-site organizations to connect branch offices to corporate LANs in a local geographic area, as well as to connect LANs within a college or corporate campus. A WLAN adapter is a *Personal Computer Memory Card Industry Association (PCMCIA)* card (commonly called PC card) for a laptop or notebook computer. It combines data connectivity with user mobility, and, through simplified configuration, enables movable LANs.

WLANs solve several problems like cabling restrictions, frequent reorganizations, and networking of highly mobile employees. While wireless connections are currently more expensive than most types of physical cabling, often their use may be cost-effective. However, it is true that most existing wireless services are less functional and offer limited coverage when compared with their wireline counterpart. But the lifetime operational costs of WLANs can be substantially less than those of wired LANs, particularly in environments requiring frequent modification.

To be able to offer broadband communication services and provide universal connectivity to mobile users, it is necessary to have a suitable standard for WLANs in place. Similarly, an approach to connecting them to the existing wired LANs and broadband networks is needed. A key design requirement for the wireless LANs is that mobile hosts must be able to communicate with other mobile and wired hosts transparently. This mobility is handled at the *Medium Access Control (MAC)* layer. In addition, it is important that the performance available to mobile users be comparable to that available to wired hosts.

To satisfy these wireless data networking needs, study group 802.11 was formed under IEEE project 802 to recommend international standards for WLANs. The mission of the study group is to develop MAC layer and physical layer standards for wireless connectivity of fixed, portable, and mobile stations within the local area. The 802.11-compliant

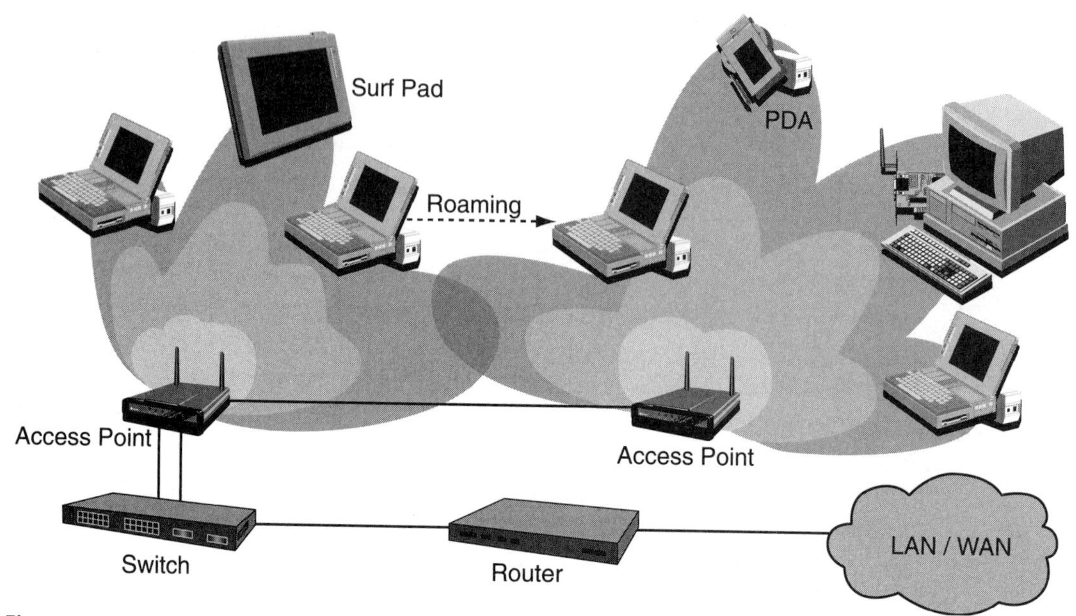

Figure 5–13 Wireless LAN configuration.

solutions consist of access points—wireless transceivers that connect directly to a wired Ethernet LAN via a built-in Ethernet port—and wireless PC cards, which allow a mobile user equipped with a notebook to connect to an access point. The access point then admits the user to the wired LAN.

The IEEE 802.11b standard for 11 Mbps WLAN connectivity on the 2.4 GHz unlicensed radio band addresses a critical issue for many organizations: wireless access to a wired LAN. Proprietary solutions for WLANs have been available for years, but the 802.11b LAN standard provides for interoperability between different manufacturers' equipment. Corporations that need faster speed migrate to an 802.11a network, which uses the less crowded 5 GHz band, as opposed to the 2.4 GHz band for 802.11b. The 802.11a has a maximum throughput of 54 Mbps and can accommodate more users due to the increase in radio-frequency channels and increased operating bandwidth. However, because of a higher absorption rate at the 5 GHz spectrum, 802.11a devices have shorter operating range of about 150 feet, compared with the 300 feet achievable by 802.11b. The range greatly depends on line-of-sight between the workstation and access point. Transmission speeds decrease as the distance between the portable workstation and the access point increases. The 802.11g, ratified in 2003, is the same high speed as 802.11a and is fully backwards compatible with 802.11b since it too uses the 2.4 GHz band. Because the IEEE only sets specifications but does not test them, a trade group called the Wireless Ethernet Compatibility Alliance runs a certification program that guarantees interoperability. Wireless Fidelity, or Wi-Fi, is a certification issued by this group, which assures compliance among manufacturers. Ongoing work is being done on the standard to accommodate higher speeds and better performance.

WLANs typically use microwaves signals, radio waves, or infrared light to transmit messages between computing devices. Subsequently, each of these technologies comes with its set of advantages and disadvantages. To select the appropriate technology, hardware, and software, one needs to consider several operational factors: data rate, number of users, interactive versus data transfer, expected response time, permissible error rate, regulatory issues, and most important of all, training, support, and maintenance of equipment and network.

Microwave LANs

Microwave technology is not really a LAN technology because it is not used to replace wired LANs within buildings. But it is classified in this category because its most important application is interconnecting LANs—bypassing T-1 circuits, providing back up against failure of the primary circuit route, connecting PBXs in a metropolitan network, and crossing obstacles such as highways and rivers. A microwave relay system is shown in Figure 5–14. Although analog and digital microwave systems are available, digital predominates in current products. A digital microwave system consists of three major components on both sides of the link: modem, RF unit, and an antenna. The RF unit is typically connected directly to the antenna, which is mounted on towers or tall buildings to transmit and collect the microwave signals, as depicted in Figure 5–15 and Figure 5–16. They must be in line-of-sight and not more than about 30 miles apart. In installations where line-of-sight cannot be obtained, passive repeaters, which redirect the signal from one path to another, are used.

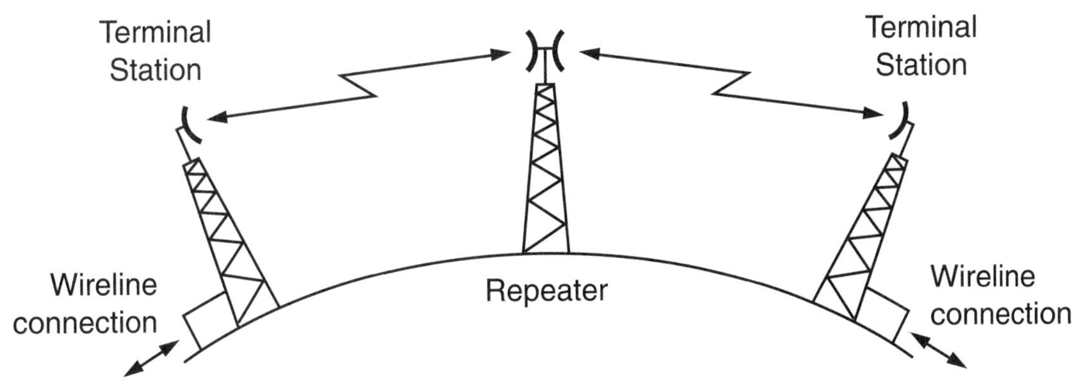

Figure 5–14 A microwave relay system.

Typical microwave systems utilize signals above 30 MHz. Generally, frequencies in the range of 300 kHz to 30 MHz are not used for communications, since there are many services located in that part of the spectrum. Additionally, these lower frequencies result in

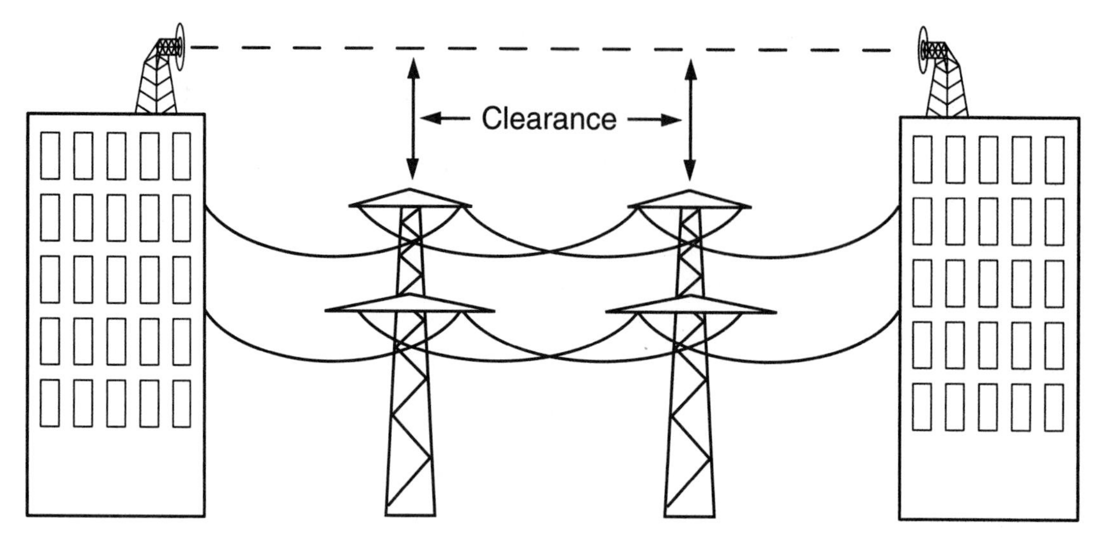

Figure 5–15 A microwave antenna must be mounted on towers or tall buildings to keep the path above electric or telephone wires.

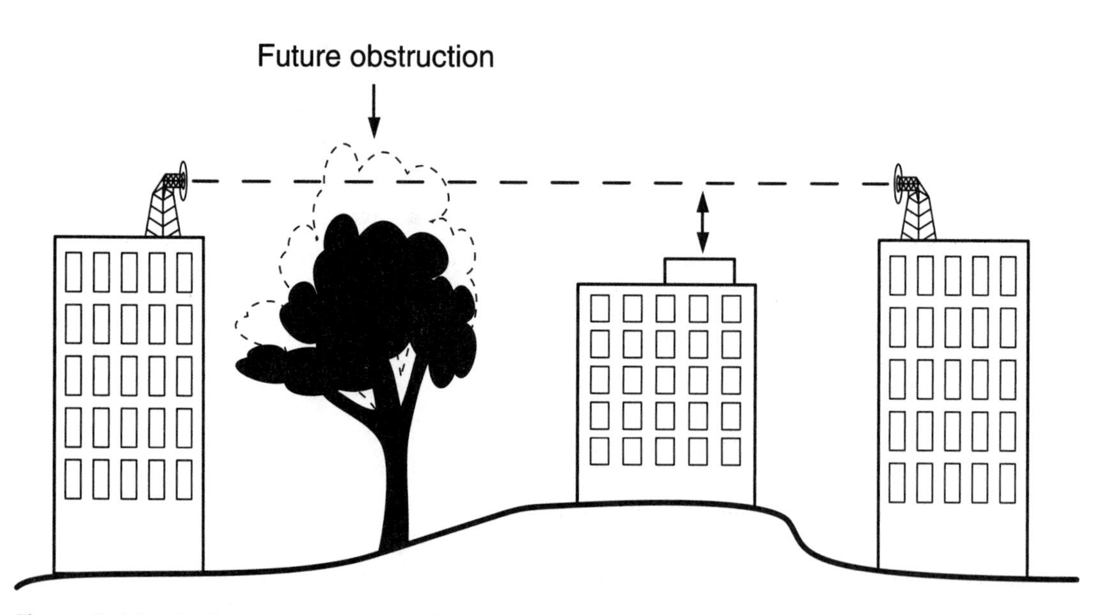

Figure 5–16 A microwave system requires a line-of-sight and any future obstructions must be taken into consideration before mounting antennas.

lower bandwidth supporting fewer channels. As a general rule of thumb, the higher the frequency, the greater the amount of bandwidth that can be carried. Therefore, the GHz bands offer the greatest value. Unfortunately, the GHz bands that are shared by both the satellite and terrestrial microwave frequencies require line-of-sight and are affected by atmospheric conditions such as rain and humidity. One major drawback to the use of microwave technology is that the frequency band used requires licensing by the FCC. Once a license is granted for a particular location, that frequency band cannot be licensed to anyone else, for any purpose, within a 17.5-mile radius.

Radio LANs

Radio LANs basically come in two forms: narrow band or spread-spectrum. Narrow band has a cost advantage but a lower data throughput rate when compared with spread-spectrum technology. However, it can still provide response within seconds to several hundred terminals, making it a preferred choice for warehousing, distribution environments, and other industrial applications. In a narrow band radio system, user information is transmitted and received on a specific radio frequency. Undesirable crosstalk between communications channels is avoided by carefully coordinating different users on different channel frequencies, while privacy and non-interference are accomplished through the use of separate radio frequencies. In addition, the radio receiver filters out all radio signals except the ones on its designated frequency. While wood, plaster, and glass are not serious barriers to radio LANs, brick and concrete walls can attenuate RF signals quite a bit. Not surprisingly, the greatest obstacles to radio transmissions commonly found in office environments are metal objects such as desks, filing cabinets, elevator shafts, and even reinforced concrete.

In business applications, however, most wireless LANs use spread-spectrum technology. It is designed to trade off bandwidth efficiency for reliability, integrity, and security. Each cell has an access point or a base station that is hard wired to the media switch. An access point is a local wireless extension that bridges a cordless LAN to existing wired LANs. A worker can senselessly move or roam from one access point to another without ever losing the network connection. This technology is called roaming-enabled access points. Of the spread-spectrum technologies, FHSS is less expensive, easier to install, and uses less power but, DSSS has wider coverage and higher throughput.

Current WLANs are a combination of basically two technologies: PC card adapters and roaming-enabled access points. A credit-card-sized PC card plugs into the portable unit and lets users create their own wireless network. The PC card has a 32- or 64-bit address and data bus and provides support for the latest 3.3 volt platform in addition to the 5 volt technology. This enables the PC card products to use less power in battery-powered PCs and PDAs.

Infrared LANs

Currently, the Infrared Data Association's (IrDA) standard for wireless data communications is being adopted by leading computer and printer manufacturers. This standard, which uses high-speed, infrared (IR) light for network connections, is the same basic technology used in remote control systems. Figure 5–17 provides a block diagram of an IR transmission system. IR systems consume very little power, and IrDA has thus become

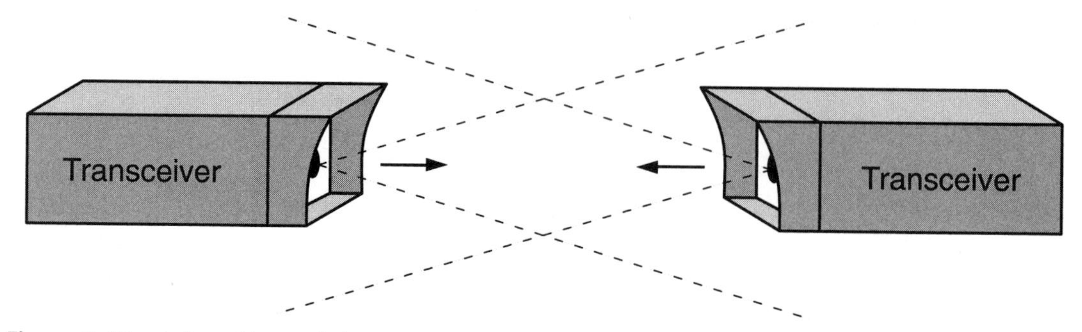

Figure 5–17 Infrared transmission system.

a natural standard for battery-operated devices. In addition, IR products have a cost advantage over spread-spectrum products and are not subject to any regulations or restrictions, worldwide. However, IR technology has not been widely used because of distance limitations. Also, unlike spread-spectrum signal, IR can only work in line-of-sight and the signal will not go through a wall.

The three types of IR systems on the market are line-of-sight, reflective, and scatter. *Line-of-sight* systems offer point-to-point, high-speed connectivity between stations, located within 100 feet. They take advantage of the built-in infrared port on the computer. Increasingly, notebook and laptop computers and other devices (such as printers) come with IrDA ports that support roughly the same transmission rates as traditional parallel ports. The only restriction on their use is that there must be a clear line-of-sight between them. Any interruption such as an object or a person walking through will disrupt the signal. Usually, the two devices are within few feet of each other. Reflective infrared systems are getting around the line-of-sight problem by bouncing the signal off walls, ceilings, and floors. Each workstation's transceiver is aimed at a spot on the wall or ceiling off which the IR signal is reflected. *Reflective* systems are best suited for areas with high ceilings where people are unlikely to disrupt the IR signal. *Scatter* systems use diffused signals similar to the manner in which light scatters. The diffused signals bounce off walls and ceilings to cover an area of up to 100 square feet. Compared with line-of-sight, reflective and scatter are low-speed systems.

Broadband Wireless Systems

Wireless Local Loop (WLL) is a broadband (voice, video, and data) system that involves a low-power digital transceiver capable of supporting bi-directional communications in a small geographic area. WLL configurations include centralized antennae that support RF connectivity to matching antennae at the customer premises, as illustrated in Figure 5–18. The WLL is being used in place of conventional wire-line connections between the local telephone exchange and customer premises. Conventionally, the local loop consists of a pair of copper wires. This is difficult to maintain, and also increasingly expensive and time-consum-

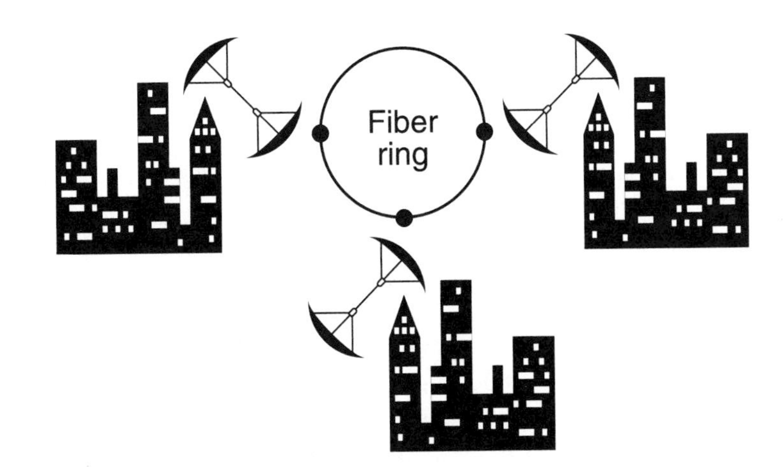

Figure 5–18
Wireless Local Loop (WLL) configurations.

ing to deploy in view of the rising cost of copper and digging. Although WLL is essentially fixed and not mobile, many WLL systems are extended to cellular or PCS systems and enable mobile wireless communications. *Local Multipoint Distribution System (LMDS)* and *Multichannel Multipoint Distribution System (MMDS)* are two popular WLL technologies.

LMDS supports signal transmission over short distances of about two miles using line-of-sight transmissions. This option uses millimeter-wave radio at frequencies of about 30 GHz. A typical installation has a central base station with an omni-directional antenna serving many residences, each of which has a directional dish aimed at the base station. Unlike many other microwave systems, LMDS can carry two-way signals from a single point to multiple points. Since LMDS systems send very high-frequency signals over short line-of-sight distances they are, by default, cellular. The radius of the cell ranges from 2 km to 5 km with a single hub transceiver in the center, which communicates with residences and businesses in the cell at blazingly fast data rates. Using Quaternary Phase Shift Keying (QPSK) modulation, LMDS supports data transfer rates of over 1 Gbps, which makes it desirable for high-speed data networks. With the use of low-powered residential transceivers, LMDS can be used for two-way communication of data, voice, and video signals.

LMDS cell layout has proven to be a very complex issue. The main factors that play into the cell size decision are line-of-sight; analog versus digital signals; overlapping cells versus single-transmitter cells; rainfall; transmission and receiver antenna height; and foliage density. Direct line-of-sight between the transmitter and receiver is essential. Locations without direct line-of-sight can occasionally fake it by using reflectors and, in some cases, amplifiers to bounce a strong signal into shadow areas. Regarding analog vs. digital, analog signals degrade faster than digital signals because they have less tolerance for noise. Additionally, many more digital signals can be squeezed onto the same amount of RF spectrum by using advanced signal modulation techniques. Also, appropriately engineered digital signals are much more robust and less susceptible to rain and foliage. So, despite the fact that analog set-tops are less expensive than digital ones, most LMDS products in the United States are considering going digital. The LMDS technology is geared more toward dense, urban environments or multitenant buildings.

The characteristics of LMDS and MMDS are identified in Figure 5–19. A MMDS system transmits microwave signals over the 2 GHz band of the radio spectrum. Unlicensed spectrum in the 2.4 GHz band shows the most promise as the expense and difficulty of obtaining licenses necessary to provision MMDS service is eliminated. Without huge up-front license costs, there is incentive to build infrastructure even in low-population areas. The low-power nature of the unlicensed spread-spectrum band makes the equipment and installation inherently inexpensive as well. In addition, MMDS is not as range-limited as LMDS, allowing it to operate cost effectively in lower-density markets. But MMDS has lower capacity when compared with LMDS. The MMDS system was initially designed as a one-way broadcast (downlink) system for TV transmission. Various two-way systems are being deployed now, but most still offer asymmetrical bandwidth with high downlink data rates and low uplink data rates. While this system may suit a residential user, it does not completely fulfill the business user's needs.

Technology	Frequency Allocation in the U.S.	Maximum Range	Characteristics
Radio LANs	902–928 MHz 2.4–2.483 GHz (most common) 5.725–5.875 GHz	25 miles	✦ Low deployment cost ✦ Low capacity ✦ Well suited for ad hoc networks
LMDS	27.5–28.35 GHz 29.1–29.25 GHz	2 miles	✦ Highest deployment cost ✦ High capacity ✦ Can be used to serve many customers
MMDS	2.5–2.69 GHz	35 miles	✦ Lower deployment cost than LMDS ✦ Lower capacity

Figure 5–19 Broadband wireless technologies and their characteristics.

SATELLITE COMMUNICATIONS

Several attractive services can be offered by utilizing the satellite technology, including PCS on a global scale, digital audio broadcasting, environmental data collection and distribution, remote sensing/earth observation, and several military applications. The International Telecommunication Satellite Organization (INTELSAT) series began with the launch of the INTELSAT-I, also known as the Early Bird, satellite in 1965. INTELSAT arose from a United Nations resolution to develop worldwide satellite communications on a nondiscriminatory basis. In the early phases of satellite communications, only two earth stations were capable of establishing communications links. INTELSAT-I system incorporated two earth antennas, four on-board antennas, two transponders, and four frequencies. The first transponder utilized the 6.30102 GHz uplink and the 4.081 GHz downlink frequencies; the second utilized the 6.3899 GHz uplink and 4.16075 GHz downlink frequencies. Since a single station was able to access and utilize the entire transponder bandwidth, INTELSAT-I was referred to as a *single-access* system, as depicted in Figure 5–20. The ever-increasing demand for more earth stations generated the need for the design of wide-band repeaters involving *multiple-access* techniques. The concept of multiple access was first introduced in the INTELSAT-II satellite systems.

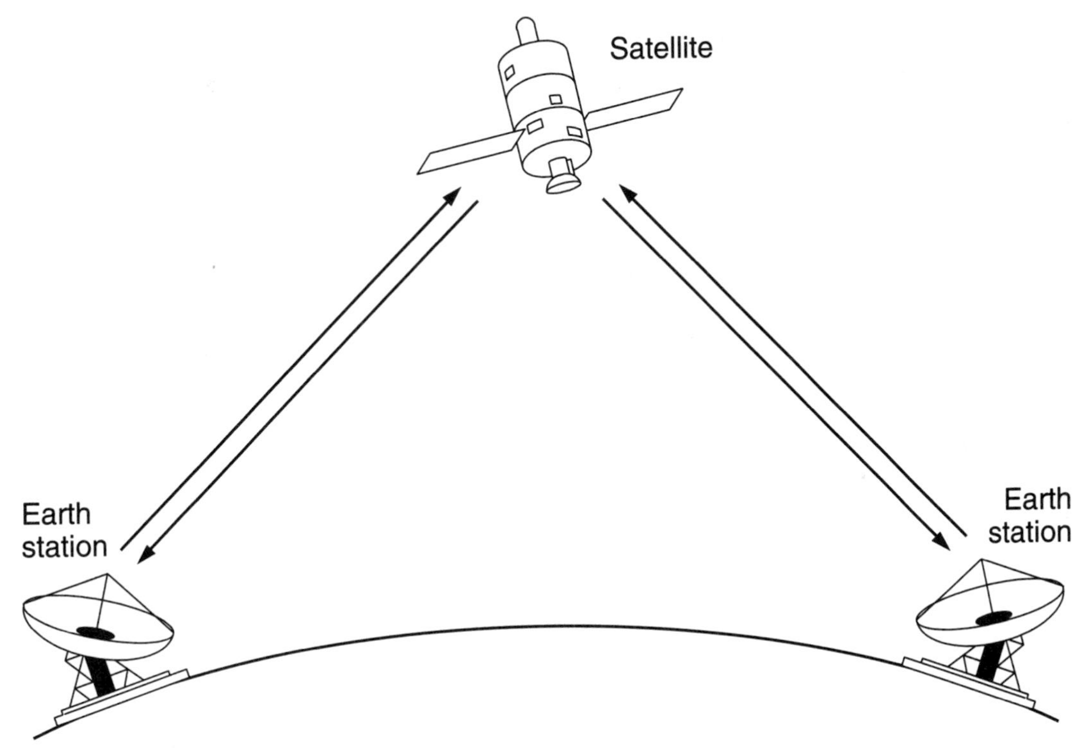

Figure 5–20 Single-access satellite system used in INTELSAT-I.

To facilitate satellite communications and eliminate interference between different systems, the ITU has divided the entire world into three regions, as shown in Figure 5–21. The same parts of the spectrum are reassigned to many nations throughout the world. The frequency spectrum allocations for satellite services are given in Figure 5–22. Each frequency band has different applications. For example, the Ku band is used for broadcasting and certain fixed satellite services. The C band is exclusively for fixed satellite services. The L band is employed by mobile satellite services and navigation systems.

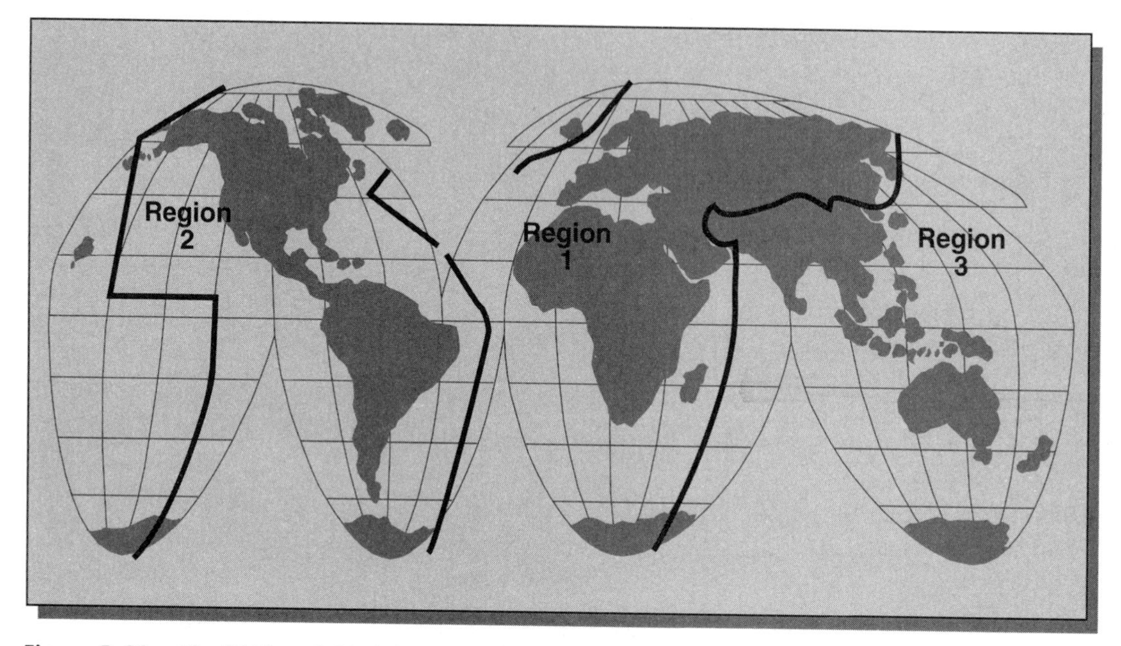

Figure 5–21 The ITU has divided the world into three regions for efficient satellite communications.

Frequency Band	Nominal Frequency Range (GHz)
L	1–2
S	1–4
C	4–8
X	8–12
Ku	12–18
K	18–27
Ka	27–40

Figure 5–22 Satellite frequency allocations.

Each satellite band is divided into separation portions: one for earth-to-space links (the uplink) and one for space-to-earth links (the downlink). Figure 5–23 provides the general frequency assignments for uplink and downlink satellite frequencies. The uplink frequency bands are slightly higher than the corresponding downlink frequency bands in order to take advantage of the fact that it is easier to generate RF power within an earth station than it is on a satellite. The uplink transmission is highly focused to a specific satellite, while the downlink transmission is focused on a particular *footprint* or area of coverage.

Uplink Frequencies (GHz)	Downlink Frequencies (GHz)
5.9–6.4	3.7–4.2
7.9–8.4	7.25–7.75
14–14.5	11.7–12.2
27.5–30	17.7–20.2

Figure 5–23 Typical uplink and downlink satellite frequencies.

A communications satellite is basically a microwave station placed in outer space. A *satellite* is a specialized wireless receiver/transmitter that is launched by a rocket and placed in orbit around the earth. There are hundreds of satellites currently in operation. They are used for such diverse purposes as weather forecasting, television broadcast, amateur radio communications, Internet communications, GPS, and the GSM. Signals are sent to the satellite by large ground station dishes and are received on the satellite by individual *transponders*. Each satellite contains about 24 transponders, which are roughly analogous to channels. These transponders then transmit the information back to earth.

The entire satellite system operates in a manner similar to that of a cellular telephone network. The main difference is that the transponders, or wireless transceivers, are in space rather than on the earth. Satellites bring affordable access to interactive broadband communication to all areas of the Earth, including those areas that cannot be served economically by any other means. The satellite network combines the advantages of a circuit-switched network (low delay), and a packet-switched network (efficient handling of multi-rate and bursty data). The most feasible use of satellite connections is in locations where high-speed wire connections are not an option for geographic or financial reasons. Very Small Aperture Terminal (VSAT) networks, which are essentially fixed satellite systems, have become mainstream networking solutions for long-distance, low-density voice and data communications because they are affordable to both small and large companies due to their lower operating costs, ease of installation and maintenance, and ability to manage multiple protocols. For instance, VSAT systems are used for batch and transaction processing, including airline reservations and credit card purchases.

Satellite Earth Station

The main mission of a **satellite earth station** is to establish and maintain continuous communication links with all other earth stations in the system through the satellite repeater. It must also provide and maintain the necessary command and control links with the spacecraft. The main system components of a satellite earth station are the antenna subsystem, transmitter section, receiver section, and command and control section. Figure 5–24 illustrates a block diagram of a satellite earth station.

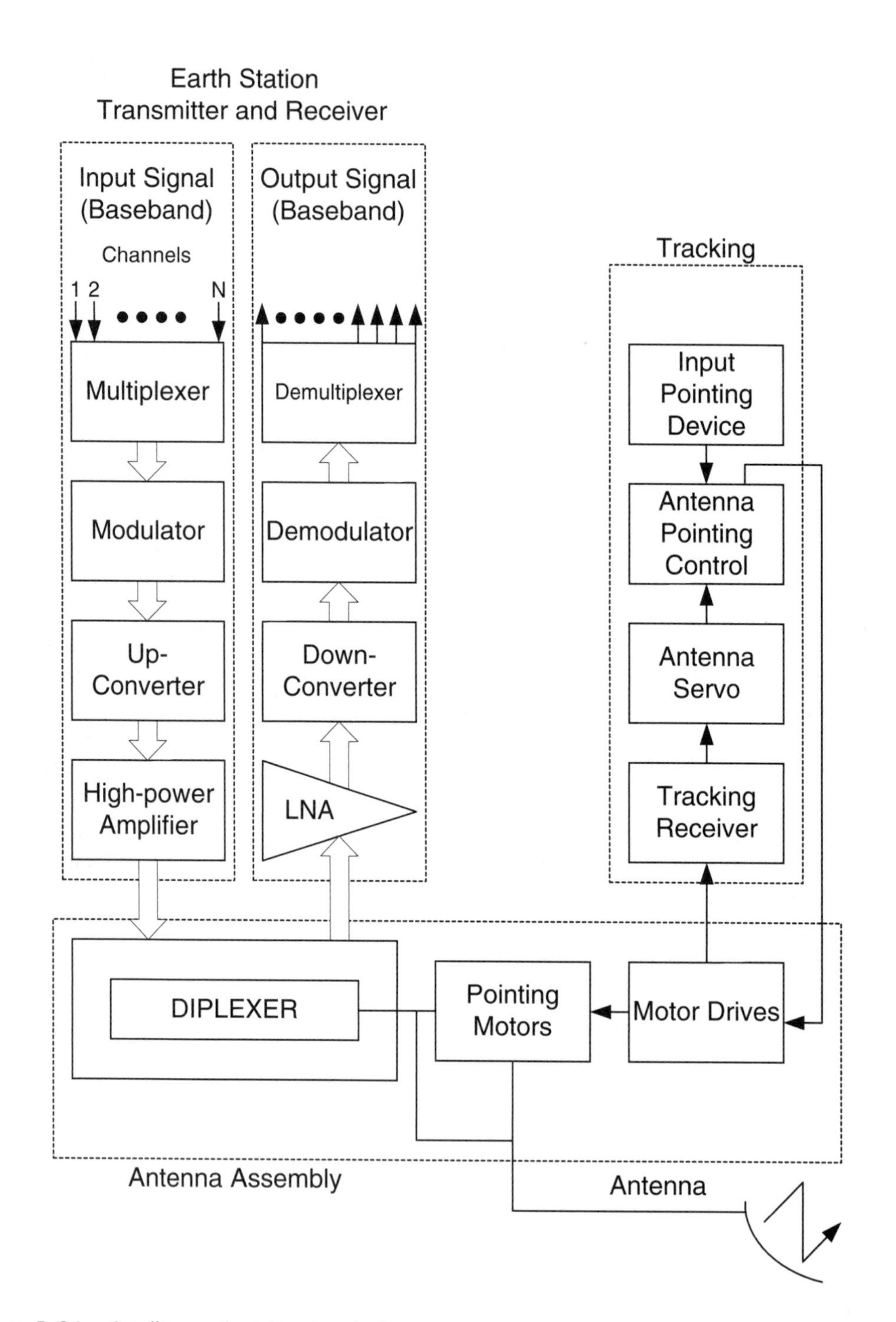

Figure 5–24 Satellite earth station block diagram.

The principal function of the earth station transmitter is to generate the composite baseband signal by multiplexing all the incoming transponder signals (usually digital), then modulating the result and up-converting it to a carrier frequency through a frequency translation circuit. The carrier frequency is then amplified through a high-power amplifier and finally transmitted to satellite transponder via the antenna subsystem.

The receiver section of a satellite earth station performs the exact opposite function of the transmitter. The microwave beam intercepted by the receiver antenna is collected at the focal point of the secondary antenna and fed to the input of a low-noise amplifier (LNA). Both antennas must exhibit a very high gain and a narrow bandwidth. Although the antenna gain is in the order of 50 dB, the output signal from the antenna is extremely weak and very compatible with the input noise power. Therefore, the presence of an LNA is absolutely necessary.

In satellite communications systems, highly directional antennas are used to provide the much-needed antenna gain both for the ground segment and on board the spacecraft. This very high signal amplification is required in order to compensate for the substantial losses the signal suffers while propagating through space. The ability of an earth station to receive satellite signals is measured by the ratio of its receiver antenna gain (G_r) to the system noise temperature (T_{sys}). This ratio is referred as **figure of merit**, which is given in Equation 5–2.

$$\text{Figure of Merit} = \frac{G_r}{T_{sys}} \tag{5–2}$$

where G_r = receiver antenna gain (dB)

T_{sys} = system noise temperature at the input of the LNA (dB)

Standards for INTELSAT systems have set the figure of merit to be equal to or higher than 40.7 dB, as shown in Equation 5–3. In order for a figure of merit to be equal to or better than 40.7, an excellent combination of receiver antenna gain and low system noise temperature must be achieved.

For INTELSAT systems,

$$\frac{G_r}{T_{sys}} = 40.7\ \text{dB} \tag{5–3}$$

Geosynchronous Satellite (GEO)

In 1963, NASA (National Aeronautics and Space Administration) set out with its Synchronous Communications (Syncom) satellite concept. The first two Syncom satellites achieved an orbit that was geosynchronous but not geostationary. In other words, their rotational period matched the Earth's own, but their orbits were inclined and eccentric.

Example 5–2

Problem

Determine the system noise temperature (T_{sys}) of a satellite receiver station in order to maintain a constant figure of merit equal to 40.7 dB with a receiver antenna gain of 55 dB.

Solution

In decibels,

$$\frac{G_r}{T_{sys}} = 40.7 \text{ dB}$$

$$G_r \text{ dB} - T_{sys} \text{ dB} = 40.7$$

$$-T_{sys} \text{ dB} = 40.7 - G_r \text{ dB}$$

$$T_{sys} \text{ dB} = G_r \text{ dB} - 40.7 \text{ dB}$$

Since $G_r = 55$ dB,

$$T_{sys} = 55 - 40.7 = 14.3 \text{ dB}$$

In degrees Kelvin,

$$T_{sys} = \text{antilog}\ (1.43) \cong 27$$

$$T_{sys} = 27°\text{K}$$

Syncom 3, launched in August 1964, circled the equator without inclination and successfully became the first geosynchronous or **geostationary orbit (GEO) satellite**.

GEOs must orbit the equator at an altitude of 22,237 miles, mostly using Ku-band (12 to 14 GHz) frequencies for television transmission. However, they were not used much for data transmission for a variety of reasons. First, most GEOs relied on a passive architecture. Essentially, these were electronic mirrors in the sky. They received signals from transceivers on Earth, amplified them, and sent them back down across their entire footprint. There were two major problems with GEOs. First, a GEO's footprint is very large since a single GEO can see about 40 percent of Earth's surface, as shown in Figure 5–25. This was useful for a broadcast medium like television but was hardly optimal for data no matter how it was multiplexed. Secondly, GEOs must be spaced far apart enough so that neither uplink nor downlink transmissions interfere with one another.

These problems were solved by the Advanced Communications Technology Satellite (ACTS) launch from the space shuttle Discovery in September 1993. It demonstrated that Ka-band (27 to 40 GHz) transmission was practical; these frequencies were previously

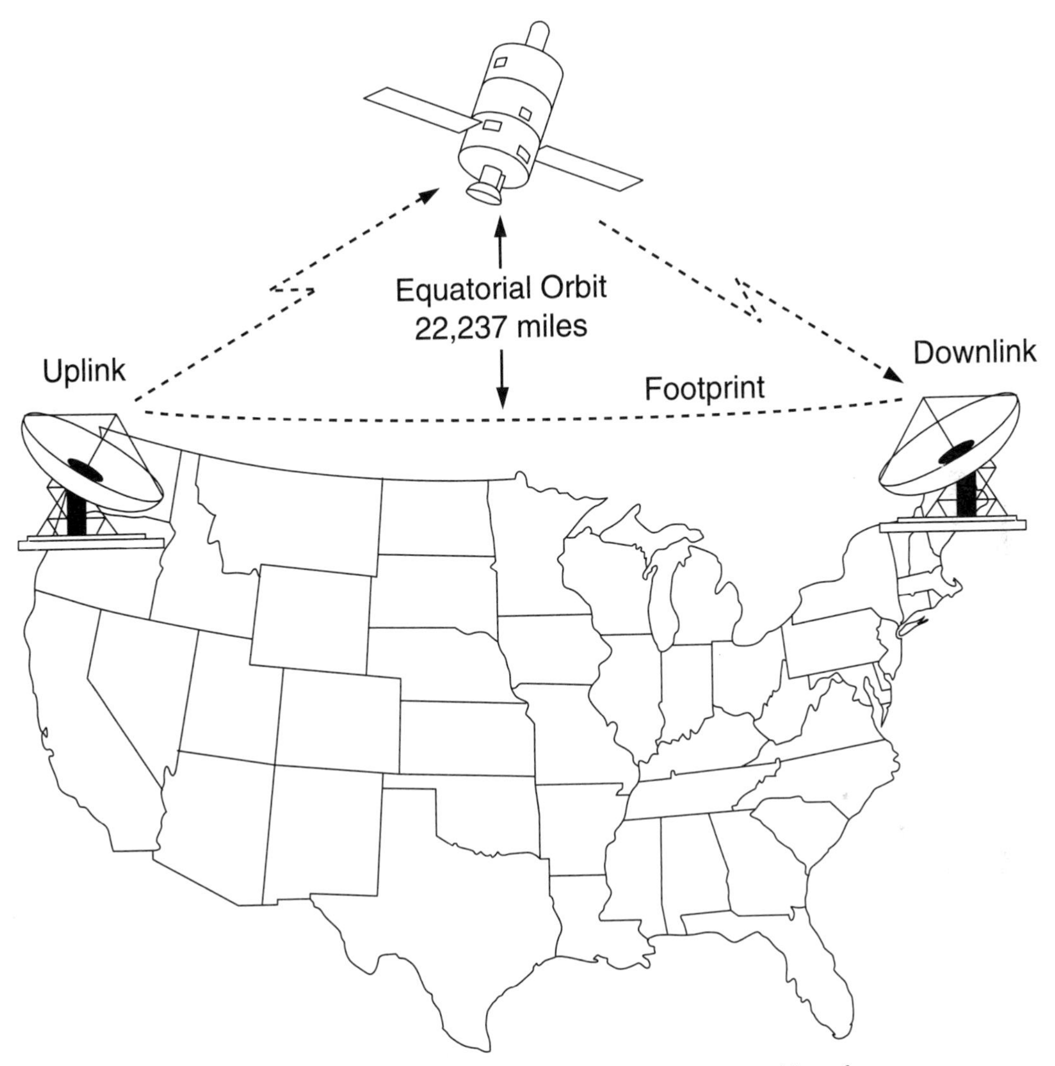

Figure 5–25 A GEO satellite's footprint covers about 40 percent of the Earth's surface.

considered unusable because of cloud cover. By unlocking a wide range of new frequencies, it made interference much less likely. The high frequencies used in the Ka-band also reduced power consumption and made smaller antennas more practical. Better still, ACTS featured antennas that could electronically divide the satellite's footprint into spot beam cells approximately 150 miles in diameter. Used in conjunction with a satellite's onboard switching, spot beam technology sends traffic to and from its destination more efficiently. Its advanced antennas proved capable of transmitting data at up to 622 Mbps to 3.5 m dishes, or 45 Mbps to 60 cm dishes. ACTS and its successors brought satellites much closer to the goal of full-duplex, broadband data transmission.

However, none of the satellites could solve the problem of latency, which was a difficulty that developed as a result of the high elevation at which GEOs must orbit the equator. While electromagnetic energy travels faster in the vacuum of space than in earthbound copper or glass, it still cannot exceed the speed of light. Therefore, it takes nearly 240 ms (milliseconds) for a radio signal to make a round trip between a ground station and a GEO directly overhead. If the ground station is at the edge of the satellite's footprint, the delay might be as long as 270 ms.

GEOs are mainly used for international and regional communications. The INTELSAT family of satellites is utilized for international communications, whereas the SATCOM, ANIK, WESTAR, COMSTAR, GALAXY, and MOLNIYA systems are used for domestic service like local television distribution and regional communications. The National Oceanic and Atmospheric Administration, (NOAA) developed and operates a system of polar orbiting satellites, as well as GEO satellites, for meteorological, oceanic, and space environmental studies. Both systems are designed to perform specific tasks; for example, the GEOs monitor weather patterns as they develop in the tropics, while the polar satellites orbiting at higher altitudes are used for data collection.

Global Positioning System (GPS)

Global Positioning System (GPS), a worldwide radio-navigation system, is funded and constantly monitored by the U.S. Department of Defense (DoD). GPS is formed from a constellation of 24 satellites at 11,000 mile altitude and their ground stations. The orbit altitude is such that the satellites repeat the same track and configuration over any point approximately each 24 hours (4 minutes earlier each day). There are six orbital planes equally spaced, and inclined at about fifty-five degrees with respect to the equatorial plane. This constellation provides the user with between five and eight satellites visible from any point on the earth. The whole idea behind GPS is to use satellites in space as reference points for locations here on earth.

A GPS receiver measures distance by timing how long it takes for the radio signal sent from the satellite to arrive at the receiver; the signal speed is same as the speed of light. Since Distance = Velocity × Time, accurate timing is the key to measuring distance to satellites. Satellites are accurate because they have atomic clocks on board, but GPS receivers with imperfect clocks must make four measurements simultaneously, as shown in Figure 5–26, and compute a correction factor for any delays. Each satellite has its own unique pseudo-random code, represented in Figure 5–27, which guarantees that a GPS receiver will not accidentally pick up another satellite's signal. Therefore, all satellites can use the same frequency without jamming each other. GPS receivers are becoming very economical and have found applications in cars, boats, planes, construction equipment, movie-making gear, farm machinery, and wireless devices including laptop computers.

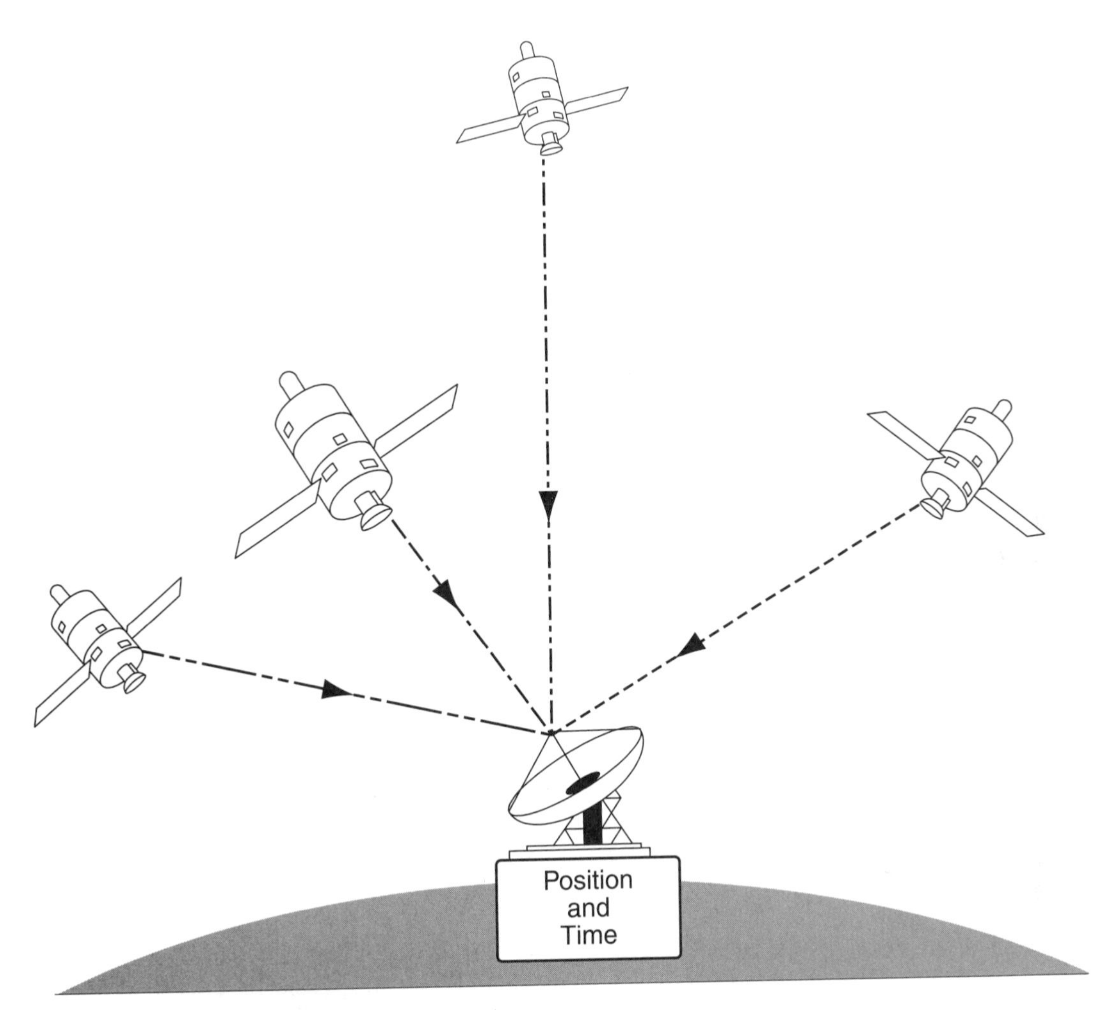

Figure 5–26 GPS navigation system.

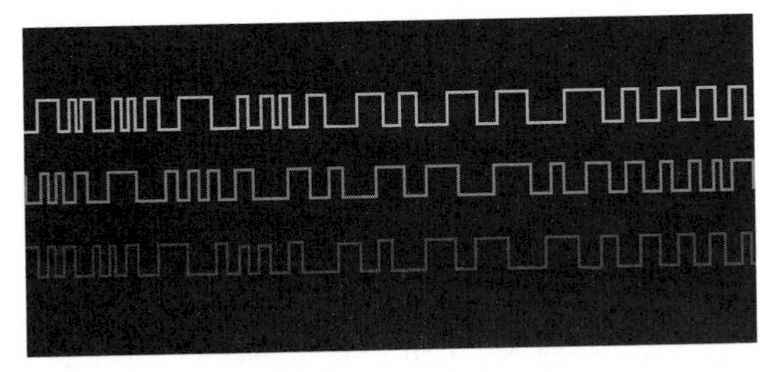

Figure 5–27 Each satellite has a unique pseudo-random code.

Low Earth Orbit (LEO) and Medium Earth Orbit (MEO) Satellites

In order to overcome the delay and to complement current terrestrial cellular networks, companies that want to provide satellite-based data communications are mostly deploying low earth orbit (LEO) or medium earth orbit (MEO) satellite constellations, as depicted in Figure 5–28 and Figure 5–29. A well-designed LEO system makes it possible for anyone to access the Internet via wireless from any point on the planet using an antenna. The system consists of a large fleet of satellites, each in a circular orbit at a constant altitude of 500 to 1,000 miles, depicted in Figure 5–30. This makes them capable of providing smaller more energy-efficient spot beams and delivering latency potentially equal to wired (transcontinental fiber optic cable) communications. The closer a satellite is to Earth, the narrower its angle of view. Therefore, instead of the eight-satellite constellation that is ample when GEOs are used, at least 48 to 288 satellites are required to provide global coverage with LEOs. Each satellite takes approximately 90 minutes to a few hours to complete one revolution.

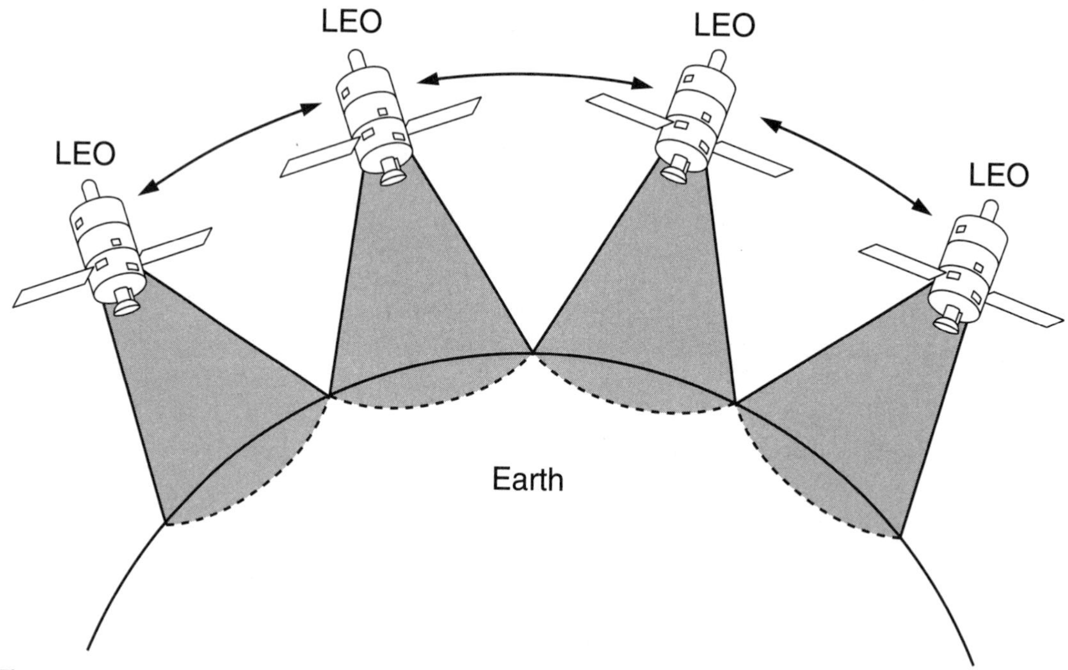

Figure 5–28 LEO satellite coverage.

Because of their low orbit, LEOs are not geostationary. Instead, they are constantly in motion with respect to any point on the Earth. Therefore, in order to create a LEO constellation, one must provide enough satellites to circle the Earth while ensuring that at least one is (or preferably, two or three are) visible to every receiving antenna at all times.

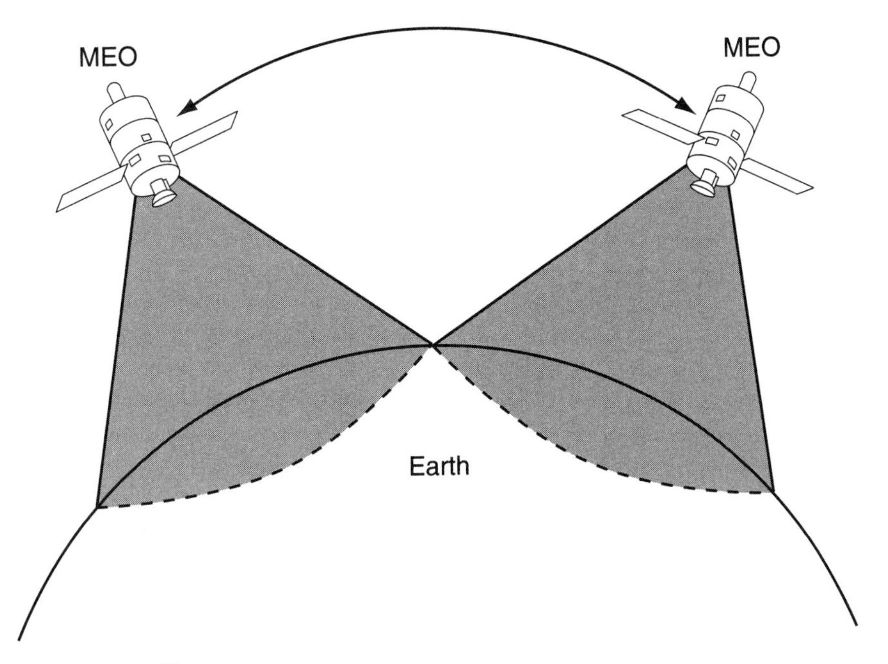

Figure 5–29 MEO satellite coverage.

Also, the antennas must be self-aiming devices. This problem has been solved by phased-array antennas, which group together many smaller antennas. By comparing the slightly different signals received by each, they can track several satellites at once without having to move physically. As a new satellite rises over the horizon, the antenna array will establish a connection, and only then will it drop the link to the satellite that is setting.

LEO constellations also face the problem of jitter, or variable latency. The distance between each LEO satellite and any particular receiver is constantly changing. In addition, a satellite may spend less than a minute over that receiver, after which time the connection will have to be handed off to the next bird. If excessive jitter is generated, it could spell problems with applications such as IP telephony or streaming video. Another physical problem with LEO constellations is the sheer number of satellites they require. Not only must a vendor launch hundreds of satellites, but it must also keep them in working order. An example of a LEO system is the IRIDIUM communications network developed by Motorola, which is represented in Figure 5–31.

There is a case, therefore, for launching MEO satellites, which, like LEOs, are not geostationary. Operating from an elevation between 1,800 and 6,500 miles, they take about two hours to pass over any point in their coverage area and experience a propagation delay of 0.1 s. Therefore, a constellation of some 16 satellites is sufficient to cover the globe. As satellite orbit heights from the Earth are reduced, the number of satellites needed to maintain constant communications increases.

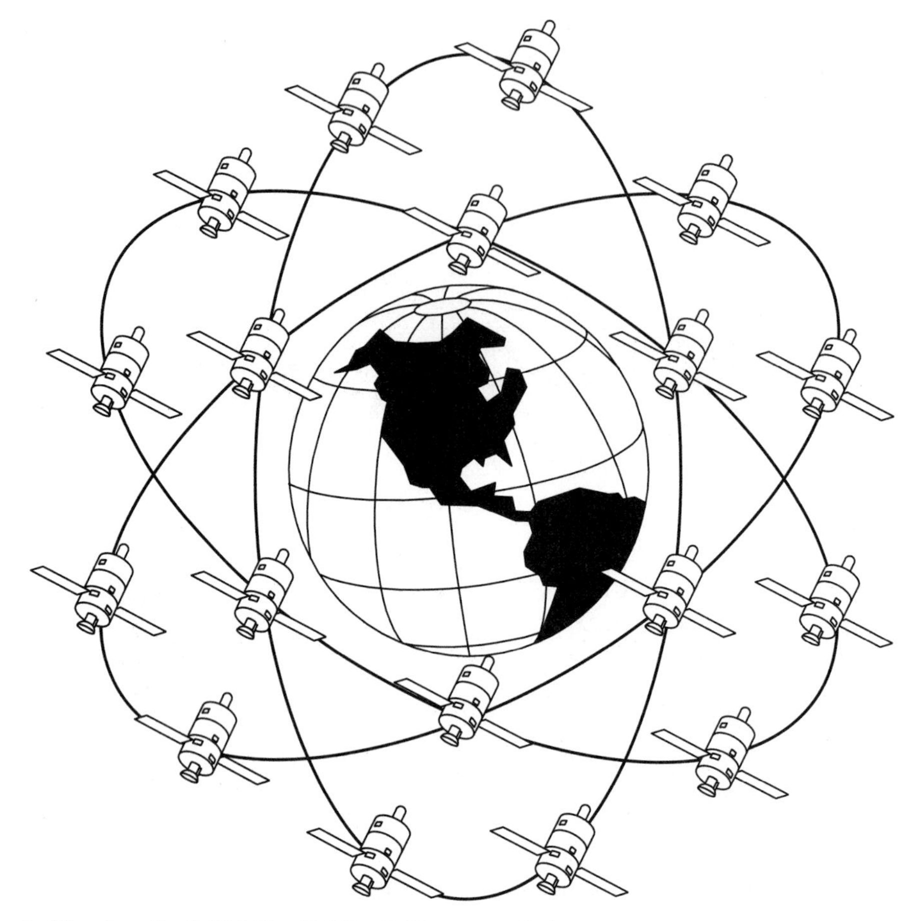

Figure 5–30 Low Earth Orbiting (LEO) satellite constellation.

INTERNATIONAL WIRELESS COMMUNICATIONS

The scope of investment and engineering efforts required to build and maintain copper-based networks has created lofty barriers that have made high penetration rates of basic telephone service possible only in industrialized nations of the world. However, advances in technology and competitive access are driving the revolution toward wireless access infrastructure for the provision of basic telephone service, especially in developing economies.

In international wireless communication systems, analog systems could be defined as first generation and digital systems as second generation (2G). These systems, however, will not be able to provide the data rates necessary for new multimedia services. There is growing need for next-generation systems that not only feature global services, but also a higher user bit rate and a transmission quality close to that of a fixed network. Thus, third generation (3G) systems combining terrestrial and satellite systems that are

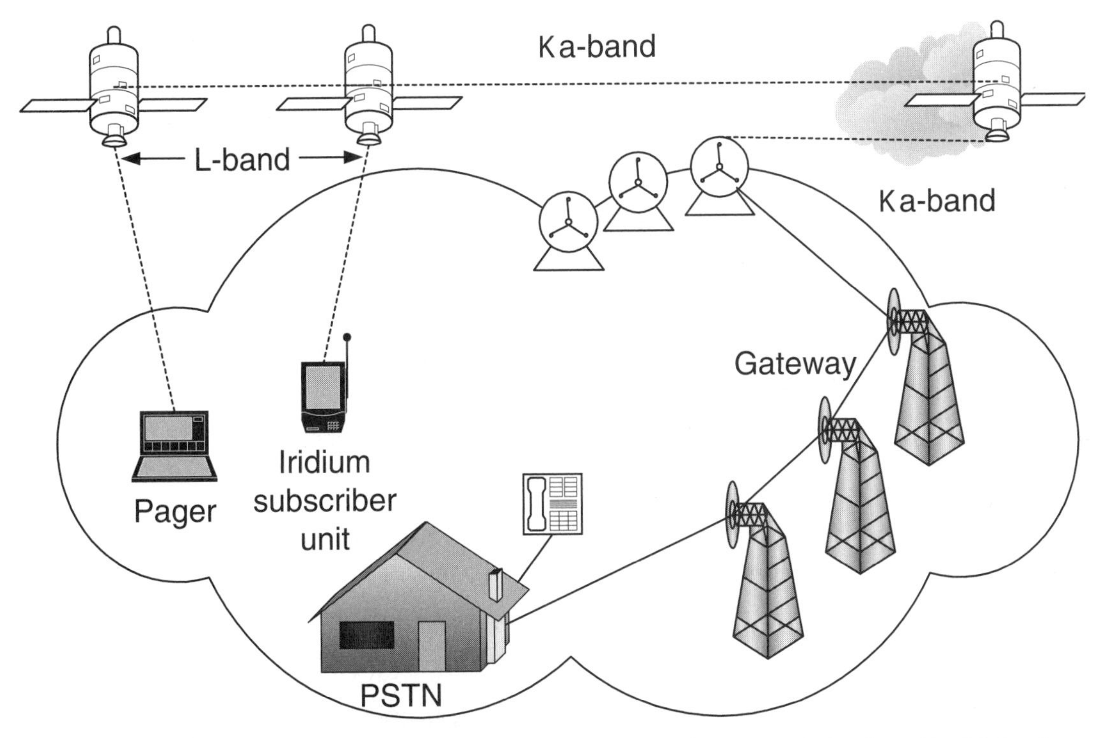

Figure 5–31 A functional diagram of the IRIDIUM communication network developed by Motorola.

expected to support global services with a high bit rate, high signal quality, and higher terminal mobility are already under development. Although its maximum supported bit rate has not yet been defined, it will be between 1 and 20 Mbps.

UMTS and IMT-2000

3G services are designed to offer broadband access at speeds of 2 Mbps, which will allow mobile multimedia services to become possible. One of the 3G systems is the Universal Mobile Telecommunications System (UMTS) developed by the CEPT and presently, main tasks for its standardization are being conducted by the ETSI. In 1998, ETSI decided on a single air interface solution for UMTS with the Wideband Code Division Multiple Access (W-CDMA) technology for wide-area applications and TD-CDMA technology for low-mobility indoor applications.

W-CDMA employs wider frequency bands. It is a promising technology because it is flexible, does not require frequency planning, and supports high data rates. A major weakness of W-CMDA, however, is its relative inefficiency in dealing with asymmetric traffic, which is envisioned to become more important as users move from speech (such as phone calls) to data (like Internet browsing). The other component of UMTS, mixed TD-CDMA, is, on the other hand, better equipped to deal with that issue because the ratio between uplink and downlink is not fixed in terms of resources. TD-CDMA seems to be a good complement to W-CDMA for 3G systems.

UMTS will provide wideband wireless multimedia capabilities over mobile communications networks. The technology will realize a range of new and innovative services including interactive video, Internet/Intranet, and other high-speed data communications services. UMTS is already a reality, with W-CDMA networks now operating commercially in Austria, Italy, Japan, Sweden and the UK. The 3G licensing process is largely completed in many parts of the world, setting the stage for the worldwide deployment of UMTS systems.

An effort similar to UMTS is underway in ITU under the name of Future Public Land Mobile Telecommunication System (FPLMTS), which was recently renamed to the more catchy International Mobile Telecommunication 2000 (IMT-2000). The system will operate in the 1.885 to 2.025 GHz and 2.11 to 2.20 GHz bands. It will provide wireless access to the global telecommunication infrastructure through both satellite and terrestrial systems and serve fixed and mobile users in public and private networks. It is expected that UMTS and IMT-2000 will be compatible so as to provide global roaming. The W-CDMA format is viewed as the leading candidate to be used for the next-generation mobile communication system to meet the needs for multimedia applications.

Global System for Mobile Communications (GSM)

The 2G **Global System for Mobile Communications (GSM)** based on TDMA technology was developed in Europe. A schematic overview of the GSM system is shown in Figure 5–32. Although GSM technology has a lot of similarities to TDMA IS-136, it developed along a very different path. Unlike in the United States, where the FCC moved the industry from a single analog standard to a new generation of multiple competing digital standards, in Europe the direction was reversed. Europe began with five incompatible analog air interfaces scattered around the continent. In the 1980s, momentum increased to build Europe's global influence as an economic block by integrating economically.

As part of that movement, in response to a European Commission directive, international agreements were devised to develop a single international open, non-proprietary digital cellular standard, with the most important goal being seamless roaming in all countries. New spectrum at the 900 MHz band was set aside for cellular service. In 1982, the CEPT held a meeting to begin the standardization process. During 1985, the CCITT created a list of technical recommendations for GSM. In 1987, all parties agreed to a compatibility specification, with an air interface based on hybrid FDMA (analog) and TDMA technologies. GSM engineers decided to use wider 200 kHz channels instead of the 30 kHz channels that TDMA used, and instead of having only 3 slots like TDMA, GSM channels have 8 slots. This allows for fast bit rates and more natural-sounding voice-compression algorithms. GSM provides data services such as email, fax, Internet

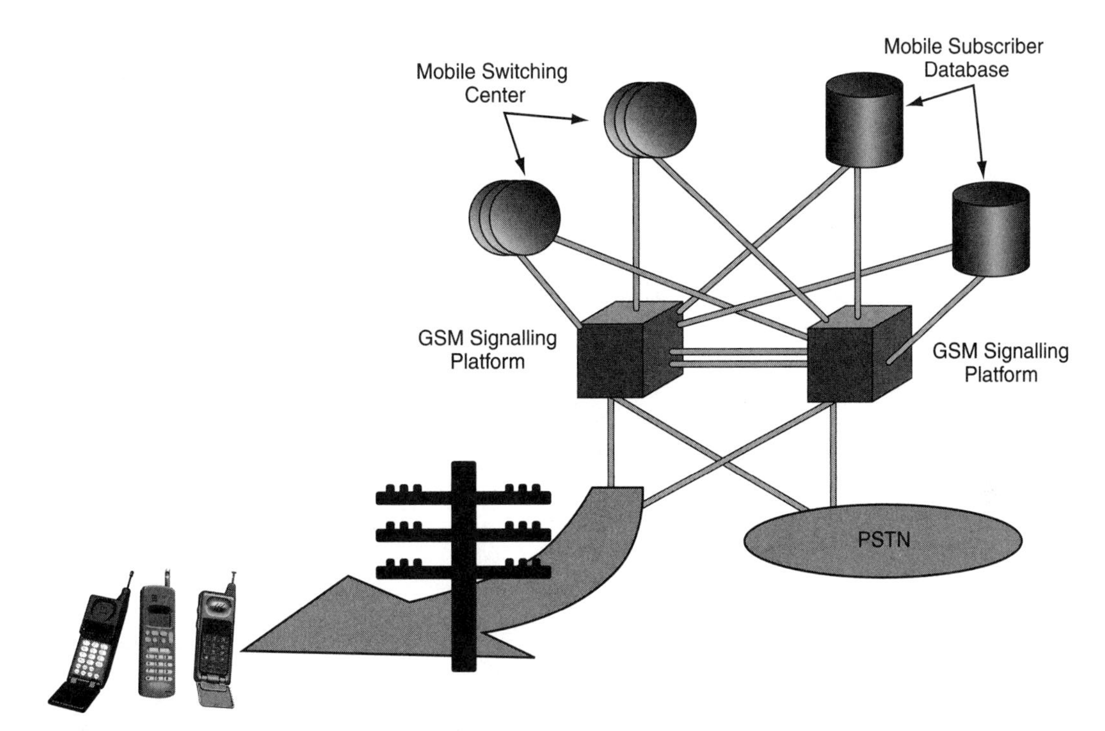

Figure 5–32 An overview of the GSM system.

browsing, and LAN wireless access and permits users to place a call from either North America or Europe.

In the United States, GSM specifications on the 1900 MHz band were developed starting in 1995. Commercial GSM 1900 MHz cellular systems have been operating in the United States since 1996. GSM networks offer transatlantic coverage for voice and data services such as fax, Internet access, and e-mail. Thus, GSM, which was first introduced in 1991, permits automatic roaming between North American, European, and Asian countries. As of 2003, some form of GSM digital wireless service was offered in over 193 countries and has become the de facto standard in the United States, Europe, and Asia.

SUMMARY

The major business trends for the 21st century—immediate response to customer needs and empowering the workforce—have industries moving employees from a traditional office to a virtual office. Several companies have realized significant savings and measurable paybacks by adopting the concept of a virtual office. The ability to access real-time information is the primary motivation for users of two-way wireless communications. The key feature of wireless communications is that individuals are able to send and receive data, audio, and video signals, *anytime, anywhere*. Users have been asking for wireless access to the same data they can access from their desktop including e-mail, calendar, workgroup documents, databases, corporate applications, and Web browsing.

A major rewrite of telecommunications regulations has removed many of the barriers preventing companies from entering new businesses. Also, the use of cellular and PCS telephones has educated the market, resulting in the growth of the wireless infrastructure. These factors have combined to create a fertile ground for several enabling technologies that have matured in the last few years. Currently, there are a number of wireless scenarios, starting with one-way paging devices and building in features and functions to full two-way data access including live audio and video. No single solution works for everyone, therefore there will be many different technologies, applications, and products available. On the international scene, we will soon realize true wireless global roaming as several countries have next generation wireless projects and programs underway. International standardization efforts will enable seamless end-to-end global service for the user. Open standards-based platforms with connections to existing telecommunications infrastructures are crucial to delivering new wireless data and telephony services.

Case Study

University X considered all of the options for wireless networking—802.11a, 802.11b and 802.11g—before deciding on an 802.11a solution. The 802.11a option won out because it provides greater bandwidth in two ways: more bandwidth per access point and closer grouping of access points, so fewer students share a given chunk of bandwidth. 802.11a provides smaller cells with more access points in a given area, which effectively divides access, roughly by classroom, and there is less overlap. Also, 802.11a enables greater number of channels as compared to 802.11b. According to the analysts for University X, 802.11g had one key deficit: it carries the inherent disadvantages of 802.11b with respect to the number of non-overlapping channels.

Questions

1. Why did University X select the 802.11a option?
2. Can you think of a scenario where 802.11b will be preferred over 802.11a?

REVIEW QUESTIONS

1. Review the development of the wireless industry with a historical perspective.
2. What is the advantage of implementing a cellular network?
3. Describe the hierarchical cell structure in a PCS network.
4. Explain how analog cellular technology works.
5. Distinguish between different digital cellular technologies and their applications.
6. Define the following terms:
   - A. Cellular
   - B. Spread spectrum
   - C. Soft handoff
   - D. Frequency Hopping Spread Spectrum
   - E. Direct Sequence Spread Spectrum
   - F. Figure of Merit
7. Discuss the specifications and applications of Bluetooth technology in wireless communications.
8. Identify the strengths, weaknesses and applications of the following wireless LAN technologies:
   - A. Microwave LANs
   - B. Radio LANs
   - C. Infrared LANs
   - D. Broadband Wireless LANs
9. Describe the role of satellite communications in the wireless world.
10. What is the function of a satellite earth station?
11. Determine the receiver antenna gain for a satellite earth station where the system noise temperature is 30°K, to maintain a minimum figure of merit of 40.7 dB.
12. Compare and contrast GEO, GPS, and LEO satellite systems.
13. Provide an overview of the current status of international wireless communications.
14. Identify five different wireless products and describe the technology(ies) implemented in their usage.

6 DATA COMMUNICATIONS

KEY TERMS

- Data Network
- Data Terminal Equipment (DTE)
- Data Communications Equipment (DCE)
- Network Architecture
- Interface
- Protocol Converter
- OSI Reference Model
- Media Access Control (MAC)
- Logical Link Control (LLC)
- Character
- ASCII Character Code
- Data Encoding
- Parity Checking
- Cyclic Redundancy Check (CRC)
- Token Passing
- Carrier Sense Multiple Access (CSMA)
- Client/Server
- Peer-to-Peer
- Repeaters
- Bridges
- Switches
- Routers
- Gateways

OBJECTIVES

Upon completion of this chapter, you should be able to:

- ✦ Discuss the meaning of data communications and related terms
- ✦ Synthesize the process of how the OSI reference model emerged as the standard network architecture with respect to the evolution of data networks
- ✦ Identify different character codes and data compression strategies
- ✦ Analyze data coding methods, error detection and correction schemes, and data link protocols
- ✦ Summarize the role of the OSI model in the development of LAN architecture
- ✦ Describe LAN access methods and the corresponding LAN technologies, primarily Token Ring, FDDI, and Ethernet
- ✦ Discuss the need for internetworking and the various devices that accomplish those functions

INTRODUCTION

Although telecommunications began with messaging and voice communications, in terms of volume, it takes a back seat to today's data communications. Back in the 1940s when the computer was invented, the data processing function was localized to the mainframe and all transactions were performed in the computer room. The dramatic increase in the performance to price ratio of computers, combined with the need for sharing information, were key factors in an evolution of data processing to *data communications*, which is defined as the exchange of digital information (0s and 1s) between two communications systems.

EVOLUTION OF DATA NETWORKS

The need for efficient data communications gave rise to **data networks** that provide worldwide access to corporate data and are collaborative in nature. A data network is a collection of devices that can store and manipulate electronic data and are interconnected in such a way that network users can store, retrieve, and share information with each other regardless of their physical location. Figure 6–1 outlines the components of a basic data communications link between two endpoints. The devices in a data network that transmit and receive data are collectively called **Data Terminal Equipment (DTE)**. These include a wide range of devices such as mainframes, simple receive-only terminals, and peripherals such as printers. DTE is coupled to the transmission medium by **Data Communications Equipment (DCE)**, which moves information by implementing communications facilities. Modems, bridges, switches, and gateways are all examples of DCE. Some devices such as multiplexers do not clearly fall into either category and must be configured depending on the specific installation requirements.

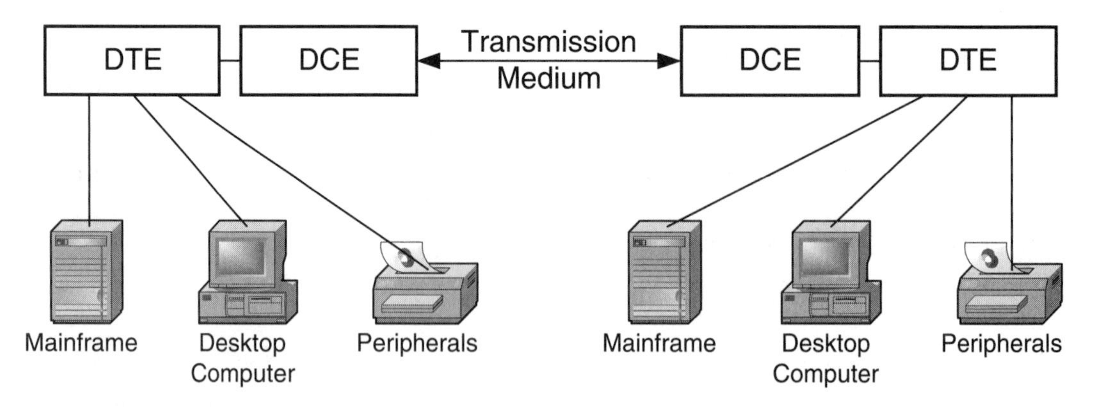

Figure 6–1 Components of a basic data communications link.

In early networks, users were connected to the mainframe by specialized wiring that was proprietary and expensive. Although smaller computer companies were competing, the major players were IBM and DEC (Digital Equipment Corporation), each with its own hardware and software platforms. IBM had everything in place: the hardware, the software, and the connections to make all the components work together in harmony—seamless communications based on the Systems Network Architecture (SNA), a seven-layered model that is illustrated in Figure 6–2. DEC introduced lower-end machines at a reasonable cost based on Digital Network Architecture (DNA), a five-layered architecture shown in Figure 6–3.The founders of DEC were ex-employees of IBM, so their approach was not very different from that of IBM; the bottom three layers of both architectures performed very similar functions.

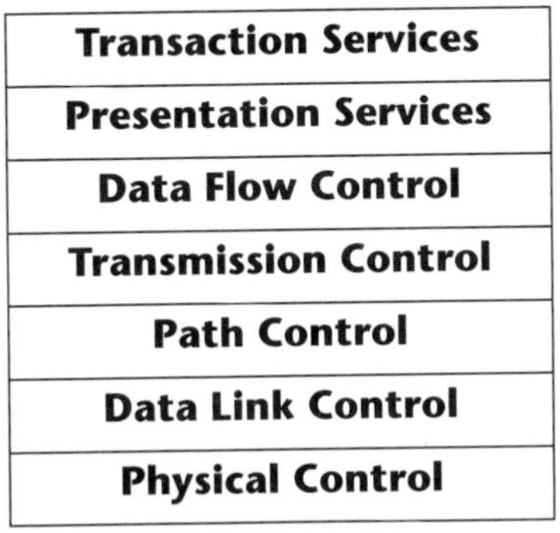

Transaction Services
Presentation Services
Data Flow Control
Transmission Control
Path Control
Data Link Control
Physical Control

Figure 6–2 IBM's SNA architecture.

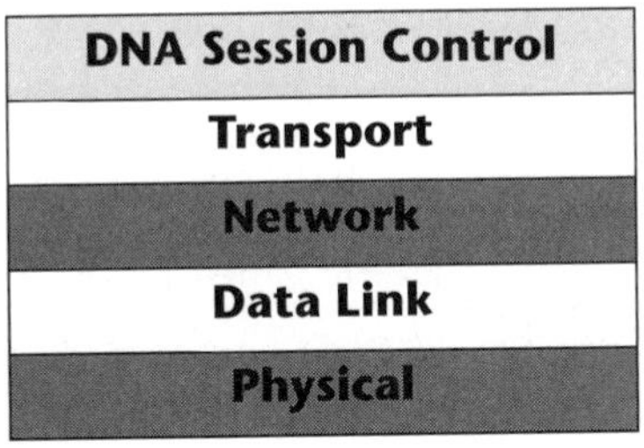

DNA Session Control
Transport
Network
Data Link
Physical

Figure 6–3 DEC's DNA architecture.

In telecommunications, **network architecture** is a coordinated set of guidelines that together constitute a complete description of one approach to building a communications environment. At first, the proprietary network architectures like SNA and DNA were not a problem because the processors were application-specific. But when users wanted to connect to both systems from a single desktop, compatibility became a big issue. This gave rise to third-party manufacturers of interfaces and protocol converters to create transparency between the two machines. The layer boundaries allow information to flow horizontally using *protocols* and vertically using *interfaces*. An **interface** provides *handshaking* from one layer to the next, for example, Layer 2 to Layer 3, as shown in Figure 6–4. Handshaking is the exchange of predetermined signals when a connection is established between two data devices and is not controlled by the operator. A protocol is a set of rules and a **protocol converter** connects two systems at the same layer (e.g., Layer 3 of one system with Layer 3 of another system) and enables two different electronic machines to communicate properly with one another, as depicted in Figure 6–5. The protocol converter box was only an interim solution before users put pressure on the coordinating companies in the industry to resolve the compatibility issue.

In 1978, the ISO accepted the challenge of finding a solution that would enable transparent data transfer between and among systems from different manufacturers. The result was a seven-layer Open Systems Interconnection (OSI) reference model.

Figure 6–4 Role of an interface.

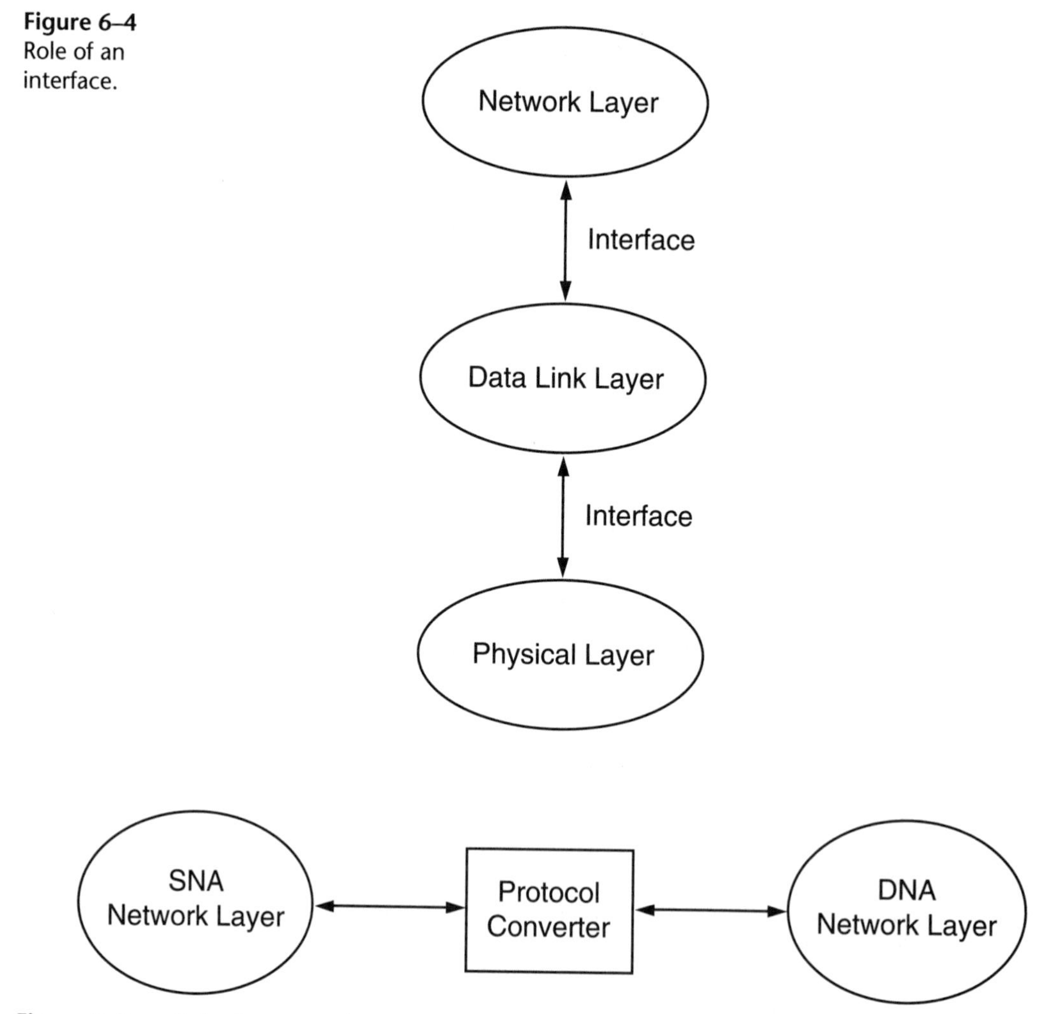

Figure 6–5 Role of a protocol converter.

OPEN SYSTEMS INTERCONNECT (OSI) MODEL

The Open Systems Interconnect, represented in Figure 6–6, is an architectural model for how digital information should be transmitted between any two points in a data communications network. The process of communication between two end users is divided into layers, with each layer adding its own set of special, related functions. So, in a given message between users, there will be a flow of data through each layer from top to bottom at the transmitting end, and when the message arrives at the other end, another flow of data from bottom to top in the receiving computer, as shown in Figure 6–7.

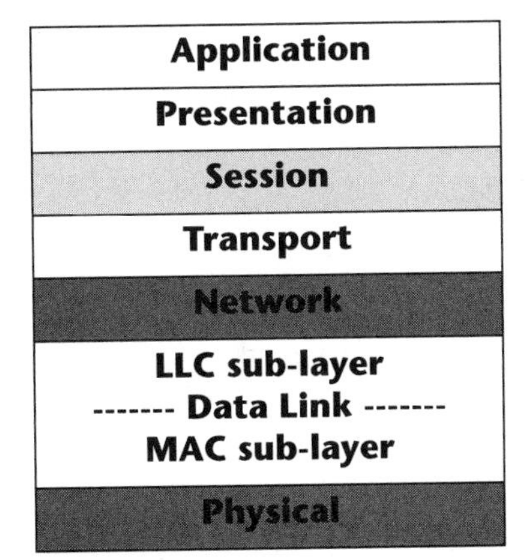

Figure 6–6 The OSI reference model.

The **OSI reference model** defines seven layers of functions that are divided in two groups. The lower three layers (physical, data link, and network) are used for a node-to-node connection, that is, when any message passes through the *host computer.* A host computer refers to the processor that performs a variety of applications including data retrieval and storage, numerical calculations, and management tasks. The host computer, which may be a mainframe or minicomputer, is a network node at the heart of a data communications network. The lower three layers deal with network-specific issues of the communications process and may be provided by a third-party communications company such as a carrier. The upper four layers (transport, session, presentation, and application) are used whenever a message passes from or to a user. Hardware considerations are relevant only at the bottom two layers, while Layer 3 and above are all software.

The primary purpose of the OSI model is to guide product implementers so that their products will consistently work with other products. Although OSI is not always strictly adhered to in terms of keeping related functions together in a well-defined layer, many products involved in data communications make an attempt to place themselves in relation to the OSI model. It is also valuable as a single reference view of communication that furnishes a common ground for education and discussion. Let us analyze the functions of each layer, beginning with the bottom layer, or Layer 1, of the OSI model.

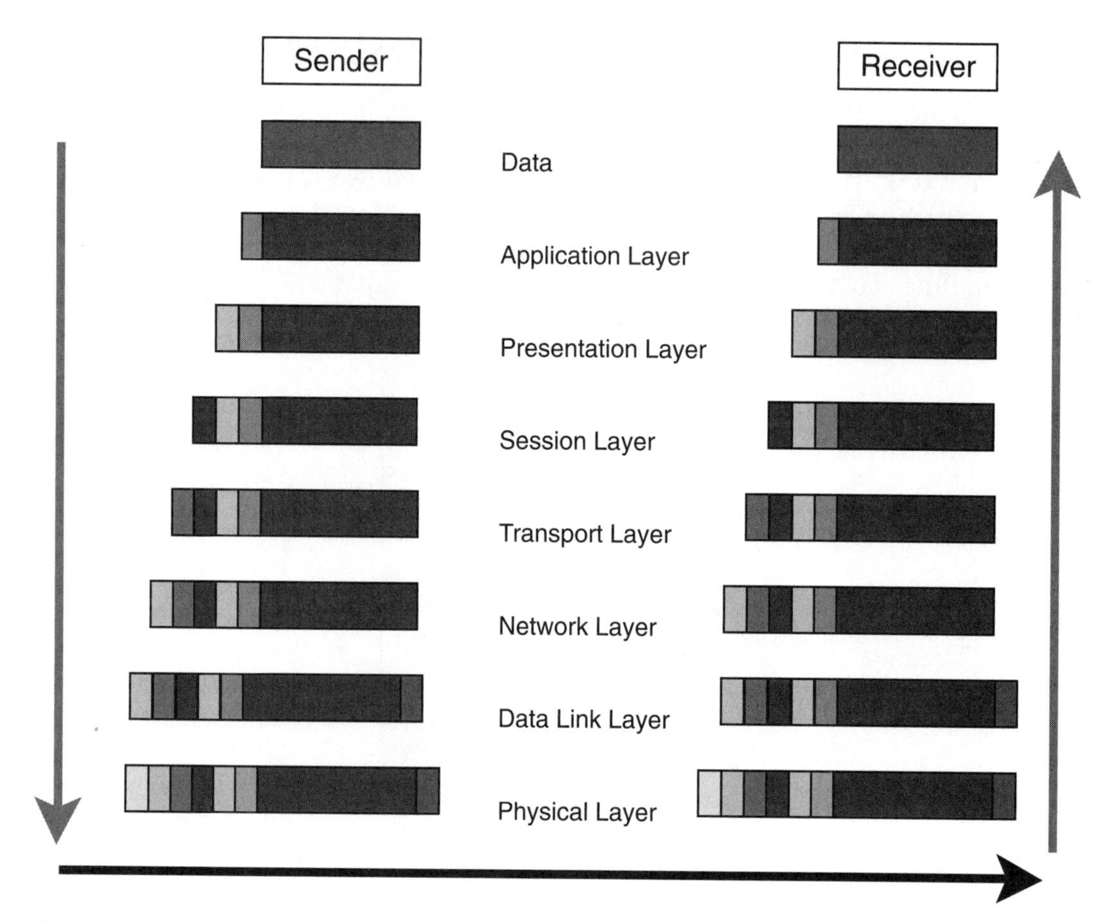

Figure 6–7 Header/trailer information is added or removed as data passes from layer to layer.

Physical Layer

The physical layer, Layer 1, interfaces the network devices with the transmission medium and provides the hardware a means of sending and receiving data on a carrier. This layer is responsible for data coding, that is, converting 0s and 1s into electrical or light pulses, and transmitting the coded pulses onto a guided medium such as copper wires, fiber optic cables, or electromagnetic waves. It is usually defined by electrical standards—electrical or optical impulses that can be interpreted as bits—as well as mechanical standards—connectors, and other physical aspects of the data path. Examples of Layer 1 include UTP cable, fiber-optic cable, and radio waves.

Data Link Layer

Layer 2, the data link layer, implements the transmission protocol and provides error control and synchronization of the bit stream at the node-to-node basis. The data link layer is subdivided into two parts: the **Media Access Control** (MAC), and the **Logical Link Control** (LLC) functions. The MAC sub-layer represents the lower portion of Layer 2 and specifies the access methods used in a network, such as the IEEE 802. The LLC sub-layer is the upper portion of Layer 2, which brings various topologies together in a common format. It also describes error checking and other functions common to all access methods.

Network Layer

The network layer, Layer 3, is responsible for addressing and delivering the packets of information, which, in effect, handles the routing and forwarding of the data. This service falls into one of two conceptual models: *virtual circuit* or *datagram*. In a virtual circuit, a connection from sender to receiver is established only on demand. Once a virtual circuit is set up, the service is comparable to a point-to-point connection where two devices communicate directly. In a datagram service, packets are delivered individually so that they can take a different route and arrive at the destination in no particular order. Each packet has control and identification information added to the data so that when all the packets arrive, they can be reassembled into a complete message. Also, a datagram is not checked at every node along the route; only the header is checked to determine the destination. An example of a datagram service is the Transmission Control Protocol/ Internet Protocol (TCP/IP).

Transport Layer

Layer 4, the transport layer, manages the end-to-end control and ensures error-free data transfer. It addresses, buffers, and multiplexes messages transmitted as packets by the network layer. The transport layer selects the most cost-efficient communication service based on the reliability needed for that transmission.

Session Layer

The session layer, Layer 5, sets up, coordinates, and terminates conversations, exchanges, and dialogs between the applications at each end. It also resolves issues such as the type of connection (half duplex or full duplex) and handles crash recovery by creating new transport layer connection. Layer 5 also authenticates and allows network access to users.

Presentation Layer

Layer 6, the presentation layer, is responsible for handling network security and architecture-independent data formats, including data conversion, encryption, and compression.

Since it is usually part of an operating system that converts incoming and outgoing data from one presentation format to another, it is sometimes referred to as the syntax layer. One example is the virtual terminal protocol, which enables a network to interface with a plethora of terminals from different manufacturers and with different capabilities. Another example is the commonly used file transfer protocol.

Application Layer

The application layer, Layer 7, provides an interface to the end user and identifies quality of service, user authentication, privacy, and other constraints on the data syntax. This layer is not the application itself, although it provides support for user programs. Some applications such as file managers, database managers, e-mail, and other network services perform application layer functions.

CHARACTER CODES

A *byte* is a string of 8 bits, which generally represents a character. Also, a byte is the shortest string of bits that a computer will manipulate as a unit. A **character** is a specific symbol, or a string of 0s and 1s that cannot be subdivided into anything smaller that retains its identity. A *character code* refers to a way of converting alphabets, numbers, punctuation marks, and other special characters into a series of 0s and 1s. One of the earliest character codes is the Morse code, which is composed of dots and dashes. Although there are several different character codes such as Baudot, EBCDIC (Extended Binary Coded Decimal Interchange Code), and Unicode, ASCII (American Standard Code for Information Interchange) is most common.

ASCII

ASCII character code, a general-purpose seven-bit binary data code developed by the ANSI, was originally defined for use by the U.S. government. With seven bits, it is possible to differentiate 2^7 or 128 different patterns. However, it is very convenient and efficient to have the number of bits in a character the same as the number of bits in a byte. Therefore, a single parity bit, which is used for error checking, is added in the highest-order or eighth-bit position. The ASCII, which consists of seven data bits and one parity bit, is the most widely-used character code. The ASCII table is represented in Figure 6–8 Part 1 and Figure 6–8 Part 2. The data bit on the far right represents the smallest place value (2°), while the data bit on the far left represents the highest place value (2^6). Some manufacturers implement an eight-bit ASCII code called Extended ASCII, in which the eighth bit is an additional data bit rather than a parity bit. The Extended ASCII is a superset of the standard ASCII—the original 128 characters are the same, while the additional 128 characters vary from vendor to vendor.

ASCII control characters			
BEL	Bell	EM	End of Medium
CAN	Cancel	ESC	Escape
DC1	Device Control 1	NUL	Null
DC2	Device Control 2	SI	Shift In
DC3	Device Control 3	SO	Shift On
DC4	Device Control 4	SUB	Substitute
Control codes			
ACK	Acknowledge	ETX	End of Text
DLE	Data Link Escape	NAK	Negative Acknowledge
ENQ	Enquiry	SOH	Start of Heading
EOT	End of Transmission	STX	Start of Text
ETB	End of Transmission Block	SYN	Synchronous Idle
Format effectors			
BS	Backspace	HT	Horizontal Tabulation
CR	Carriage Return	LT	Line Feed
FF	Form Feed	VT	Vertical Tabulation
Information separators			
FS	File Separator	RS	Record Separator
GS	Group Separator	US	Unit Separator

Figure 6–8 ASCII character code (part 1 of 2).

DATA ENCODING METHODS

There are innumerable variations in how best to transmit data (0s and 1s) across various media. The physical layer or Layer 1 protocol defines the **data encoding** and decoding methods. *Shannon's Channel Coding Theorem* states that if the transmission rate is equal to or less than the channel capacity, then there exists a coding technique which enables transmission over the channel with an arbitrarily small frequency of errors. There are various ways in which bits are represented by electrical or light pulses.

Bits 7654321	Character	Bits 7654321	Character	Bits 7654321	Character	Bits 7654321	Character
0000000	NUL	0100000	SP	1000000	@	1100000	`
0000001	SOH	0100001	!	1000001	A	1100001	a
0000010	STX	0100010	"	1000010	B	1100010	b
0000011	ETX	0100011	#	1000011	C	1100011	c
0000100	EOT	0100100	$	1000100	D	1100100	d
0000101	ENQ	0100101	%	1000101	E	1100101	e
0000110	ACK	0100110	&	1000110	F	1100110	f
0000111	BEL	0100111	'	1000111	G	1100111	g
0001000	BS	0101000	(	1001000	H	1101000	h
0001001	HT	0101001	)	1001001	I	1101001	i
0001010	LF	0101010	*	1001010	J	1101010	j
0001011	VT	0101011	+	1001011	K	1101011	k
0001100	FF	0101100	,	1001100	L	1101100	l
0001101	CR	0101101	-	1001101	M	1101101	m
0001110	SO	0101110	.	1001110	N	1101110	n
0001111	SI	0101111	/	1001111	O	1101111	o
0010000	DLE	0110000	0	1010000	P	1110000	p
0010001	DC1	0110001	1	1010001	Q	1110001	q
0010010	DC2	0110010	2	1010010	R	1110010	r
0010011	DC3	0110011	3	1010011	S	1110011	s
0010100	DC4	0110100	4	1010100	T	1110100	t
0010101	NAK	0110101	5	1010101	U	1110101	u
0010110	SYN	0110110	6	1010110	V	1110110	v
0010111	ETB	0110111	7	1010111	W	1110111	w
0011000	CAN	0111000	8	1011000	X	1111000	x
0011001	EM	0111001	9	1011001	Y	1111001	y
0011010	SUB	0111010	:	1011010	Z	1111010	z
0011011	ESC	0111011	;	1011011	[	1111011	{
0011100	FS	0111100	<	1011100	\	1111100	|
0011101	GS	0111101	=	1011101	]	1111101	}
0011110	RS	0111110	>	1011110	^	1111110	~
0011111	US	0111111	?	1011111	_	1111111	DEL

Figure 6–8 ASCII character code (part 2 of 2).

Non-Return to Zero (NRZ)

Non-Return to Zero (NRZ), depicted in Figure 6–9 (a), is the simplest representation of digital signals—high for 1 and low for 0. Two of the biggest shortcomings of NRZ are its DC component and its inability to carry synchronization information along with the data. First, if an NRZ signal has a sequence of 1s, the signal cannot pass through such electrical components as transformers and capacitors, which only conduct when the signal is changing. Second, if a series of 1s appears in an NRZ transmission, the receiver will require an additional synchronization signal to determine how many 1s there are. NRZ has applications in RS-232 links and in data storage on hard disk drives. A slight variation of NRZ, the Non-Return to Zero Inverted (NRZ-I) encoding method is used in Gigabit Ethernet to send pulses down to the fiber.

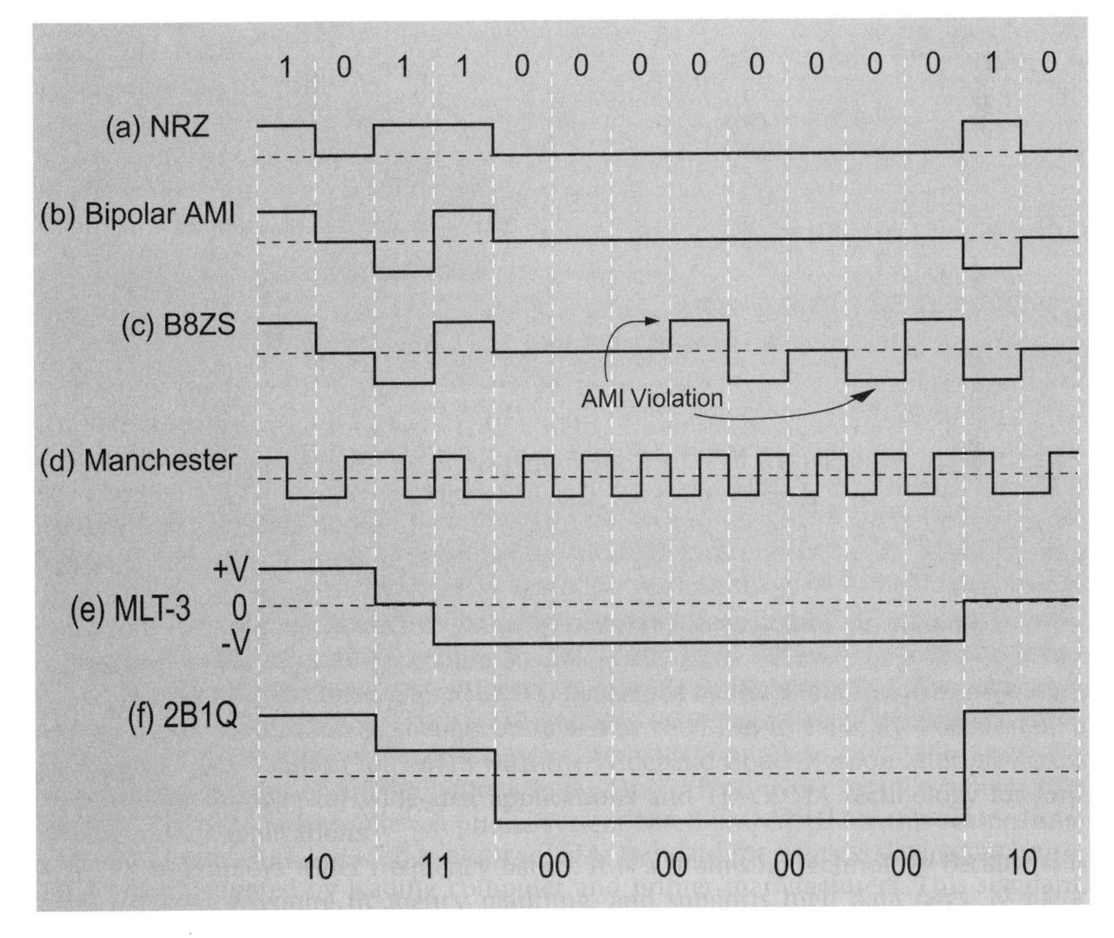

Figure 6–9 Data encoding methods.

Bipolar Alternate Mark Inversions (Bipolar AMI)

In Bipolar Alternate Mark Inversions (AMI), represented in Figure 6–9 (b), marks are analogous to 1s and spaces are analogous to 0s. It solves the DC component problem of NRZ by alternating the polarity of 1s: the first 1 is a positive signal, the second 1 is a negative signal, and the signal value of subsequent 1s alternate; 0s are represented by no signal. However, with a long string of zeros, Bipolar AMI signals can lose self-synchronization.

Bipolar with 8 Zero Substitution (B8ZS)

A variant of Bipolar AMI, Bipolar with 8 Zero Substitution (B8ZS) takes care of the self-synchronization problem by breaking the alternation rule when it comes across a sequence of eight consecutive 0s. As shown in Figure 6–9 (c), it puts 1s in the places of the fourth and fifth 0s and in the places of the seventh and eighth 0s. With the first substitute 1 incorrectly having the polarity of the previous 1, and the third substitute 1 incorrectly having the polarity of the second substitute 1, the receiver recognizes an intentional violation and concludes that there is, in fact, a sequence of eight 0s. This coded violation ensures that there will never be a sequence of more than seven successive no-signal bit times. Non-coded violations indicate spoiled bits. These rules add a degree of physical-layer error detection to the B8ZS encoding method, which is often used in T-1 lines.

Manchester Encoding

Represented in Figure 6–9 (d), Manchester encoding or its variation is implemented in Ethernet and Token Ring LAN technologies. A 1 is indicated by a high to low transition in the middle of a pulse, while a 0 is indicated by a low to high transition in the middle of a pulse. As a result, Manchester encoding has no DC component and is fully self-synchronizing. If there is no transition in a bit time, it indicates a physical-layer error. The drawback of this protocol is that its bandwidth requirement is twice the baud rate. For example, 16 Mbps Token Ring requires a signaling of 32 MHz, and to achieve 10 Mbps on Ethernet, one needs a bandwidth of 20 MHz. For 100 Mbps transmission, the waveform frequency would peak at 200 MHz, but CAT 5 UTP is only rated at 100MHz, so Fast Ethernet would be impossible to implement. Two forms of waveform encoding have been implemented as alternatives to Manchester encoding: Multi-Level Transition (MLT-3) and 4B/5B encoding.

Multi-Level Transition (MLT-3)

MLT-3 represents a coding scheme used to support high-speed LANs that operate at 100 Mbps and beyond. As shown in Figure 6–9 (e), a binary bit is encoded as one of three levels: a positive voltage +V, a negative voltage –V, and a zero voltage. Encoding occurs based upon the following rules: 1) if the next input bit is 0, the next output value is the same as the preceding value, and 2) if the next input bit is a 1, then the next output value results in a transition. The output process examines the preceding output value: if the preceding value was either +V or –V, then the next output value is zero, and if the preceding output was zero, then the next output will be non-zero and have the opposite sign to the last non zero output. In MLT-3, since coding bits are encoded as one of three voltage levels the signaling rate (baud rate) is one-third of the transmission rate. Thus, a signaling rate of 33 MHz will support a LAN transmission rate of 100 Mbps.

4B/5B Encoding

In 4B/5B, every possible 4-bit pattern is assigned a 5-bit code. Instead of sending the actual 4 bits across the wire, a 5-bit code is transmitted. This is referred to as 4B/5B encoding. Since there are 16 possible 4-bit patterns and 32 possible 5-bit patterns, it is possible to pick symbols that ensure that every valid 4-bit representation has at least two transitions; enough transitions ensure proper synchronization. Using the encoding chart in Figure 6–10, the 8-bit sequence 10110110 would actually be transmitted as 10 bits, 1011101110. The main advantage of 4B/5B is that an encoded stream of bits will never contain more than three 0s in a row. This helps with synchronization and leads to higher transfer rates than the basic Token Ring or Ethernet protocols. Besides Fast Ethernet, Fiber Distributed Data Interface (FDDI) also uses 4B/5B combined with MLT-3. Gigabit Ethernet over copper uses 8B/10B encoding, where each 8-bit byte is encoded into a 10-bit stream, combined with NRZ-I.

2 Binary 1 Quaternary

Illustrated in Figure 6–9 (f), 2 Binary 1 Quaternary (2B1Q) has found wide applications in broadband communication technologies such as Integrated Services Digital Network (ISDN), Symmetric Digital Subscriber Line (SDSL), and High-Bit Rate Digital Subscriber Line (HDSL). A line with 2B1Q encoding uses four distinct signaling levels, with data represented in two-bit units. By encoding two bits with each signal transition, 2B1Q represents the distinction between bits per second and baud rate. As a case in point, in ISDN, the signal occupies a bandwidth of 80 kHZ, the baud rate is 80 kbaud, but the raw data rate is 160 kbps.

ERROR DETECTION AND CORRECTION

A key requirement of a data communications network is ensuring the integrity of the information. This is typically accomplished in two steps: Detecting when errors occur at the receiver during transmission, and triggering retransmission or performing an error correction in the event that an error is detected. Data-link-level protocols use one of two approaches to recover lost data. In a Stop and Wait protocol, the sender transmits a packet and waits for acknowledgement (ACK) from the receiver before sending a new packet. If no ACK is received within a specified time, the packet is resent. This method is

Four Data Bits				Five-bit Code				
A3	A2	A1	A0	Q4	Q3	Q2	Q1	Q0
0	0	0	0	1	1	1	1	0
0	0	0	1	0	1	0	0	1
0	0	1	0	1	0	1	0	0
0	0	1	1	1	0	1	0	1
0	1	0	0	0	1	0	1	0
0	1	0	1	0	1	0	1	1
0	1	1	0	0	1	1	1	0
0	1	1	1	0	1	1	1	1
1	0	0	0	1	0	0	1	0
1	0	0	1	1	0	0	1	1
1	0	1	0	1	0	1	1	0
1	0	1	1	1	0	1	1	1
1	1	0	0	1	1	0	1	0
1	1	0	1	1	1	0	1	1
1	1	1	0	1	1	1	0	0
1	1	1	1	1	1	1	0	1

Figure 6–10 4B/5B encoding.

relatively inefficient when compared to the Sliding Window protocol where the sender transmits multiple frames before receiving ACK. This keeps the pipe full; upper limit on the number of frames outstanding (unacknowledged) is determined by the channel bandwidth and other transmission parameters.

There are two types of errors: single-bit errors and burst errors. As opposed to a single-bit error, a burst error affects several bits in a single character or transmission frame. Upon detecting a burst error, the receiver requests a retransmission. There are different error-checking mechanisms such as echo checking, parity checking, longitudinal redundancy check, checksums, hamming code, and cyclic redundancy check. Some commonly used methods are discussed below.

Parity Checking

Parity checking is one of the simplest error-detection schemes. It is appropriate for use in asynchronous transmission because it checks one character at a time. An example that represents the transmission of an ASCII character using parity checking is depicted in Figure 6–11. The two popular types are *odd parity* and *even parity.* In the odd parity scheme, the parity bit is such that the total number of 1s in a single character is odd; in the even parity scheme, the parity bit is such that the total number of 1s in a single character is even. Either of these two schemes may be used as long as both transmitter and receiver use the same one. If this cannot be confirmed, the parity function may be turned off, which in effect ignores the parity bit, allowing a device implementing parity checking to communicate with a device incapable of that function.

ASCII Character	Even Parity	Odd Parity
A	[0]1000001	[1]1000001
D	[0]1000100	[1]1000100
	↕ even parity bit	↕ odd parity bit
8	[1]0111000	[0]0111000
DEL	[1]1111111	[0]1111111

Figure 6–11 Parity checking (parity bits are shown in boxes).

In parity checking, a parity generator at the sending station inserts a parity bit preceding each character's first stop bit. A parity generator circuit is made up of several levels of Exclusive-OR (XOR) circuits, as shown in Figure 6–12. At the receiving station, the total number of binary 1s in the data character and the parity bit are counted and compared with the selected parity scheme. If they match, there is no error, while a mismatch indicates a transmission error. Parity checking has two major limitations. First, it is only an error-detection scheme and performs no error correction so that if an error is detected, the character has to be retransmitted. Second, it can detect only an odd number of bit errors in the transmission of a byte, because an even-bit change in a single character causes the parity condition to be the same as if an error did not occur.

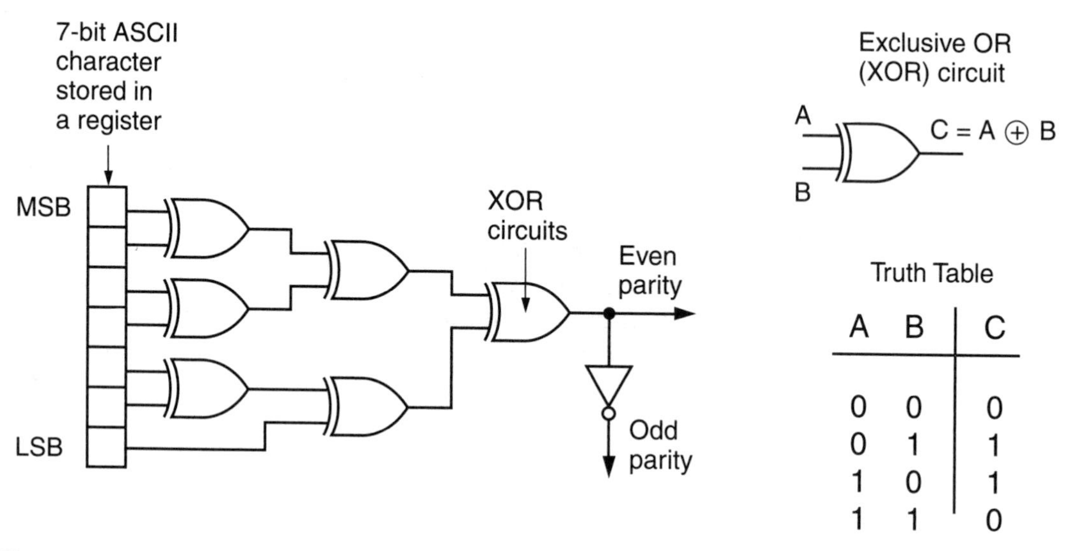

Figure 6–12 A parity generator circuit.

Longitudinal Redundancy Check (LRC)

Following directly from parity checking, Longitudinal Redundancy Check (LRC) operates on a group of bytes. It is used to create a cross-grid matrix pattern to pinpoint a bad bit. In addition to generating a parity bit for each character, LRC produces a *Block Check Character (BCC)* to provide extra error-detection capabilities for a block of data. It transmits one BCC for eight characters, where each character comprises seven data bits and one parity bit. The BCC is computed as the data is transmitted, and it is then appended to the message. At the receiving end, the computer computes its own version of the BCC on the received data and compares it to the received BCC. The two should be the same for error-free transmission.

Let us discuss the example shown in Figure 6–13, where data is transmitted in eight-character blocks. Using the odd-parity scheme, a parity bit is added for each seven-bit ASCII character. In addition, a horizontal parity check is conducted on each row to produce the BCC. This process is referred to as longitudinal redundancy check. The individual character parity bits and the BCC bits provide a form of coordinate system that allows a particular bit error in one character to be identified. Once it is identified, the bit is simply complemented to correct it. The advantages of LRC are that it is simple and improves the odds of detecting errors, but its significant overhead is a major disadvantage.

Hamming Code

Hamming code is an error-detection-and-correction scheme designed for single bit errors. If more than a one-bit error occurs, this method will not be able to detect the errors accurately. Hamming code requires generation of several parity bits that are then interspersed

Character	D	A	T	A		C	O	M	BCC
(LSB)	0	1	0	1	0	1	1	1	0
	0	0	0	0	0	1	1	0	1
ASCII	1	0	1	0	0	0	1	1	1
	0	0	0	0	0	0	1	1	1
Code	0	0	1	0	0	0	0	0	0
	0	0	0	0	1	0	0	0	0
(MSB)	1	1	1	1	0	1	1	1	0
Parity (odd)	1	1	0	1	0	0	0	1	1

LSB: Least Significant Bit
MSB: Most Significant Bit
BCC: Block Check Character

Figure 6–13 Longitudinal redundancy check.

with data in a specific pattern. Since one data bit will affect more than one parity bit, the software can calculate which bit has flipped its value and will change it back to its correct value. When an error is corrected by the receiving device, it is referred to as *forward error correction*. The error-correction capability is an important feature of hamming code, as it eliminates the need for retransmission in case of single-bit errors.

Cyclic Redundancy Check (CRC)

Adopted by the ITU (formerly CCITT), **Cyclic Redundancy Check (CRC)** is one of the most widely-used, reliable, and efficient error-detection schemes. The CRC method detects multiple errors within any length of message, and requires far less transmission overhead as compared with parity checking. Therefore, it is used in synchronous transmission on blocks of data that can be several thousand bytes and where single bit errors occur less frequently as compared with multiple or burst errors. CRC is used in Ethernet, Token Ring, ISDN, FDDI, and other network technologies.

CRC uses a unique mathematical algorithm called a polynomial (P), which is known to both the transmitter and the receiver. This scheme generates an additional bit pattern called a Frame Check Sequence (FCS), which is appended to the data block as either a 16-bit or 32-bit value. As depicted in Figure 6–14, the process begins with adding the same number of binary zeros as the bits in the desired FCS sequence, and dividing the resultant binary number by the P. The remainder becomes the FCS, which overlays the zeros added in the first step. At the receiver, the original data plus the FCS are concatenated together and then divided by the P. If there is no remainder, the CRC algorithm

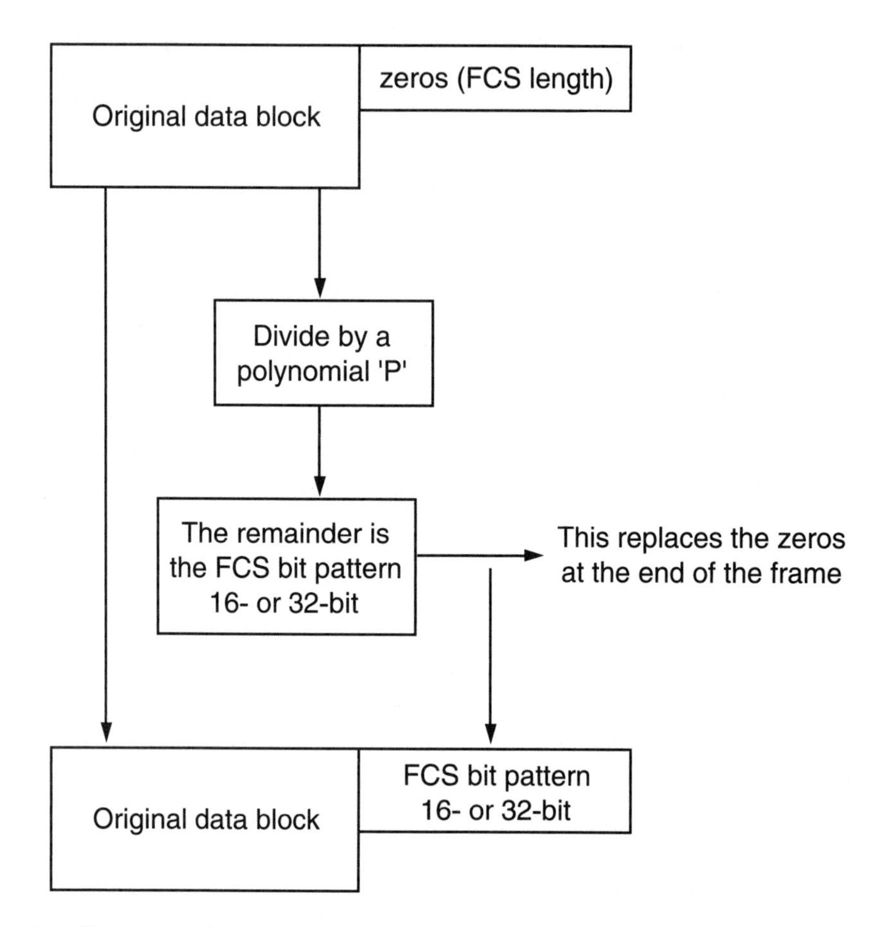

Figure 6–14 CRC process flowchart.

assumes that there are no errors. Standardization ensures that a CRC-32 (an FCS of four bytes) protocol on one system is able to communicate seamlessly with a CRC-32 protocol on another system.

If there are no errors in the block or frame, it is acknowledged by the receiving device through the transmission of an ACK, which cues the sending device that the block of data can be erased from buffer memory and the next data block can be sent. But a data block in error is negatively acknowledged with a NAK, which prompts the transmitting device to retransmit that specific block, which has been stored in buffer memory. The result is an integrity factor of 10^{-14}; in other words, the possibility of an undetected error is 1 in 100 trillion. While CRC is memory- and processor-intensive and therefore relatively expensive to implement, it is frequently used in most computer communications because it ensures virtually error-free data transmission.

DATA LINK PROTOCOLS

The data link layer deals with how data is logically packaged to cross from one user to another. Bit patterns acquire meaning at this layer. In asynchronous transmission (sometimes called start-stop transmission) each character is sent independently, and the start bit is used by the receiver for synchronization. In synchronous transmissions, data link protocols are divided into two broad categories: *byte-oriented protocols*, which use special characters to signal the beginning and end of the message and *bit-oriented protocols*, which use a single special character (flag character) to mark both the beginning and end of the message, and the header and following message are always the same defined length.

The most widely-used byte-oriented protocol is IBM's Binary Synchronous Communications (BISYNC or BSC). The two of the most widely-used bit-oriented protocols are High-Level Data Link Control (HDLC) and IBM SNA's Synchronous Data Link Control (SDLC) which may be considered a subset of HDLC. HDLC is specified as the standard protocol for Layer 2 in the OSI model. It allows every device to both send and receive information without being controlled by any other device, so that all devices have equal right to use the communication facility. Ethernet, Token Ring, ISDN, and FDDI are some of the popular network technologies that are based on the HDLC frame structure.

High-Level Data Link Control (HDLC)

The HDLC frame format, shown in Figure 6–15, consists of raw data or information and an overhead that envelopes the raw data. An opening flag is a sequence of eight bits, which marks the beginning of a packet, followed by 16- or 32-bit source and destination station addresses. One or two bytes of control information describe the type of HDLC frame, routing parameters, and other packet identifiers. The variable data field or raw data is now inserted, followed by a 16- or 32-bit FCS and a closing flag, which marks the end of the HDLC frame. The HDLC frames are created by equipment that transmits the packets across the network.

	Opening Flag	Station Address	Control Information	Data Field	FCS	Closing Flag
Bytes	1	2 or 4	1 or 2	Variable	2 or 4	1

Figure 6–15 HDLC frame format.

LAN ACCESS METHODS

Soon after the ISO developed the OSI reference model in 1978, the IEEE formed the 802 Committee in 1980. It was charged with developing LAN standards for the physical and data link connections between devices. A LAN access method describes how the devices access the network and share the transmission facilities. By definition, a shared-media LAN has only one path to handle high-speed data. However, the total capacity of the path exceeds the transmission speed of any single station, so stations are usually unaware that they are sharing the medium. *Polling* refers to the process of a host computer asking an intelligent terminal if it has any data to send to the host computer. This task is typically accomplished by a *front-end processor (FEP),* which handles all the routine communications procedures for the host computer. *Selecting* occurs when a host computer or a FEP sends data to a terminal after the terminal indicates that it is ready to accept data. The *polling and selecting* access method is not very common because it requires the use of a central controller to execute and monitor the process.

Popular LAN access methods are classified as *noncontention* or *contention.* The IEEE 802 committee adopted three incompatible LAN standards: 802.3, a bus or star contention network for light-duty applications, 802.4 (bus) and 802.5 (ring) token passing noncontention networks for use in situations where assurance of network access is needed, and 802.11 for wireless networks. Ethernet is based on 802.3, while Token Ring and FDDI are based on 802.5. Choosing the most appropriate access method for an organization requires an accurate and comprehensive look at present and projected data traffic patterns.

Token Passing

The noncontention-based **token passing** is a deterministic system; network access is orderly and response time is fairly predictable even under conditions of high load. In token passing, depicted in Figure 6–16, each station receives and then regenerates the token. When a station wishes to add a data packet to the token, it transforms the token to a different bit pattern called a connector, which alerts all stations to the fact that a data packet is following. When the addressed station receives the data, it indicates that in the packet's control field. When the source receives the connector and data packet, and sees that the data has been received, it does not regenerate the data packet and changes the connector to a token. The main disadvantage of token passing is the delay while a token circulates. In a token-passing network, theoretically, if any station or node goes down, the entire network collapses. But this problem is eliminated by bypass software that automatically reroutes traffic past a dead station. Token passing is implemented primarily on bus topology in Token Bus, and ring topology in Token Ring and FDDI networks.

Carrier Sense Multiple Access (CSMA)

The contention-based **Carrier Sense Multiple Access (CSMA)** defined in IEEE 802.3 was designed for light industrial use as the response time can be unpredictable, especially under conditions of heavy load. It is a *multiple access* method, meaning any of the network devices can transmit data onto the network at will; there is no central controller. It is also a *broadcast* network where all stations see all frames, regardless of whether they represent an intended destination. Each station must examine received frames to deter-

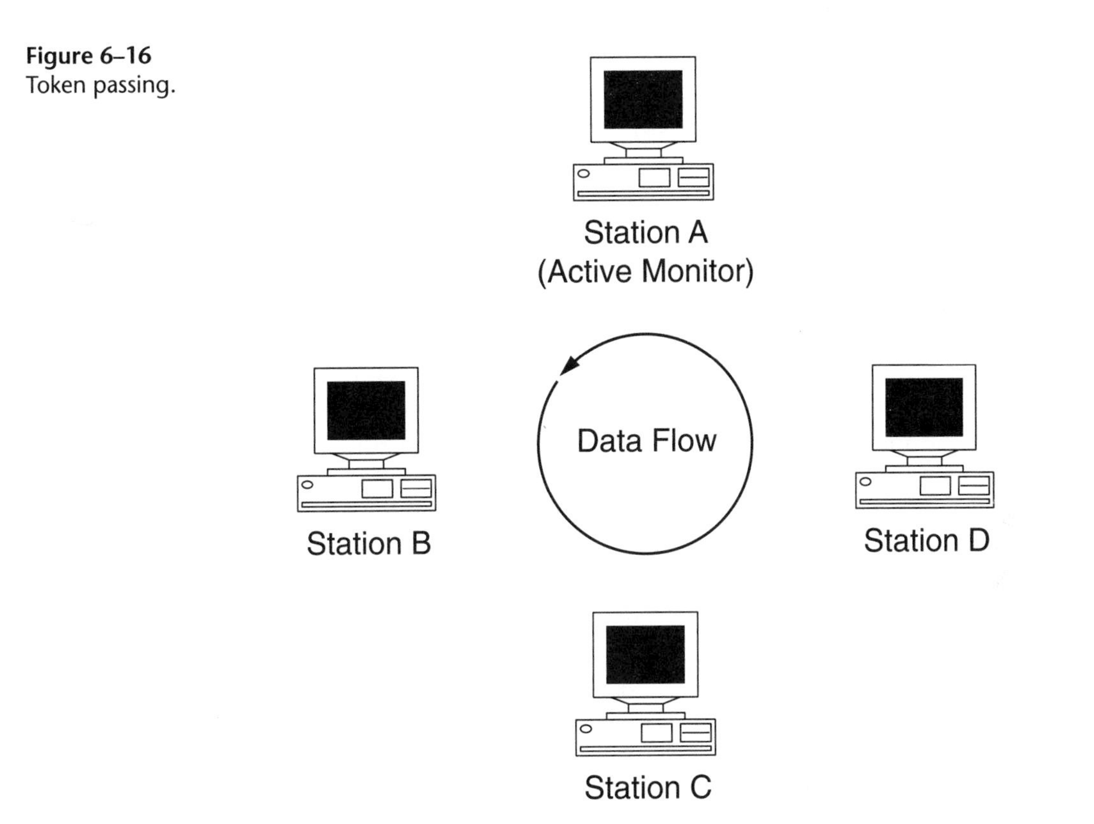

Figure 6–16
Token passing.

mine if the station is a destination. If so, the frame is passed to a higher protocol layer for appropriate processing.

A system employing CSMA is in one of three states: transmission, contention, or idle. Network efficiency is the result of a combination of *carrier sensing,* same as the contention state, which prevents transmissions that otherwise would result in collisions, and *collision detection* or *collision avoidance,* which initiates the retransmission of transmissions that have collided. A small amount of idle time is required before and after each transmission.

Carrier Sensing

Carrier sensing reduces the probability of collisions. An interface cannot transmit when another one is transmitting. Since signal propagation takes time, some time elapses between the beginning of the transmission and the time a collision might occur from an interface that had not yet begun receiving the transmitted signal. Both interfaces believe they were the first to begin transmitting, and they are both correct in their own frame of reference. This period of uncertainty is known as the *contention time.* In order to avoid collisions, it is set at twice the propagation time between the most distant pair of interfaces.

CSMA with Collision Detection (CSMA/CD)

CSMA with Collision Detection (CSMA/CD) is the dominant standard since it is specified for IEEE 802.3 Ethernet LANs. In CSMA/CD, a station wishing to transmit listens to the network to see if it is already in use. If in use, the station waits, otherwise the station transmits. A collision occurs when two stations listen for network traffic, hear none, and transmit simultaneously. In this case, both transmissions are damaged, and the stations must retransmit at some later time. The process is represented in Figure 6–17 (a), (b), and (c). CSMA/CD stations can detect collisions, so they know when they must retransmit.

Collision detection, upon detecting a collision, immediately stops the transmission of a frame. Consider a system that broadcasts frames of the same size. The amount of time to transmit a frame is the ratio of the frame length and the bit rate. A collision will result with a given frame if any other frame is broadcast within one frame time of the start of the given frame. This window of vulnerability is two frame times in duration, since the collision may be either with a frame that started before the given frame or after it. To recover from the collision, the network interface waits a short while and then retransmits. If a collision occurs on the second attempt, it waits again before retransmitting. This process is called *backoff*. The wait times and increases in wait times are random because collisions will continue to occur if wait times are equal for all the frames that collide. The retransmission is attempted about six times, and then the data to be transmitted is simply discarded. Hence, a busy CSMA/CD network, such as Ethernet, will drop packets.

CSMA with Collision Avoidance (CSMA/CA)

In collision avoidance, the sender sends a request to send (RTS) frame to receiver and indicates the time needed to complete data transmission. The receiver sends clear to send (CTS) frame, indicates time to complete data transmission and reserves channel for the sender. The sender transmits the data, and the receiver responds with an ACK frame. The RTS and CTS let other stations know of the data transmission so that collision is avoided; sender backs off if CTS is missing. ACK facilitates reliable transmission at data link level. CSMA/CA is used in IEEE 802.11 wireless LANs.

LAN TECHNOLOGIES

The ability to link a wide range of computers using a vendor-neutral network technology is an essential feature for today's LANs. Most LANs must support a wide variety of computers purchased from different vendors, which requires a high degree of network interoperability. Popular baseband LAN technologies include Token Ring and Ethernet.

Token Ring

Token Ring, defined in IEEE 802.5, is a 4 or 16 Mbps ring-architectured baseband system, which uses a token-passing access method with predictable response time even under

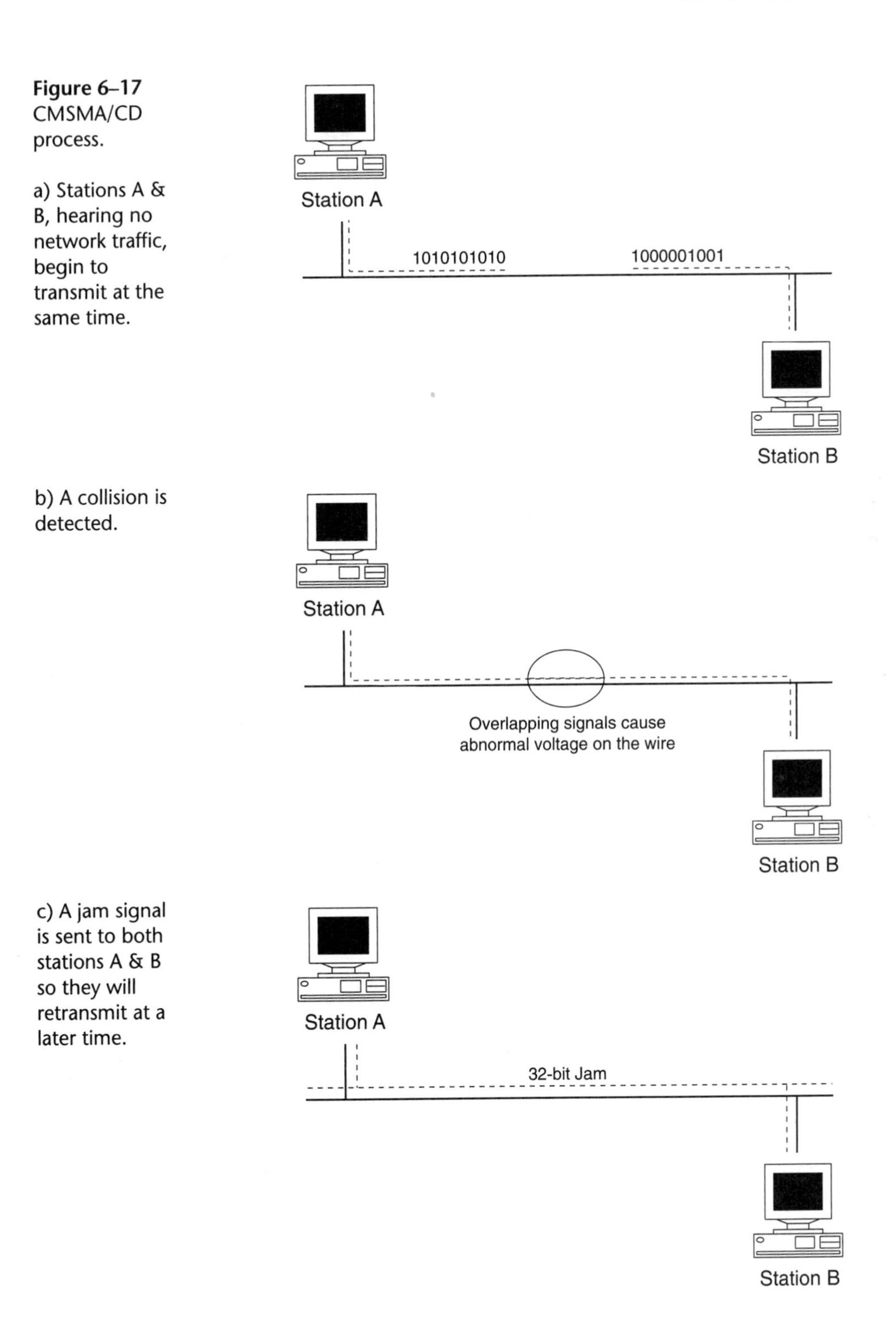

Figure 6–17 CMSMA/CD process.

a) Stations A & B, hearing no network traffic, begin to transmit at the same time.

b) A collision is detected.

c) A jam signal is sent to both stations A & B so they will retransmit at a later time.

heavy load. The sequence of nodes is governed by the physical order in which the nodes appear on the ring. Token Ring runs on different wiring types—UTP is supported for 4 and 16 Mbps but STP is required for the backbone. Although Token Ring networks are wired as a ring, the physical wiring is configured as a star. Two pairs of wires from each station terminate in a *Multistation Access Unit (MAU)* located in the center of the network, as shown in Figure 6–18. The MAU routes the token from port to port. Token Ring is a more robust network compared with Ethernet, but it is also more expensive—the Token Ring NICs are more costly and the MAUs are more expensive than Ethernet hubs.

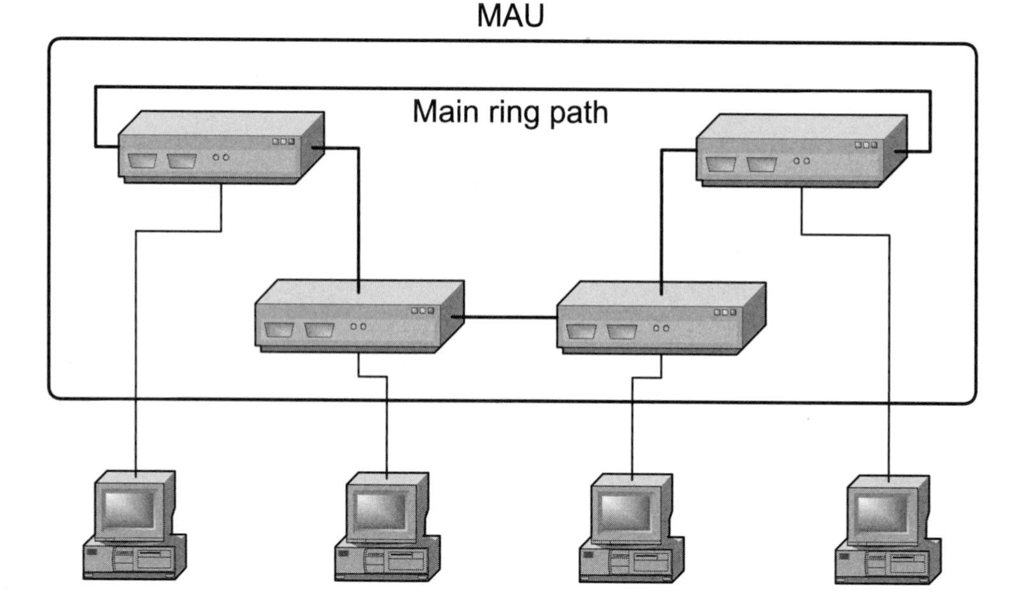

Figure 6–18 Token Ring MAU (Multistation Access Unit).

Fiber Distributed Data Interface (FDDI)

Fiber Distributed Data Interface (FDDI) covers Layers 1 and 2 of the OSI Reference Model, as shown in Figure 6–19. In FDDI, stations are connected in a dual ring implemented with optical fibers using the token-passing concept but operating at a higher speed. The dual-ring architecture provides a 100 percent redundancy. The transmission rate is 100 Mbps, the total end-to-end diameter of the network cannot exceed 200 km, and FDDI specifications permit a maximum of 500 nodes. There are two major differences between Token Ring and FDDI. First, in contrast to Token Ring, multiple frames can circulate simultaneously on an FDDI network. When a station completes sending a frame, it reinserts the token on the network. A second station on the ring can seize the token and send a frame while the first frame is still making the trip back to the originating station.

Second, the active monitor provides a master clock in Token Ring, whereas in FDDI each station provides its own clocking.

At the physical layer, the FDDI is broken into two separate sublayers. The first, Physical Media Dependent (PMD), which is at the bottom, is responsible for specifying the physical cables and connectors used; the second, Physical Layer Protocol (PHY), is media-independent, as it defines the data coding methods. There are four different kinds of PMD sublayers, according to the kind of transmission medium being used:

- PMD defines the use of multimode fibers
- SMF-PMD defines the use of Single Mode Fibers (SMF)
- LCF-PMD defines the use of multimode Low Cost Fibers
- TP-PMD defines the use of Twisted Pair cabling—CAT 5E UTP or Type I STP cable

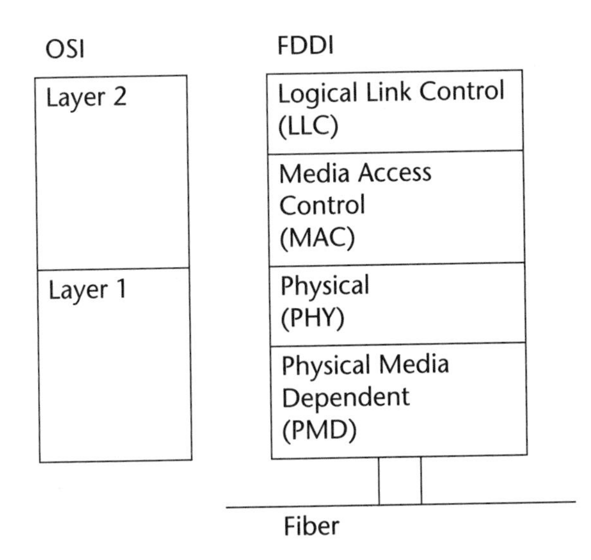

Figure 6–19 FDDI stacks up against Layers 1 and 2 of the OSI model.

The LCF-PMD is less expensive to implement because of cheaper connectors that lower the cost of horizontal distribution of fiber optic cabling. The primary application of FDDI is in a MAN or a LAN backbone, where this high-speed backbone is used to connect a group of lower-speed LANs such as Token Rings and Ethernets. An example is illustrated in Figure 6–20.

Ethernet

Ethernet was invented in the 1970s by Dr. Robert M. Metcalfe at the Xerox Palo Alto Research Center. The first Ethernet system ran at approximately 3 Mbps and was known as experimental Ethernet. Formal specifications for Ethernet were published in 1980 by a

Figure 6–20
FDDI implemented in a MAN.

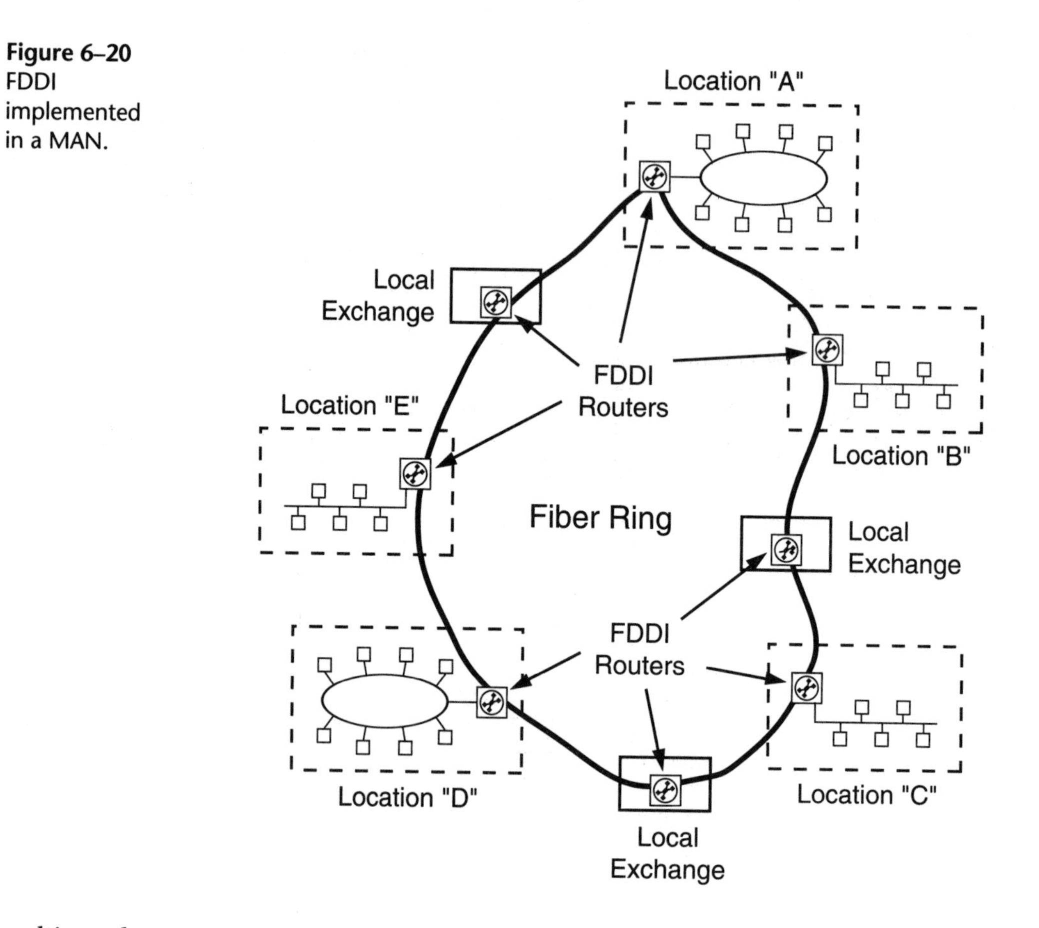

multi-vendor consortium that created the DEC-Intel-Xerox (DIX) standard. It marked a turning point in the evolution of experimental Ethernet into an open, production-quality Ethernet system that operates at 10 Mbps. Ethernet technology, with some variation, was then adopted as IEEE 802.3 in 1985 by the IEEE LAN standards committee.

As both Ethernet and IEEE 802.3 implement CSMA/CD, differences between the two are subtle. The frame structure is somewhat different, and the transmission media they support are not identical. Ethernet provides services corresponding to Layers 1 and 2 of the OSI reference model, while IEEE 802.3 specifies the physical layer (Layer 1) and the media access control portion of the data link layer (Layer 2), but does not define a logical link control protocol. Frequently, the terms Ethernet and IEEE 802.3 are used interchangeably despite the differences.

An Ethernet frame varies between 64 to 1518 bytes, depending upon the size of the data field. It is illustrated in Figure 6–21 and is described as follows:

- ✦ Preamble: The alternating pattern of 1s and 0s tells receiving stations that a frame is coming. An additional byte serves to synchronize the frame-reception portions of all stations on the LAN.

- Destination and Source Addresses: Each address comprises six bytes: the first three bytes called Organizationally Unique Identifiers (OUIs) are specified by the IEEE on a vendor-dependent basis, and the last three bytes are specified by the Ethernet vendor that wishes to build Ethernet interfaces. This six-byte address is also known as the physical address, hardware address, or MAC address. A pre-assigned node address vastly simplifies the setup and operation of the network. The source address is always a unicast (single-node) address, while the destination address can be unicast, multicast (group), or broadcast (all nodes).
- Length: The length indicates the number of bytes of data that follows this field.
- Data: This is the frame's *payload* or actual data, which is the reason the frame is being transmitted. The data field can vary from 46 to 1500 bytes. After the physical-layer and link-layer processing is complete, the data contained in the frame is sent to an upper-layer protocol. Ethernet expects at least 46 bytes of data. If data in the frame is insufficient to fill the frame to its minimum size, padding bytes are inserted to ensure at least a 46-byte data frame.
- Frame Check Sequence (FCS): This sequence contains a four-byte CRC value, which is created by the sending device and is recalculated by the receiving device to check for damaged frames.

Ethernet is a star-architectured baseband system in which each physical link is associated with a unique MAC-addressed port. When frames or packets are sent into the network, the first step is to pick one of the MAC-addressed ports within the larger logical trunk port over which to send the data. This selection is done by the MAC selection algorithm, which plays another vital role of making sure that frames being transmitted into the network arrive in order at the destination network device. To do this, the algorithm creates a set of associated frames known as a session. The common characteristic of a session is that all frames in it have the same source and destination addresses. By controlling the session, the MAC selection algorithm makes sure all frames going between a particular source and destination travel on the same physical link within the trunk group, thereby arriving at the destination in correct order.

	Preamble	**Source Address**	**Destination Address**	**Length**	**Data (payload)**	**FCS**
Bytes		6	6	2	Variable (46 to 1500)	4

Figure 6–21 An Ethernet frame format.

10BaseT

10BaseT, where the T stands for twisted-pair wire, has been one of the most popular Ethernet standards in a LAN environment and is illustrated in Figure 6–22. It supports a 10 Mbps transmission rate and uses a star topology, which is implemented through a **hub**, a wiring concentrator. Since all devices are connected centrally to the hub, the hub

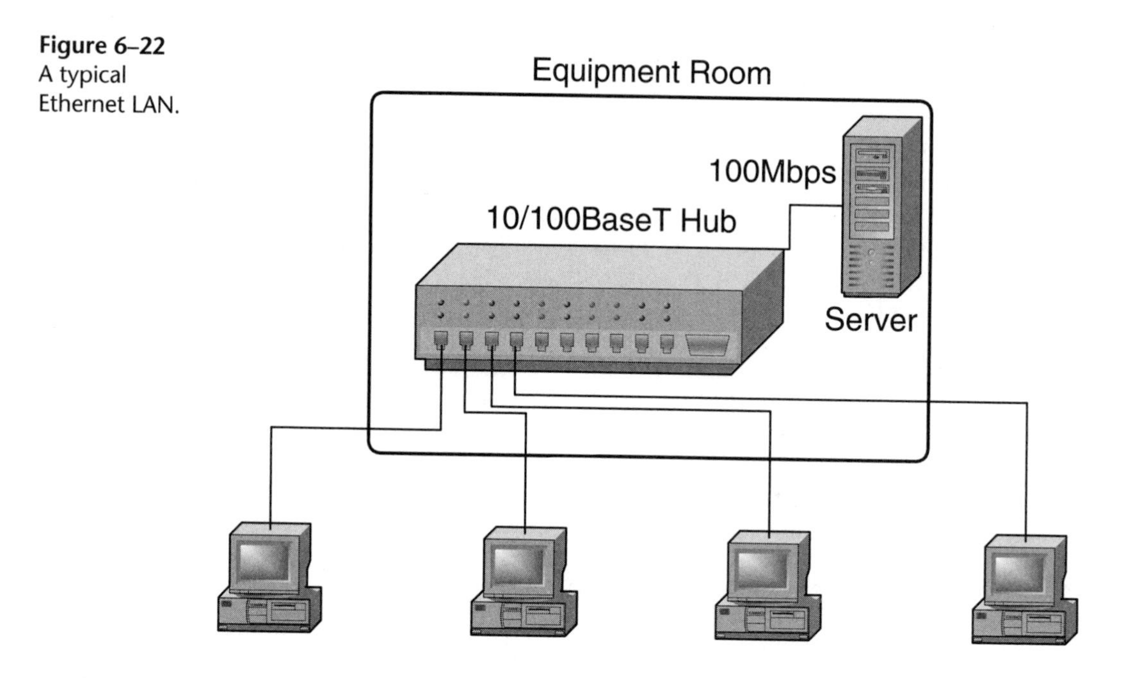

Figure 6–22
A typical Ethernet LAN.

becomes the focal point of additions, moves, or changes to the network. Port-switching hubs allow network managers to move users from one segment to another without reconfiguring the hub. Before a new device is inserted into the loop, at minimum, the hub verifies valid signal quality. A device with poor signal quality or inappropriate clock speed is left in bypass mode, which allows other devices to continue without disruption. These features enable a much more dynamic environment because devices can be hot-inserted or removed, and problems can be more readily identified. Usually, each port on a hub has a corresponding LED, which gives at-a-glance status information of insertion, bypass, and bad-link states. For example, a blinking LED may represent traffic, but if it is off, it may indicate no link or a bad link.

In a busy Ethernet, several interfaces may attempt to transmit when they sense an idle network, resulting in a collision. In 10BaseT Ethernet, stations up to two kilometers apart could sense when a collision occurs. The distance limitation results from the relationship between the time required to transmit a minimum-sized Ethernet frame (64 bytes) and the ability to detect a collision. When a collision occurs, the MAC layer detects it and sends a jam signal, which tells the transmitting stations to back off and retry.

Ethernet is described as a light-duty network because as the number of hosts increase, the share each gets of the total available bandwidth decreases, and efficiency begins to drop. With a sufficiently small number of hosts contending for the total network bandwidth, one can hope to see about 8 Mbps transmitted on a 10 Mbps network. In the extreme case, an Ethernet can achieve a condition called collapse in which it does not leave the contention state for any significant amount of time. In theory, it can be triggered by a bad combination of timing with just a handful of interfaces all initiating a transmission as soon as the network appears idle, but the probability is astronomically small that such an event would occur. Typically, collapse is triggered by a failure in the circuitry of an interface when it sends a continuous stream of bad frames often without stopping to do carrier sensing or collision detection. An inability to detect collisions leads to network instability.

100BaseT

100BaseT, defined by IEEE 802.3u in 1994, is commonly referred to as Fast Ethernet. It uses the CSMA/CD specification and retains the Ethernet frame format, size, and error-detection mechanism. In addition, it supports all applications and networking software running on 10BaseT networks. The upgrade comes with a minimal amount of investment, training, and reconfiguration of existing hardware and applications. A typical 10BaseT and 100BaseT LAN configuration is depicted in Figure 6–23. 100BaseT hubs must detect dual speeds and respond accordingly much like Token Ring 4 or 16 Mbps MAUs, and NICs can support 10 Mbps, 100 Mbps, or both. The main difference between 100BaseT and 10BaseT, other than the obvious speed differential, is the network diameter. Increasing the clock rate tenfold means that the time needed to transmit a frame is reduced by a factor of 10. That, in turn, directly affects network diameter, shrinking it from 2 kilometers for 10BaseT to 200 meters for 100BaseT.

1000BaseT

1000BaseT, also known as Gigabit Ethernet, is defined by the IEEE 802.3z, which was published in 1997. It is an extension of the Ethernet standard and uses the same frame format, frame size, and management objects used in earlier Ethernet networks. Gigabit Ethernet offers 1000 Mbps of raw-data bandwidth while maintaining backward compatibility with Fast Ethernet and Ethernet network devices. It provides for full-duplex operating modes for switch-to-switch and switch-to-end-station connections, as well as half-duplex operating modes for shared connections. Since Gigabit Ethernet represents another tenfold increase in clock speed over Fast Ethernet, it should require another tenfold reduction in network diameter. But a 20-meter network is clearly impractical, so the 802.3z committee redefined the MAC layer for Gigabit Ethernet.

In general, Gigabit Ethernet can operate over fiber-optic cabling, Category 5 UTP, and coaxial cabling as well. The IEEE standard supports a single-mode fiber-optic link with a maximum length of 5 km; a multimode fiber-optic link with a maximum length of 500 meters, and a coaxial copper-based link with a maximum length of 25 meters, such as within a wiring closet. To accommodate CAT 5 UTP cable at distances up to 100 meters, IEEE 802.3ab specifies a logical media-independent interface between the MAC layer and Layer 1, but the silicon technology and digital-signal processing involved in chip design is extremely complicated. Gigabit Ethernet over copper sends and receives simultaneously on all four pairs, while 10BaseT and 100BaseT only use two of the four

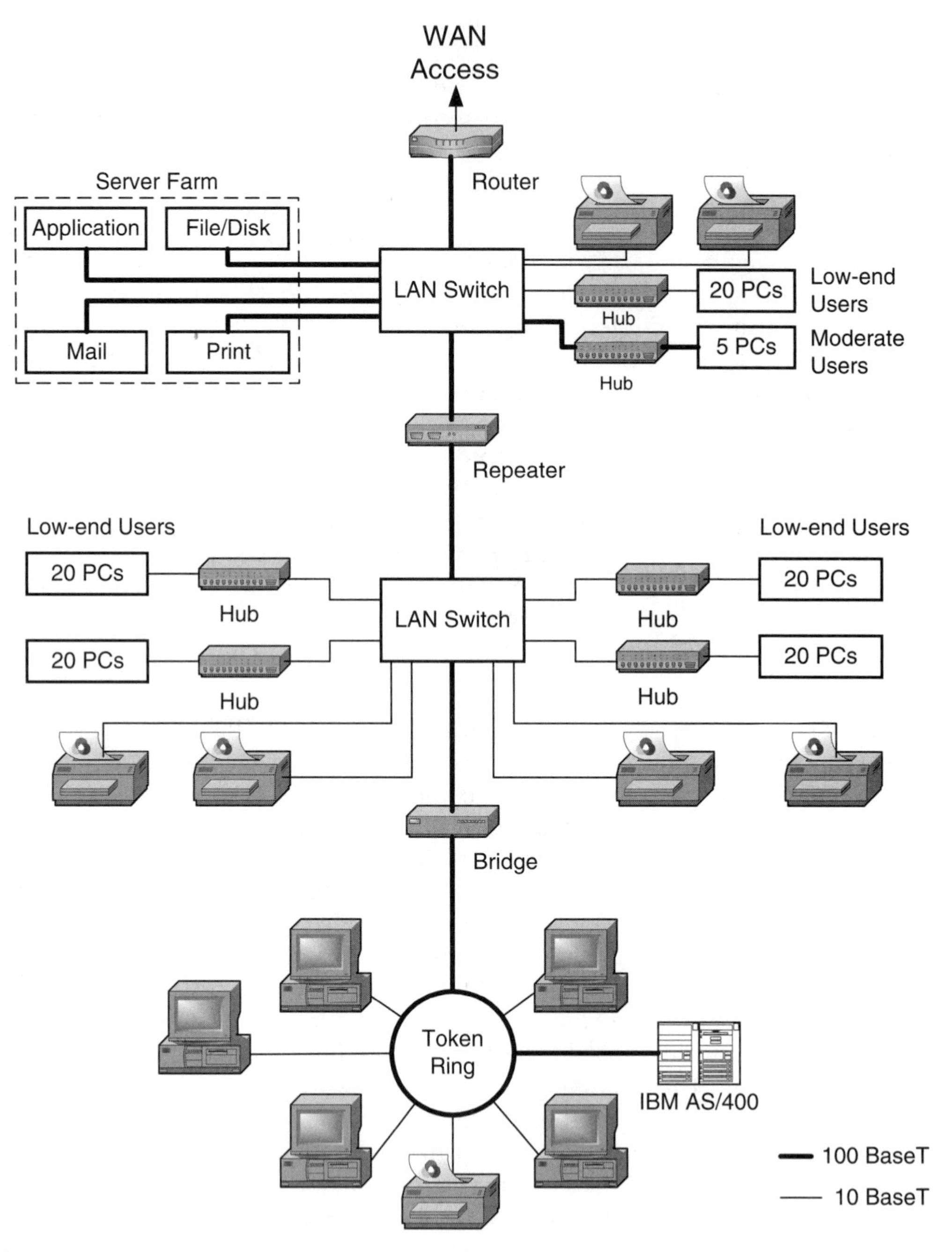

Figure 6–23 Typical 10BaseT and 100BaseT LAN configuration.

pairs. Thus, in Gigabit Ethernet, it is critical to comply with Category 5 requirements across all four pairs, which requires highly accurate test equipment.

With Gigabit Ethernet, network managers can accommodate increasingly complex applications with bigger appetites for bandwidth. These include data-intensive applications in science, engineering, and medicine, as well as multimedia Web traffic. The overall result is a gigabit-scaled campus network with a balanced network design that breaks down the conventional limits of Ethernet. Migration to Gigabit Ethernet will occur gradually, and initial implementation has been in the backbone of existing Ethernet LANs—switch-to-switch links, and switch-to-server links. Eventually, upgrades will reach server connections and the desktop as well. A Gigabit Ethernet network can support a greater number of segments, more bandwidth per segment, and hence a greater number of nodes per segment.

LAN CONFIGURATIONS

By the late 1980s the personal computer revolution was in full force, and desktops were capable of running applications themselves. But users wanted more than stand-alone operation; they wanted to share resources—expensive peripherals and files—with other users. New technologies like Ethernet and Token Ring were becoming available, and LANs began to appear. There are two generic approaches to configuring LANs: *Client/Server* and *Peer-to-Peer.*

Client/Server

Client/server computing, today's most popular networking strategy, was developed in the early 1990s. The model is depicted in Figure 6–24. A special dedicated machine, a server, is used to manage and provide special services. Servers are computers on the network that are accessible to network users. The client part is any other network device or process that makes requests to use server resources and services. There are essentially five basic components of a LAN:

- Network devices include servers, workstations and shared devices such as printers
- Network Interface Cards (NICs) required by each network device to access the network
- Cable or electromagnetic waves as a physical transmission medium
- Network Operating System (NOS) to control the network
- Network communication devices such as hubs, switches and routers

Servers contain resources that they serve to users who request the services. The most common types of servers are a file server and a print server. File servers are often set up so that each user on the network has access to his or her personal directory, along with a range of public or shared directories where applications and data are stored. The client

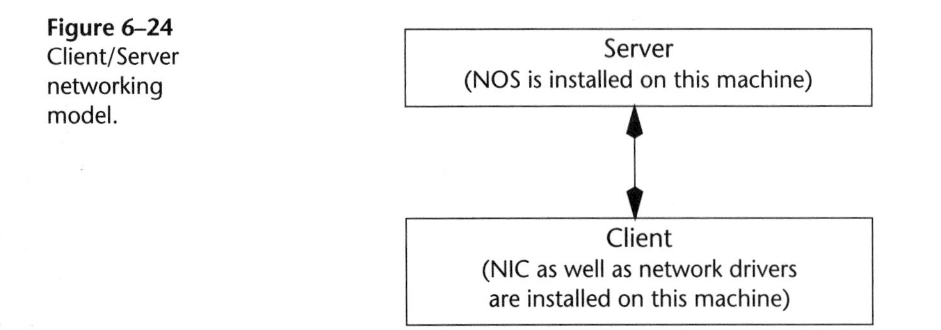

Figure 6–24 Client/Server networking model.

does the processing on most LANs, but on some, the file server and the client can share the processing duties. A print server stores the print jobs that are sent from clients to network printers in a queue before they are printed.

A NIC, which is a piece of hardware and part of Layer 1, is a computer circuit board or card that is installed in a computer to provide a dedicated, full-time connection to a network. Personal computers and workstations on LANs contain a NIC specifically designed for the LAN transmission technology, such as Ethernet or Token Ring, as illustrated in Figure 6–25. The address of each node (any device on the LAN) is provided by the NIC installed in the device. The speed of transmission will depend on the type of medium, the capabilities of the NIC, the client and the server. Typical speed ranges for LANs are from 10 to 1000 Mbps, and a few are even higher.

In a client/server network, the NOS software resides on the server and operates between the user application and the data link layer of the 802 standards. The open architecture enables any NOS to operate with any 802 standards by changing the NIC and network driver software. The NOS monitors, and at times, controls, the exchange and flow of files, email, and other network information. The NOS provides centralized control and administration of the entire network; it enables the sharing of printers, files, databases, and applications and provides directory services, security, and housekeeping aspects to manage a network. It enables differential access to files on network disks; that is, it controls which files a user can access, as well as how the user accesses those files. For example, a user may have access to a word processor file, but only to read it, not to modify it. At the same time, another user may have access to the same file and be able to modify it.

Peer-to-Peer

A LAN that does not make a distinction between a user workstation and a server is called a **peer-to-peer** network. It allows you to connect two or more computers in order to pool their resources, as depicted in Figure 6–26. Individual resources such as disk drives, CD-ROM drives, scanners and even printers are transformed into shared resources that are accessible from each of the computers. Each computer acts as both a client (a worksta-

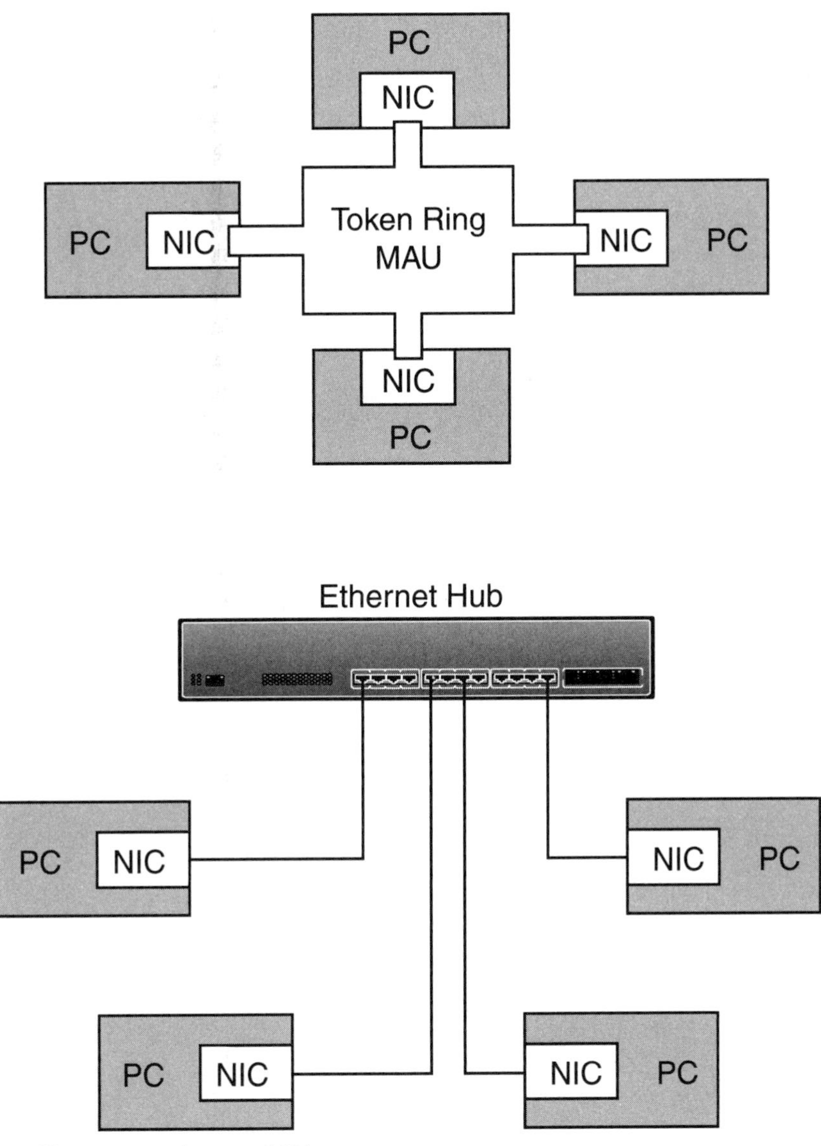

Figure 6–25 Client connection to a LAN.

tion) and a server (used for information storage). Network Operating Systems are classified according to whether they are client/server or peer-to-peer. AppleTalk and Windows for Workgroups are examples of peer-to-peer NOS. A NOS must be installed on every machine that is a member of the peer-to-peer network.

A peer-to-peer workstation is a microcomputer which is used in a fashion similar to a microcomputer in a stand-alone mode. The main difference is that the network provides additional locations from which application programs and files can be retrieved. You can

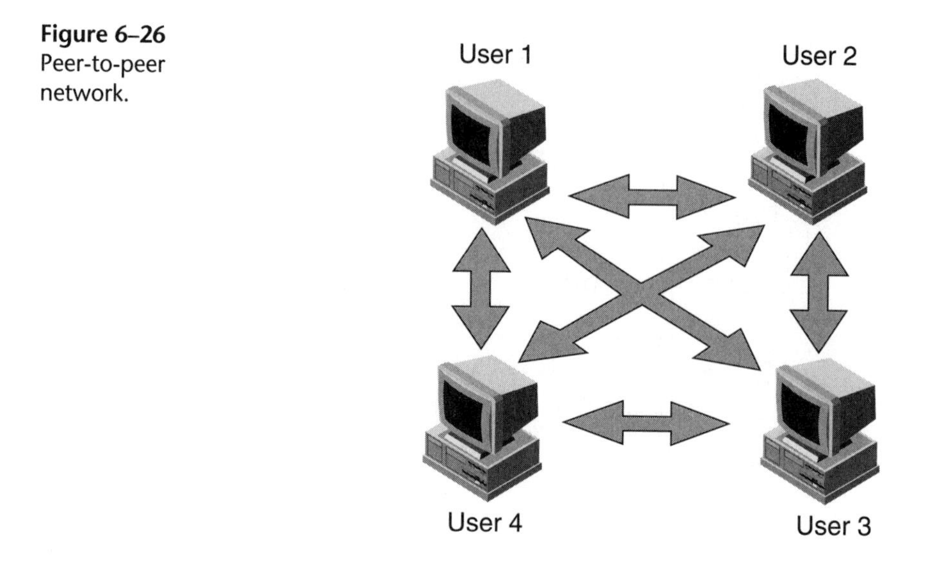

Figure 6–26 Peer-to-peer network.

network workstations from different vendors, such as IBM and Apple. Workstations are advanced machines typically used by engineers, architects, graphic designers, and other individuals who require a faster microprocessor, a large amount of RAM, special features such as high-speed graphics adapters, and a relatively fail-safe system. One instance in which such a peer-to-peer network may be useful is for a product design group that might want to share the files.

Most NOS software allows each peer-to-peer computer to determine which resources will be available for use by all other users. When one computer's disk has been configured so that it is being shared, it will usually appear as a new or additional drive to the other users. As an example, consider users X and Y are on a peer-to-peer network. User X has a DVD-R/W drive (specified as drive D), while user Y does not. If user X configures the D drive on his machine so that it is shared, then user Y can map to the user X's D drive, and have access to the DVD-R/W, in addition to the drives on his own computer. Because drives can be easily shared between peer-to-peer computers, data only needs to be stored on one computer. The advantages of peer-to-peer over client-server networks include:

- Network is easy and inexpensive to setup and maintain
- No need for a network administrator
- Makes efficient use of computing resources in a home or small office environment

INTERNETWORKING

Internetworking refers to the equipment and technologies involved in connecting either LANs to LANs, WANs to WANs, or LANs to WANs. When designing a network, it is important to remember that the capacity of the internetworking devices tends to limit the overall network capacity. Typical examples of internetworking devices are repeaters, bridges, switches, routers, and gateways. Each device operates at all layers below a particular layer of the OSI model, as shown in Figure 6–27.

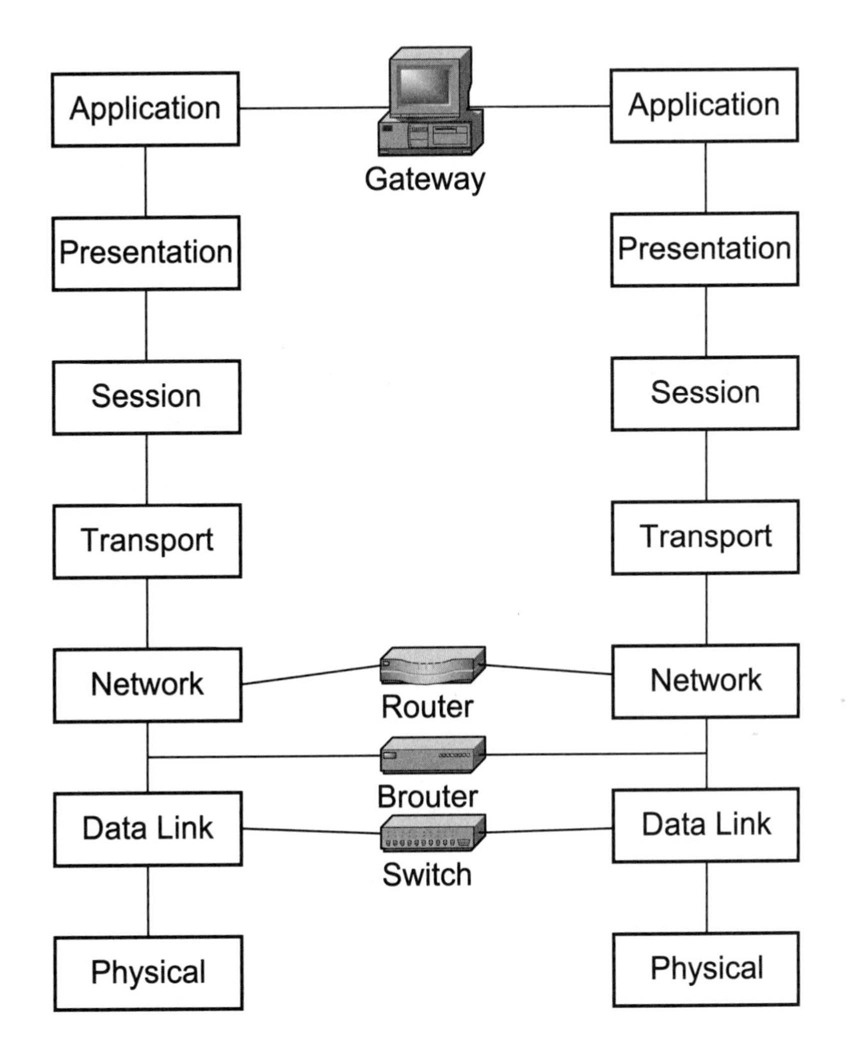

Figure 6–27 Network interconnection can occur at various layers of the OSI model.

Repeaters

Repeaters operate at Layer 1 of the OSI model and simply act as transceivers that receive, amplify, and retransmit information. They are unintelligent devices connected between two LAN segments of the same type, so that a Token Ring repeater will transmit only Token Ring traffic, as shown in Figure 6–28. Repeaters are used to either extend a LAN

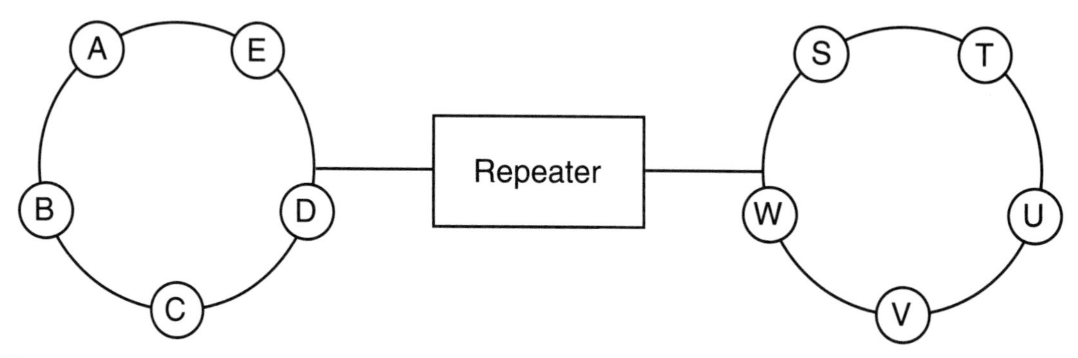

Figure 6–28 Interconnecting two Token Ring networks with a repeater.

beyond its usual distance limitation, or to provide electrical isolation, or to accomplish media conversion. A typical hub is a multi-port repeater. The signal received at the backbone is regenerated and transmitted to all other ports. Repeaters may introduce undesired side effects where they add not only cost to the circuit, but also distortion as a result of limited bandwidth, additional noise, and other undesirable changes to the signal they amplify. Designers attempt to minimize the use of repeaters to the greatest extent possible.

Bridges

Bridges are intelligent devices that operate at Layer 2 and are used to interconnect LANs of the same type, as illustrated in Figure 6–29. A basic bridge has ports connected to two (or more) otherwise separate LANs. Packets received on one port may be retransmitted or forwarded on another port. Unlike a repeater, a bridge will not start retransmission until it has received the complete packet. As a consequence, stations on either side of a bridge may be transmitting simultaneously without causing collisions. A bridge, like a repeater, does not modify the contents of a packet in any way, and a simple bridge retransmits every packet whether or not this is necessary.

Unlike repeaters, bridges may provide *filtering* and *forwarding* services across the link. A *learning bridge* examines the source field of every packet it sees on each port and builds up a picture of which addresses are connected to which ports; therefore, a packet will not be transmitted to a network segment that has no need to see the information. Filtering means that if the destination of a packet is the same side of the bridge as its origin, the bridge ignores it, but if the address is on the other segment, the bridge lets it across or forwards it. If a bridge sees a packet addressed to a destination that is not in its address table, the packet is retransmitted on every port except the one it was received from. Bridges also address table entries; if a given address has not been communicating in a specified period of time then the address is deleted from the address table. The learning bridge concept works equally well with several interconnected networks, provided that there are no loops in the system. Bridges use a method known as the spanning tree algorithm to con-

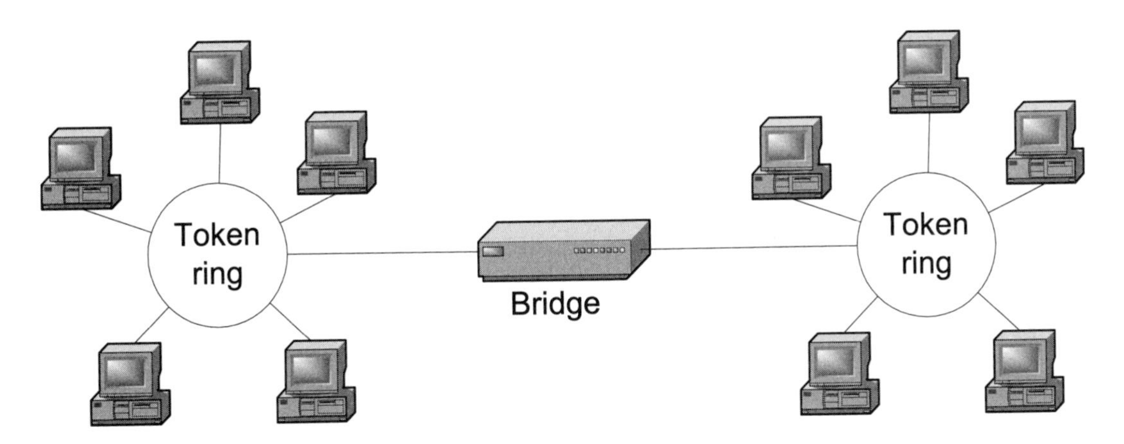

Figure 6–29 Bridge interconnects two Token Ring LANs.

struct a non-looping topology by deciding not to use certain links in the network. The links are still there and may come into use if the network is reconfigured.

Switches

Switches are fast multiport bridges that provide an economical way to resolve network congestion. A switch establishes point-to-point connection between ports at the full wire speed of the network and eliminates collisions between segments, as depicted in Figure 6–30. All the available bandwidth is dedicated to the two ports that are connected during the transmission of a packet. A switch port can be connected to a single station such as a server, or to a multistation network segment. They have low latency and a high filtering and forwarding rate compared with other alternatives. Latency is synonymous to delay, which is the amount of time it takes for a packet of data to get from one designated point to another.

Over the last few years, LAN switching has greatly increased network performance by replacing shared media with dedicated bandwidth. Users benefit from direct access to their networks, and the bottlenecks of shared Ethernet or Token Ring disappear as point-to-point switching is deployed. Because of their speed and simplicity, network switches are replacing hubs and routers as the dominant form of internetworking. In addition, switch vendors are introducing Web interfaces to make them easier to manage, and policy-based networking capabilities to add intelligence and business prioritization to network switching. Switches have two major weaknesses: Under heavy load conditions, some switches may drop packets, and they lack the ability to filter out or block unwanted or unauthorized traffic.

Spanning Tree Protocol

Spanning Tree Protocol, officially part of the IEEE 802.1D standard for MAC, is a link management protocol used by Layer 2 devices. In a network with redundant connections between bridged LANs, one connection is always in the forwarding position, passing all

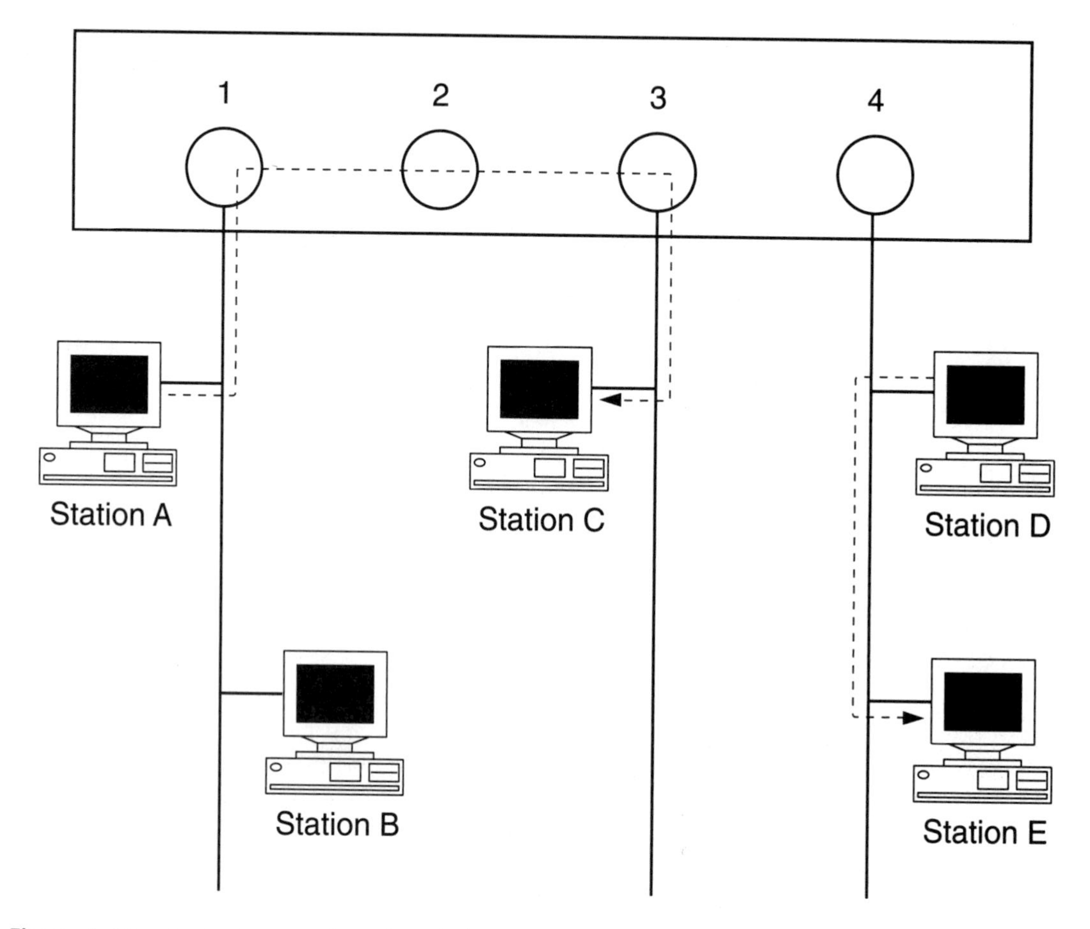

Figure 6–30 An implementation of a switch.

traffic. The other is in a standby, or blocking, position. If the first connection goes down, spanning tree is the algorithm that learns about the disruption and ensures the backup connection kicks in. Without spanning tree in place, it is possible that both connections may be simultaneously live, which could result in an endless loop of traffic on the LAN, as illustrated in Figure 6–31. A weakness of spanning tree is that it is a slow protocol and often cannot keep up with the speed of today's networks. For example, broadcast traffic on a bridged network always has the potential of slowing down the network when protocols such as spanning tree react too slowly. One can get around spanning tree by using Layer 3 routers or switching routers capable of port-level routing, but they are more expensive than a Layer 2 switch or bridge.

There has been a rise in the installation of hybrid switching routers that combine the speed and simplicity of the Layer 2 switch with the intelligence of the Layer 3 router by operating on both layers. Multilayer switching improves network performance by providing

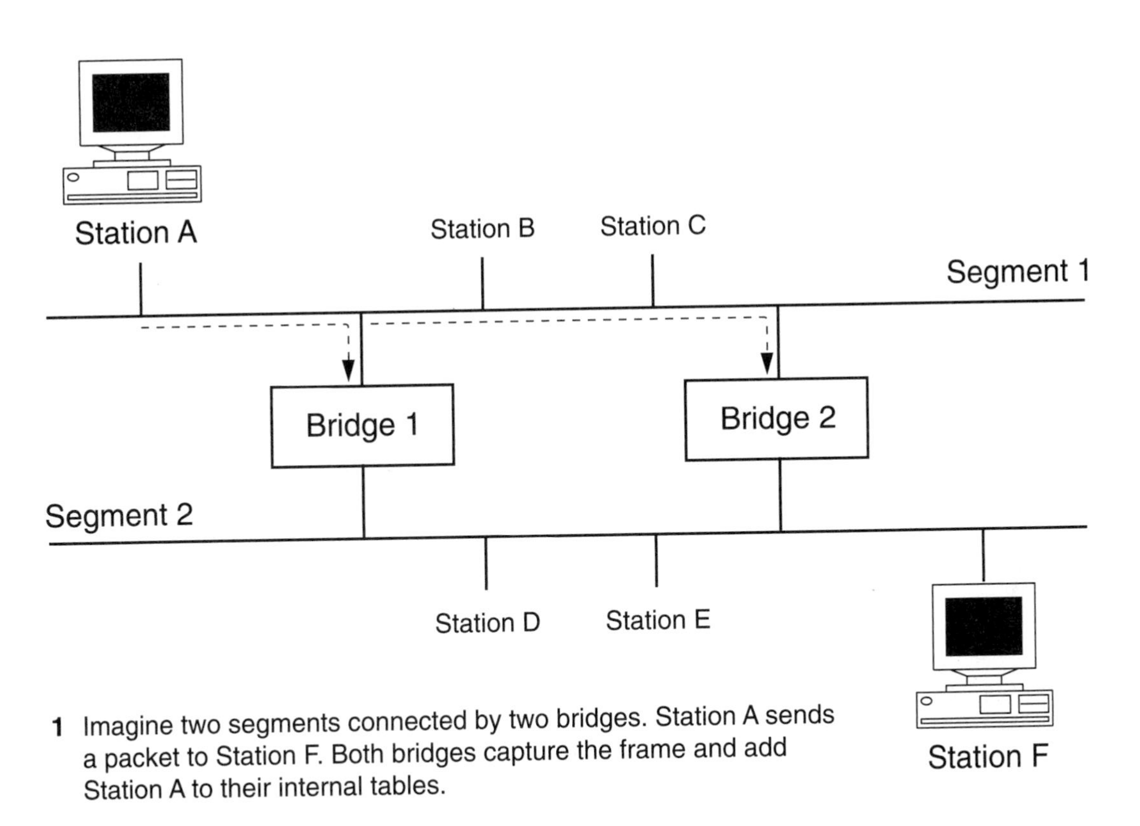
Station A
Station B
Station C
Segment 1
Bridge 1
Bridge 2
Segment 2
Station D
Station E
Station F
1 Imagine two segments connected by two bridges. Station A sends a packet to Station F. Both bridges capture the frame and add Station A to their internal tables.

Figure 6–31 Use of spanning tree protocol to prevent infinite traffic loops (part 1 of 3).

multiple capabilities, including the ability to offload Layer 3 traffic from traditional routers, to support multiple protocols, and to switch and route on a per-port basis. This ensures maximum flexibility of network design. It removes the scalability and throughput restrictions that limit network growth, while building the foundation for an emerging generation of high-speed networked applications.

Routers

Routers operate on Layer 3, the network layer that routes data to different networks. Routing is important when multiple segments are connected in such a way that there is more than one possible path between one station and another on the network; it provides intelligent redundancy and security required to select the optimum path. A router determines the next network point to which a packet should be forwarded toward its destination. It is located at any juncture of networks, and it decides which way to send each information packet based on its current understanding of the state of the networks to which it is connected. A router creates or maintains a table of the available routes and their conditions and uses this information along with distance and cost algorithms to

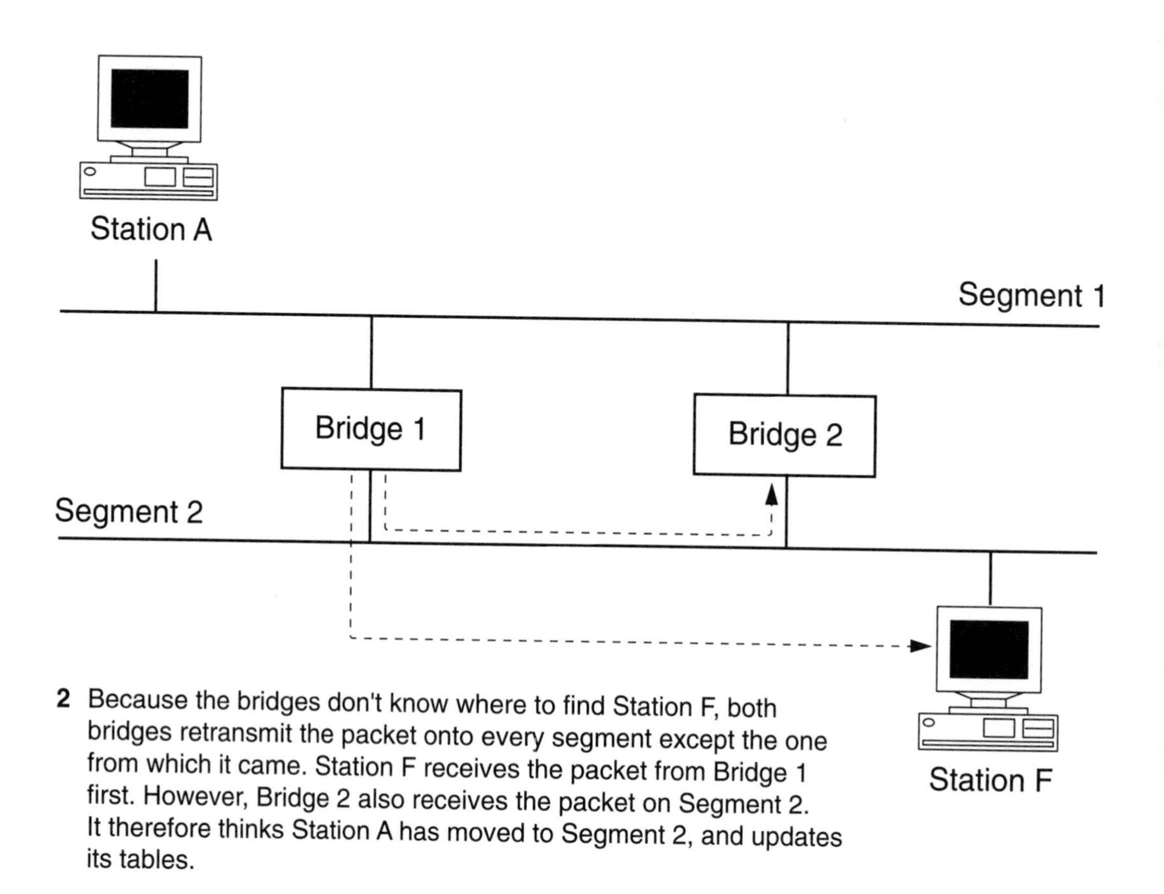
Station A
Segment 1
Bridge 1
Bridge 2
Segment 2
Station F
2 Because the bridges don't know where to find Station F, both bridges retransmit the packet onto every segment except the one from which it came. Station F receives the packet from Bridge 1 first. However, Bridge 2 also receives the packet on Segment 2. It therefore thinks Station A has moved to Segment 2, and updates its tables.

Figure 6–31 Continued (part 2 of 3).

determine the best route for a given packet. Typically, a packet may travel through a number of network points with routers before arriving at its destination. Usually routers are used for connecting remote networks. They have the following drawbacks:

- ✦ They are complex and difficult to install, configure, and manage
- ✦ They have a lower packet-filtering and forwarding rate as compared to switches
- ✦ They are protocol dependent, and may be unable to handle some protocols without network reconfiguration. Furthermore, some protocols are not routable

Router performance is measured by its packet-forwarding rate, which is the number of packets transferred per second from input to output ports. This rate depends on packet size, and how many protocols are being supported. The larger the packet, the greater the router throughput because each forwarded packet requires reading the packet header, which consumes time. Routers have protocol intelligence, which enables them to handle multiple protocols. Many organizations continue to rely upon traditional multiprotocol

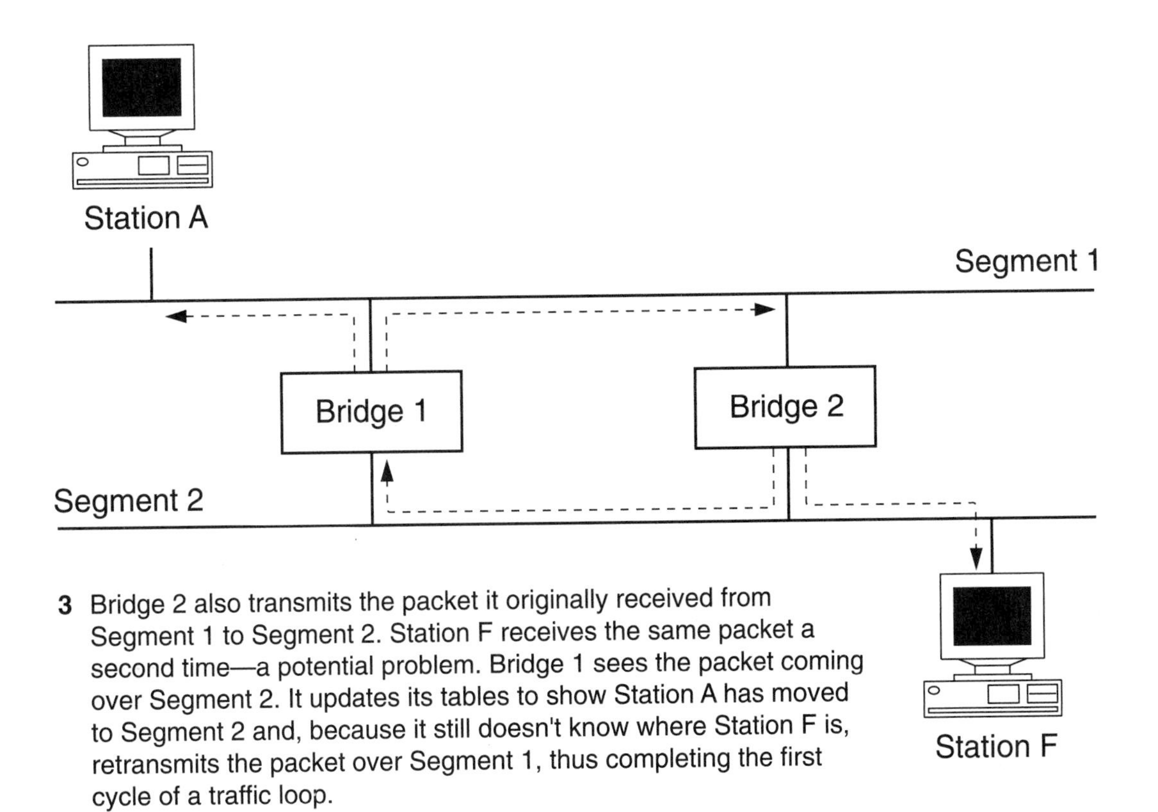

Station A
Segment 1
Bridge 1
Bridge 2
Segment 2
3 Bridge 2 also transmits the packet it originally received from Segment 1 to Segment 2. Station F receives the same packet a second time—a potential problem. Bridge 1 sees the packet coming over Segment 2. It updates its tables to show Station A has moved to Segment 2 and, because it still doesn't know where Station F is, retransmits the packet over Segment 1, thus completing the first cycle of a traffic loop.
Station F

Figure 6–31 Continued (part 3 of 3).

routers to provide the foundation for their networking infrastructure. In that case, the router, or more specifically the backplane of the router, is the collapse point for the entire enterprise as the total network response time depends upon how the router manages all the WAN and LAN connectivity, as illustrated in Figure 6–32.

Sometimes, a router is included as part of a network switch. Switching routers provide the ultimate flexibility and investment protection by allowing users to switch or route on per port basis. Such an architecture enables users to maximize the backbone design by deploying either switching or routing wherever it is needed in the network. A Brouter is a network bridge combined with a router. Integrated Routing and Bridging (IRB) allows users to both route and bridge a protocol in the same router, with connectivity between all the interfaces.

Gateways

A **gateway** operates at Layer 7 and spans all seven layers of the OSI model. As a result, the activity of the gateway is processor-intensive, and the device can become a bottleneck. It

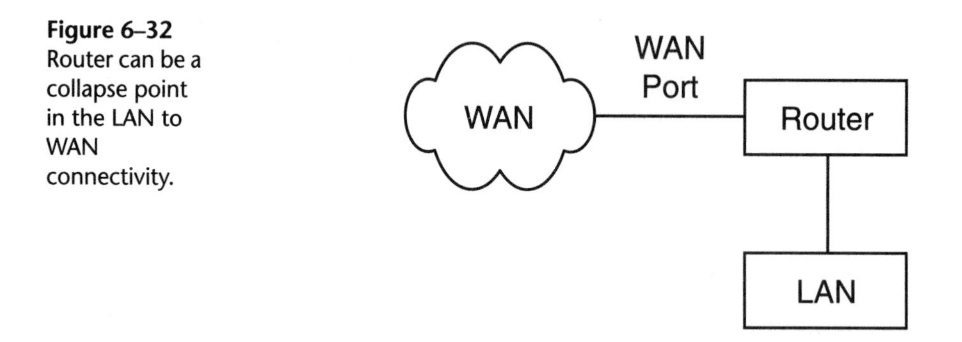

Figure 6–32 Router can be a collapse point in the LAN to WAN connectivity.

is designed to link incompatible networks—for example, Ethernet to SNA—by protocol conversion. A gateway typically handles three different protocols: source, destination, and transmission path protocols. Today, router is used to describe such a device; the term gateway refers to special-purpose devices that perform application layer conversions of information received from various protocols.

Channel Service Unit (CSU)/Data Service Unit (DSU)

Channel Service Unit (CSU)/Data Service Unit (DSU) come in either stand-alone units or combination CSU/DSUs. As shown in Figure 6–33, at least one of these devices is required for any digital line termination, but combination CSU/DSU is most common. A CSU is used to connect a switched digital line, such as a T-1, to a DCE device. It protects the switched line from electrical damage and provides a way for the common carrier to test the circuit through loopback. A DSU is used to access dedicated lines and works like a repeater. The purpose of the CSU/DSU is to encapsulate information into proper framing and to ensure proper timing before connecting to the WAN.

Trunking

Trunking is a technique that allows a networking device to bond multiple physical links into a group that works like one logical link, as shown in Figure 6–34. In a LAN environment, trunking can be used in any type of network capable device such as a Layer 2 switch, a Layer 3 switching router, or a computing device such as a server or desktop computer. The networking device is equipped with multiple physical links that become part of a larger entity known as the trunk port, which is viewed as one big logical link. For example, four Fast Ethernet ports can be trunked to work like one big 400 Mbps pipe. Trunking provides a cost-effective way to pour bandwidth into critical network points.

It is efficient to internetwork in a pure IP (Internet Protocol) environment so that there is no need to make disparate systems communicate. The more protocols you run, the more bandwidth you utilize in simply handshaking. IP is fast becoming the protocol of choice in many LAN to WAN environments; it is the middleware for interconnectivity.

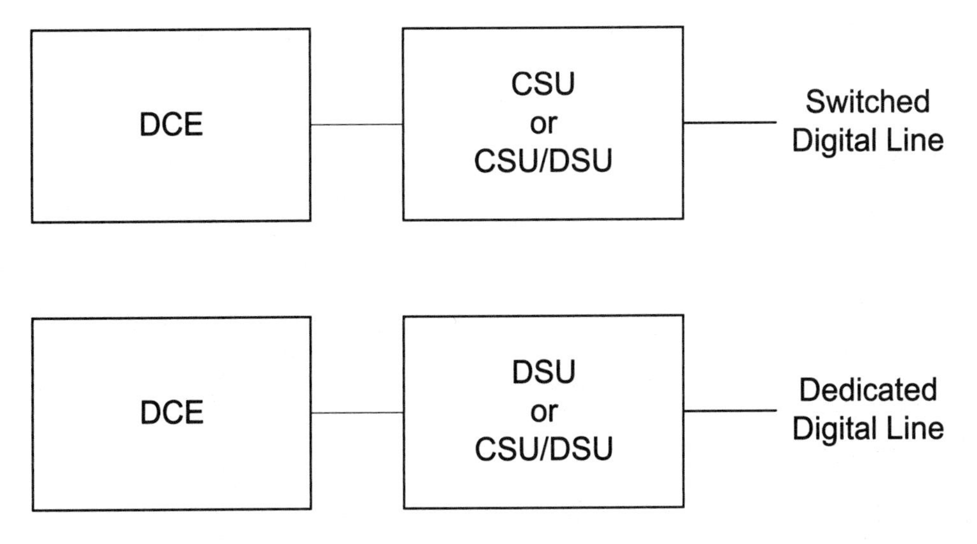

Figure 6–33 CSU/DSU implementation.

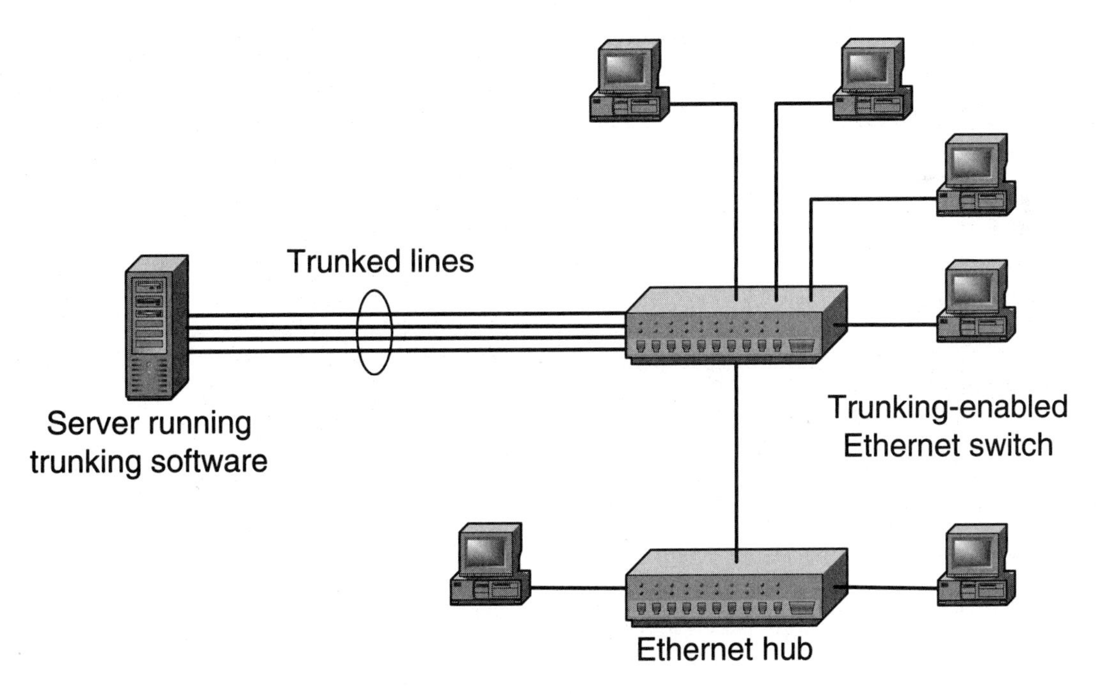

Figure 6–34 Trunking lets companies increase the bandwidth from server to switch.

Although IP has been around since the 1970s, its popularity as a platform for business development is recent.

SUMMARY

The data communications infrastructure evolved as computer networks were designed to enable the crucial technologies of the information age. The developments can be traced from a mainframe environment to that of distributed computing, with client/server computing becoming the technology of choice in today's networks. The proprietary nature of earlier networks was changed to an open environment by the seven-layer OSI reference model, which provided for interoperability among vendors. The network and higher layers are entirely software products, while hardware considerations are relevant only at the bottom two layers.

There are several LAN technologies in use today, but Ethernet has become one of the most flexible and pervasive technologies within the enterprise and beyond. From the time of the first Ethernet standard, the specifications and the rights to build Ethernet technology have been made easily available to anyone. This openness, combined with the ease of use and robustness of the Ethernet system, resulted in its widespread implementation. An essential feature of Fast Ethernet and Gigabit Ethernet implementations is the backward compatibility, allowing network managers to easily plan and deploy higher speed networks. However, Ethernet performance could suffer under heavy load conditions, therefore stable technologies such as Token Ring and FDDI have been used in the backbone.

REVIEW QUESTIONS

1. Define the following terms:
   - A. Data communications
   - B. Interface
   - C. Protocol
   - D. Virtual Circuit
   - E. Datagram
   - F. Byte
   - G. Character
   - H. Character code
   - I. Forward Error Correction
   - J. Stop and Wait protocol
   - K. Sliding Window protocol
   - L. Client/Server Computing
   - M. Polling and Selecting
   - N. Network Interface Card
   - O. Network Operating System
   - P. Trunking
   - Q. Latency
2. Synthesize the process of how the OSI reference model emerged as the standard network architecture with respect to the evolution of data networks.
3. Describe the function of each layer of the OSI model.
4. Distinguish between different data coding methods, and state the applications for each.
5. Compare and contrast error detection and correction schemes: Parity Checking, Longitudinal Redundancy Check, and Cyclic Redundancy Check.
6. Summarize the role of the OSI model in the development of LAN architecture.
7. Describe the LAN access methods: Token Passing, CSMA/CD, and CSMA/CA.
8. Explain the HDLC frame format.
9. Analyze the following LAN technologies and their applications:
   - A. Token Ring
   - B. FDDI
   - C. Ethernet
10. What are the essential components of a client/server LAN? Describe the role of each component.

11. Identify the similarities and differences between a client/server and peer-to-peer network.

12. Discuss the functions of the following:

    A. Hub
    B. Repeater
    C. Bridge
    D. Switch
    E. Router
    F. Gateway
    G. CSU/DSU

7 WIDE AREA NETWORK AND BROADBAND TECHNOLOGIES

KEY TERMS

Quality of Service (QoS)
Latency
Jitter
Packet Assembler/Disassembler (PAD)
Packet Switching Exchange (PSE)
Payload
X.25
Frame Relay
Permanent Virtual Circuit (PVC)
Committed Information Rate (CIR)
Network Downtime
Switched Multimegabit Data Service (SMDS)
Integrated Services Digital Network (ISDN)
Basic Rate Interface (BRI)
Primary Rate Interface (PRI)
Synchronous Optical Network (SONET)
Automatic Protection Switching (APS)
Asynchronous Transfer Mode (ATM)
Packetization Delay
Segmentation and Reassembly
Switched Virtual Circuits (SVC)
Digital Subscriber Line (DSL)
Cable Modem (CM)
Fixed Wireless

OBJECTIVES

Upon completion of this chapter, you should be able to:

- Analyze quality of service metrics for transporting information over a network
- Discuss the process of transmitting a message over a packet switching network
- Determine the significance of CIR and SIR
- Explain the frame format in Frame Relay
- Discuss the segmentation and reassembly process in ATM
- Describe the SONET protocol stack and STS-1 frame structure
- Evaluate the strengths, weaknesses, and applications of X.25, Frame Relay, SMDS, ISDN, SONET, ATM, PoS, DTM, DSL, and cable modems

INTRODUCTION

The basic objective of WAN technologies is to provide a means of LAN interconnects over long distances, that is, from few km to virtually around the world. Organizations want to connect LANs across the boundaries of a wide area and provide high-speed data transfer at a reasonable cost. Considering the bursty nature of data traffic, the LAN-to-LAN or LAN-to-WAN connection is not required all the time. *Bursty* refers to non-constant or variable rate information transfer with peak activity in spurts, followed by periods of idle time. Therefore, a meshed leased line network is an expensive proposition unless the volume of data justifies the connection full time. Prior to the arrival of WAN technologies and services, the only way to connect LANs was through dedicated leased lines. This facility provides reliable, high speed services starting as low as 2.4 kbps and ranging as high as 45 Mbps (T-3 service), to link two or more sites for a fixed monthly charge. Leased lines can be either fiber optic or copper and provide a consistent amount of bandwidth for data, voice, and video links between sites. Although this is still the way some WANs are connected, businesses are taking advantage of more flexible, and in many cases, more economical alternatives offered by network providers.

As multigigabit traffic in a variety of formats—data, voice, and video—migrates upward from the LANs and MANs to the WAN, there is a growing need for broadband technologies to efficiently integrate heterogeneous protocols across the WAN. The most crucial considerations for successfully transporting voice and video over a data network are the **Quality of Service (QoS)** metrics. QoS refers to a set of characteristics that define the delivery behavior of different types of network traffic and provide certain guarantees. The characteristics most often defined are throughput, **latency** or transit delay, **jitter** or variation, availability of service, and acceptable error rates or packet loss. Latency is the end-to-end delay that a signal element experiences as it moves across the network; jitter is the variability (in effect, the standard deviation) of the latency in the network. Although some latency and jitter is expected, excessive amounts can cause degradation in error performance and distortion of the resulting analog signal after decoding at the receive end of the circuit. This chapter discusses high-speed communications services that carriers offer in order to enable organizations to transmit digital signals over a WAN.

PACKET SWITCHING NETWORKS

Packet switching is considered to be one of the most efficient means of sharing both public and private network facilities among multiple users. Each packet contains all of the necessary control and routing information to deliver the packet across the network. Packets can be routed either independently or as a series of packets that must be maintained and preserved as an entity. Let us trace the process of transmitting a message from user A to user B in a packet-switching network, depicted in Figures 7–1 and 7–2:

- The data is sent from user A DTE to a host computer. This is typically a serial asynchronous connection.
- The host computer is connected to a **Packet Assembler/Disassembler (PAD)**, which can be a software package or a piece of hardware that either receives data and breaks it down into packets or reassembles the received packets into a continuous data stream. The PAD assembles the data into packets when one of three events occurs: 1. A predetermined number of bytes are received; 2. A specified time elapses; or 3. A particular bit sequence occurs.
- The PAD functions like a statmux as it interleaves the packets from user A with packets from several other ports, based on an algorithm that allows each port to appear as though a dedicated link is available to it.
- The PAD creates packets for delivery after adding necessary overheads and then routes the packets to a network node, specifically called **Packet Switching Exchange (PSE)**. A PSE is a computer that analyzes each packet to verify that it was received without errors, reads the address, frames the information, and routes the packet to the next appropriate port.
- There is a synchronous connection between one PSE and another. A virtual connection is created, allocating the time slots for the packets from user A to travel on the virtual circuit to user B. When a user sends data in a burst, multiple packets will be generated and routed across the network based on the addresses contained in the packets.
- The packet is buffered (or stored) at every node until the next node checks and acknowledges the packet.
- At user B's DTE, packets are checked, acknowledged, stripped of overheads, and the **payload** is delivered to the receiving DTE. Payload is the actual information that needs to be transmitted.

X.25

X.25, one of the first packet-switching technologies, was built on the OSI model. It is a Layer 3 protocol that specifies the interface between a DTE and a packet-switched network. The data link layer or Layer 2 uses a derivation of HDLC called balanced link access procedure to control errors and packet transfer, and to establish the data link. The X.25 network layer establishes logical channels and a virtual circuit from each source to each destination to keep track of the packets on each connection. It uses a form of statistical time-division multiplexing that analyzes past users to allocate more interleaved packet slots to heavier users and less interleaved slots to lighter users.

In the initial rollout of X.25 services, the original network connections were on analog lines at up to 9.6 kbps. Since analog circuits have error rates of 10^{-5} or 10^{-6}, X.25 was designed to ensure guaranteed data delivery and integrity. Therefore, this technique involves error checking at every node and continual message exchange regarding the

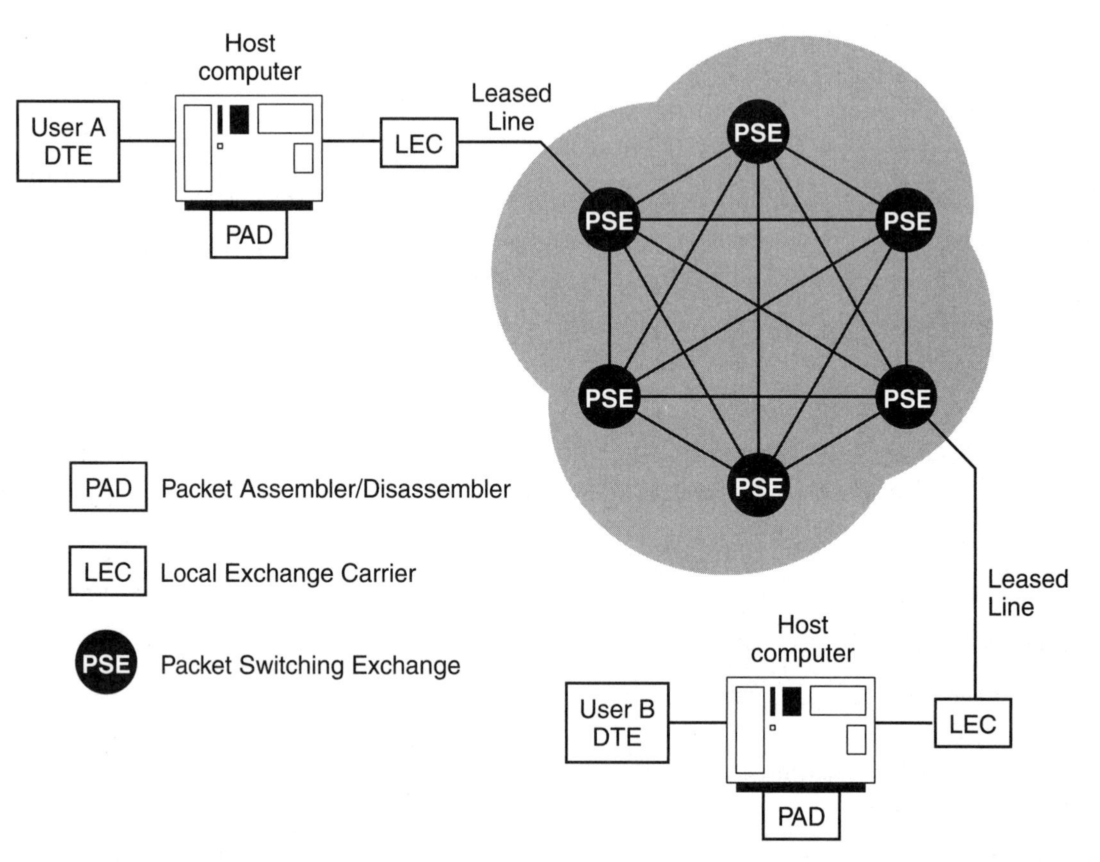

Figure 7–1 Users are connected to a packet switching network cloud via a leased line to a PSE, and hardware or software PAD function.

progress of packets, from node to originator and from node to destination. As a result, X.25 is slow and tedious. Today's fiber-optic networks deliver error performance of at least 10^{-9}, and often better. In such high-quality circuits, the X.25-intensive processing for every link imposes excessive latency that is rather unnecessary. Also, X.25 networks provide access speeds of up to 56 kbps, which is slow for interconnecting LANs. One way to create a faster service is to reduce overhead associated with the network by eliminating some of the processing, mainly in the error detection and correction schemes employed at every node. This concept was implemented in *Frame Relay*.

FRAME RELAY

Frame Relay is a fast packet-switching technique that provides a cost-efficient means of connecting an organization's multiple LANs, as shown in Figure 7–3. Each node on the network only checks the address but never verifies the payload and passes the *frame*, or a

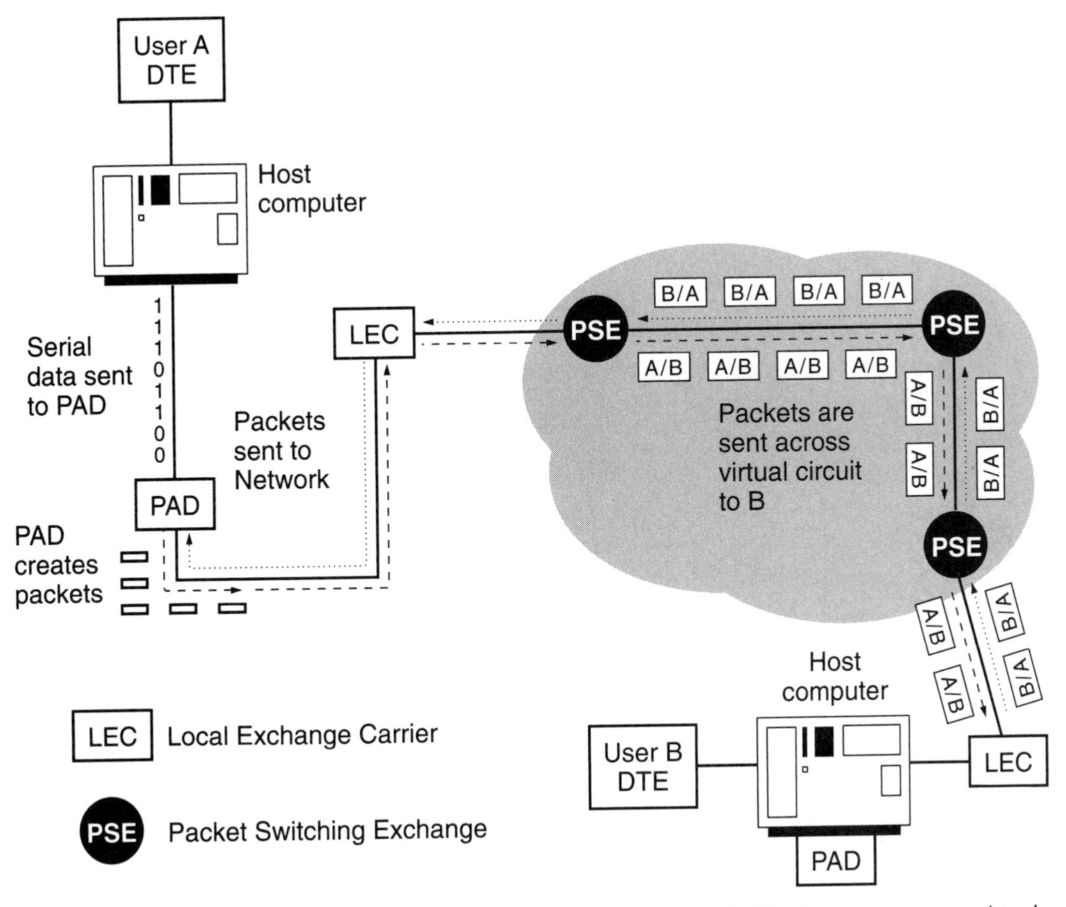

Figure 7–2 Serial data sent from user A to the PAD get packetized (A/B), then sent across a virtual circuit to user B. B can be returning responses B/A on a duplex circuit.

block of data, along a predefined path so that data integrity is assumed to be an end-user responsibility. This is a major digression from X.25, where each node is responsible to perform error checking on every packet. While the X.25 was originally designed to handle customer's asynchronous traffic, Frame Relay is designed to take advantage of the network's ability to transport data on a low-error, high-performance digital network and to serve the needs of intelligent synchronous applications. Another advantage of Frame Relay is that it can support the same traffic as X.25 while facilitating bandwidth on demand requirements for bursty traffic.

In Frame Relay, access from each site is provided into the network cloud, requiring only a single connection point. The data transmission is done over a shared packet switched cloud with connection-oriented paths. Data transported across the network is interleaved on a frame-by-frame basis so multiple users can be connected on the same link concurrently, as shown in Figure 7–4. Communications from a single site to any of

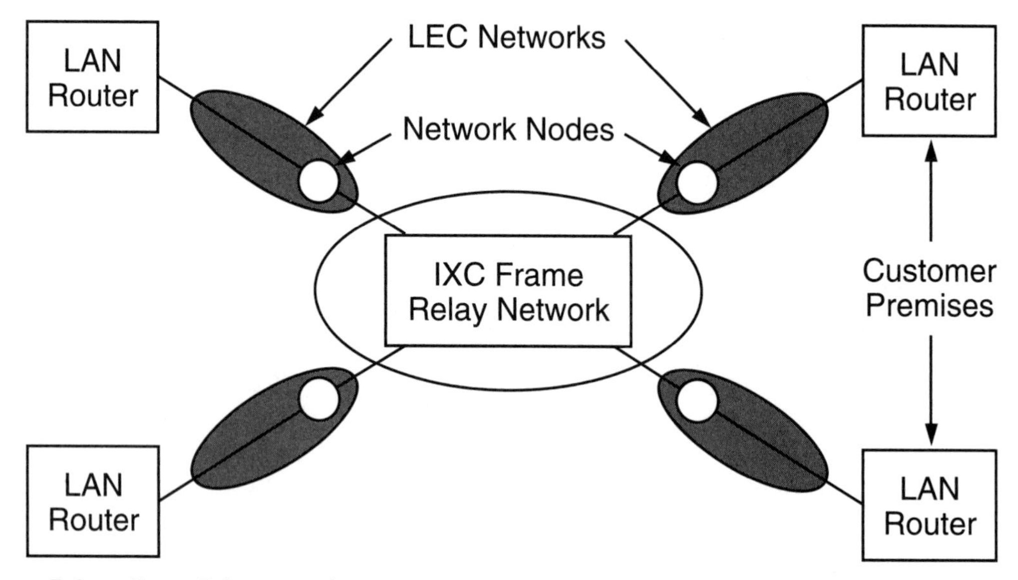

Figure 7–3 Frame Relay network connects an organization's multiple LANs.

the other sites is accomplished using a pre-defined network connection of virtual circuits, called **Permanent Virtual Circuit (PVC)**. A PVC connects two sites just as a leased line would, except that the bandwidth is shared among multiple users. Since the PVCs are pre-defined for each end-to-end connection, a network path is always available. This eliminates the call set-up time associated with dial-up lines and X.25. All major carriers offer interLATA Frame Relay service, and most LECs offer it within the LATA. Access to the Frame Relay network is usually offered through fractional T-1 or T-1 lines. The service is more flexible and the cost is less distance-sensitive than in dedicated leased lines. Also, capital equipment cost within the network, support, and most diagnostic tasks are off-loaded to the carrier.

In a Frame Relay service, the access rate or delivery rate is prespecified. The customer selects a **Committed Information Rate (CIR)**, which is a guaranteed rate of throughput when using Frame Relay. CIR is assigned to each of the PVCs selected by the user. Because the service is full duplex, a different CIR may be assigned in each direction, which gives a customer additional flexibility. Since LANs are prone to bursty traffic, the transmission rate may burst over the CIR. The burst rate (Br) is usually limited to twice the CIR. The burst excess rate (Be) is in addition to Br. The total throughput is a sum of CIR, Br, and Be.

CIR + Br + Be = Total Throughput

It should be pointed out that there are no standard offerings in the industry. Although the total throughput can go up to the channel access rate, the allocation of Br and Be depends on the carrier's networks and the service agreement between the customer and

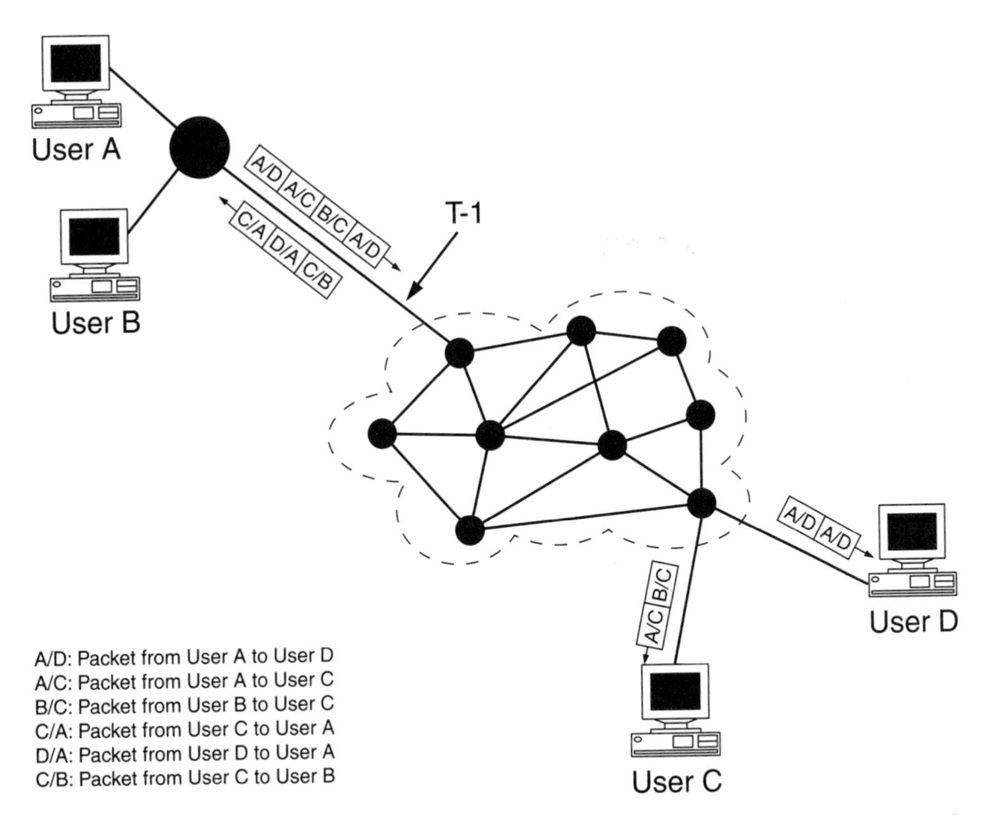

Figure 7–4 In Frame Relay, multiple users can be connected on the same link concurrently.

the carrier. There are some carriers that will not allow any bursting across the network, limiting the maximum throughput to the CIR. If allowed, bursting is typically limited to two seconds or less. When using the Br or the Be, the network will make its best attempt to deliver the frames, but there are no guarantees. As the network begins to get congested, the frames that are riding the network beyond the CIR can be discarded, while frames that are within the CIR will have priority. Because the network will only make its best attempt to deliver the data, the end-user equipment must be intelligent enough to recognize that frames have been discarded.

A Frame Relay frame, shown in Figure 7–5, comprises an opening flag followed by a frame header, which has control information such as discard eligibility setting and congestion notification. The variable data field can extend up to 1610 bytes. Finally, a CRC Frame Check Sequence is followed by a closing flag.

Advantages of Frame Relay

Most importantly, Frame Relay supports interconnection of LANs running multiple protocols, including Appletalk, SNA, DecNet, X.25, IPX, and TCP/IP, which provides fairly

	Opening Flag	Frame Header	Variable DATA Field	FCS	Closing Flag
Bytes	1	2 or 4	up to 1610	2	1

Figure 7–5 The frame format used in Frame Relay.

robust interoperability between various switching platforms. Other significant advantages include increased utilization and resultant savings because of network consolidation, network flexibility and scalability, since PVCs can be reconfigured or added without much difficulty, and improved network uptime. The risk of **network downtime** is always a major consideration in network design. A single failure in a dedicated leased line can bring the entire network to a halt, and the use of redundant links can be expensive. Frame Relay provides network connections into a virtual cloud so that there is automatic rerouting of network links within the cloud. This robustness of the delivery and recovery mechanism is the responsibility of the network provider. The only single point of failure that remains is in the local loop, which can be redundantly protected at a greatly reduced price.

SWITCHED MULTIMEGABIT DATA SERVICE (SMDS)

Switched MultiMegabit Data Service (SMDS) does not require PVCs for each pair of nodes, and it is available at higher rates than is Frame Relay. As shown in Figure 7–6, SMDS is a public, packet-switched service aimed at enterprises that do not want to commit to predefined PVCs but need to exchange large amounts of data with other enterprises over a WAN on a bursty basis. This high-speed connectionless data service is also deployed for linking LANs within a metropolitan area using the IEEE 802.6 MAN protocol. Like Frame Relay, SMDS is available from many LECs and long-distance carriers.

SMDS is a service, not a technology, so it can use any transport mechanism, including ATM. SMDS uses a technique called *Distributed Queue on a Dual Bus (DQDB)*, which distributes the network service requests of users into queues, to handle the transfer of information on unidirectional buses, as shown in Figure 7–7. The goal of SMDS is to provide high-speed data transfer on a switched, as-needed basis. Similar to CIR in Frame Relay, there is a *Sustained Information Rate (SIR)* in SMDS. The SIR is based on one of five classes of service, which provides control over how much information a node can place on the network, which in turn controls network congestion. The access class of service and the corresponding SIR is shown in Figure 7–8. The current use of SMDS includes the following:

- LAN-to-LAN interconnection in a metropolitan area
- Intra- and inter-company document transfer and sharing
- Host-to-host transfers

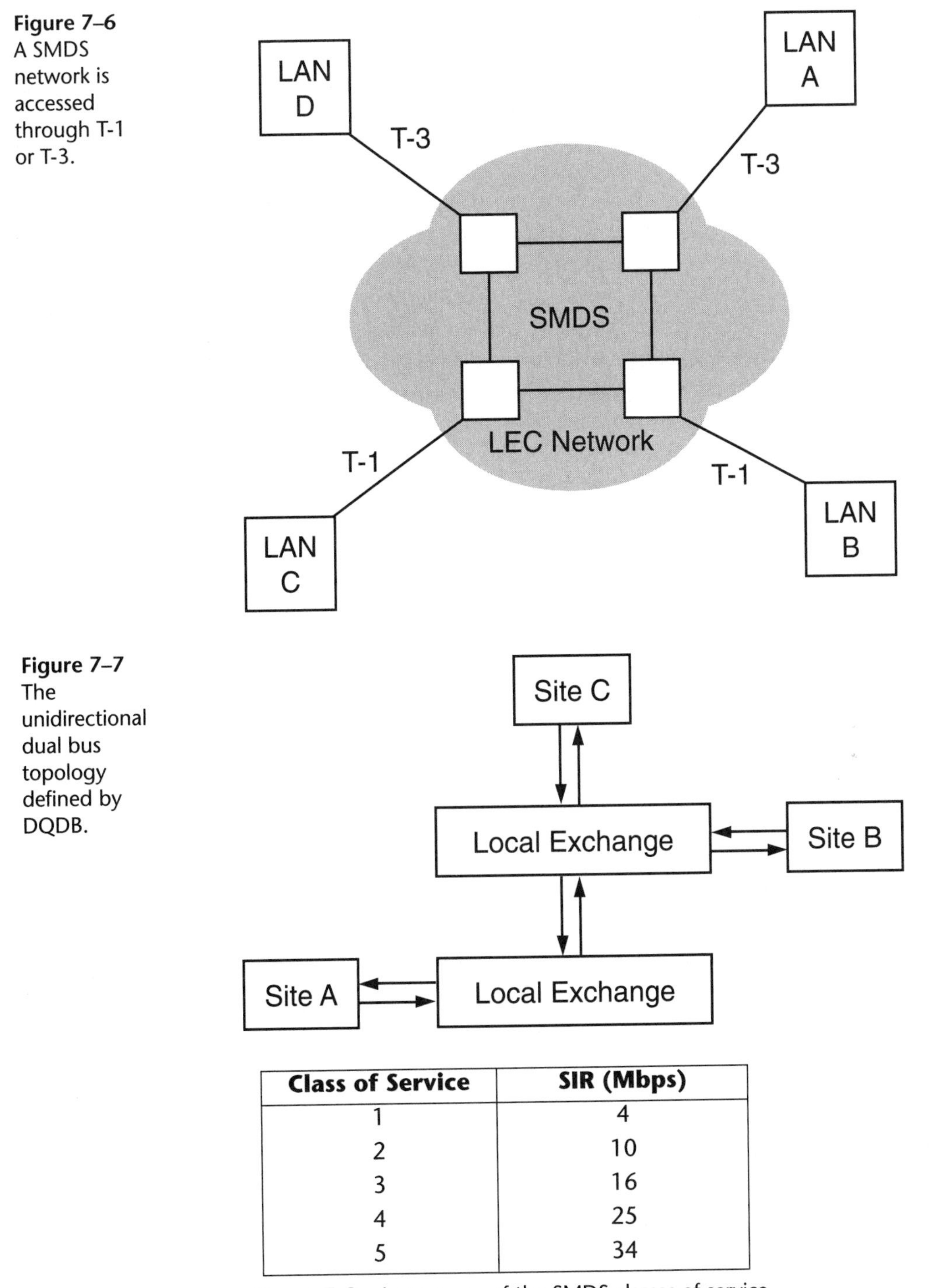

Figure 7–6 A SMDS network is accessed through T-1 or T-3.

Figure 7–7 The unidirectional dual bus topology defined by DQDB.

Class of Service	SIR (Mbps)
1	4
2	10
3	16
4	25
5	34

Figure 7–8 A summary of the SMDS classes of service.

INTEGRATED SERVICES DIGITAL NETWORK (ISDN)

Of late, **Integrated Services Digital Network (ISDN)** has overcome considerable negative perceptions rooted in shortcomings of the technology's initial design in North America more than 20 years ago. It has suffered for years as a result of a lack of standards, and even though ISDN standards have improved, the whole technology has been stigmatized. ISDN was developed as a way for telecommunications companies to support data and voice transmission over a single line, which was not available using the old telephone system. Although ISDN has been widely implemented in Europe, it is not as prevalent in North America, and the vast majority of subscriber loops in the United States are still analog. An *ISDN line* uses regular phone lines but provides end-to-end digital connectivity, and it distinguishes itself from other services by guaranteeing bandwidth and allowing users to simultaneously utilize voice and data applications.

The ISDN User-to-Network Interface (UNI) can be divided into two categories: **Basic Rate Interface (BRI)** and **Primary Rate Interface (PRI)**. The BRI is appropriate for a single two-wire subscriber loop, typically for an advanced user or home office application. It is a 2B+D interface, with two 64 kbps bearer (B) channels that can carry voice, data or video, and one 16 kbps data (D) channel, which provides intelligent line management. The D channel carries out-of-band signaling information to facilitate the establishment, maintenance, and clearing of ISDN channels, and specifies how a user connects to the carrier network. The phone company replaces the conventional analog circuit with a BRI circuit and a Network Terminating unit 1 (NT1), which provides the entry termination at the customer's home or business. NT1 is a device that accepts a two-wire signal from the phone company and converts it to a four-wire signal that sends and receives to and from devices within the home or business. Some ISDN equipment such as an ISDN terminal adapter (the ISDN equivalent of a modem) may already have a built-in NT1.

The North American PRI is appropriate for a business that utilizes a T-1 line, as it offers an economic alternative for connecting digital PBXs, LANs, host computers, and other devices to the network, as shown in Figure 7–9. It uses 23 B channels and one D channel, all at 64 kbps yielding a 1.536 Mbps line (equivalent to T-1), while the international PRI standard uses 30 B channels plus 2 D channels, all at 64 kbps yielding a 2.048 Mbps line (same as E1). For example, Internet Service Providers (ISPs) use ISDN PRI to speed subscriber connections and reduce their costs. PRI aggregates both digital ISDN BRI and analog modem dial-up connections with one link and phone number. This eliminates the need for separate hunt groups and the cost of additional modem racks and T-1 lines to support analog service.

At the customer end, the ISDN architecture includes applications such as voice, telemetry services, facsimile, video teleconferencing, text messaging, electronic mail, and wideband services, all on a single line. As represented in Figure 7–10, at the ISDN network node, the integrated customer access is separated into components, either physically or logically, and diverted to the appropriate type of network:

- Channel-switched network
- Circuit-switched network
- Packet-switched network
- Common channel signaling network (SS7)

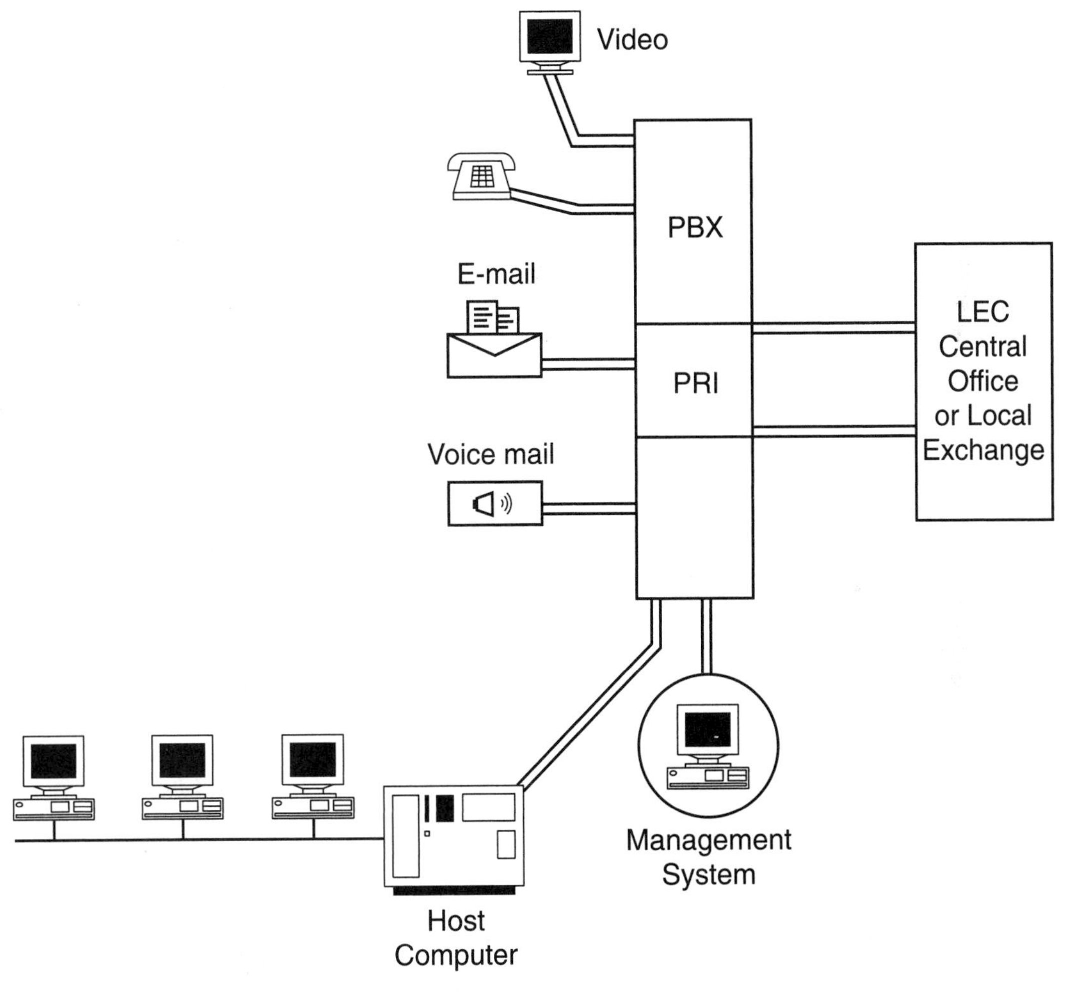

Figure 7–9 ISDN Primary Rate Interface (PRI) applications.

Advantages and Disadvantages of ISDN

The end-to-end digital connectivity in ISDN results in higher throughput and, in effect, reduced call lengths. From the carrier's perspective, the shorter call duration enables the company to provide service to a greater number of users without adding circuit capacity.

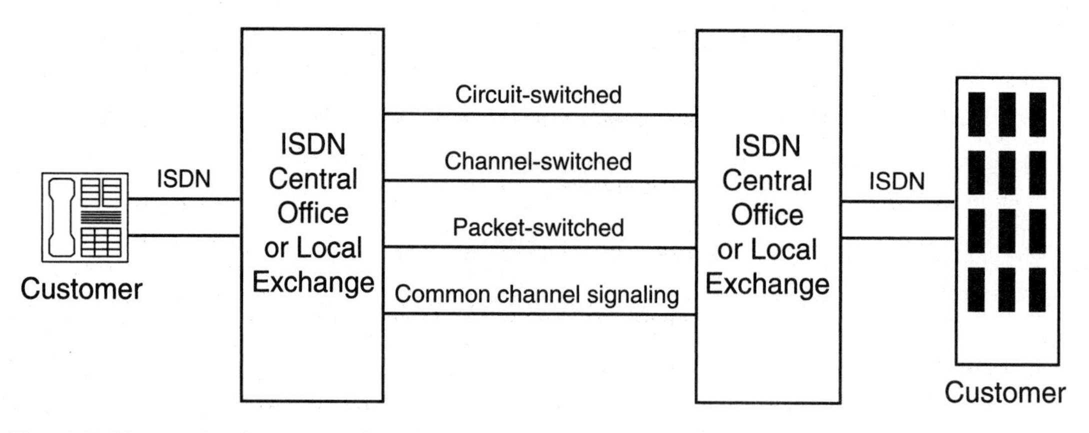

Figure 7–10 At the ISDN network node, integrated customer access is separated into components.

This improved efficiency reduces operating costs for the carriers. From the user's perspective, ISDN offers enhanced calling features such as digital voice quality, speed dialing, call return, caller ID, call forwarding, and local number portability, but its real value is the availability of multiple channels. It is possible to use one B channel for voice communications while simultaneously using the second B channel for data transmission, or the two B channels can be combined to provide data transmission speeds of up to 128 kbps (or up to 512 kbps with compression). An ISDN network layout is shown in Figure 7–11.

For example, it is possible to use both channels to surf the Internet at 128 kbps, and if a call comes in, the ISDN line will drop the Internet connection from 128 kbps to 64 kbps (one B channel) for the duration of the call. When the call is completed, the second B channel will reconnect to the Internet for a combined 128 kbps connection. Technology has made it possible for ISDN modems to use the D channel as an Always On conduit that enables a third call. Using Always On/Dynamic ISDN (AO/DI), consumers can conduct three calls simultaneously, for example, a phone call, a fax, and a data connection. This flexibility, coupled with the ability to guarantee bandwidth, positions ISDN as today's state-of-the-art media for small/home offices fulfilling all of their communication needs.

ISDN's high-speed performance, secure transmissions, and throughput make it a viable solution for high-speed remote access for telecommuters. Also, ISDN is exceptionally fitted to videoconferencing—to the desktop or the conference room. Using PRI or multiple BRI lines, channel bonding enables dynamic bandwidth allocation to support the high transmission rates (typically 64 kbps to 384 kbps) that video requires. And when not used for video, channels can be assigned to other data or voice uses. Using PRI's D channel, users can exchange supporting information (for example, text and graphic files) while videoconferencing. Despite the benefits of greater speed and flexibility, the rate of ISDN installations is far less than the number of analog lines that are installed each year. The main reason is that ISDN is not available in some areas. Also, the cost of ISDN is somewhat prohibitive; most ISDN services require additional charges such as line instal-

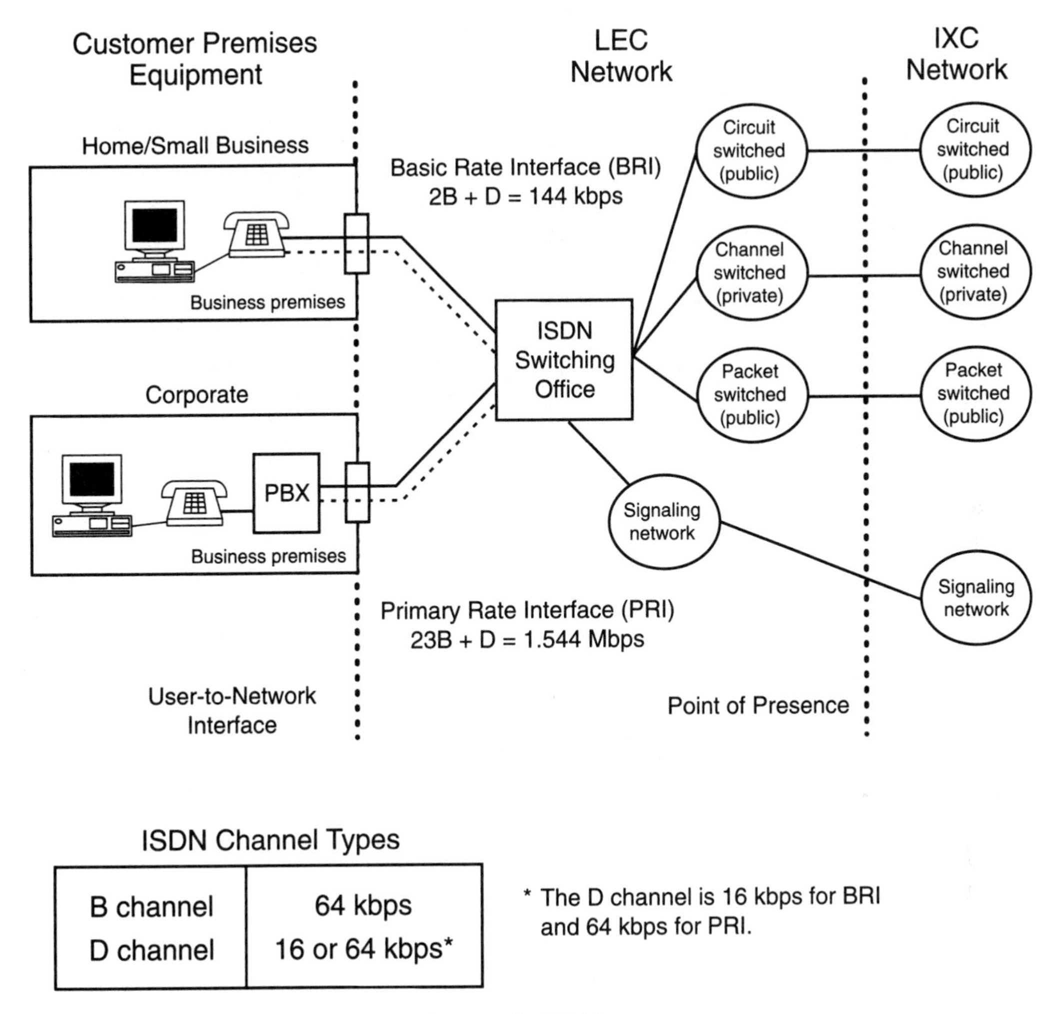

B channel	64 kbps
D channel	16 or 64 kbps*

Figure 7–11 Integrated Services Digital Network (ISDN).

lation and usage fees. Beyond pricing issues, ISDN is typically harder to set up and configure than a traditional analog modem.

SYNCHRONOUS OPTICAL NETWORK (SONET)

North American **Synchronous Optical Network (SONET)** is a physical layer or Layer 1 technology first conceived in the mid 1980s by MCI Communications. The term synchronous refers to the fact that data transmission is tightly bound to a clock signal, which made time division multiplexing much simpler than it had been and highly suitable for its initial telephony applications. SONET is organized as a master-slave relationship, with clocks of the higher-level nodes feeding timing signals to clocks of the lower-level nodes. All nodes can be traced to a primary reference source, a Stratum One atomic clock with extremely high stability and accuracy. Although SONET could theoretically be run over copper, it was designed from the start to hasten the move to fiber optic cable by ensuring interoperability. SONET networks are optimized for voice and do not inherently provide efficient bandwidth utilization for data.

SONET, a way of transmitting data in frames over WAN fiber-optic lines, is widely used by telecom companies. As fiberoptic systems were introduced, a new Synchronous Digital Hierarchy (SDH) evolved. SDH was accepted in other parts of the world except the United States, which adopted the term SONET. Although the international and U.S. versions of SDH/SONET are very close, they are not identical. SONET is based upon multiples of a fundamental rate of 51.840 Mbps, called STS-1 (Synchronous Transport Signal, Level 1). After the information has been converted into an optical signal, it is referred to as an OC-1 (Optical Carrier, Level 1). Some typical rates are given in Figure 7–12.

OC Level	STS Electrical Level	Transmission Rate (Mbps)
OC-1	STS-1 electrical	51.84
OC-3	STS-3 electrical	155.52
OC-12	STS-12 electrical	622.08
OC-24	STS-24 electrical	1244.16
OC-48	STS-48 electrical	2488.32
OC-192	STS-192 electrical	9953.28
OC-768	STS-768 electrical	39,813.12

Figure 7–12 SONET transmission rates.

STS-1 Transmission Rate

Each STS-1 frame is transmitted every 125 μs, or there are 8,000 frames transmitted per second, with each frame consisting of 810 bytes. That works out to a rate of 51.840 Mbps, as shown below:

$$\begin{aligned}\text{STS-1 Transmission Rate} &= (8{,}000\ \text{frames/s}) \times (810\ \text{bytes/frame}) \times (8\ \text{bits/byte}) \\ &= 51{,}840{,}000\ \text{bps} \\ &= 51.840\ \text{Mbps}\end{aligned} \tag{7–1}$$

This is known as the STS-1 signal rate—the electrical rate used primarily for transport within a specific piece of hardware. The optical equivalent of STS-1, which is known as OC-1, is used for transmission across the fiber. Multiple STS-1 (or OC-1) tributaries can easily be combined to increase bandwidth as higher-level signals are integer multiples of the base rate. For example, STS-3 (or OC-3) is three times the rate of STS-1 (3 x 51.84 = 155.52 Mbps). An STS-12 rate would be 12 x 51.84 = 622.08 Mbps. SONET can carry signals with speeds below STS-1 when it needs to by dividing its payload into smaller segments known as virtual tributaries. The International SDH system is based upon a fundamental rate of 155.520 Mbps, three times that of the SONET system. This fundamental signal is called STM-1 (Synchronous Transport Module, Level 1).

Advantages of SONET

SONET is based on ring topology, as illustrated in Figure 7–13. It defines the properties of an optical bit stream and specifies the characteristics of optical fiber that let data move across the medium. There are many valuable aspects of SONET that have made it the primary protocol for long haul, high-speed optical fiber communications.

First, every type of standard communications traffic can be multiplexed into SONET. SONET defines data units as frames, which make it easy for different traffic streams—traditional synchronous telephone streams such as T-1 and T-3 circuits, as well as ATM and other asynchronous data sources—to share the same bit stream.

Second, SONET is scalable and generally the first technology to implement a new round of high-speed interfaces. As OC-3 (155 Mbps) becomes a mere tributary technology, and OC-48 (2.4 Gbps) becomes readily available on high-speed switches and routers, and OC-768 (40 Gbps) is no longer a figment of imagination, SONET is consistently the first, and often the only way these ever-increasing throughput rates are implemented.

The third important aspect of SONET is its standardization. Before SONET, each maker of fiber optic equipment had its own implementation methods. The current ferment in worldwide communications probably could not have happened without the standardization of fiber transport protocols. At present, there is no standards-based alternative to SONET for high-speed optical data transmission.

The fourth valuable facet of SONET is its built-in fault tolerance called **Automatic Protection Switching (APS)**, which is widely implemented by major carriers and has prevented many communications services from being laid low by wayward backhoes. SONET rings with APS use redundant strings of fiber, so that if a fiber breaks, traffic can be switched to another within microseconds.

SONET Protocol

As shown in Figure 7–14, SONET has its own protocol stack with four layers: photonic, section, line, and path. The photonic layer describes how STS electrical data is converted into light pulses and vice versa. The section layer operates between optical repeaters, and its functions include scrambling, framing, and error monitoring. It deals with the transport of bits and the conversion from STS electrical pulses to OC light pulses. The line layer operates between line terminating equipment, providing synchronization, and

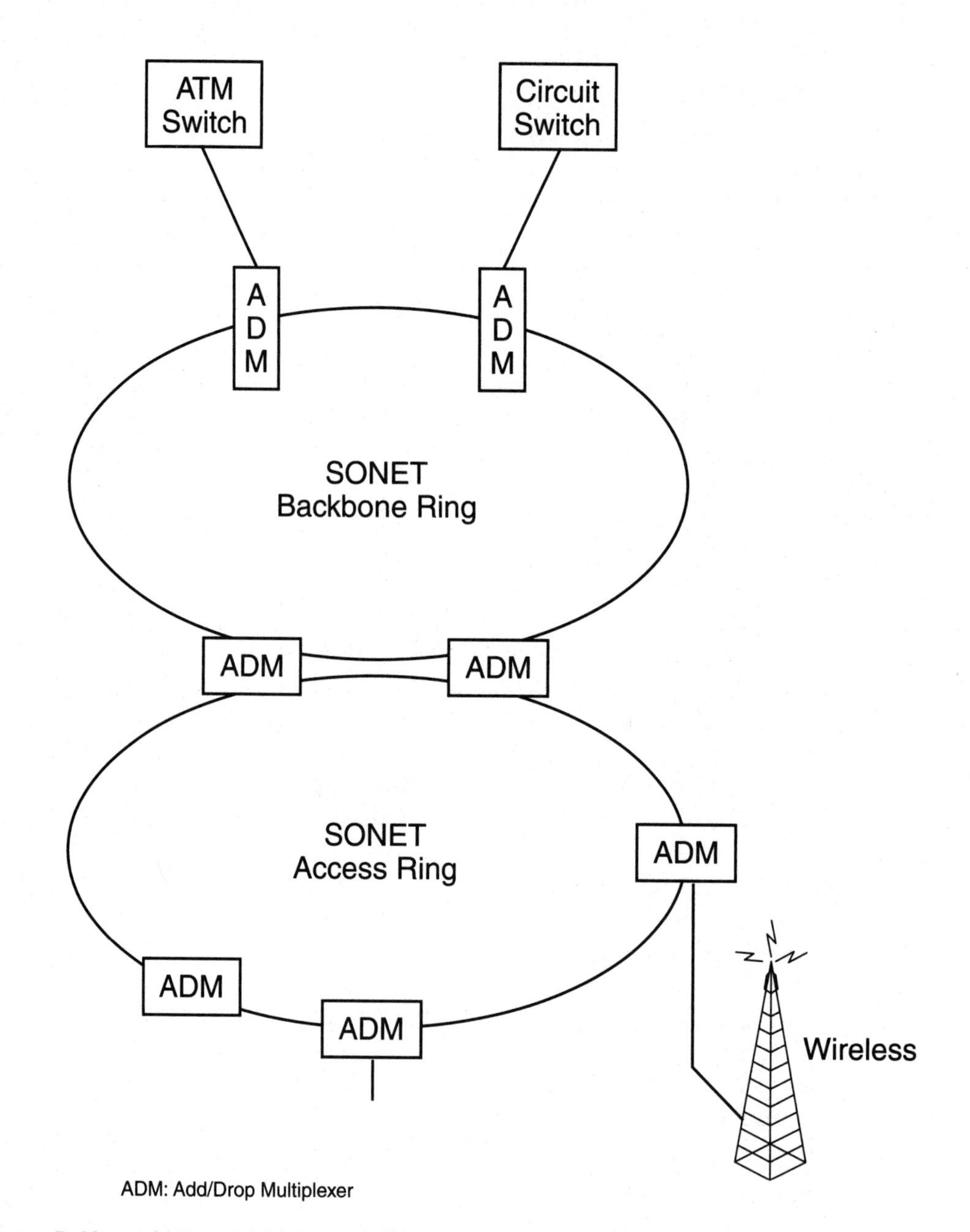

Figure 7–13 Add/Drop Multiplexers (ADM) at nodes provide access to the SONET Ring.

invoking APS when required. Finally, the path layer is used for end-to-end communications and control. The section, line, and path layers are embodied as overhead in the basic STS-1 frame.

Path	✦ Control end-to-end communication
Line	✦ Synchronization ✦ Invoke APS when required
Section	✦ Scrambling ✦ Framing ✦ Error Monitoring
Photonic	✦ STS electrical data is converted into light pulses and vice versa

Figure 7–14 The four layers of the SONET protocol.

STS-1 Frame Structure

As illustrated in Figure 7–15, each STS-1 frame is a 9-row by 90-column structure, for a total of 810 bytes, where each column contains 9 bytes. The frame can be divided into two main areas: transport overhead, and the Synchronous Payload Envelope (SPE). The first three columns of each frame make up the transport overhead, while the remaining 87 columns make up the SPE. The 27-byte transport overhead is divided into two pieces: section overhead (9 bytes), and line overhead (18 bytes). The 783-byte SPE can be further divided into two parts: the STS Path Overhead (POH), and the payload. The STS POH consists of 9 bytes and is used to communicate diagnostic information from the point where a payload is mapped into the STS-1 SPE to where it is delivered. The order of transmission of bytes is row-by-row from top to bottom and from left to right (with the most significant bit first).

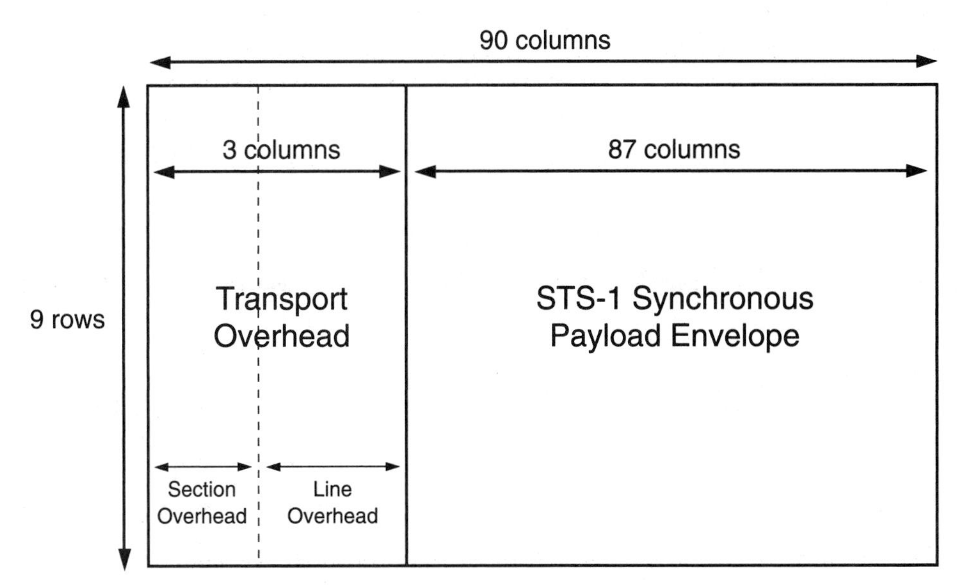

Figure 7–15 STS-1 frame format.

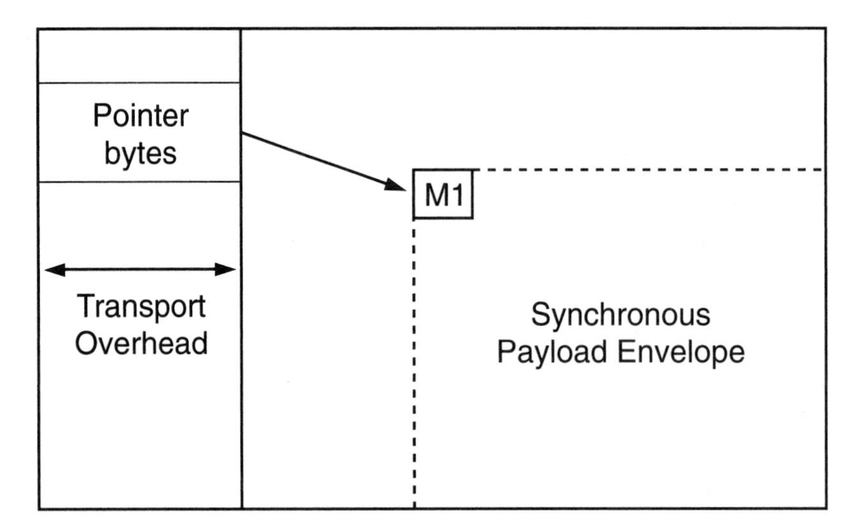

Figure 7–16 A payload pointer indicates the location of the SPE.

SONET provides substantial overhead information, allowing simpler multiplexing and greatly expanded Operations, Administration, Maintenance, and Provisioning (OAM&P) capabilities. The transport overhead of each individual STS-1 frame is aligned before interleaving, but the associated STS SPE does not have to be aligned because each STS-1 transport overhead has a payload pointer to indicate the location of the SPE, as shown in Figure 7–16. The SPE can have any alignment within the frame, so that it may begin in one STS-1 frame and end in the next. The pointer bytes in the STS transport overhead indicate where the STS-1 SPE begins. Once the payload is multiplexed into the SPE, it can be transported and switched through SONET without having to be examined or demultiplexed at intermediate nodes. The pointers also compensate for frequency and phase variations as they allow the transparent transport of SPEs between nodes with separate network clocks but having almost the same timing. For these reasons, SONET is said to be service-independent or transparent.

ASYNCHRONOUS TRANSFER MODE (ATM)

Asynchronous Transfer Mode (ATM) is a cell relay transport mechanism that evolved from the development of the Broadband ISDN (B-ISDN) standards. ATM is a telecommunications concept defined by the ANSI and the ITU standards committees for the transport of a broad range of user information—voice, data, and video communication. It can aggregate user traffic from multiple applications onto a single User Network Interface (UNI). Instead of using the processor-intensive slow-speed services of X.25 packet switching or the speedier Frame Relay, a mix of packet/frame technology evolved using a fixed-size *cell*. The term cell is just a different name for a group of data bits, like the terms "packet" and "frame" used in other standards. Both X.25 and Frame Relay use variable-length packets or frames.

This causes some latency within a network because the processing equipment uses special timers and delimiters to ensure that all of the data is encapsulated as a single unit.

In ATM, data is broken down into fixed-size units called cells and transported through the network. The system operation can be characterized as being similar to Conventional TDM because it transmits ATM cells synchronously and continuously, whether or not any data is being sent. The term asynchronous is used to describe an ATM network because traffic does not need to wait for any particular cell or time slot. When a user sends data, the data is allocated to cells dynamically. ATM, which is designed to be carried on SONET, inserts fixed-size cells into the STS payload, which allows for the fluid and dynamic allocation of the bandwidth available. ATM starts at 50 Mbps (equivalent to one OC-1) and to achieve the 155 Mbps rate, three OC-1s are kept together as a single data stream. The rate can escalate to 622 Mbps and go into the 1.2 Gbps class.

Cell Relay combines the high throughput and bandwidth optimization of Frame Relay with the predictability of Time Division Multiplexing, making it suitable for both bursty data traffic and isochronous voice/video traffic. At the same time, ATM gets around the inefficiency of the fixed time slot, rates, and formats of the TDM world and the unreliable packet-switching world by dynamically allocating whatever is necessary to the user whenever the user wants it.

ATM Layer and Cell Format

ATM is much more than a Layer 2 or data link layer technology as it provides services associated with network and transport layers, Layers 3 and 4. It performs routing functions as well as intelligent queuing and scheduling mechanisms to cope with IP or other network layer protocols. Also, it allows multiple data streams such as IP, voice, SNA, video, and others to share the same link. On the other hand, ATM clearly fits the Layer 2 definition when it interfaces with the physical layer. Its Layer 1 options include many fiber optic and copper transmission media, including SONET (at practically any throughput rate), T-1 lines, and Category 5 UTP copper cabling. ATM provides extraordinary scalability of both bandwidth and distances over which the physical network can extend, and the fixed-length cells make for fast hardware-based switching. An ATM cell is like a typical Layer 2 frame in that it has error-correction capabilities and addressing information with significance only for the local data link.

An ATM cell consists of 53 eight-bit bytes (also called octets). The 53-byte cell size results from a compromise between voice and data communications at the design stage. Engineers who were primarily interested in voice applications, which have short, repetitive frames, had wanted ATM cells to be made even smaller to keep the packetization delay low enough to handle digitized voice in real time. **Packetization delay** is defined as the time it takes to fill a cell. On the other hand, the 53-byte cell size is too small for efficient data transfer. The 53 bytes consist of 48 bytes of payload or actual message bits called the Protocol Data Unit (PDU) and 5 bytes of routing, priority and other useful information called the header, as shown in Figure 7–17.

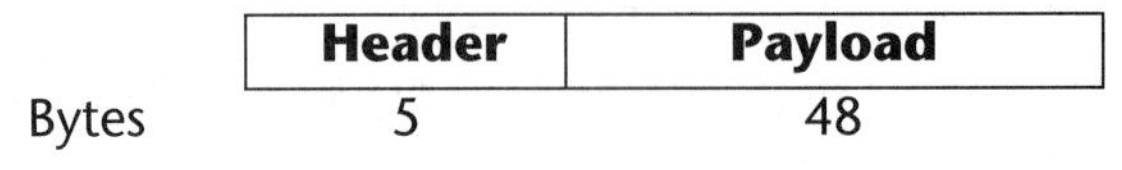

Figure 7–17 An ATM cell structure.

At the transmitting end, the ATM Adaptation Layer (AAL), which sits right above the ATM layer, formats data into 48-byte cell payloads in a process known as *segmentation.* The ATM layer then adds a 5-byte header with connection and QoS information to each cell and passes the cells, now 53 bytes, to the physical layer. At the destination, the ATM layer receives the cells, removes their headers, and passes them to the AAL. The AAL reconstructs them into high-level data in a process known as *reassembly,* and transmits the data to the destination devices. This **segmentation and reassembly** process is also known as packet-to-cell conversion. Figure 7–18 shows how the bottom three layers of the ATM protocol stack correspond to the bottom two layers of the OSI model.

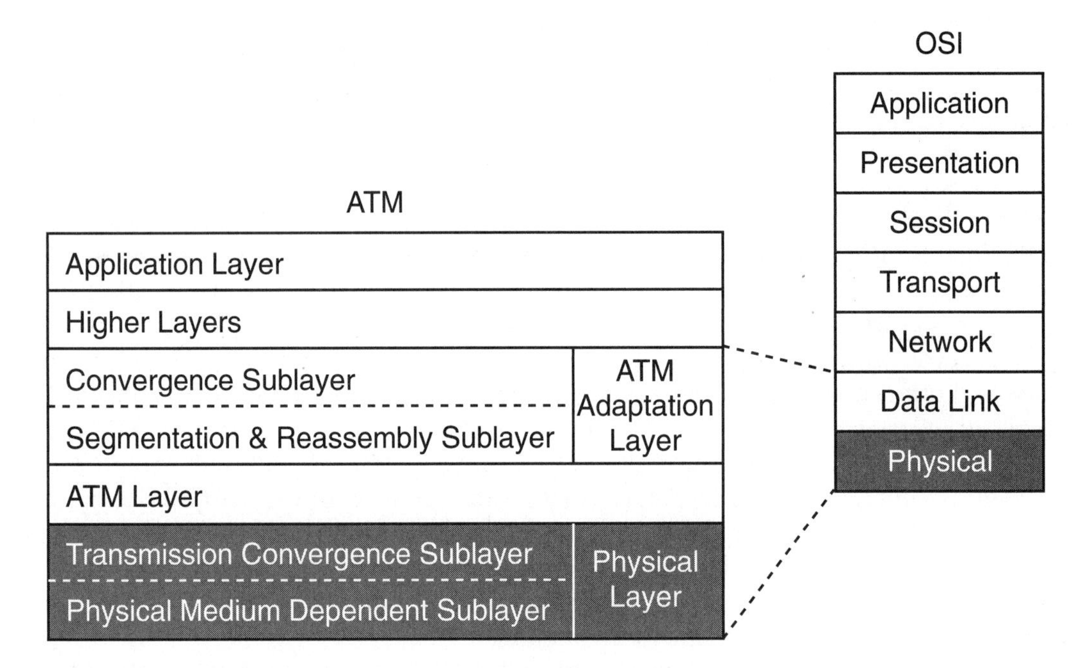

Figure 7–18 Correspondence between the bottom three layers of the ATM protocol stack and the bottom two layers of the OSI model.

ATM is a connection-oriented protocol that requires a virtual path, which is a collection of virtual circuits, to be set up for each flow. A virtual circuit, which is the basic unit, carries a single stream of cells in order from node to node to create an end-to-end connection across an ATM network with all cells routed the same way through the network. This results in faster recovery in case of major failures. Virtual circuits may be point-to-point or point-to-multipoint and can be statically configured as PVCs or dynamically controlled via signaling as **Switched Virtual Circuits (SVCs)**, providing a rich set of service capabilities. SVCs are the preferred mode of operation because they minimize reconfiguration complexity.

Advantages of ATM

In a conventional circuit-switched voice network, quality is not an issue since bandwidth is guaranteed, and transmission delay (latency) and delay variation (jitter) simply does not occur. On the other hand, in conventional packet-switched data networks, delay has been acceptable because data transmissions have not been as time-sensitive as voice. Until recently, guaranteed bandwidth has not been a concern in data networks. The practice was if data reached its destination sooner, so much the better, but it if it got there a few minutes later, there was little impact. As the criticality of data has increased, so too has the need for guaranteeing both bandwidth availability and timely delivery.

ATM's primary advantage is its ability to accept real-time traffic, such as voice and video, without introducing latency and jitter, while also serving as a statistical multiplexer for a number of streams of non-real-time data. Traditional synchronous networks can support real-time traffic by provisioning a circuit for its exclusive use. Their drawback is that when there is no traffic, the capacity of that circuit is wasted, which is an expensive proposition. Datagram networks, such as the Internet, use their carrying capacity very efficiently by leveling out bursts of traffic, using queues on intermediate nodes, and forcing end nodes to retransmit lost packets. The price of this bandwidth efficiency is unpredictable performance, intermittent high latency, and at times, unacceptable levels of jitter for consumers of real-time data.

Let us assume that there is substantial congestion on the network. Switches and routers manage congestion by creating queues. Queues hold excess traffic and dole it out in an orderly way when the congestion abates. With multiple queues, the switch or router can implement priorities based on some form of traffic differentiation. Queues must be managed carefully to prevent them from creating unpredictable latencies and, as a result, severe jitter. An important characteristic of ATM is that the network will not give traffic access to a virtual circuit unless it can ensure a contracted QoS standard from end to end. Voice and video traffic get the same latency and jitter metrics on an ATM network that they get with a network made up of TDM circuits using leased lines. A data stream will get the equivalent of a busy signal if the ATM network is too congested.

ATM's ability to create virtual paths and channels means that multiple customers' traffic can be carried on a single network without security concerns or encryption overhead. The virtual circuits ensure true QoS on a per-connection basis, allowing for the convergence of voice, video, and data traffic. Using statistical multiplexing, ATM guarantees the necessary bandwidth and latency characteristics by protecting each call from the effects of every other call on the network. ATM data switches are becoming increasingly popular because they support QoS, which is vital for real-time applications such as video conferencing and other mission-critical applications.

At the same time, data that is not time-sensitive can make do with the leftover capacity of the channel and potentially pay a lower fare for sacrificing guaranteed service quality. ATM provides four service classes that are identified in Figure 7–19.

ATM Class of Service	Characteristics	Applications
Constant Bit Rate (CBR)	✦ Specifies fixed bit rates, so it is analogous to a leased line	✦ Voice ✦ Other latency-sensitive applications
Variable Bit Rate (VBR)	✦ Provides a specified throughput but data is not sent evenly	✦ Voice ✦ Video conferencing
Available Bit Rate (ABR)	✦ Provides a guaranteed minimum capacity but allows data to burst over when the network is available	✦ Electronic document transfer
Unspecified Bit Rate (UBR)	✦ Does not guarantee any throughput levels	✦ File transfer ✦ Electronic mail

Figure 7–19 Summary of ATM classes of service.

The Constant Bit Rate (CBR) and the Variable Bit Rate real-time (VBR-rt), are both appropriate for latency-sensitive applications such as voice. The VBR-rt requires a timing relationship while the VBR non-real-time (VBR-nrt) is designed for most LAN-to-LAN applications. For instance, a videoconferencing signal can be mapped to the VBR-rt class, while a file transfer can be mapped to an Available Bit Rate (ABR) or an Unspecified Bit Rate (UBR) connection. Thus, ATM creates a single network for voice, video, and data, where voice and video streams can maintain the low latency and low jitter they require.

Drawbacks of ATM

ATM's major drawback is the cell tax, which refers to the overhead for converting IP traffic to ATM. The segmentation and reassembly results in wasted bandwidth when compared with pure IP throughout the entire transfer. The 10 percent (5-byte header for every 48-byte payload) overhead that results is fairly high. In addition, the 53-byte ATM cells are too small for efficient data transfer since a minimum IP data packet is 64 bytes. For transmission over ATM, this must be split into two cells. Fitting irregular-length packets into cells often leaves them only partially filled. Furthermore, a large packet will have to be segmented into many cells. As an example, for a 1,518-byte packet, there will be as many as 32 cells, and the loss of any one of these cells requires retransmitting the entire packet.

Because of these factors, IP running over ATM may not realize more than 80 percent of the underlying SONET line rate. While this may be acceptable when there is a mix of voice and data traffic, pure IP networks resent wasting 20 percent of their bandwidth.

Another disadvantage is that the technology can be disruptive for most existing LAN connections. Its connection-oriented behavior and heavy protocol overhead require different expertise, management techniques, service requirements, and training. Although ATM was originally designed to run all the way from the core of the network to the desktop, the speed of Ethernet interfaces keeps increasing without costing nearly as much as the ATM interfaces they compete with. Therefore, as illustrated in Figure 7–20, while ATM dominates service-provider core-data networks and has presence in enterprise backbones, it has only been adopted for end-user systems in niche applications, where the utmost performance is worth the high price. Integrating ATM at the network core with Ethernet at the edge of the network has become a reality, although meshing ATM with Ethernet networks is fairly complex. There have been initiatives to overcome these difficulties. Classical IP (CIP) over ATM, LAN Emulation (LANE), and Multiprotocol (MPOA) over ATM let ATM operate like a connectionless, routed network. However, these solutions add complexity and expense to an environment that is not inexpensive in the first place.

While companies may like the benefits of ATM, in a significant number of cases, their networks will not require the quality of service features for which ATM was designed. Very often, the logical answer will be evolutionary rather than revolutionary: extending their current Ethernet capabilities with upgrade to Gigabit Ethernet. This will preserve users' financial and training investment in their Ethernet-based networks, while enabling them to adapt to future requirements. However, Gigabit Ethernet lacks ATM's resiliency, and it does not support load sharing. Without load sharing, even 1000 Mbps backbones can become overwhelmed when supporting multiple Fast Ethernet feeds. Nor does Gigabit Ethernet support true, end-to-end QoS. Although Class of Service (CoS) traffic *prioritization* schemes can be added to the technology, they increase the cost and complexity of Gigabit Ethernet and do not provide ATM's jitter and latency guarantees.

An extension of QoS, prioritization is the means by which certain frames are given preferential treatment over others. This technique enables Layer 2 switches and Layer 3 routers to differentiate between packets and handle them differently based on their tags. Prioritization is meaningless unless a congestion level of 100% is reached on a given output port. For example, if a switch is operating at 80% port capacity, prioritization services do not kick in because there is no contention for bandwidth.

ATM Standards

ATM is a set of international interface and signaling standards defined by the ITU, with the ATM Forum playing a pivotal role. The Forum, established in 1991, is an international voluntary organization composed of vendors, service providers, research organizations, and users. Its purpose is to accelerate the use of ATM products and services through

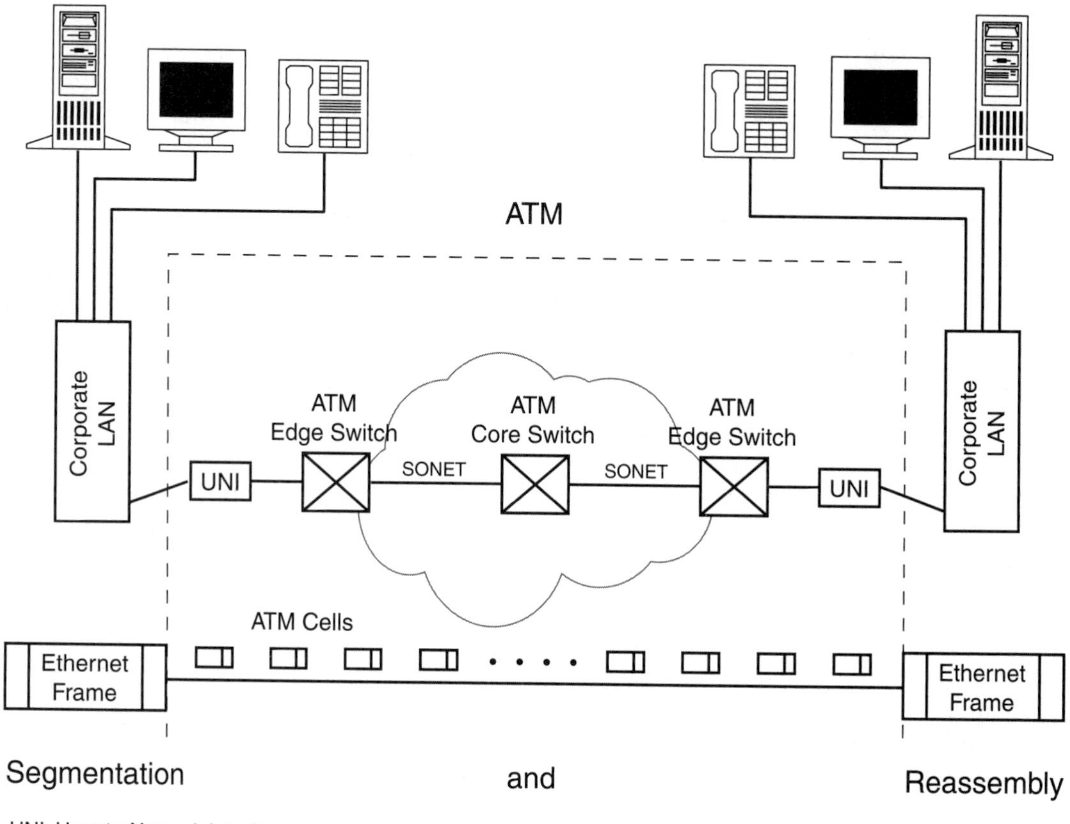

Figure 7–20 An ATM network.

the rapid convergence of interoperability specifications, promotion of industry cooperation, and the development of a cohesive set of specifications that provide a stable ATM framework. One key benefit of standardized architecture as compared with a proprietary switch is that when the signaling software on the NIC is completely standards based, any switch developed with this technology can be seamlessly integrated with any other, similarly open, platform. As a result, service providers can achieve true vendor independence as they expand their distributed multimedia switch sites.

The two standards that provide connectivity between networks are Broadband Inter-Carrier Interface (BICI) and Public or Private Network-to-Network or Node-to-Node Interface (PNNI). PNNI is more feature-rich than BICI and supports routing and bandwidth reservation sensitive to class of service. It uses a multilevel hierarchical routing model, providing scalability to large networks. Parameters used as part of the path computation process include the destination ATM address, traffic class, traffic contract, QoS requirements, and link constraints. The Forum's latest efforts, in cooperation with the Internet

Engineering Task Force (IETF), have been to bring end-to-end QoS for IP traffic crossing ATM backbones by defining new methods for transporting IP data and video over WANs. The goal is to go beyond simple transport to a Guaranteed Frame Rate (GFR) specification, which is a new ATM class of service that lets users mix Ethernet and IP traffic more effectively on an ATM network. For example, Real-Time Multimedia Over ATM (RMOA) specification is aimed at bringing reliability to IP voice and video applications. Figure 7–21 depicts perceived transmission quality versus latency for voice applications.

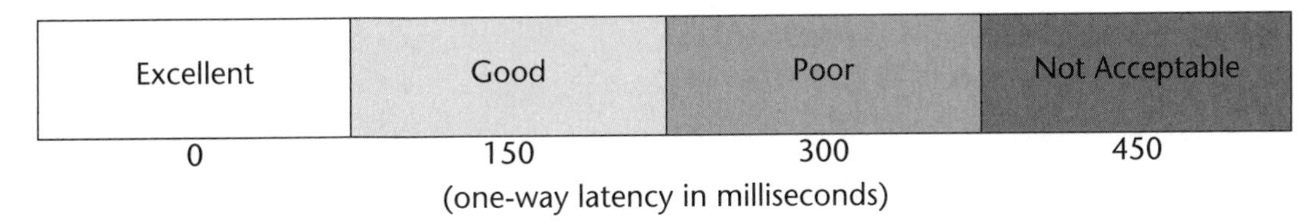

Figure 7–21 Perceived quality versus latency.

PACKET OVER SONET (POS)

A new technology called Packet over SONET (PoS), also known as IP over SONET, was devised in 1994 and adopted a year or two later. PoS is designed specifically for high-speed, high-volume IP packet traffic and lends itself most comfortably to a data-only network. IP packets have no particular requirement for service quality, at least for traditional IP networks. ATM's statistical-multiplexing function can be performed just as well by a router when there is no critical real-time traffic. These circumstances resulted in PoS emerging as a strong contender for connecting high-speed backbone routers as it eliminates the need for ATM by mapping IP packets directly onto SONET frames, slashing the cell tax. PoS protocols are optimized for variable-length packets rather than fixed-length ATM cells.

For example, assume that a typical IP packet for Internet browsing is about 576 bytes. If you transfer this over ATM on an OC-3 link, you will realize an efficiency of about 80% of the bandwidth for the actual IP transfer. The other 20% is a combination of ATM and SONET overhead. Now, if the 576 bytes are transmitted using PoS technology, you will find that you get almost 95% efficiency, which is a significant savings.

Consider the network shown in Figure 7–22 that uses a combination of T-1 and T-3 lines for voice and data, which makes it a prime candidate for ATM over SONET or PoS redesign. The ATM approach, depicted in Figure 7–23, adds capacity and simplifies network design by using OC-1 and OC-3 links, but there is no protection against circuit failure. In the PoS approach, depicted in Figure 7–24, the network is built around an OC-12 SONET ring with APS, making it a more robust alternative. Although one cannot do away with the data link layer, it can be substituted with the HDLC-based Point-to-Point Protocol (PPP).

One special example of PoS application is the undersea cable, where bandwidth is at a premium. There are PoS links running beneath the Atlantic and the Pacific oceans. PoS

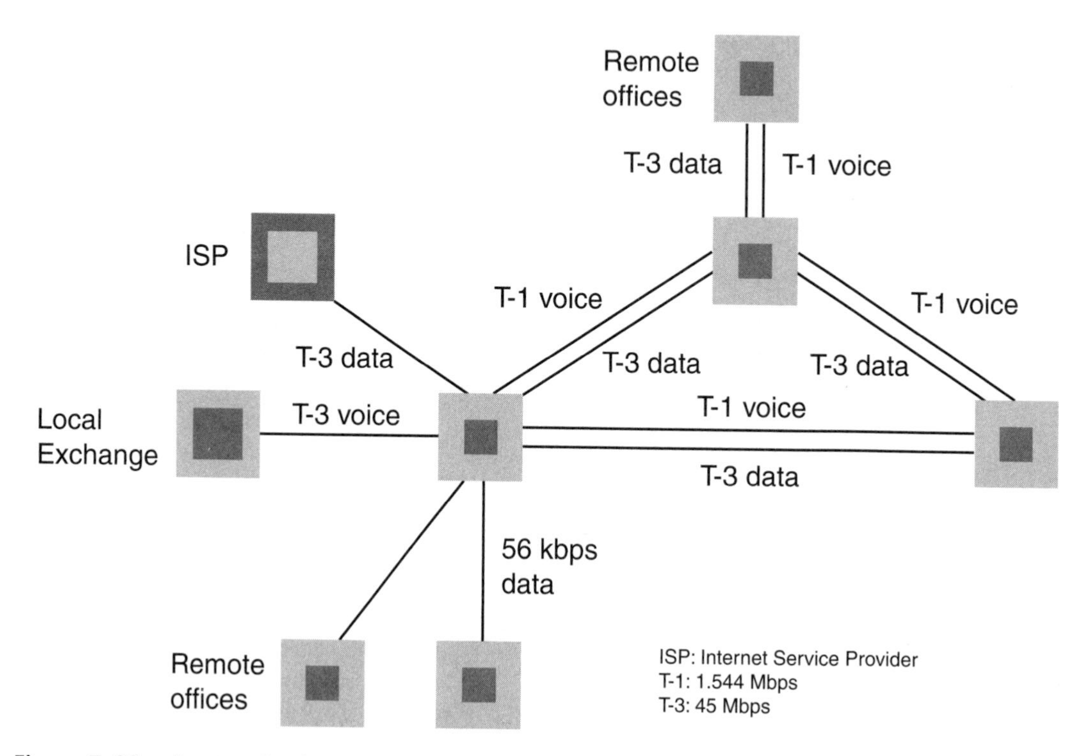

Figure 7–22 A network using a combination of T-1 and T-3 lines for voice and data is candidate for ATM or PoS.

can also operate as a high-speed WAN technology over OC-3 leased lines, making it an alternative for enterprise applications such as corporate Intranets over PPP. There are various relevant standards for employing PPP instead of ATM at Layer 2, namely, a combination of two IETF recommendations: RFC1619 (PPP over SONET/SDH) and RFC1662 (PPP in High-Level Data Link Control-like framing). With no ATM, QoS controls are added at Layer 3 by implementing MultiProtocol Label Switching (MPLS), which includes ATM-like features, such as differentiated CoS and QoS management. Although MPLS enables PoS to deliver QoS that compares well with that of ATM, there are concerns over the processing power MPLS requires in routers when more than two or three classes of service have been defined.

DYNAMIC SYNCHRONOUS TRANSFER MODE (DTM)

Dynamic Synchronous Transfer Mode (DTM) is a new broadband network technology that helps enterprise networks efficiently carry voice, data and streaming video on a single, integrated network. It combines the advantages of circuit- and packet-switching. DTM can automatically set up connections and assign a certain number of channels or

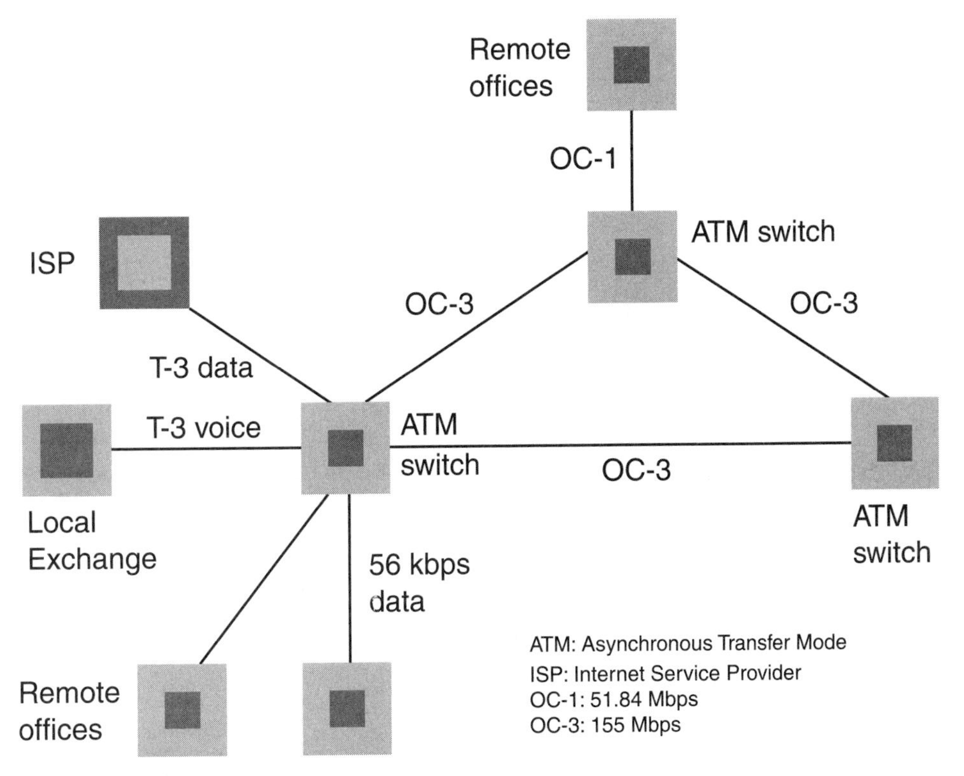

Figure 7–23 Framework for ATM over SONET network.

bandwidth to each connection, depending on traffic type and bandwidth requirements with little or no delay. Specifically, DTM divides fiber pipe capacity into frames of 125 µs, which are further divided in slots, depending upon the bit rate. For example, with a bit rate of 2.5 Gbps, the number of slots is around 4,800. Each DTM node in the network maintains a status table that contains information about free bandwidth in other nodes. When more bandwidth is needed, a node consults its status table to decide which node to ask for more bandwidth. Once established, a DTM channel provides guaranteed service, and, depending on the channel's traffic load, the capacity of the channel can be altered during operation. DTM also provides a means by which to log resource usage for billing, management, and administration purposes.

As a Layer 2 technology that works directly on fiber, DTM may be used on top of, or in parallel with, existing SONET infrastructures. While it was primarily developed to efficiently handle IP traffic, its synchronous properties also allow it to carry legacy telephony and leased-line traffic effectively over the same infrastructure as IP and streaming video. DTM maintains full isolation between the different transmission streams and produces little overhead. These synchronous properties are more important today since IP is also carrying more real-time applications. DTM combines the reliability and guaranteed

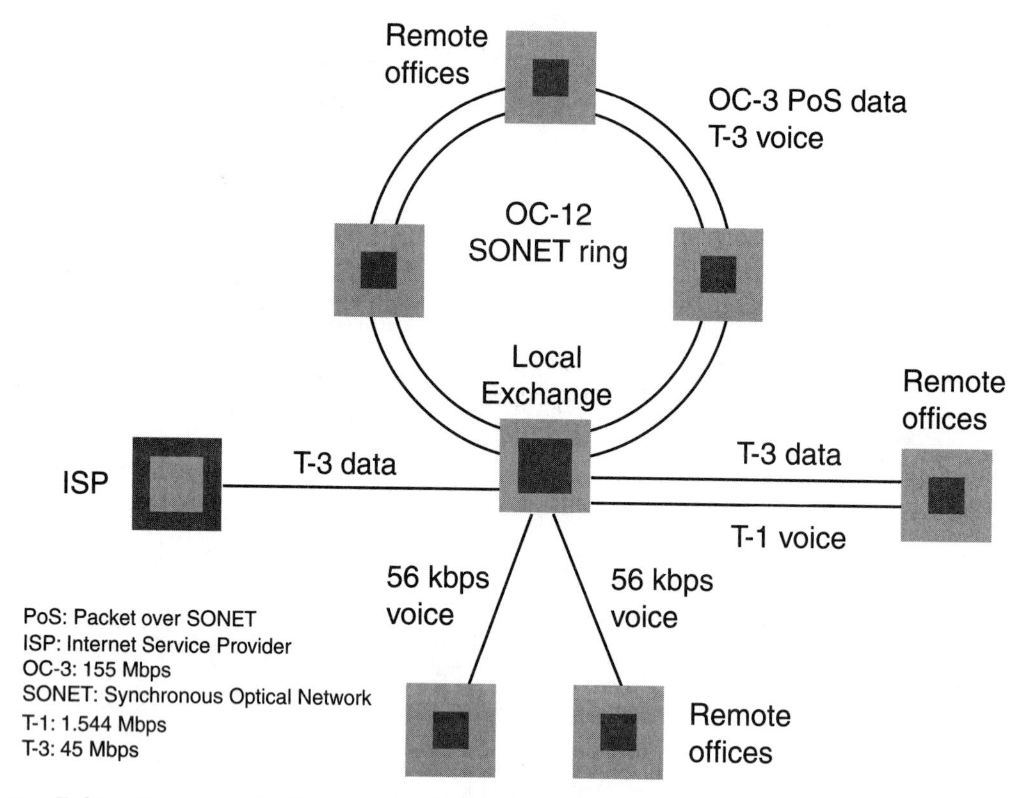

Figure 7–24 Framework for Packet over SONET network.

service of SONET and the flexibility and traffic-engineering capabilities of ATM, but at a lower cost and with less complexity. The protocol also works over a DWDM infrastructure, where DTM switches provide add-drop support between nodes that share a wavelength and switch traffic between nodes situated on different wavelengths. Its major disadvantage is that it is still a new technology, so it has not been carefully scrutinized and lacks the international standardized architecture of ATM.

RESIDENTIAL OR SMALL BUSINESS ACCESS TECHNOLOGIES

The dial-up model emerged in the late 1980s to meet the needs of smaller businesses, branch offices, residential users, and emerging telecommuters. Dial-up access can be either analog or digital. The 19.2 kbps WAN connection that eventually evolved into the V.90 56 kbps connection is an asynchronous analog connection. At the receiving end, a Remote Access Server (RAS) located at the organization's office or at the Internet Service Provider's (ISP) POP terminates the connection. Providing high-speed data service that is always on is relatively new ground for cable and telephone companies. While cable companies have an early lead, the ubiquity of telephone lines gives telephone companies an edge. The public perception, as well as the business mergers between the cable and telephone industries, promises to make it an interesting contest.

Digital Subscriber Line (DSL)

Digital Subscriber Line (DSL) access technology, illustrated in Figure 7–25, is a new platform for delivering broadband services to homes and small businesses, thus bringing the information highway to the mass market. The technology is also used for connecting remote corporate offices to company headquarters located in the same metropolitan area. It can be implemented on most of the existing copper infrastructure, enabling the rapid and near-ubiquitous offering of new high-speed data access services with minimal expense. DSL supports a wide variety of high-bandwidth applications, such as high-speed Internet access, telecommuting, virtual private networking, and streaming multimedia content. In addition to fast Internet access, the technology accommodates ordinary analog telephone calls over the same wires. The line-sharing technique is a sophisticated form of old-fashioned frequency-division multiplexing, with each service—analog voice telephony, uplink data and downlink data—occupying its own frequency band but using advanced modulation techniques to overcome the shortcomings of twisted-pair cabling. The analog telephone line has all the traditional features such as call waiting, call forwarding, and others.

Speeds for DSL are dependent on loop links, with substantial decrease in throughput as users travel farther away from the local switching exchange. The largest drawback to DSL is that it is a relatively new technology. Given that universal industry standards have yet to be ratified and that the equipment is based on proprietary solutions, there is slight reluctance to release this technology on a large scale until cohesiveness is established.

Prominent DSL related technologies, identified in Figure 7–26, include Rate Adaptive DSL (RADSL), High data-rate DSL (HDSL), Symmetric DSL (SDSL), and Asymmetric DSL (ADSL), collectively known as xDSL. RADSL adjusts rates based on the quality and length of the line, while HDSL uses advanced modulation techniques for efficient transmission of signals over T-1 lines up to 12,000 feet long and uses less bandwidth and no repeaters. ADSL is intended to complete the high-speed connection with the customer's premise. It transmits two separate data streams with much more bandwidth devoted to the downstream leg going to the customer than to the returning leg. The speed depends on the length and condition of the local loop, where the uplink rates can go up to 768 kbps and downlink rates can vary between 1.5 to 60 Mbps under ideal conditions. ADSL succeeds because it takes advantage of the fact that most of its target applications, such as Internet access and remote LAN access, function perfectly well with a relatively low upstream data rate. For example, MPEG movies require 1.5 or 3 Mbps downstream but need only between 16 kbps and 64 kbps upstream. The protocols controlling Internet or LAN access require somewhat higher upstream rates but in most cases can get by with a 10-to-1 ratio of downstream-to-upstream bandwidth.

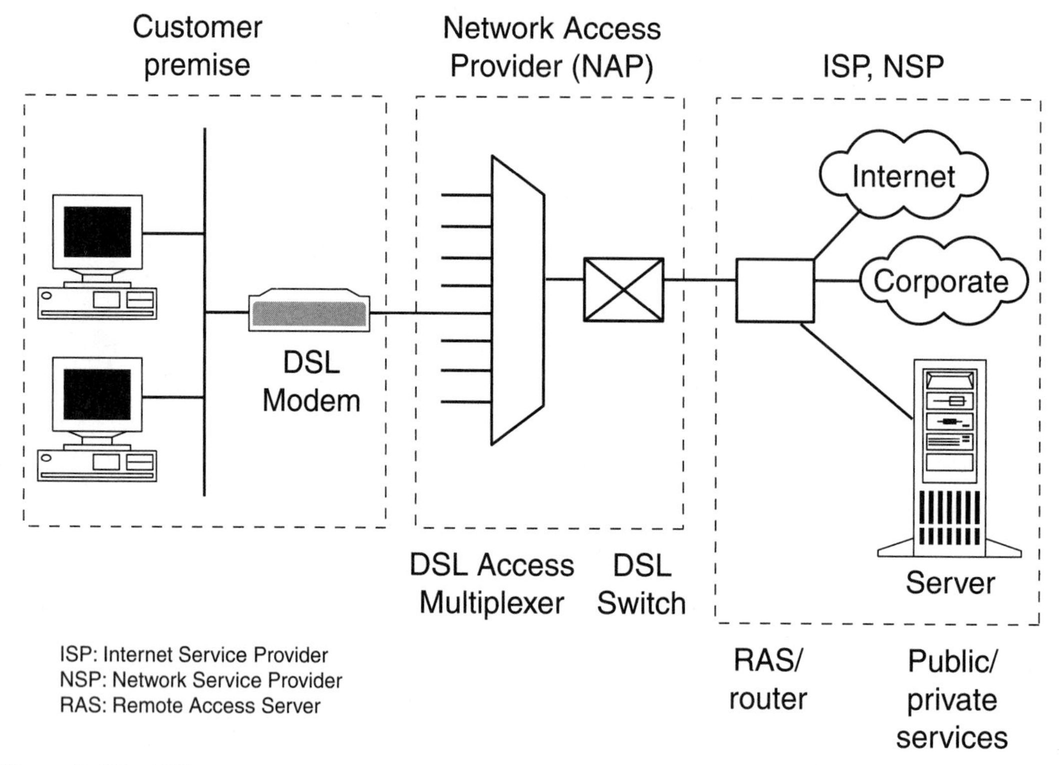

Figure 7–25 DSL access.

Cable Modems (CMs)

The coaxial cables that bring TV channels to about 90% of American homes offer tremendous potential for high-speed data connections to the Internet. In theory, a **Cable Modem (CM)** provides speeds up to 35 Mbps downstream and 10 Mbps upstream, more than 100 times as fast as most modems and ISDN adapters. However, in most cases, neighbors have to share the cable bandwidth to the headend, making practical speeds slow down to 8.5 or 9 Mbps—a problem not shared by ADSL. With cable modems, the available bandwidth per person decreases as more people log on.

A CM has two connections: one to the cable wall outlet and the other to a PC or to a set-top box for a TV set. As shown in Figure 7–27, the Cable Modem Termination System (CMTS) can talk to all the CMs, but the CMs can only talk to the CMTS. So if two CMs need to talk to each other, the CMTS will have to relay the messages. One of the limitations to the proliferation of CMs is that since the current coaxial cable network is not designed for two-way transmissions, cable companies have to expend significant effort and expense to facilitate the handling of two-way transmissions.

Today's increasing demand for bandwidth has created the need for a Hybrid Fiber/Coaxial (HFC) cable network. Older CATV systems were provisioned using only coaxial

DSL Option	Maximum Speed to Customer	Maximum Speed from Customer	Characteristics
Integrated Services Digital Network DSL (ISDN/DSL)	144 kbps	144 kbps	Uses mature ISDN technology
Rate-adaptive DSL (RADSL)	Up to 6.3 Mbps	Up to 640 kbps	Service provider must activate the desired bandwidth up to the capability of the loop
High data-rate DSL (HDSL)	1.544 Mbps	1.544 Mbps	Works well at short distance from local switching exchange. Very sensitive to condition of copper plant
Symmetric DSL (SDSL)	768 kbps	768 kbps	Supports equal transmission in both directions for IP telephone calls
Asymmetric DSL (ADSL)	9 Mbps	640 kbps	Maximizes amount of data that can be sent from Internet to PC, consistent with POTS

Figure 7–26 Prominent DSL technologies.

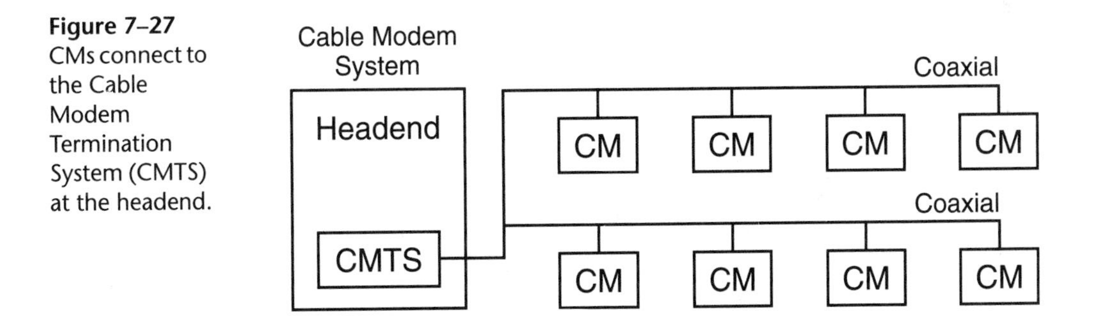

Figure 7–27 CMs connect to the Cable Modem Termination System (CMTS) at the headend.

cable, but modern systems use fiber transport from the headend to an optical node located in the neighborhood to reduce system noise. Coaxial cable runs from the node to the subscriber so that the fiber plant is generally a star configuration with all-optical-node fibers terminating at a headend. Although HFC can provide 30 to 40 Mbps total capacity downstream using a single 6 MHz analog channel spectrum (typically shared by 100 to 250 homes), it has very limited upstream capacity.

Fixed Wireless

Fixed wireless, utilizing a Multi-channel Multi-point Distribution System (MMDS), is not a new technology. However, its utilization toward developing high-speed Internet services, to spread broadband networks to areas too expensive to reach using DSL or cable modems, is relatively young. It operates over a licensed spectrum in the 2.5 though 2.7 GHz band. As depicted in Figure 7–28, fixed wireless broadband gets its name from the antennas that need to be "fixed" high above the ground so they can broadcast broadband access to homes or businesses within a 35-mile radius. Unlike mobile wireless, fixed wireless uses a stationary digital transceiver that is at the home or business receiving the service. Basic customer premise equipment (CPE) includes an externally mounted antenna and a small wireless modem for a computer connection; CPE costs vary with the location of the site.

Carriers ignored early generations of fixed wireless equipment because a connection could only be made if the antenna was within line of site of a receiver, but new generation of equipment does not require an antenna to be within line of sight. Telephone companies could also use the technology to give customers DSL access without the big investments now required to wire remote neighborhoods. Fixed wireless operates at speeds comparable to DSL and cable modem services; typical downstream transmission rates are 1 Mbps, scalable to 10 Mbps with average upstream speeds around 512 kbps. Pricing is comparable to DSL and cable modem broadband service offerings, less CPE and setup outlays.

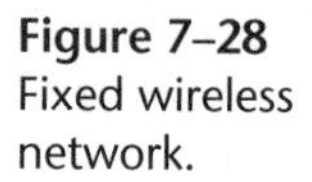

Figure 7–28
Fixed wireless network.

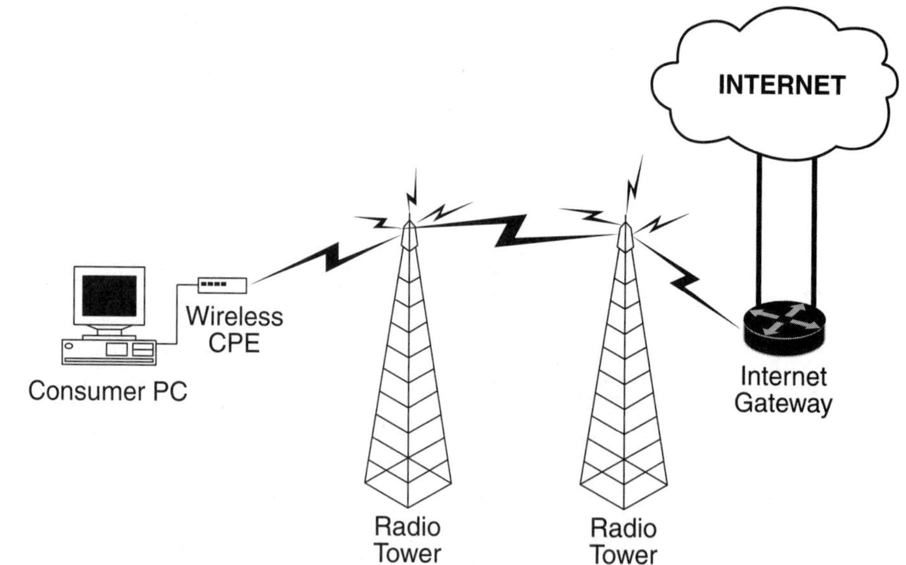

VSAT (Very Small Aperture Terminal)

VSAT (Very Small Aperture Terminal) is a satellite communications system that serves home and business users. A VSAT end user needs a box (transceiver) that interfaces between the user's computer and an outside small-diameter antenna dish (0.6 to 3.8 meter). The transceiver receives or sends a signal to a satellite transponder in the sky. The satellite sends and receives signals from an earth station computer that acts as a hub for the system. Each end user is interconnected with the hub station via the satellite in a star topology. For small businesses, VSAT technology represents a cost effective, secure and reliable solution for an independent communications network connecting a large number of geographically dispersed locations. VSAT networks offer value-added services capable of supporting data, voice, and video signals over LAN. Generally, these systems operate in the Ku-band and C-band frequencies.

Passive Optical Network (PON)

Passive Optical Network (PON) technology, still under experimentation, promises to facilitate the rollout of broadband services by cable television operators, incumbent telcos, data CLECs, and wireless carriers. PONs will break the bandwidth bottleneck, leading to a local access network that is digital, broadband, and interactive. In the short term, the most promising application for PONs is as an optical feeder for telcos, cable TV companies, and wireless providers. PONs complement emerging broadband access technologies such as xDSL (copper), HFC (coax) and LMDS (wireless), rather than competing against them, by allowing for shorter access drops and higher bandwidth. Because it provides a low-cost, high-bandwidth, fault-tolerant solution for carriers interested in delivering lucrative revenue-generating broadband services, PON is a promising solution for the next-generation broadband local loop.

The most significant benefit that PON offers to end-users is a vast increase in the amount of bandwidth delivered to homes and businesses. It opens the door to more bandwidth-intensive applications such as teleconferencing, video-on-demand, electronic commerce, and digital broadcasting, and it eliminates any network-induced wait time experienced when surfing the Internet. PON provides a dedicated optical connection that is secure all the way to the central office because there are no active electronics in the outside plant—just optical fiber to the splitter and then to the home. This is particularly important to users with a home-office connection to the corporate LAN. PON also offers subscribers quicker service delivery, as new subscribers can be easily added to an existing facility by interconnection at an existing splitter or by installing a second-tier splitter for increased drop capacity.

ATM-Passive Optical Network (ATM-PON) is cited as potentially the most effective broadband access platform. It is a point-to-multipoint, cell-based, optical-access architecture that facilitates broadband communications over a purely passive optical-distribution network between an Optical Line Terminal (OLT) at the central office and multiple remote Optical Network Units (ONUs) on customer premises within a 20-km radius. It was first proposed as a standardized solution in the early 1990s by the Full-Service Access Network (FSAN) initiative, which comprised 14 telephone companies from around the world. FSAN is an effort to set common requirements for full-service optical-access networks among all operators globally, and it works closely with the ITU-T.

In an ATM-PON system, a maximum of 64 ONUs can share the capacity of a single fiber using ATM transport and passive optical splitter/combiner technology. Full-duplex transmission via a single fiber facility can also be achieved using independent wavelengths (1310/1550 nm) for each direction. Due to the physical convergence of the upstream burst transmissions from each ONU via one or more passive optical splitter/combiner elements, the timing of each ONU transmission must be precisely synchronized with delay compensation to account for unequal distances between the OLT and each individual ONU. To accomplish transmission-delay equalization among all ONUs, the OLT performs a ranging procedure to measure the logical reach distance to and from each ONU and assigns a specific equalization delay adjustment to each ONU.

Ethernet PONs are in the early phases of commercial development. Although ATM PONs have a slight head start in the marketplace, current industry trends including the rapid growth of data traffic and the increasing importance of Fast Ethernet and Gigabit Ethernet services favor Ethernet PONs.

SUMMARY

Since the 1990s, there have been rapid technological advancements in wide-area broadband technologies. Early WAN access interconnecting corporate offices was confined to private, leased-line connections which were expensive, inefficient, and required dedicated back-up circuits to improve reliability and fault-tolerance. As a result, smaller offices did not have a viable way to communicate over the WAN. But the bursty nature of the way we conduct our business allowed the sharing of resources among many users, thereby sharing the bandwidth available. WAN technologies such as X.25, Frame Relay, SMDS and others replace fully meshed networks of leased lines with substantial cost savings, and extend the performance and efficiencies of a company's LANs over a wide area on a switched, as-needed basis. Figure 7–29 summarizes the available spectrum of networking technologies and data rates and Figure 7–30 provides an overview of the strengths and weaknesses of popular WAN technologies.

Among Layer 2 technologies, Frame Relay and SMDS are very efficient for data communications, but they cannot adequately support voice or video traffic because of variable end-to-end delivery times. Voice and video transmissions are of a constant bit rate nature and intolerant of delays. ATM was designed for handling multimedia applications (voice, video, data, and a variety of other services) without degradation by providing full QoS and traffic management guarantees on end-to-end connections. Large mission-critical infrastructures that demand reliable delivery of high-bandwidth time-sensitive appli-

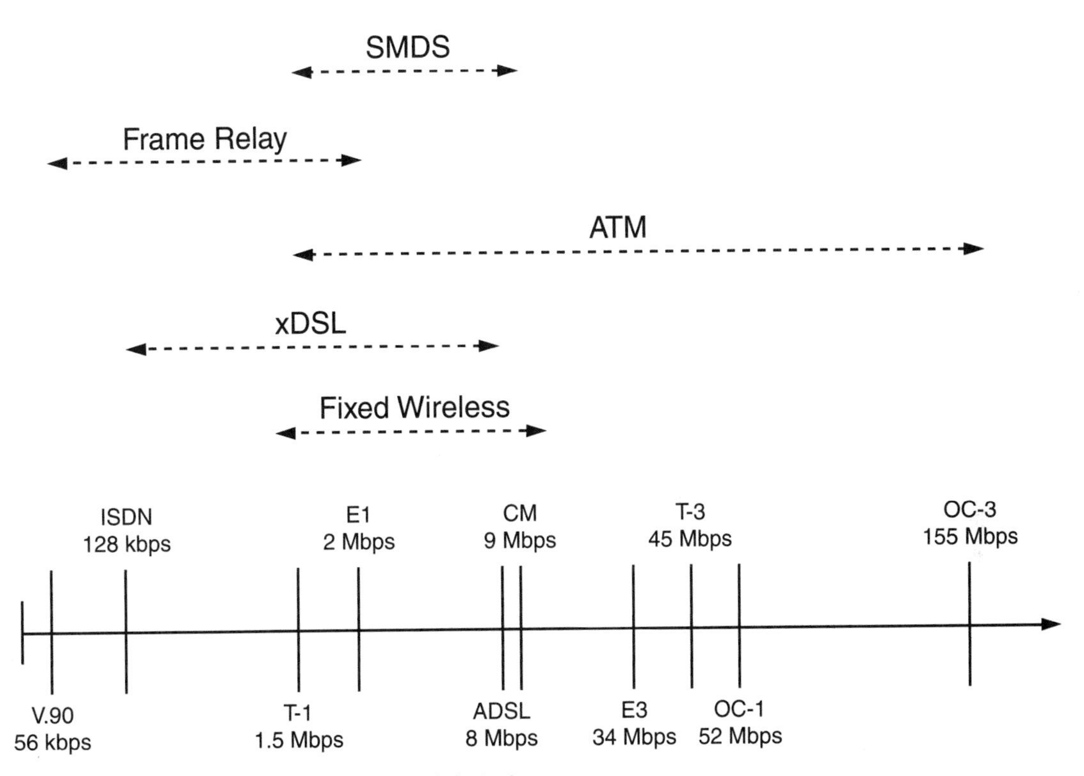

Figure 7–29 Network technologies and their data rates.

cations have been built around high-speed ATM switches interconnected with Gigabit SONET connections, and, most recently, Terabit routers connected over SONET links. But in most cases, the ATM core has been integrated with Ethernet at the edge of the network. At Layer 1, SONET continues to be the transport medium of choice for service providers as it has matured into a stable technology familiar to installation, maintenance, and operations personnel. Its ring architecture provides rapid restoration, easy access to lower-bandwidth circuits at intermediate network points, and circuit visibility.

Regarding access technologies for homes or small businesses, DSL, CM, fixed wireless, and ISDN offer an always-on connection so users do not have to worry about dial-up issues. With DSL, the bandwidth is delivered over the same copper wiring used by the POTS, but users must live within approximately a three-mile radius of each DSL-equipped switching office. CM service can provide more bandwidth for end users' money as compared with ISDN and ADSL services, but it does not offer the same secure and deterministic environment. Although DSL and CM are high-speed low-cost alternatives, an ISDN connection allows the greatest flexibility as it provides access to multiple channels (voice, data and fax) over a single line. PON is a new technology and the scope of its implementation depends upon telecommunications carriers and equipment vendors.

Popular WAN Technologies	OSI Layer	Strengths	Weaknesses
SONET	Layer 1	✦ Every kind of communications traffic can be multiplexed into SONET ✦ Easily scalable ✦ Standardized ✦ Built-in fault tolerance	
X.25	Layer 3	✦ Guaranteed data delivery ✦ High data integrity	✦ Slow and tedious because it involves error checking at every node
Frame Relay	Layer 2	✦ Robust interoperability between various switching platforms ✦ Improved network uptime because of virtual circuits ✦ Widely available	✦ Users have to commit to predefined PVCs ✦ Mostly baseband applications
SMDS	Layer 2	✦ Access rates higher than those available for Frame Relay ✦ No commitment to PVCs	✦ Not very flexible or scalable ✦ Mostly baseband applications
ATM	Layer 2	✦ Supports true QoS for broadband applications ✦ Easily scalable ✦ Standardized	✦ Cell tax for purely data traffic like IP traffic ✦ Can be disruptive for existing Ethernet LANs
ISDN	Layer 2	✦ End-to-end digital connectivity ✦ Broadband applications such as video conferencing are much faster and clearer	✦ Cost is somewhat prohibitive ✦ Typically a little more difficult to set up and configure

Figure 7–30 Strengths and weaknesses of popular WAN technologies.

The economics of different competitive alternatives reveal significant differences in how the future might unfold. Some broadband networks, such as cable and satellite systems, require huge upfront payments before the network ever becomes operational. But once they achieve coverage, their plant is scalable through cell splitting, as has been done with cellular, or by reducing the number of homes per node, which is the cable analog of cell splitting. On the other hand, solutions such as xDSL are more heavily weighted to incremental costs. For instance, for each new subscriber, the telco will have to ensure that the line is properly conditioned, install subscriber equipment, and also install equipment at the end office serving the subscriber. This appears to be a much less risky strategy if demand is expected to be low or slow to increase. Although network convergence has been happening deeper in the PSTN for trunking applications between data switches and PBXs, as well as in toll bypass applications in the long-distance network, not much attention has been paid to delivering combined data and voice in the local loop.

Case Study

ISDN provided an ideal solution for improving customer service and cutting costs in a large regional bank's mortgage department. Each of the bank's 100 branch locations had just one dedicated mortgage loan officer. When the loan officer in a branch was not available, customers couldn't be served, even though loan officers in other branches often were idle. Bank managers decided to centralize the loan specialists and place videoconferencing kiosks in each branch. Dial-up ISDN service (BRI in small branches, PRI in larger) connects kiosks to the central location. Customers can interact face-to-face with a mortgage loan expert via these terminals. More customers are served, and the mortgage department needs fewer loan officers because workload is evenly distributed.

Questions

1. What alternate technologies can be used in the situation described above?
2. Compare these technologies with ISDN with respect to cost, efficiency, and scalability.

REVIEW QUESTIONS

1. Define the following terms:
   - A. Bursty
   - B. Quality of Service
   - C. Latency
   - D. Jitter
   - E. Packet Assembler/Disassembler
   - F. Packet-Switching Exchange
   - G. Payload
   - H. Permanent Virtual Circuits
   - I. Committed Information Rate
   - J. Network Downtime
   - K. Distributed Queue on a Dual Bus
   - L. Sustained Information Rate
   - M. Basic Rate Interface
   - N. Primary Rate Interface
   - O. Automatic Protection Switching
   - P. Packetization Delay
   - Q. Switched Virtual Circuits
   - R. Cell Tax
   - S. Point-to-Point Protocol
2. Discuss the process of transmitting a message over a packet-switching network.
3. Explain the frame format in Frame Relay.
4. Describe the SONET protocol stack and STS-1 frame structure.
5. Compare and contrast:
   - A. X.25 and Frame Relay
   - B. Frame Relay and ATM
   - C. Packet over SONET and ATM
   - D. DSL, CM, and fixed wireless
6. Evaluate the significance of each:
   - A. Committed Information Rate
   - B. Sustained Information Rate
   - C. 53-byte cell size in ATM
7. Explain the frame format in Frame Relay.
8. Discuss the "segmentation and reassembly" process in ATM.
9. Describe the SONET protocol stack and STS-1 frame structure.

10. Evaluate the strengths, weaknesses, and applications of the following:

- **A.** X.25
- **B.** Frame Relay
- **C.** SMDS
- **D.** ISDN
- **E.** SONET
- **F.** ATM
- **G.** Packet over SONET
- **H.** DTM
- **I.** DSL
- **J.** CM
- **K.** Fixed Wireless
- **L.** PON

8

INTERNET AND CONVERGED NETWORKS

KEY TERMS

Convergence
Transmission Control Protocol (TCP)
User Datagram Protocol (UDP)
Connection-oriented
Connectionless
Internet Protocol (IP)
Address Resolution Protocol (ARP)
IP Version 4 (IPv4) Addressing
Dotted Decimal Notation
Subnets
IP Version 6 (IPv6)
Slow Start
Spoofing
Virtual Private Network (VPN)
Tunneling
Intranet
Extranet
Converged Data/Voice Networks
Voice over IP (VoIP)
Assured Quality Routing (AQR)
Streaming Mode
Buffered Mode
Lossless Data Compression
Lossy Data Compression

OBJECTIVES

Upon completion of this chapter, you should be able to:

- Outline the TCP/IP model and evaluate its role as the de facto standard for global communications with reference to the OSI model
- Differentiate between TCP and UDP transport protocols
- Apply different IPv4 addressing schemes and the Dotted Decimal Notation
- Evaluate the advantages of Multicast, Subnets, and Classless Addressing
- Explain the updates in IPv6 as compared to IPv4
- Discuss prominent TCP/IP applications and protocols

- Analyze the shortcomings of TCP transmission via satellite
- Describe the role of Internet2
- Compare and contrast SNA with TCP/IP
- Analyze the need for Virtual Private Networks and describe their components
- Determine the significance of Intranets and Extranets for businesses
- Develop an argument for converged networks
- Compare and contrast between Voice over IP and traditional PSTN
- Discuss popular protocols for multimedia transmission over IP
- Explain lossless and lossy data compression techniques and their implementation

INTRODUCTION

The existing telecommunication network evolves to match the paradigm shift from a voice network to a data-centric one to a data/voice/video converged network. Integrated communication includes information building blocks necessary for doing business in the 21st century: voice, data, video, and Internet access. Traditionally, only large corporations with extensive financial and technical resources could utilize this broad base of communications capabilities. Most small- and medium-sized businesses relied solely on telephones, fax machines, and mail. This disparity was accepted by smaller organizations since they lacked the resources necessary to build and support complex information networks. But as LANs become increasingly pervasive in smaller companies, the demand for integrated communications, including computer-telephony integration, grows stronger.

The term **convergence** has taken on multiple meanings depending on the perspective and the context. For consumers, convergence may mean pervasive computing—the merger of the computer, telephone, TV, Internet, and every imaginable appliance. For businesses, it refers to the merging of computers and communications, or packet-switched data and circuit-switched voice. Converged data/voice networks can utilize one of many options such as Voice over Frame Relay (VoFR), Voice over ATM, and **Voice over IP (VoIP)**, in which the digitized voice is encapsulated in frames, cells, or packetized in IP packets and is then transported over data networks. The Internet protocols can be used to communicate across any set of interconnected networks, as they are equally well suited for LAN as well as WAN communications.

TCP/IP MODEL

About the same time that the OSI reference model was designed, the TCP/IP protocol suite emerged from research under the auspices of DARPA. In the mid-1970s, DARPA and other government organizations understood the potential of packet-switched technology and were just beginning to face the problem that virtually all companies with networks had dissimilar computer systems. With the goal of heterogeneous connectivity in mind, DARPA funded Stanford University and Bolt, Beranek, and Newman (BBN) to create a series of communication protocols. The result of this development effort was a protocol suite, with the *TCP* and the IP at the heart of that protocol suite.

Although originally designed for operation on the Internet, TCP/IP is equally adaptable for a closed network such as a LAN. Unlike OSI, TCP/IP is not a true international standard, but it has emerged as a de facto standard for global communications. A drawback of the OSI is by virtue of its open architecture—within each layer of the OSI, a user may select among many different protocols, resulting in uninteroperability, or incompatibility, between different networks. As a result, protocol converters remain a necessary overhead.

The TCP/IP has become the widest accepted set of protocols in the telecommunications industry and is implemented in both LAN and WAN environments. It has many benefits that support it as one of the best choices for a company protocol suite, including:

- Ease with which it can be configured, managed, maintained, and scaled
- Higher flexibility than any other protocol
- Good error-detection and recovery mechanisms
- Hardware and software products are competitively priced
- Broad appeal, especially because of the growing popularity of the Internet

Let us see how the TCP/IP set of protocols fit into the OSI network architecture, as shown in Figure 8–1. The TCP/IP sits atop the physical and data link layers that are in actuality not part of this family of protocols, enabling it to seamlessly transport data between different LANs. The IP is analogous to the OSI network layer, while the TCP is analogous to the OSI transport layer. The TCP is **connection-oriented**, while the User Datagram Protocol (UDP) is a **connectionless** version of TCP. Connection-oriented means a connection must be established prior to actual data transfer occurring, while connectionless references transmission occurring on a best-effort basis, with an acknowledgment flowing back only after transmission was initiated. Some Internet applications have been developed to use TCP, while others have been developed to use UDP. Thus, it is important to note the differences between each transport protocol. The session and presentation layers are not used for most standard TCP/IP services; protocols at the application layer interface directly with the TCP or the UDP at the transport layer.

Transmission Control Protocol (TCP)

Transmission Control Protocol (TCP) is a reliable, **connection-oriented**, unicast (point-to-point), guaranteed delivery protocol that performs end-to-end error checking, correction, and acknowledgement. It ensures that data is delivered error-free with no loss or duplication. The format of a TCP packet is shown in Figure 8–2. It can be used with a variety of Layer 3 protocols, not just IP. At the transmitting end, its function is to receive

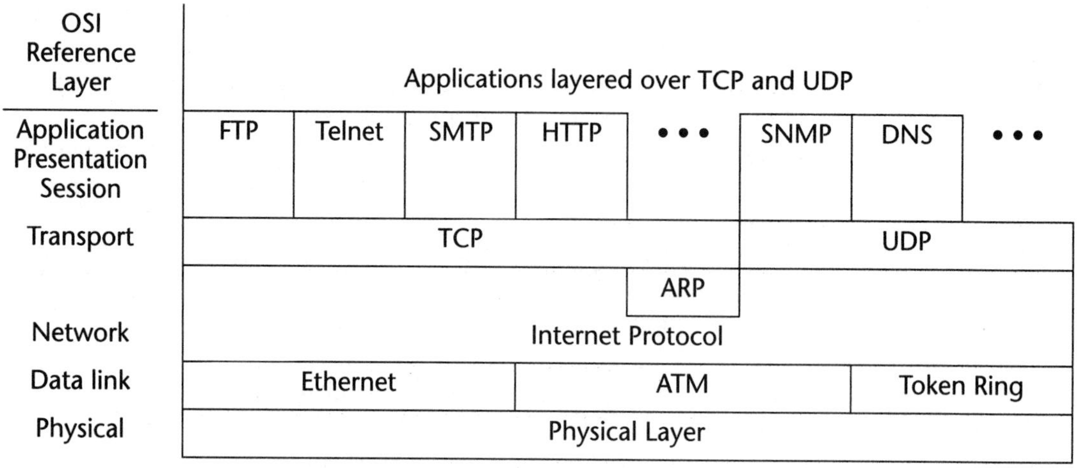

OSI Reference Layer	Applications layered over TCP and UDP							
Application Presentation Session	FTP	Telnet	SMTP	HTTP	• • •	SNMP	DNS	• • •
Transport	TCP					UDP		
					ARP			
Network	Internet Protocol							
Data link	Ethernet			ATM			Token Ring	
Physical	Physical Layer							

Legend: FTP = File Transfer Protocol

SMPT = Simple Mail Transport Protocol

HTTP = HyperText Transmission Protocol

SNMP = Simple Network Management Protocol

DNS = Domain Name Service

ARP = Address Resolution Protocol

Figure 8–1 Correlation between TCP/IP and OSI layers.

	Source Address	Destination Address	Control Information	CRC	Urgency	Options	Payload
Bytes	2	2	12	2	2	variable	variable

Figure 8–2 TCP packet format.

messages from the application program and break them into packets for transmission. At the destination, it resequences packets that arrive out of sequence and communicates directly with the application program in the host. It should be noted that TCP supports only data reliability and is not suitable for transport of multimedia streams, which require consistent time delivery at the receiver and only need to be semi-reliable. Applications that use TCP include FTP (File Transfer Protocol), HTTP (HyperText Transmission Protocol), TELNET (Terminal Emulation), and SMTP (Simple Mail Transfer Protocol).

User Datagram Protocol (UDP)

User Datagram Protocol (UDP), in comparison with TCP, is an unreliable, **connectionless** protocol, but it has less overheads. Connectionless means that the UDP protocol does not require a session to be established between two devices, and transmission occurs on a best-effort basis. If reliability is ensured at a higher layer such as the application layer, then UDP, which is more efficient than TCP, can be used for transport. Applications such as SNMP (Simple Network Management Protocol) and RTP (Real-time Transport Protocol) use UDP.

Internet Protocol (IP)

The **Internet Protocol (IP)**, equivalent to Layer 3, segments and packets data for transmission and then places a header for delivery. The IP header is in addition to the TCP or UDP header appended to the application data. The IP header includes the source and destination addresses, enabling an end-to-end data flow. Each network node reads the header information and routes the packet accordingly, but there is no assurance of delivery. As shown in Figure 8–3, at Layer 2, which commonly represents a LAN transmission facility, a LAN header such as those formed by Ethernet or Token Ring will prefix the IP header. A LAN trailer that includes FCS characters for error checking will be added as a suffix.

There is no relationship between a Layer 2 address and an IP address, which means that when IP data is delivered to a LAN, another mechanism called the **Address Resolution Protocol (ARP)** is required to enable data to reach its destination that uses a different addressing scheme. The physical machine address is also known as a MAC address. ARP provides the rules for making the correlation and providing address conversion from IP to MAC, and vice versa.

IP routing specifies that IP datagrams travel through internetworks one hop at a time. The entire route is not known at the outset of the journey. Instead, at each stop, the next destination is calculated by matching the destination address within the datagram with an entry in the current node's routing table, which is referred to as dynamic routing. Each node's involvement in the routing process consists only of forwarding packets to the next node based on internal information, regardless of whether the packets get to their final destination. In other words, IP lacks end-to-end error checking and acknowledgement; therefore, it is described as an unreliable, connectionless, best-effort, datagram protocol. The format of an IP version 4 (IPv4) packet is depicted in Figure 8–4.

IP Version 4 (IPv4) Addressing

The **IP version 4 (IPv4) addressing**, officially standardized in September 1981, requires a unique, 32-bit address to be assigned to each host connected to an IP-based network. The basic addressing scheme is a two-level hierarchy, as illustrated in Figure 8–5, where all

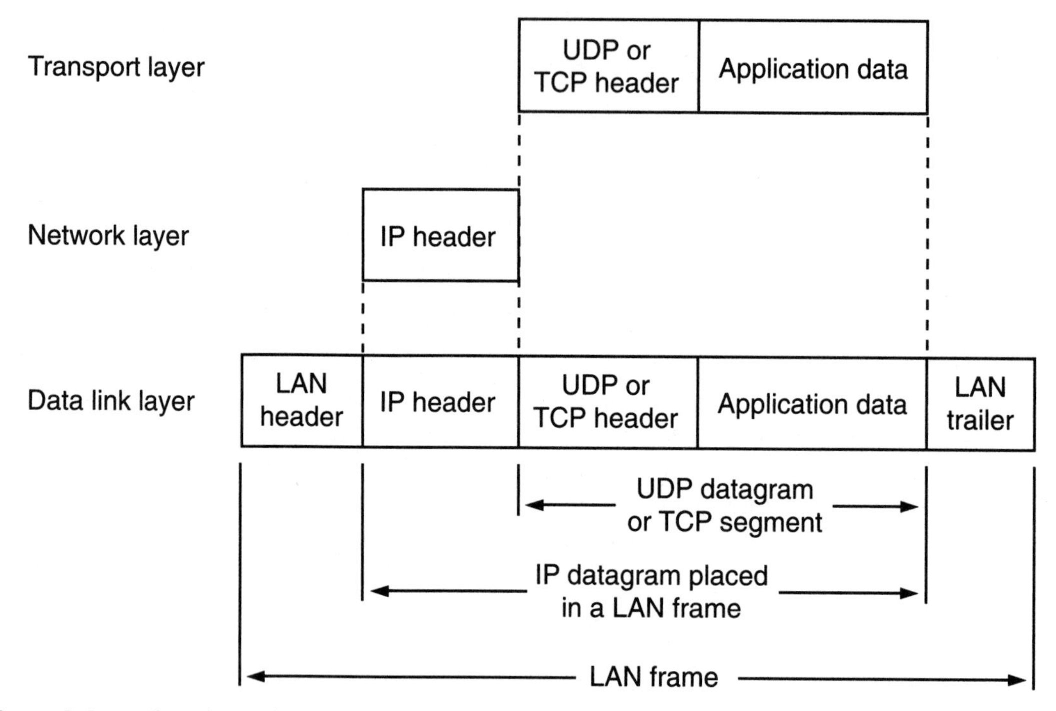

Figure 8–3 The relationship of headers at the transport, network, and data link layers.

	Header and Control Information	CRC	Source Address	Destination Address	Options	Payload
Bytes	10	2	4	4	variable	variable

Figure 8–4 IPv4 packet format.

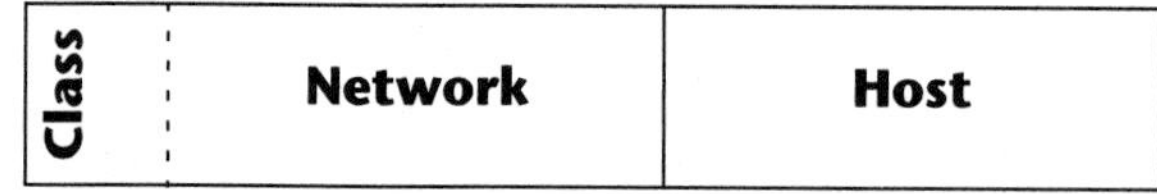

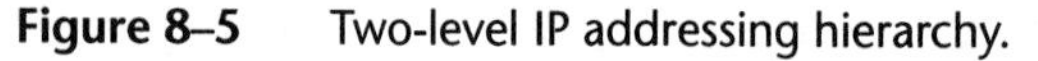

Figure 8–5 Two-level IP addressing hierarchy.

hosts on the same network must be assigned the same network prefix, but they must have a unique host address to differentiate one host from another. Similarly, two hosts on different networks must be assigned different network prefixes; however, the hosts can have the same host address. IP addresses are composed of three parts: class, network, and host, where the lengths of these fields are variable. IPv4 addressing supports five different network classes, represented in Figure 8–6. The bits to the far left indicate the network class.

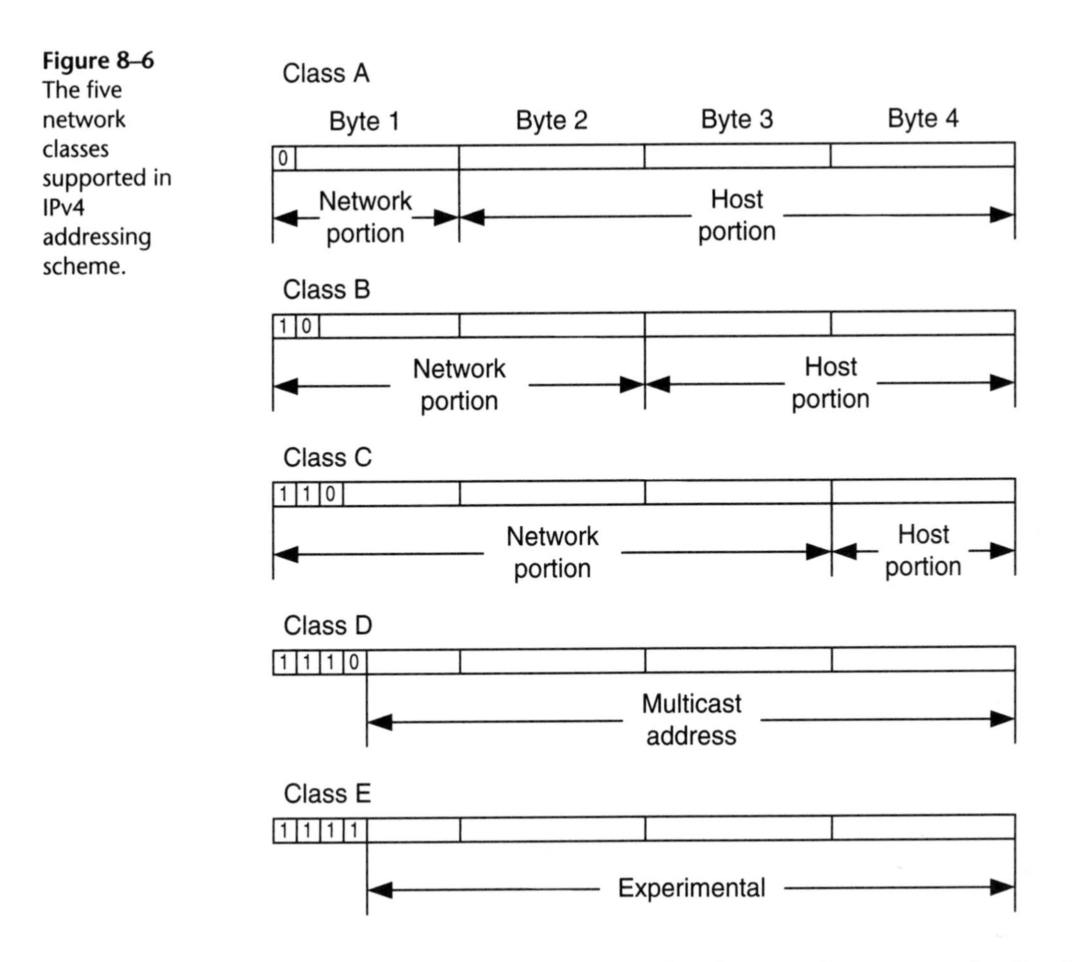

Figure 8–6 The five network classes supported in IPv4 addressing scheme.

Class A networks are intended mainly for use with a few very large networks. The first bit, set to the value 0, identifies the class; the next seven bits are reserved for network address, and the remaining 24 bits are assigned to hosts. Therefore, only a maximum of 128 Class A networks are available, each of which can have 16.78 million hosts. Typical examples of Class A networks are very large organizations and countries that have national networks.

Class B networks allocate two bits for class, set to the value 10, 14 bits for network addresses, and 16 bits for host addresses. This scheme offers a good compromise between network and host address space. The Class B network can have a maximum of 16,384 networks, each able to support up to 65,536 hosts. Class B addresses are normally assigned to relatively large organizations with tens of thousands of employees.

Class C networks allocate the first three bits for class, set to the value of 110, 21 bits for a network field, but provide only 8 bits for a host field. So the number of hosts per network may be a limiting factor for Class C networks. It is quite common to have multiple Class C addresses assigned to a single organization that supports more than 256 hosts but is not large enough to justify a Class B address. With 21 bits available for a network field, there are approximately 2 million Class C networks.

Class D addresses are reserved for multicasting, which is an addressing technique that allows a source to send a single copy of a packet to a specific group through the use of a multicast address. Since TCP is a unicast transport protocol, all multicast applications must run on top of UDP or, alternatively, interface directly with IP and provide their own customized transport layer. A Class D address is defined by assigning the value 1110 to the first four bits, with the remaining 28 bits used for multicast address. Until recently, the use of multicast addresses was relatively limited; however, its use is increasing considerably as it provides a mechanism to conserve bandwidth.

To understand how Class D addressing conserves bandwidth, consider a digitized video presentation routed from the Internet onto a private network, where ten users wish to receive the presentation at their respective terminals. Without a multicast transmission capability, ten separate audio and video streams would be transmitted onto the private network. In comparison, through the use of a multicast address, a single data stream is routed to the private network. Since an audio/video stream can require a relatively large amount of bandwidth, the ability to eliminate multiple data streams via multicast can prevent networks from being saturated.

Class E addresses are reserved for experimentation and future use. The first four bits have the value of 1111, which results in 28 remaining bits that are capable of supporting approximately 268.4 million addresses.

Dotted Decimal Notation

Dotted Decimal Notation is a technique used to express IP addresses via the use of four decimal numbers separated from one another by decimal points. Internet Assigned Numbers Authority (IANA) was responsible for three things: 1. assigning IP addresses, that is, the four octets such as 206.135.10.2, that are used to identify every Internet router, server and workstation; 2. running the root name servers that provide the essential base for the Domain Name System (DNS); and 3. acting as final arbiter and editor for key standards developed by the Internet community. The DNS server provides translation between a domain name that is a meaningful and easy-to-remember and an Internet address. For example, the domain name *www.ilstu.edu* has an IP address of 138.87.4.3. The last identifier in the domain name, which is the *edu* part of the domain name, reflects the purpose of the organization or entity. In the United States, classical domain name identifiers are:

- *com* for commercial organizations
- *edu* for educational institutions
- *gov* for governmental organizations
- *int* for organizations formed under international treaty
- *mil* for military units

- *net* for network access providers
- *org* for nonprofit organizations

Other countries assign similar domain names but also append a two-letter country code, such as *.ca* for Canada. A typical Australian Internet address might be hmobson@dccomp.com.au, where *au* indicates Australia. The new organization, called the Internet Corporation for Assigned Names and Numbers (ICANN), is starting to shape Internet administration such that it is representative of the broad range of users and organizations that are a reason for the success of the Internet. New suffixes have been under consideration since mid-1990, but there were disputes over how many, which ones, and how they would be registered. ICANN was designated by the Commerce Department in 1998 as the overseer of domain names and online addresses. In 2000, ICANN approved:

- *.aero* for the aviation industry
- *.biz* for business
- *.coop* for business cooperatives
- *.info* for general use
- *.museum* for museums
- *.name* for individuals
- *.pro* for professionals

Creating the new suffixes is similar to adding area codes to the national phone system to accommodate growth. They could make addresses more simple and Web sites easier to find. A computer user, for example, could someday type ua.aero to reach the United Airlines website instead of www.united.com. The new suffixes could also begin a new Internet land rush, with speculators and trademark holders competing to claim the best names first.

Dotted-decimal notation divides the 32-bit IP address into four 8-bit (1-byte) fields, with the value of each field specified as a decimal number. That number can range from 0 to 255 (binary 0 to 11111111) in bytes 2, 3, and 4. In the first byte of an IP address, the setting of the first few bits for the address class limits the range of decimal values that can be assigned to that byte. For example, a Class A address is defined by the setting of the first bit position in the first byte to 0. Thus, the maximum value of the first byte in a Class A address is 127 (binary 01111111). Figure 8–7 provides the decimal values of the bit positions in a byte. As represented in Figure 8–8, address classes can be recognized from the first decimal number: 0 to 127 are Class A, 128 to 191 are Class B, and 192 to 223 are Class C. The dotted-decimal number 96.104.62.131 is equivalent to binary 01100000.01101000.00111110.10000011; 96 is equivalent to the first eight bits 01100000, 104 is equivalent to the next eight bits 01101000, and so on. Obviously, it is easier to work with four decimal numbers separated by dots than with a string of 32 bits.

2^7	2^6	2^5	2^4	2^3	2^2	2^1	2^0
128	64	32	16	8	4	2	1

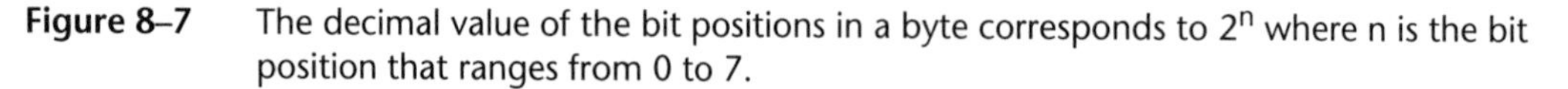

Figure 8–7 The decimal value of the bit positions in a byte corresponds to 2^n where n is the bit position that ranges from 0 to 7.

Class	First Number Range (decimal)
A	0–127
B	128–191
C	192–223

Figure 8–8 Address classes can be recognized from the first decimal number.

Subnets

One of the limitations associated with the use of IPv4 addressing is the necessity of assigning a distinct network address to each network. This can result in the waste of many addresses as well as a considerable expansion in the use of router tables. For example, assume that a Class C network supports 29 workstations and servers. After adding an address for a router port, it would use only 30 out of 255 available addresses. Thus, the assignment of two Class C addresses to an organization that needs to support two networks with a total of 60 devices would result in 450 (255 x 2 – 60) available IP addresses being wasted. In addition, routers outside the organization would have to recognize two network addresses instead of one. Because of this problem, the RFC 950, which became a standard in 1985, defines a procedure to **subnet** or divide a single Class A, B, or C network into subnetworks.

Through the process of *subnetting*, the two-level hierarchy of Class A, B, and C networks is turned into a three-level hierarchy. In doing so, the host portion of an IP address is divided into a subnet portion and a host portion. Figure 8–9 provides a comparison between the two-level hierarchy initially defined for Class A, B, and C networks and the three-level subnet hierarchy. Since the network portion of all the subnet addresses remains the same, the route from the Internet to all subnets that belong to a particular IP network address stays the same. This means that routers outside the organization have only one entry for the network, but routers within an organization must have the ability to differentiate between different subnets. This is illustrated in Figure 8–10 and Figure 8–11.

Classless Addressing

As explained earlier, the use of individual Class A, B, and C addresses can result in a significant amount of unused address space. The use of *classless addressing* is increasing as a mechanism to both extend the availability of IP addresses and to enable routers to oper-

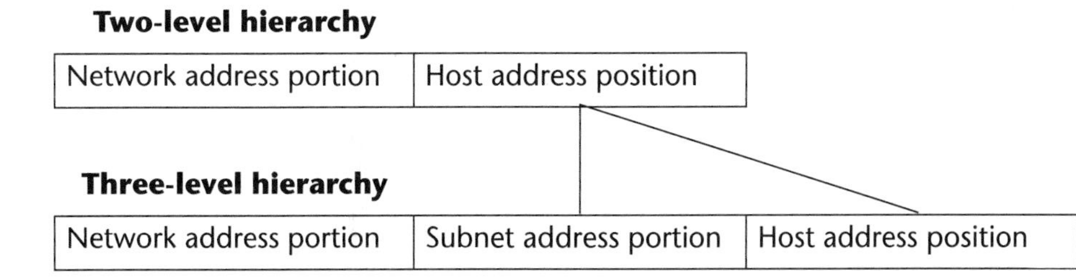

Figure 8–9 Comparison between the two-level and the three-level hierarchy.

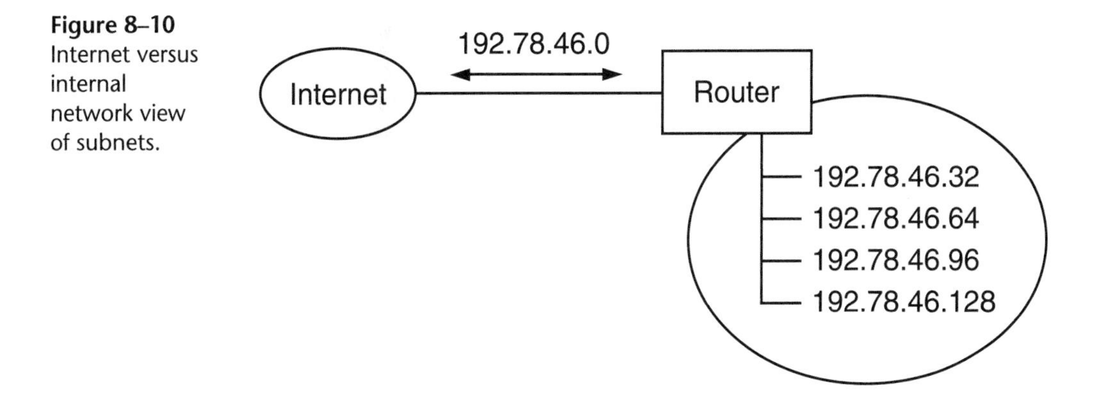

Figure 8–10 Internet versus internal network view of subnets.

Base Network	11000000.01010000.00101110.00000000 = 192.78.46.0
Subnet #1	11000000.01010000.00101110.00100000 = 192.78.46.32
Subnet #2	11000000.01010000.00101110.01000000 = 192.78.46.64
Subnet #3	11000000.01010000.00101110.01100000 = 192.78.46.96
Subnet #4	11000000.01010000.00101110.10000000 = 192.78.46.128

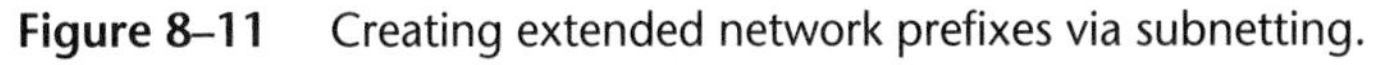
Figure 8–11 Creating extended network prefixes via subnetting.

ate more efficiently. In classless addressing, an organization is assigned a number of bits to use as the local part of its address corresponding to its needs. For example, if an organization requires 4000 IP addresses, it would be given 12 bits, or 2^{12} (4096) distinct addresses to use as the local part of its address. The remaining 20 bits in the 32-bit address space are used as a prefix to denote what is referred to as a supernetwork. It provides a more efficient mechanism because a router outside the organization has only one entry for the supernetwork, which is the IP address of the router within the organization that interfaces with the outside world. This organization router may then list the 4000 IP addresses in its routing table. But with 4000 hosts, it is more likely that the organization would have subnets in order to improve the efficiency, and the router would then list only the subnet addresses. Figure 8–12 lists few examples of classless address blocks that can be assigned from available Class C address space.

Prefix Bits	Local Bits	Equivalent Number of Class C Addresses	Distinct IP Addresses
24	8	1	256
23	9	2	512
22	10	4	1,024
20	12	16	4,096
18	14	64	16,284
17	15	128	32,768

Figure 8–12 Classless addressing.

IP Version 6 (IPv6)

IP version 6 (IPv6) is a new version of IP, which is designed to be an evolutionary step from IPv4. It is a natural increment and can be installed as a normal software upgrade in Internet devices, and it is interoperable with the current IPv4. Network nodes and clients with either IPv4 or IPv6 can handle packets formatted for either level of the Internet Protocol. In other words, users and service providers can update to IPv6 independently without having to coordinate with each other. IPv6 is designed to run well on high-performance networks, for example, Gigabit Ethernet, OC-12 and ATM, and at the same time still be efficient for low-bandwidth networks like wireless.

IPv6 has been developed to extend source and destination addresses and provide a mechanism to add new operations with built-in security. Although IPv4 is still widely used, over the next few years, the IPv4 32-bit address will be replaced with the IPv6 128-bit address. This extension anticipates considerable future growth of the Internet and provides relief for what was perceived as an impending shortage of network addresses. In addition to retaining the unicast and multicast addresses of IPv4, the introduction of an *anycast* address in IPv6 provides the possibility of sending a message to the nearest of several possible gateway hosts with the idea that any one of them can manage the forwarding of the packet to others. Anycast messages can be used to update routing tables along the line.

An IPv6 packet format is represented in Figure 8–13. A fixed-length header allows for faster processing of the IP datagram. The header includes extensions that allow a packet to specify a mechanism for authenticating its origin, for ensuring data integrity, and for

	Header and Control Information	Source Address	Destination Address	Payload
Bytes	8	16	16	variable

Figure 8–13 IPv6 packet format.

ensuring privacy. Options are specified in an extension to the header that is examined only at the destination, thus speeding up overall network performance. Flow labeling and priority allows packets to be identified as belonging to a particular flow so packets that are part of a multimedia presentation that need to arrive in real-time can be provided a higher QoS relative to other customers. One reason for the slow adoption of IPv6 is that it is enormously difficult to change network-layer protocols; it is like replacing the foundation of a house. On the other hand, introducing new application-layer protocols is like adding a new layer of paint to a house. Thus, the Internet has witnessed rapid deployment of new application-layer protocols: HTTP, Web, audio and video streaming, and chat.

TCP/IP-based Applications

Some TCP/IP-based applications layered over TCP and UDP include FTP, HTTP, TELNET, SMTP, SNMP, RTP, and UNIX Remote Login, which caters specifically to UNIX-based host computers. Each of these Layer 7 or application protocols is designed for a specific function: FTP is used for moving data files; HTTP is used by browsers to connect with Web servers; TELNET is used to access a remote host computer; SMTP provides a universal format for exchanging electronic mail; and SNMP provides the mechanism to transport status messages and statistical information about the operation and utilization of TCP/IP devices. Some frequently used applications are explained as follows.

Simple Mail Transfer Protocol (SMTP) is a mechanism for sending standard, interoperable text messages from one computer system to another. The *Post Office Protocol* provides a store-and-forward mechanism that allows for clients to be disconnected from the network and still receive mail. The process is illustrated in Figure 8–14. The simplicity of the protocols means that they can be widely implemented at a low cost on almost any platform, but it also exposes a fundamental limitation. Pictures, media, and large binary files are too unwieldy for SMTP to handle. With no method of sending binary files, the Internet Engineering Task Force (IETF) adapted SMTP to handle attachments. The protocol that describes how to handle attachments using standards-based mail is called *Multipurpose Internet Mail Extensions (MIME)*. It permits mail clients to send attachments of any kinds and marks them with a label that indicates how the information was encoded and what application created the attachment. The richness of MIME allows pictures, audio, and video to be sent as part of electronic mail, as shown in Figure 8–15.

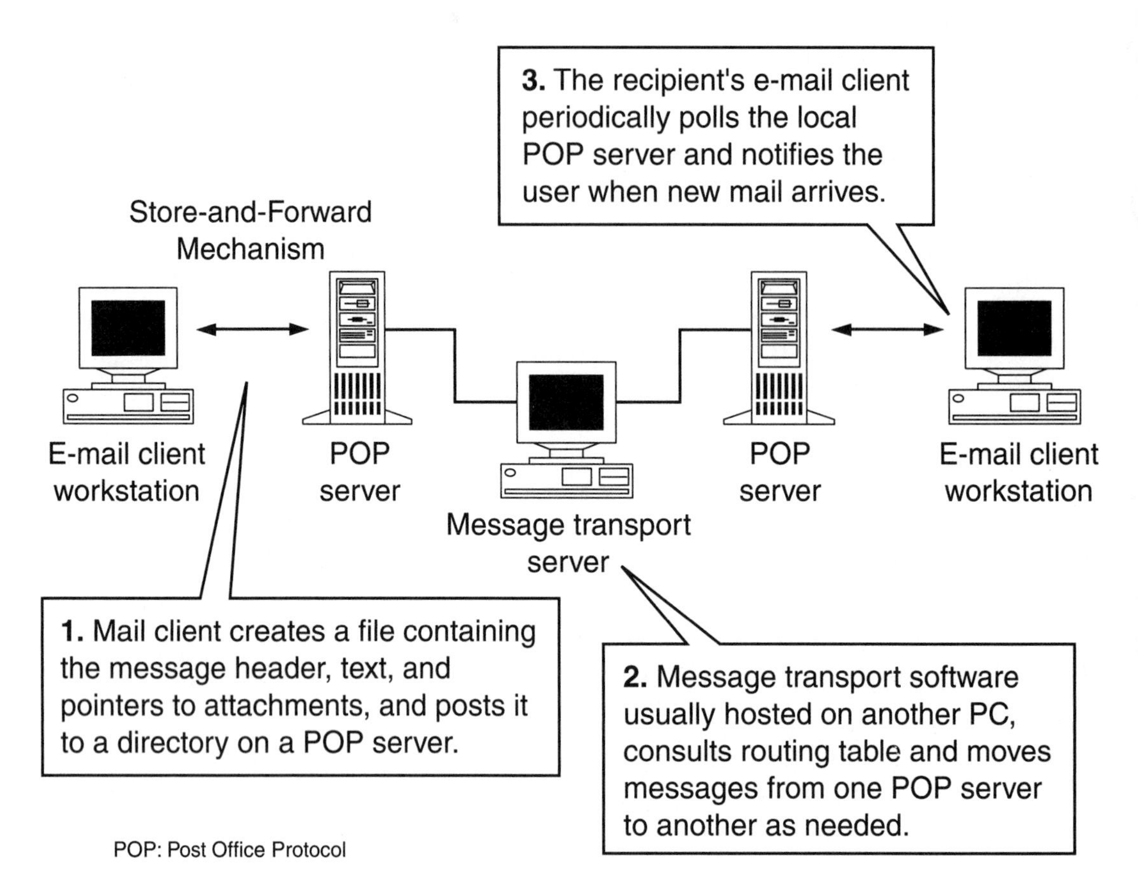
3. The recipient's e-mail client periodically polls the local POP server and notifies the user when new mail arrives.
Store-and-Forward Mechanism
E-mail client workstation
POP server
Message transport server
POP server
E-mail client workstation
1. Mail client creates a file containing the message header, text, and pointers to attachments, and posts it to a directory on a POP server.
2. Message transport software usually hosted on another PC, consults routing table and moves messages from one POP server to another as needed.
POP: Post Office Protocol

Figure 8–14 Store-and-forward mechanism provided by the Post Office Protocol.

Content	Description
Application	Contains application or binary data, such as word-processing documents and spreadsheets
Audio	Contains audio or voice data
Image	Contains a still image, such as a picture
Message	Contains another encapsulated mail message
Multipart	Contains several body parts, possibly of different types, within the same message
Text	Various character sets and formatted text
Video	Contains video or moving image data

Figure 8–15 MIME message types.

Despite the flexibility, MIME and the underlying mail delivery protocols suffer from a fundamental disadvantage compared with proprietary mail systems. Group scheduling and collaboration tools that are taken for granted under client/server applications cannot be implemented using existing standards-based mail. In response to the demands, new protocols are being crafted that can make interoperable mail rich in functionality as proprietary systems. The first of these specifications, the *Internet Message Access Protocol (IMAP)* is already well established in mainstream messaging software. It permits true mail-to-client/server interaction by allowing the mail to stay at the server and to be accessed and managed by the clients. While the Post Office Protocol is easy to implement at the server, IMAP requires careful consideration for additional disk space, disk quota administration, and server performance. In case of digital mobile devices, the WAP is an open standard for providing Internet communications and telephony services.

Point-to-Point Protocol (PPP) was designed by the IETF to route multiple protocols over dial-up and dedicated point-to-point links. For example, when a student connects to a college LAN over a modem, PPP enables the communication over a serial link. The Link Control Protocol operates above the HDLC at Layer 2, and is responsible for link negotiation and authentication between devices—the simplest method uses password. Once a link is established, the Network Control Protocol (NCP) provides a framework to enable the network layer protocol, such as IP, to establish a connection. Once a connection is established, data transfer can take place.

First used in router-based data networks as a protocol running on top of T-1 access lines to ensure multivendor router interoperability, PPP is not designed to blend or multiplex multiple streams of data, and it cannot prioritize data streams or respond gracefully to congestion. But if all one needs to do is to provide a nearly error-free, point-to-point link for data, PPP can handle the job. With its security provisioning, multiprotocol support, IP address dynamic assignment, and support for asynchronous and synchronous connections, PPP has become the de facto standard for Internet access.

Serial Line Internet Protocol (SLIP) can also be used for carrying IP over an asynchronous dial-up or leased line, but it does not support error detection or correction. Therefore, in almost all cases, the use of PPP is preferable.

TCP VIA SATELLITE

Transmission Control Protocol is not well suited for satellite transmission because it employs an algorithm known as **slow start**. When a connection first starts up, it determines the available bandwidth on the network by starting with an initial window size of one segment (usually just 512 bytes) and then increases the window size only when packets are delivered successfully and acknowledgements arrive. This process is enabled by a *sliding-window* protocol, which must contain adequate buffering to resequence packets between two hosts. Its advantage is that it avoids stressing congested networks. Its disadvantage is that it can take a significant amount of time on a high-latency network before full bandwidth is made available. Web pages are particularly difficult to deal with, as a single one may require many different TCP connections, each of which has to go through the slow-start process. Throughput is limited by the formula given in Equation 8–1:

$$\text{Throughput} = \frac{\text{Window size}}{\text{Round-trip time}} \tag{8–1}$$

Example 8–1

Problem

Use the standard maximum window size of 65,535 bytes and GEO satellite latency of 500 ms to find the maximum throughput.

Solution

$$65{,}535 \text{ bytes} = 524{,}280 \text{ bits}$$

$$\begin{aligned}\text{Throughput} &= \frac{\text{Window size}}{\text{Round-trip time}}\\ &= \frac{524{,}280}{500 \text{ ms}}\\ &= 1.05 \text{ Mbps}\end{aligned}$$

Therefore, it may be desirable to tune the TCP stack for a larger maximum window size. Finally, TCP normally uses a cumulative acknowledgment scheme. If a segment is lost, the sender must restart transmission over again from the lost segment, even if subsequent segments have been successfully received. This means that the window size gets reduced and the slow-start mechanism is invoked yet again.

One way around slow start is TCP **spoofing**, where the spoofing box provides premature confirmation that a TCP segment has been received and then keeps track of actual acknowledgments and asks for retransmittals when needed. Although spoofing is effective, it may cause problems resulting from incompatible protocols when satellite and terrestrial links are combined.

INTERNET2 (I2)

Internet2 (I2) is an outcome of collaborative efforts to address the bandwidth issue while creating and sustaining a leading-edge network capability vital to the research and instructional teaching missions of higher education. I2 has been developed under the direction of the University Corporation for Advanced Internet Development (UCAID), a nonprofit consortium of more than 163 universities working in partnership with government entities and 60 corporations and affiliate members. The goal of the I2 project is not to replace the existing Internet infrastructure but to add a backbone that can handle the traffic anticipated in the coming decade.

I2 members are using advanced network capabilities to explore applications, most of which require guaranteed end-to-end network performance that is not possible with current Internet technology. The following is an example of what I2 can accomplish. A data-heavy high-definition TV signal has been transmitted in real-time over more than 1,000 miles from Stanford University in California to the University of Washington, which required a bandwidth of 270 Mbps. I2 is designed to help alleviate online traffic jams through the creation of a limited number of regional hubs, called GigaPOPs, which serve as access points for distributed, scalable, high-speed, and high-performance network computing applications.

SNA VERSUS TCP/IP

To understand why companies are exploring the option of moving from Systems Network Architecture (SNA) to TCP/IP, we must first understand the history of SNA and TCP/IP environments and what features have made them popular protocol choices. SNA was originally developed by IBM for enterprise use and has proved to be a reliable and manageable way of running large WANs. Developed for the mainframe marketplace, SNA is a hierarchical networking scheme that describes unique roles for different components of the system, providing efficient service to networked applications.

With the explosion of the Internet in the early 1990s and the widespread use of the Unix operating system, TCP/IP has matured into its current position as the widely accepted standard for WANs, evolving from a limited range of interoperability services to a full protocol suite for enterprise networks. The main benefit of TCP/IP is that it allows users to standardize on a single networking protocol across any processor platform within a heterogeneous network, in addition to gain in terms of cost, management, and usability.

As described in Figure 8–16, SNA is a centralized architecture with a mainframe computer executing tight control over peripheral network elements, while TCP/IP is a distributed architecture with no central host. SNA does not allow any data loss and is well-suited for extremely critical data such as a bank's financial information. On the other hand, TCP/IP is appropriate for graphics files where some data loss is tolerated. TCP/IP is a relatively simple protocol and has an open-source code that is readily available, while SNA is complex and proprietary since only IBM has all the code.

The popularity of the personal computer has lead to the slow demise of centralized host networks. In response, IBM developed the next generation of SNA, referred to as Advanced Peer-to-Peer Networking (APPN). It has inherited many of its predecessor's features and is now a standard feature of the midrange AS/400 environment. This process of evolution has developed SNA into a networking architecture that embraces both traditional, centralized

	SNA	TCP/IP
Similarities	✦ WAN protocols	✦ WAN protocols
Differences	✦ Hierarchical network architecture ✦ Proprietary (only IBM has all the source code) ✦ Centralized computing with the mainframe executing tight control over peripheral network elements	✦ Peer-to-peer network architecture ✦ Open source code ✦ Distributed computing with no central host

Figure 8–16 Similarities and differences between SNA and TCP/IP.

mainframe networking and modern, peer-oriented networking. In addition, the introduction of TCP/IP into the AS/400 operating system has facilitated the integration of midrange systems into the corporate LAN. The challenge that AS/400 users face is the continuing need to support SNA-based legacy applications and hardware over the new TCP/IP-centric networks they create.

VIRTUAL PRIVATE NETWORK (VPN)

One major movement forward in the use of the Internet for business applications is the implementation of **Virtual Private Networks (VPNs)** using a service provider's IP, Frame Relay, or ATM infrastructure, depicted in Figure 8–17. VPNs are encrypted tunnels through a shared private or public network that forward data over the shared media rather than over dedicated leased lines. *Encryption* involves scrambling of data by use of a mathematical algorithm so that the scrambled information is undecipherable and meaningless. **Tunneling** refers to the process where the source end encrypts its outgoing packets and encapsulates them in IP packets for transit across the Internet, while at the receiving end, a gateway device removes and decrypts the packets, forwarding the original packets to their destination. VPNs are characterized by connections that occur between point-to-point secure clients and are controlled by software and protocols during the connection. After the data transmission session is terminated, the connection between the locations is abandoned. A VPN uses the Internet's structure of routers, switches, and transmission lines while providing security for the users.

The most compelling argument for replacing a private network by a VPN is cost-effectiveness. Sharing the cost of leased lines in a public network, such as the Internet, can cut monthly recurring costs by an order of magnitude. There are significant differences between a private network and a VPN. First, a VPN is dynamic: VPNs do not maintain permanent links between the end points that make up the corporate network. Second, rather than depending on the security of data associated with private lines, VPNs

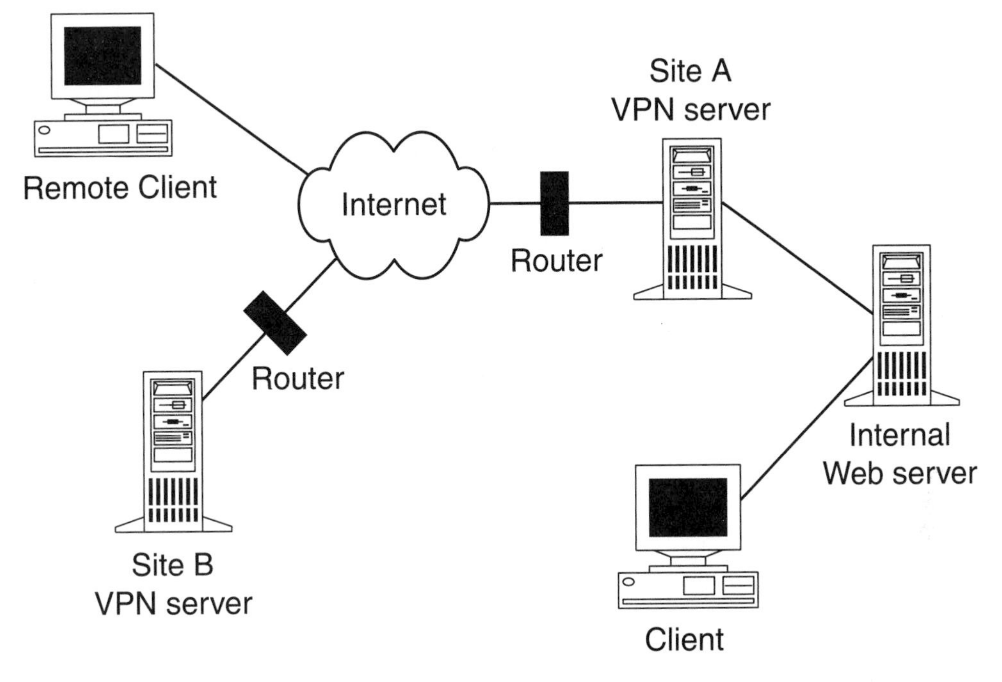

Figure 8–17 An example of a VPN.

have to maintain the privacy of communications, mainly by encrypting and tunneling the data that is forwarded between sites, as illustrated in Figure 8–18.

Tunnels can consist of two types of end points: a client, which is an individual computer, or a LAN with a security gateway. In LAN-to-LAN or site-to-site tunneling, a security gateway, router, or firewall serves as an interface between the tunnel and the private LAN. In such cases, usually corporate environments, users on either LAN can use the tunnel transparently to communicate with one another. Client-to-LAN tunnels, on the other hand, are usually set up to connect either a mobile user or a small branch office to a corporate LAN. The client must first run special software to initiate the creation of a tunnel and then exchange traffic with the corporate network.

Virtual Private LAN Service (VPLS) has emerged to meet a growing need to connect geographically dispersed locations with a protocol-transparent, any-to-any, full-mesh service. This is difficult for service providers to achieve with existing network architectures such as ATM and Frame Relay. VPLS is a class of VPN that supports the connection of multiple sites in a single bridged domain over a managed IP/Multi-Protocol Label Switching (MPLS) network. All services in a VPLS appear to be on the same LAN, regardless of location, which removes complexity from enterprise networks and lets carriers scale the networks. A VPLS uses edge routers that are connected by a full mesh of MPLS label switched path tunnels,

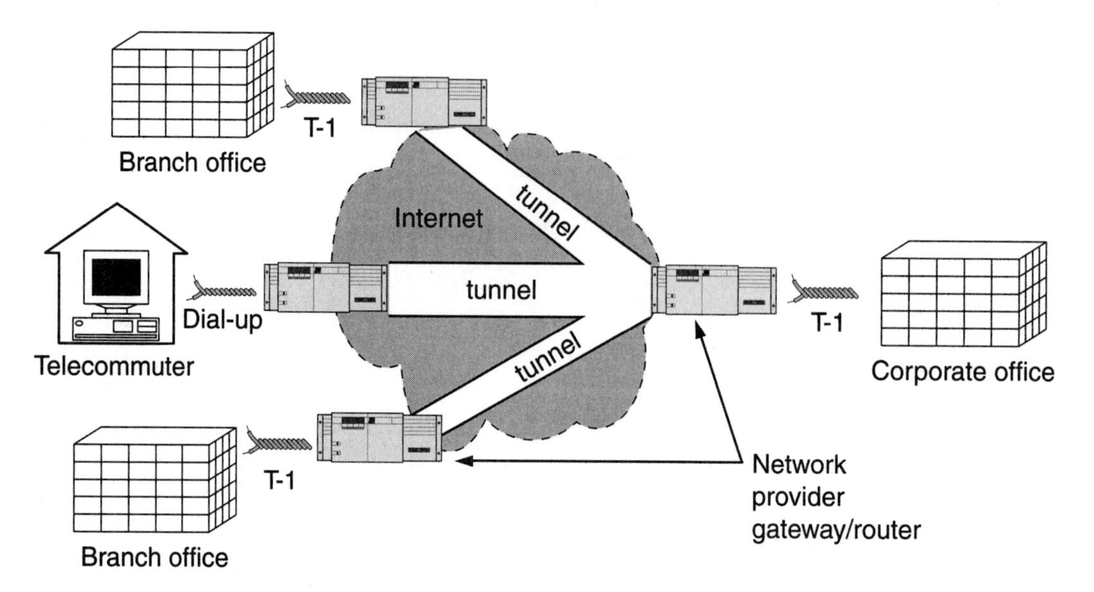

Figure 8–18 The use of tunneling in a VPN.

enabling any-to-any connectivity. The IETF standardization process for VPLS is moving forward, and it is using MPLS as a way to offer Ethernet connectivity from one site to many, versus the one site-to-one site limit of point-to-point service.

VPN Protocols

Different methods used to access a VPN are illustrated in Figure 8–19. VPN protocols support a variety of security measures for protecting the data, although the process varies from protocol to protocol. The leading protocols are: Point-to-Point Tunneling Protocol (PPTP), Layer 2 Transport Protocol (L2TP), and Internet Protocol Security (IPSec). The Layer 3 IPSec, developed by the IETF, is a broad-based open solution that is a part of IPv6. It is often the preferred VPN solution for IP environments because its standards include some of the strongest security measures and management systems. Moreover, security arrangements can be handled without requiring changes to individual user computers. A disadvantage of IPSec is that it can only work with IP traffic, while other protocols encapsulate packets at Layer 2 and are therefore capable of handling traffic other than IP, such as IPX (Internetwork Packet Exchange).

VPNs normally occur in one of three scenarios:

- *Intranet VPNs* link a corporate main office with remote or branch offices. The corporation generally maintains each end of the network; access is provided through a firewall.

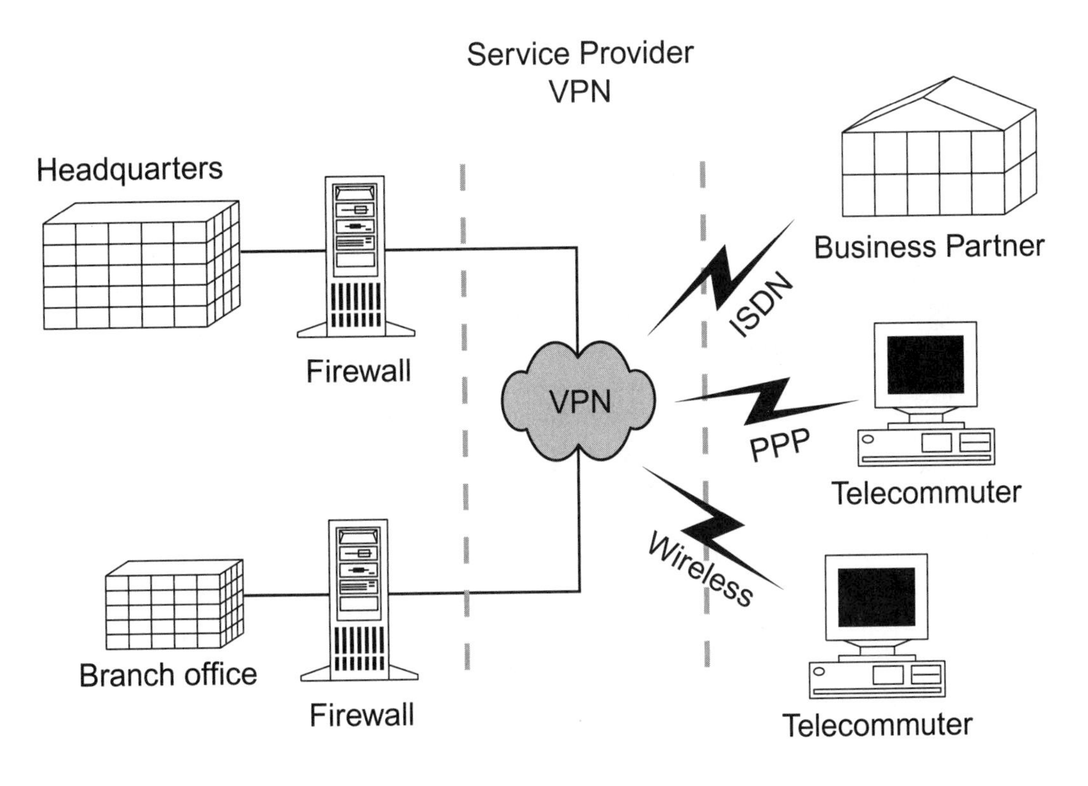

Figure 8–19 Different methods used to access a VPN.

- *Remote Access VPNs* link telecommuters or traveling employees with the corporate network, and these are becoming increasingly important; access is provided via digital methods such as ISDN or wireless medium.
- *Extranet VPNs* link corporate partners, suppliers, customers, and investors together to share and distribute information worldwide via the Internet, which has allowed companies to work seamlessly across time zones and continents without maintaining expensive, dedicated communication networks; access is provided through a firewall.

INTRANET AND EXTRANET

An **Intranet** is a private network that uses TCP/IP, HTTP, and other Internet protocols but is contained within an enterprise. In general, it looks like a private version of the Internet. Its main purpose is to share a company's information and computing resources among its employees and facilitate teamwork across multiple platforms. Intranets can be linked with one another or with LANs over a wide area by VPNs, creating enterprise-wide Intranets. VPNs, based on IP, can also extend the ubiquitous nature of Intranets to remote offices, mobile users, and telecommuters. Enterprises may allow users within their Intranet to access the public Internet through one or more gateways and firewall servers that have the ability to screen messages in both directions so that company security is maintained. *Firewall,* a piece of hardware and software, is a security device that allows limited access out of and into one's network from the Internet.

An **Extranet** is an Intranet that allows controlled access by authenticated outside parties to enable collaborative business applications across multiple organizations. Extranets are therefore more private than the Internet, yet more permeable than an Intranet because they allow access to third parties through authentication. There are two basic configurations: a direct leased line, where an enterprise can have full physical control over the line from Intranet to Intranet, and a secure link over the Internet defined only by access privileges and routing tables establishing a VPN. The second configuration is most popular because it provides an economical mechanism for constructing an Extranet.

The Intranet/Extranet model may be illustrated best within the construct of the classic three-tier client/server architecture, as shown in Figure 8–20. These three tiers include the data and system services tier, the application tier, and the presentation tier. The data and system services tier contains core technologies such as servers, databases, security, multiplatform interoperability, connectivity, and scalability. Key business activities that are conducted throughout a collaborative environment take place via the application tier. One key application is Groupware, which is a class of software tools that enable collaboration on projects within a corporation, or across business enterprises. The presentation tier includes the interface layers between users and back-end resources they wish to access, so that the interface can be customized to each individual user according to the user's profile. A group of users can literally be accessing the same data simultaneously, but each user would see a different subset of that data with an entirely different view. In this way individual users and groups can be assured of relevant and useful experiences as they work, collaborate, and communicate within the Intranet or Extranet environment.

The explosive growth of Intranets and Extranets can be attributed to the almost universal acceptance of the IP protocol as a preferred networking protocol standard. Through widespread adoption of this standard, enterprises can connect LANs over a wide area or link Intranets together for business-to-business interoperability relatively easily and inexpensively, whereas to build those kinds of connections in the past was difficult and also very expensive. For example, prior to open architecture, the prospect of linking the product supply chain—manufacturers, suppliers, dealers, off-site contractors, and customers—was virtually impossible. The only options were expensive proprietary networks that were custom developed to overcome the heterogeneous platforms and networks within each organization. Web-based Intranets and Extranets open up these closed systems by being able to achieve integration across distributed, cross-platform environments, as depicted in Figure 8–21.

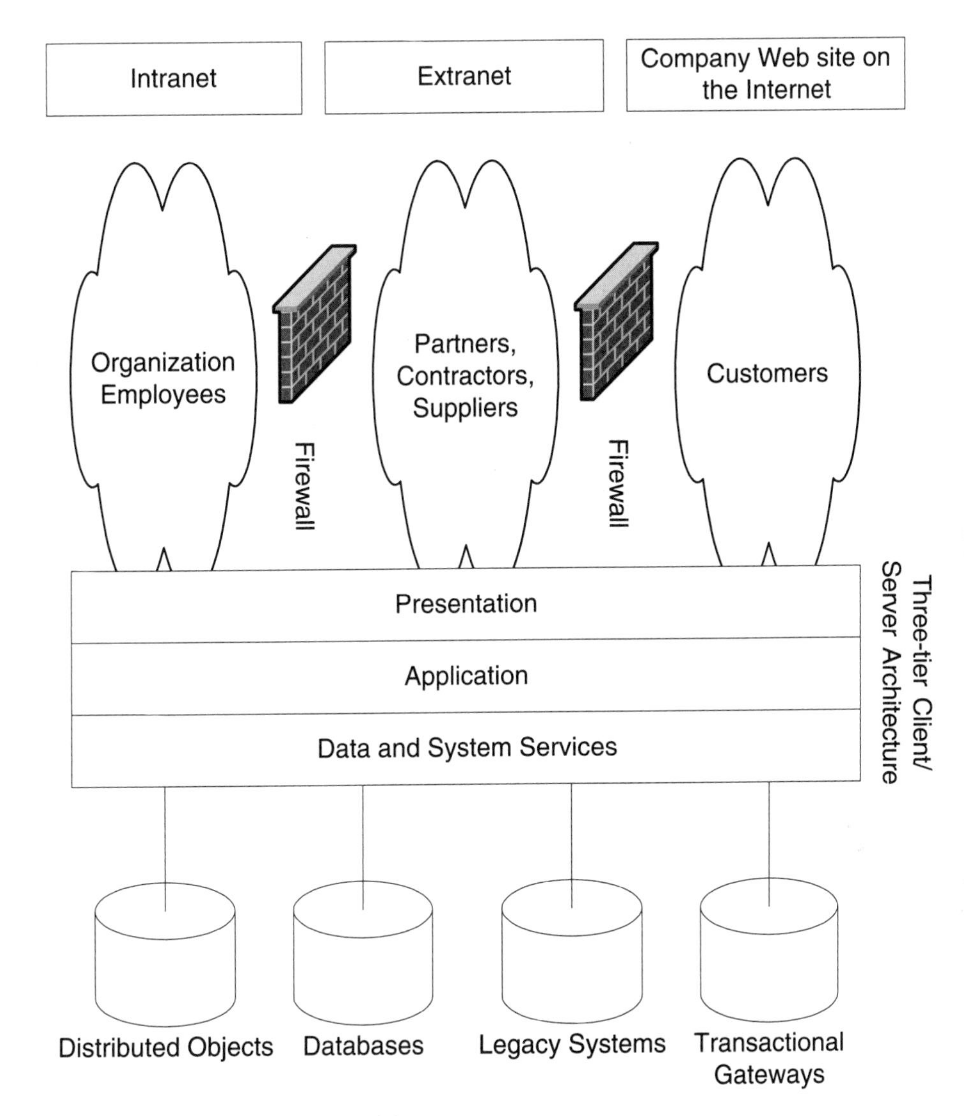

Figure 8–20 An Intranet/Extranet model.

CONVERGED NETWORKS

One of the goals of the telecommunications industry is to bring the open systems model of the data world to a voice network that has thus far been based on proprietary switches and software. The move is toward high-capacity voice switches that can communicate with both the traditional circuit-switched telephone networks and the packet-switched data networks so that the LAN can carry voice and video, along with data, as illustrated in Figure 8–22. The Class 5 switch still enjoys a position of prominence

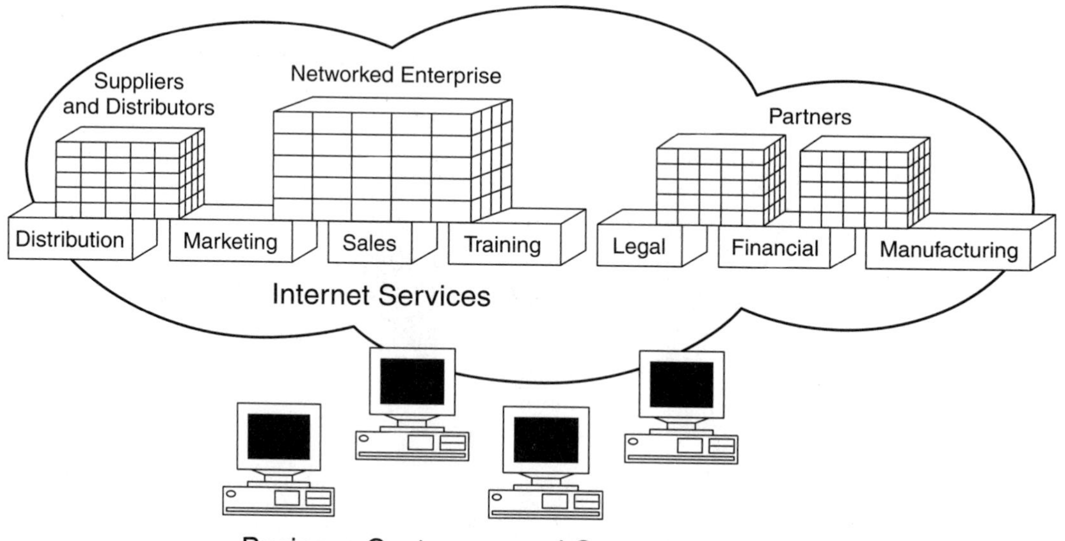

Figure 8–21 One depiction of the use of Intranets and Extranets.

because its features add substantially to revenues. The first challenge is to connect the Class 5 switch to the broadband access network.

The telephone network provides one level of QoS: you either get a 64 kbps voice channel or you do not. Voice-digitization modules lower the operating rate of a voice conversation to 4 or 8 kbps per voice channel, depending upon the compression techniques used. So you can transmit between 16 and 8 voice conversations on one 64 kbps line, or a mixture of voice and data over a single connection. **Converged Data/Voice Networks** can be described as the application of voice digitization and compression schemes through a variety of hardware and software products to enable voice to be transported on public and private networks originally developed to transport data. It fulfills one of the most fundamental aspects of network management: to integrate the transport of information in a cost-effective manner.

Converged networks must fulfill certain system requirements; the most important requirement is the ability to treat different data types differently. A network must have low delay for voice telephony, the ability to handle megabit data streams with ease for video, and low error rates and strong security for mission-critical data such as financial transactions. In ensuring QoS for converged networks, latency and jitter, however important, are only two of the issues. A provider must also address call-completion ratio, post-dial delay, and echo cancellation. Further, the solution must accommodate a wide variety of useful features, including automatic attendant capability, directory lookup, hunt groups, teleconferencing, and caller ID, and it must be able to work with standard tele-

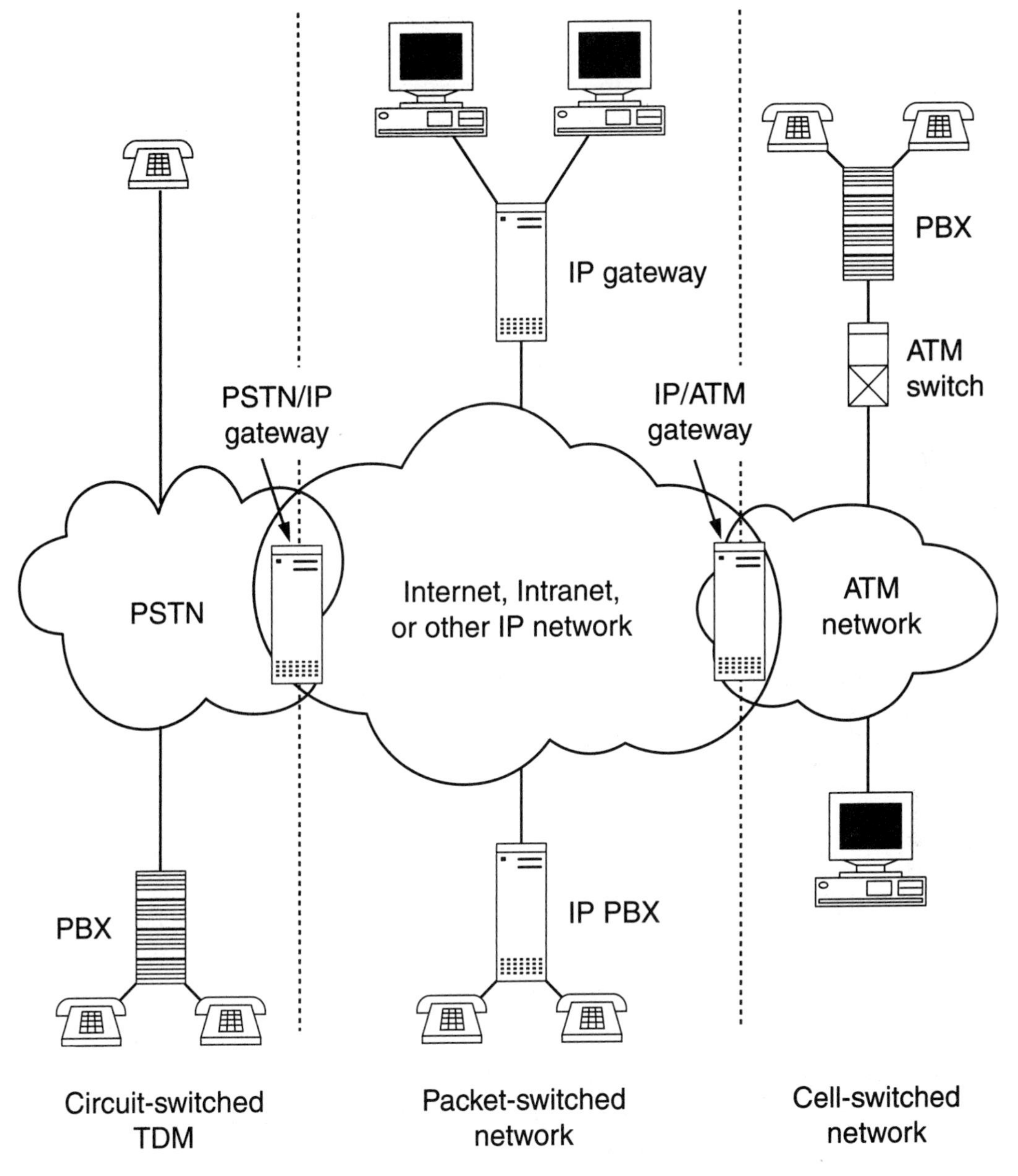

Figure 8–22 Converged networks.

phone sets. Only when a network is engineered for all of these parameters can the claim of toll quality truly be made.

Many organizations operate separate voice and data networks based on a traditional separation of the two technologies for two primary reasons: economics and technology.

In some situations, the cost associated with integrating the two might exceed the potential savings or require a period of time that would result in a relatively poor return on investment. As for technology, until recently, equipment was not available to effectively and efficiently transport voice over data networks. Voice compression and the development of techniques to enable compressed voice to flow effectively and efficiently through a packet network are two key technologies for transporting voice over data networks.

VOICE OVER IP (VoIP)

One of the latest technological developments has been in the area of **Voice over IP (VoIP)**, a term that refers to the process of transmitting telephone calls over the Internet rather than through the traditional telephone system. TCP is used by applications where the integrity of data is a primary concern, but it can result in unacceptable delays when transporting digitized voice and is rarely, if ever, used for voice transmission. Its connection-oriented capability results in its use for establishing communications between network devices and call-control operations, while UDP is used for the flow of digitized voice. Although UDP is used to transport datagrams, the actual transfer occurs via the use of virtual circuits with an IP header added to the UDP header for routing purposes. There are a number of voice compression algorithms currently in use for VoIP; G.723 is the standard recommended by the ITU for low bit-rate voice transmission over the Internet. An integration of the disparate technologies of SS7 and IP is a critical step before VoIP becomes a mainstream technology.

PSTN and IP Internetworking

As shown in Figure 8–23, a VoIP gateway bridges calls between the PSTN and an IP network as it performs the internetworking function between different signaling protocols and media streams carried on the Internet and PSTN. Gatekeepers work in conjunction with a network of gateways. While gateways transfer calls between the PSTN and IP networks, gatekeepers provide the intelligence for a VoIP network. Among other things, gatekeepers provide an interface to billing systems, as well as network security by preventing unauthorized usage. The network management station enables the control and monitoring of the whole network from changing call routing tables to monitoring gateway and gatekeeper activities. **Assured Quality Routing (AQR)** is a set of processes and procedures that marries packet and circuit switching to automatically reroute calls to the PSTN when parameters on a given call exceed developed acceptable ranges. AQR ensures high standards such as high call-completion ratios, low post-dial delay, and, most importantly, high voice quality.

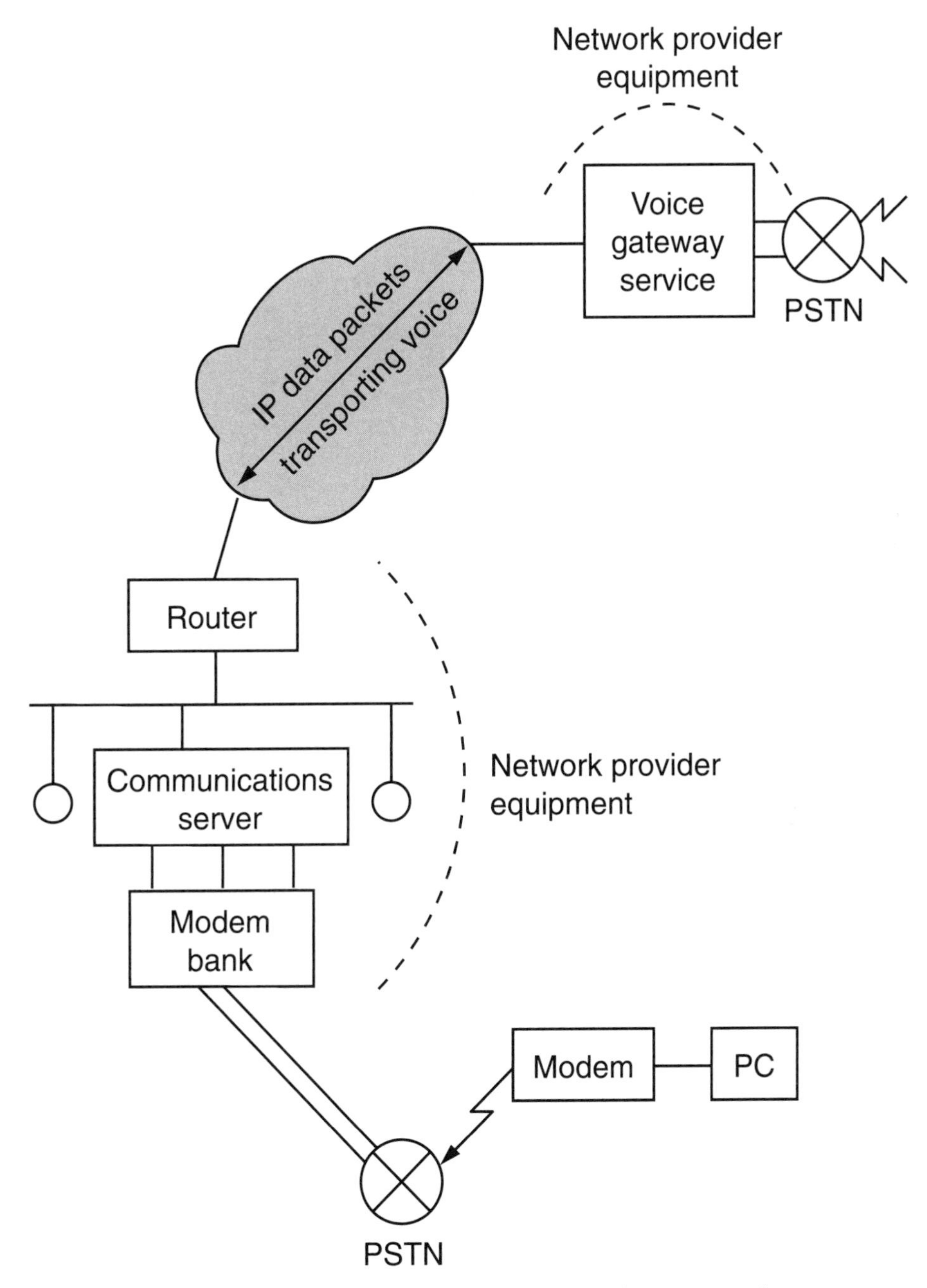

Figure 8–23 A VoIP gateway bridges calls between the PSTN and an IP network.

VoIP Call Process

Several scenarios can be used for implementing VoIP where the cost is significantly lower than long-distance telephone rates over the circuit switched network.

1. In a computer-to-computer service, a destination telephone number is entered into a special software client. The software connects over the Internet to the service provider's gatekeeper, gets authorization to use a gateway, and calls that specific gateway, which then calls the destination phone number and bridges the users' sound cards so that the call completely bypasses the telephone network.
2. In a phone-to-phone service, users on both ends utilize the traditional telephone. The caller dials a local phone number that connects the phone call to a local interface between the telephone system and data network. This interface could be provided by, for example, an ISP, which routes the call through the data network to another interface that is located close to the caller. At that point, the call rejoins the regular telephone network and proceeds to its destination. In this scenario, the caller pays only for the two local call segments, plus any charges associated with Internet use and the interface between the telephone system and the Internet.
3. In a computer-to-phone service, the originating phone call is placed by a computer and transmitted through a data network to the destination telephone handset.

VoIP QoS

Enterprise network managers who manage VoIP services are concerned that there is often a difference between service-quality levels reported by traditional network-management tools and the opinions expressed by service users. The main reason for this difference is that network behaviors that can significantly impact VoIP performance include jitter buffer *discards* and *bursts* (varying periods of high and low packet loss), but current data applications simply do not address those behaviors. The purpose of the jitter buffer is to add a small amount of delay to each packet to compensate for congestion-induced variations in packet arrival times or jitter at the receiver. Packets arriving outside the jitter buffer's delay window will be discarded. The discards can cause significant artifacts in the audio heard by the listener. Thus, accounting for discards is crucial to gaining insight into call quality.

Packet loss concealment (PLC) capabilities in modern vocoders (voice coder/decoder) can conceal packet loss and discards. VoIP call quality is usually measured against PSTN toll-quality voice, focusing on intelligibility to the *listener* and *conversational* smoothness. To score *listener* quality, parameters that must be taken into account include the vocoder type, whether PLC is enabled, whether packets were lost in burst or gap states, the number, density and duration of bursts, and at what point in the call losses and discards occurred. To score *conversational* quality, end-to-end delay must be measured and included in the calculations to understand the potential effects of double talk. A VoIP project should begin with a comprehensive traffic-study and network usage-pattern analysis to help you know what kind of trunking and bandwidth you need, and it should include analysis of bandwidth usage, availability, calling rates, latency, and jitter.

Contemporary IP networks offer a best effort service, that is, the network tries to deliver all packets to their destinations to its best ability. This means that as traffic load approaches and exceeds the network's capacity, there will be competition between packets for the network resources. The goal of much of the QoS work in the IP community is to provide a network that offers service differentiation. With the onset of congestion (competition) in the network, some packets must be given preference over others. The IETF continues work on two protocols that deal with IP QoS: *Differential Services (DiffServ)*, and the *MultiProtocol Label Switching (MPLS)*.

DiffServ establishes mechanisms to prioritize certain data streams, though it does not guarantee particular service levels for any particular stream. At Layer 3, routers can affect service quality through variable queuing. This ensures that in the presence of congestion, a high-priority packet will transit the router faster than a low-priority packet, but it cannot guarantee latency.

MPLS makes it easy to manage a network for QoS by setting up a specific path for a given sequence of packets identified by a label put in each packet, thus saving the time needed for a router to look up the address for the next node to forward the packet to. MPLS is called multiprotocol because it works with IP, ATM, Frame Relay, and Ethernet network protocols by forwarding packets at Layer 2 rather than at Layer 3.

Voice Over Frame Relay (VoFR)

A growing number of vendors are now delivering *Voice over Frame Relay (VoFR)* solutions, and standards are evolving at a rapid pace. To converge voice and data traffic onto a single Frame Relay trunk, *Frame Relay Access Devices (FRADs)* employ complex queuing algorithms that can process frames by traffic priority and maximum elapsed time in queue. Since queuing and transmission latency in FRADs are directly dependent on frame size, mixing frames of widely varying sizes will result in variable latency, or jitter for voice transmissions. *Frame Relay segmentation* alleviates this problem by segmenting all voice and data transmission to a fixed frame or cell size, as depicted in Figure 8–24. But data efficiency is maintained by taking multiple voice samples (usually three to five, depending on trunk speed), and packing them into a single larger frame, which improves data performance. Also, most FRADs can detect and weed out silence patterns, reducing bandwidth requirements. In Frame Relay networks, one or more frames occasionally get discarded because of a number of possible conditions, such as congestion, excessive bursting, or PVC rerouting. To minimize the clipping effect caused by frame loss, some of the more advanced FRADs employ a lost speech interpolation technique that blends speech samples that precede and follow the lost frame(s) to mask the lost speech elements.

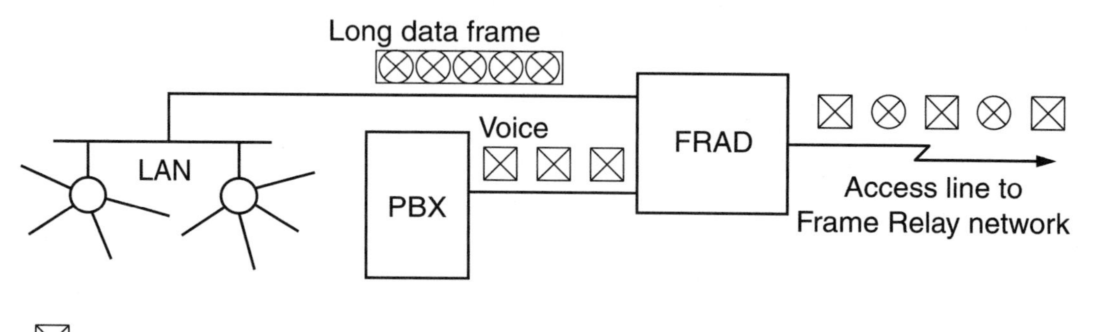

Figure 8–24 Frame Relay Access Devices (FRADs).

Voice Over ATM

Voice over ATM supports multiple classes of service to obtain the predictability and reliability required for end-to-end transmission of voice, data, and video. Each traffic class has predefined characteristics, as well as varying levels of service guarantees, and is based on the use of three key attributes:

1. Timing relationship between the source and destination
2. Variability of the bit rate
3. Connection mode

The timing relationship between the source and destination defines the ability of the receiver to receive the original data stream at the same rate at which it was originated. For example, a voice conversation digitized at 64 kbps via PCM must be read by the receiver at that data rate in order for it to be correctly interpreted. Regarding the bit rate, the CBR was designed primarily for supporting voice communications. It allows the amount of bandwidth, end-to-end delay, and the delay variation to be specified during call setup. The third attribute, connection mode, can be either connection-oriented or connectionless. Examples of connection-oriented applications include voice calls and SNA data sessions, while examples of connectionless applications include Ethernet transmission and applications that use UDP.

MULTIMEDIA OVER IP PROTOCOLS

Popular multimedia transmission over IP protocols include: Real-time Transport Protocol (RTP), which involves sequencing of packets; Resource Reservation Protocol (RSVP), which involves reservation of bandwidth; Session Initiation Protocol (SIP), which represents an IETF proposed standard for control of multimedia call sessions; Open Settlement

Protocol (OSP), which is designed to handle authorization, call routing, and call detail billing between ISPs; and various videoconferencing standards and their applications.

Real-time Transport Protocol (RTP)

Implemented in real-time applications that are bandwidth intensive and sensitive to latency and jitter, RTP can compensate for packet loss. For example, streaming video sent from a server to a client often experiences packet loss, but the client can detect the loss and fill in the gaps. RTP supports unicast as well as multicast transmissions. RealAudio and RealVideo applications allow either streaming mode or buffered mode operation. **Streaming mode** is extremely sensitive to congestion as it plays the audio/video in real-time across the network. The information is sent in compressed form over the Internet in a continuous stream and is played as it arrives. **Buffered mode** downloads the entire file to the client and then plays the audio/video in the client environment rather than across the network.

Resource Reservation Protocol (RSVP)

RSVP is the centerpiece of Layer 3 and Layer 4 mechanisms to ensure QoS for real-time IP data streams. It lets applications reserve a path with certain metrics across an IP network. Like ATM, it implements an admission policy—if the resources of the network are inadequate to provide the required service levels, the connection will be refused. RSVP has been successfully implemented on a local scale but has had problems scaling to larger traffic volumes typical of the Internet backbone for the following reason. Routers are designed to consider one packet at a time, processing each one on its individual merits. Router speed can therefore be maximized and its computational resources minimized because it does not have to store and look up pathway information for the duration of a stream of data. However, RSVP requires establishing paths from source to destination. On a customer-premises network, this overhead may be easily accommodated, but on the Internet, where the paths traverse the Internet core and traffic volumes are enormous, RSVP's tracking of paths is uneconomical and difficult to implement.

Open Settlement Protocol (OSP)

OSP involves a dedicated settlement server designed to handle authentication, authorization, call routing, and call detail over IP networks. OSP inherently supports a multi-service, multi-provider environment, and the following key transactions occur when a call is initiated:

- Authentication: Once it is determined that the call will be terminated outside the originating network based on dialed destination digits, the caller must first be authenticated to use the terminating network's facilities.

- Authorization: The settlement server ascertains whether the call can be terminated on a peer network and at the same time identifies a set of possible terminating endpoints. A token is then sent to a specified chosen target terminating endpoint.
- Billing: Basic call rating information needs to be communicated and may be established beforehand through pre-rating. In cases where a guaranteed level of QoS is required, it can also be negotiated.
- Usage: A record detailing the service including information such as the endpoint IP address, caller ID information, level of QoS, and protocol is sent to the settlement server for reconciliation.

H.323

The H.323 standards architecture is broad—it specifies gateways and gatekeepers that enable connections among LAN-based desktop videoconferencing (DVC) units, ISDN-connected H.320 units, analog telephone-connected H.324 devices, and ISDN and POTS telephones. H.323 is a recommended standard from the ITU for multimedia transmission on LANs and via packet networks like IP and Frame Relay that have non-guaranteed bandwidth. Initially, H.323 was not devised for voice but for multimedia videoconferencing, and it was typically only concerned with a few highly intelligent endpoints. Later versions feature SS7 integration for transfer of enhanced services across the PSTN border, bandwidth management, and QoS issues. New capabilities include call hold, call park, call waiting, some multimedia broadcasting capability, and address translation across telephone number, URL ID, e-mail ID, as well as IP address.

A collection of related protocols is described in Figure 8–25. The new G.723 audio-encoding and H.263 video-encoding protocols ensure high-quality audio that does not degrade video performance. G.723 requires about 16 kbps for audio and leaves the rest of the available bandwidth for video. On the other end of a connection, a gateway usually is confronted with an H.320 system that uses H.261 video and a 64 kbps G.711 audio standard. Therefore, a gateway must transcode the protocols without introducing noticeable delays. The multiplex standards H.221, H.223 and H.225 combine audio, video, and data in one bitstream for different suites of videoconferencing standards. In addition, standards such as the Media Gateway Control Protocol (MGCP) and associated gateway functions continue to be developed. The effective and widespread deployment of multimedia and telephony services will depend on the successful implementation of these protocols over SS7 and over IP networks.

DATA COMPRESSION

Data compression is arguably one of the most important aspects of the Internet and multimedia. This technique is applied to data prior to its transmission, or dynamically to a

Standard	Application
H.320	A suite of multimedia conferencing standards on switched digital networks, such as ISDN.
H.323	A suite of multimedia conferencing standards on traditional packet-switched LANs with no guaranteed quality of service (QoS)
H.324	Standard for multimedia conferencing on analog phone lines (POTS)
H.261	Video compression standard specified for H.320, H.323, and H.324
H.263	Video compression standard for H.324 (low bit-rate communications)
G.711	Basic audio compression at 48 kbps to 64 kbps (ISDN)
G.723	Audio compression for H.324 POTS standard videoconferencing
H.221	Multiplex standard for H.320
H.225	Multiplex standard for H.323
H.223	Multiplex standard for H.324
H.245	Standard for multimedia system control in H.323, H.324, and H.310; includes features such as capability exchange and signaling
H.GCP	ITU's extension of H.323 to enable IP gateways to speak to SS7 systems
JPEG	Joint Photographic Experts Group of ISO produced still-image compression standard
MPEG	Motion Picture Experts Group of ISO produced video standards for broadcast television or playback on multimedia PCs
RTP	Real-time Transport Protocol involves sequencing of packets and compensates for packet loss in either streaming or buffered mode
SIP	Session Initiated Protocol, speaks between user and IP/media gateways
V.80	Specified for H.324 systems that convert synchronous data streams to asynchronous, enable rate adjustments during a call, and notify client software of lost packets

Figure 8–25 Multimedia standards or protocols and their applications.

data stream being transmitted. It is used to reduce the number of bits that must pass over the communications medium by a ratio of 4:1, which reduces transmission time, yields higher throughput rate and improved response time. **Lossless** data compression used for text transmission refers to techniques in which no data is lost, while **lossy** data compression results in some amount of data loss. FTP and TELNET are examples of traditional applications that do not tolerate packet loss. Most image compression technologies, such as JPEG and MPEG, use a lossy technique.

Run-Length Encoding

Run-Length Encoding (RLE) is a lossless compression technique that compacts redundant data. It is used in some common Windows file formats and in fax modems. As transmitting fax modem scans a document, it quickly senses a run of *white space* and encodes it in few bits. Once some *real* data appears, the modem notes this fact and begins to send corresponding bits to the internal buffer. After a specific number of bits are stored in the buffer, the modem packs them into a frame for transmission. As runs of *white space* and runs of *real* data reoccur, the modem recognizes that fact and adjusts accordingly. Therefore, a modem can transmit a document with a lot of white space quicker as compared with a document of dense text. You may want to try faxing a white sheet of paper and then a document of very dense text; you will immediately see the difference.

Huffman Code

The basic principle behind the lossless Huffman code is that certain symbols or combinations of symbols appear in text more frequently than do others. Huffman coding is based upon a simple philosophy: The most common data should be noted with the least number of bits, while the least common data should be noted with the most number of bits. The objective of the Huffman Code is to compress the actual number of bits sent from the source to the destination by using a variable-length code, where the code for each character has a unique prefix. A code table is organized in a tree-like structure with frequently occurring characters closer to the root of the tree and represented by shorter codes. A technique related to Huffman coding is Shannon-Fano coding; the algorithm used to create the Huffman codes is bottom-up, and the one for the Shannon-Fano code is top-down.

As shown in Figure 8–26, if the characters to be encoded are arranged in a binary tree, a code for each character is found by following the tree from its root to the character in the leaf so that the encoding is a string of bits. In Figure 8–26, E is represented by 00, T is represented by 10, and S is represented by 011. Therefore, the word SET would be encoded as 011 00 10. It is expected that a message coded in Huffman code would contain fewer bits than any other code that uses a fixed number of bits for every symbol. However, the quality of the compression depends on the code table, which can be either static (fixed) or dynamic (changes according to the data being transmitted). If the code

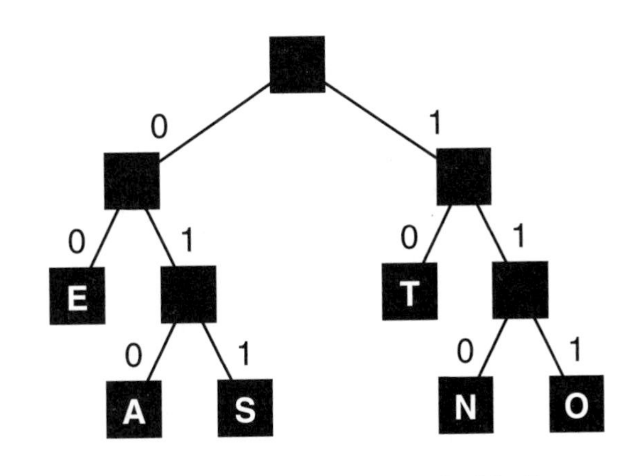

Figure 8–26 Huffman code, where characters to be coded are arranged in a binary tree.

table is well adapted to the data being compressed, Huffman coding can reach the theoretical limit of compression. One very popular compression algorithm, MNP 5, is a combination of Huffman coding and RLE. It is implemented in the hardware of many modems to effectively increase the transmission speed.

Transform Coding

Lossy image coding techniques normally have three components: image modeling, which defines such things as the transformation to be applied to the image; parameter quantization, whereby the data generated by the transformation is quantized to reduce the amount of information; and encoding, where a code is generated by associating appropriate code words to the raw data produced by the quantizer.

Transform coding, a lossy technique, can be generalized into four stages:

- ✦ image subdivision
- ✦ image transformation
- ✦ coefficient quantization
- ✦ Huffman encoding

The first step, image subdivision, is to subdivide an n×n image into smaller n×n blocks. In image transformation, the image is represented in a new domain where a reduced number of coefficients contains most of the original information. This is where certain characteristics of the human visual system can be exploited to achieve higher compression ratios. For example, our eyes are less sensitive to the color blue than they are to green. Using Fourier transform on frequency, we can take advantage of this fact when compressing a color image by throwing away some of the blue information that our eyes probably would not detect. The next step, quantization, reduces the amount of data used to represent the information within the new domain. Quantization in most cases is not a reversible operation, and therefore it belongs to the lossy method. The final step of encoding is usually a lossless technique like Huffman code, which may further reduce the bit rate. The JPEG and MPEG standards are based on transform coding.

Video signals consist of a number of still picture frames composed each second, for example, 30 still pictures per second in a high-quality video transmission. In such cases, although the first picture is sent in its entirety, subsequent frames are analyzed and only the changed pixels are sent to the receiver rather than the entire frame. The receiver combines this change information with its memory of the previous frame, and locally builds new versions of the frames for display. In this scheme, compression algorithms lose some of the resolution of video images, but it may be acceptable as long as the communication keeps up with the speed of the moving image.

SUMMARY

Telecommunications are experiencing cost-performance breakthroughs as a result of a combination of competition and technology. Packet voice can be implemented, but the carriers are being conservative because of the large installed base of circuit-based switches and the peripheral operation support systems, billing systems, network management systems, intelligent-network call-feature systems, and signaling systems. The major convergence issue is not packet versus circuit switching but operational systems-level mediation and the ultimate replacement and upgrade of all network support systems. Mediation equipment and software will allow coexistence with, and then migration to, a packet-voice world from the existing circuit infrastructure at a controlled pace.

There are many instances when Internet users need to access the PSTN and PSTN users need to access Internet telephony services. The IETF PSTN/Internet Inter-Networking (PINT) Working Group addresses technologies, architectures, and the arrangement through which Internet applications can request PSTN services. Another IETF Group addresses the opposite arrangement, in which PSTN users request services that require an interaction between the PSTN and the Internet. Services that PINT addresses often appear on Web pages—click-to-dial, click-to-fax, and voice call-back services. For example, users visiting a shopping Web site may need more information about a product they want to buy. A click-to-fax-back button initiates a sequence that sends information to the user's fax machine through the PSTN. The common denominator of the enriched PINT services is that they combine the Internet and PSTN services in such a way that the Internet is used for non-voice interactions, while voice (and fax) is carried entirely over the PSTN. The PINT Gateway is the PSTN node that interacts with the Internet.

The technology movement adds new capacities for human interaction and constitutes a resource that changes the way we live and work together. Integrated business communications play a key role in the convergence of voice and data by connecting people with other people and the information they need and enabling them to obtain it or act upon it anytime, anyplace. Convergence spans all forms of computer and communications technology and is becoming a key force in productivity improvement and cost reduction. Providing concurrent voice and data management enables network managers to plan their networks in response to changing traffic requirements, and it equips them with the diagnostic tools to isolate problems and restore service levels rapidly. As networks become valuable assets, network security becomes a prime concern.

Case Study

The nation's prominent long-distance carrier is converting its circuit-switched telephone network to packet-switched technology akin to that used on the Internet. The company claims the project is the first time a national telecom carrier has set out to totally revamp its systems, converting them from a technology used by most phone companies to carry voice calls for more than 100 years to a technology better suited to carry data. When completed, the packet-switched network will allow voice or data traffic to be chopped into packets and sent across networks using multiple routes, the same way the Internet handles IP traffic. The company is replacing its old switches with ATM switches and gateways. Customers will not be forced to change equipment to use the upgraded network, but upgrading the packed-based communications gear would let them gain access to IP-based products and services, including multimedia streams, IP technology, IP Centrex, and yet-to-be-invented services. Businesses should save money on capacity increases, combined voice and data traffic on one network, and lower maintenance costs. Because packet-switched networks are more reliable and redundant, service quality should improve as well. Analysts say the shift from circuit-switching to packet-based telephony could be as important for the telecommunications world as the paving of dirt roads was for the automobile industry, because it will result in an all-digital packet network that can easily handle a mix of voice, data, video and other types of traffic.

Questions

1. Discuss the potential advantages and disadvantages of a major carrier's move from circuit-switched to packet-switched technology.
2. Why do you think only one carrier has proposed such a move, while other carriers are cautious and closely watching the transformation?

REVIEW QUESTIONS

1. Define the following terms:
   - A. Connection-oriented
   - B. Connectionless
   - C. Slow Start
   - D. Spoofing
   - E. Tunneling
   - F. Assured Quality Routing
   - G. Streaming mode
   - H. Buffered mode
   - I. Frame Relay Segmentation
   - J. Lost Speech Interpolation
   - K. Lossless data compression
   - L. Lossy data compression
2. Outline the TCP/IP model and evaluate its role as the de facto standard for global communications with reference to the OSI model.
3. Differentiate between TCP and UDP transport protocols.
4. Discuss IPv4 addressing scheme and provide an example for each network class.
5. Explain the use of the Dotted Decimal Notation with an example.
6. Evaluate the advantages of multicast, subnets, and classless addressing.
7. Describe the updates in IPv6 as compared with IPv4.
8. Discuss prominent TCP/IP applications and protocols.
9. Analyze the shortcomings of TCP transmission via satellite.
10. Compare and contrast:
    - A. SMTP, MIME, and IMAP
    - B. SNA with TCP/IP
    - C. RTP with RSVP
    - D. Intranet with Extranet
    - E. DiffServ with MPLS
    - F. RLE with Huffman Code
11. Explain the four steps in Transform Coding.
12. Describe the role of Internet2.
13. Analyze the need for virtual private networks and their characteristics.
14. Determine the significance of Intranets and Extranets for businesses.
15. Develop an argument for converged networks.
16. Provide a framework for PSTN and IP internetworking; identify the necessary components and include call quality measures.
17. Compare and contrast Voice over Frame Relay and Voice over ATM.

KEY TERMS

Network Management
Network Manager
Policy Management
RAID Levels
Platform
Network Operating System (NOS)
Network Administration and Maintenance
Virtual LAN (VLAN)
Network Utilization
Computer and Network Security
Passive Threats
Active Threats
Virus
Worm
Vulnerability Management
Security Policy
Authorization
Encryption
Authentication
Intrusion Detection System (IDS)
Firewall
Direct Attached Storage (DAS)
Network Attached Storage (NAS)
Storage Area Network (SAN)
Two-tier Versus Three-tier Architecture
Business Continuance

OBJECTIVES

Upon completion of this chapter, you should be able to:

- Outline the responsibilities of a network manager
- Evaluate the importance of policy management
- Assess the need for a network operations center
- Differentiate between network administration and maintenance, and management tasks
- Evaluate different RAID levels
- Design an ergonomically-appropriate workstation

- ✦ Determine the characteristics of the major components of LAN Hardware
- ✦ Analyze the specifications and characteristics that should be considered when evaluating a network operating system
- ✦ Assess different security threats and the need for a security policy
- ✦ Evaluate different security measures such as access control, encryption, authentication, Intrusion Detection System (IDS), and firewalls
- ✦ Estimate the significance of performance monitoring
- ✦ Determine the appropriateness of different storage strategies like DAS, NAS, and SAN

INTRODUCTION

The telecom industry is at the brink of combining three types of management—system, network, and services—to achieve totally integrated **network management.** In telecommunications, it is not unusual to find computer specialists, engineers, technologists, and service managers using the same vocabulary to mean different things and pursuing different goals. Indeed, even the term "network management" has different meanings. In telephony, it means the control of congestion in the network, while in data communications, several additional factors are involved, such as maintaining network availability, ensuring security, managing storage, and balancing server and network performance. The common threads are routing, switching, transport, network applications, and intelligent network services.

Baseband network managers track network traffic and error rates because the quality of the physical and logical networks is measured by bit errors and dropped frames. The performance of each network element is assessed and accounted for in determining total network performance. On the other hand, a broadband network manager has those standards to maintain, but also tracks service performance against contracted goals and customer management of network assets. However, there are some fundamental network management concepts that are common to all telecom networks.

Network management encompasses both human and automated tasks that support the creation, operation, and evolution of a network. For a network to be effective and efficient over a long period of time, a network management plan must have two goals: prevent problems where possible and prepare for problems that will most likely occur. The job of a **network manager** keeps evolving with time and developments in technology. Looking at the big picture (as depicted in Figure 9–1), the responsibilities of the job include:

- ✦ Policy Management
- ✦ Evaluation of Hardware and Software
- ✦ Network Administration and Maintenance
- ✦ Network Security
- ✦ Configuration Management

Responsibilities of a Network Manager

Policy Management	Evaluation of Hardware and Software	Network Administration and Maintenance	Network Security	Configuration Management

Figure 9–1 Responsibilities of a network manager.

POLICY MANAGEMENT

A pivotal function of network management is **policy management**, which is an implementation of a set of rules or policies to dictate user connectivity and network resource priorities. Policy management includes three fundamental functions:

1. Provisioning or configuring of the network switches and routers,
2. Verification (or auditing) of network operation, and
3. Enforcement of the policies, especially technology standards.

Technology standards are the foundation of policy management. Here is an example of what can happen when standards policy is not enforced: An organization wanting to assess its IT infrastructure distributed 2,500 surveys via internal e-mail. When users attempted to open the e-mail, the survey caused some desktops to lock up and crash, disrupting productivity and generating mountains of hate mail to the IT department. The department thought that it was adhering to standards, but unfortunately, the enterprise actually had three e-mail systems—two office suites and eight versions of one suite—in operation.

EVALUATION OF HARDWARE AND SOFTWARE

The two major components that are evaluated when designing and upgrading a network are the hardware and the software.

Hardware

In recent years, with the increased use of computers in the workplace, a greater emphasis has been placed on workstation ergonomics, and much has been published about the total workstation environment. Having information about ergonomic design and how the arrangement of the equipment can impact comfort, health, and productivity is useful because network administrators are often called on to advise others in the company about terminal use and environment. Information about an ergonomically designed workstation appears in Figure 9–2.

In a client/server environment, a user is presented with a standard interface from any client workstation within the organization. The network administrator has complete control over the applications and the operating system versions. Therefore, any delay experienced by the user is blamed on the entire network rather than the limiting component. In the past when users needed faster processing, a new desktop would generally solve the problem, which is not the case for a client/server model.

Several factors determine the response time of a server. The primary one is number of users that are attached to the system—if a specific application has a large number of users, one server may be dedicated to that single application, in which case it is referred to as a *dedicated server.* Other servers could be used to distribute the load of other programs on the system. A *remote-access server,* which provides a bridge between the network, a modem, and a telephone line, allows users to dial into a network from anywhere in the world. A downside of client/server is that server downtime adversely affects multiple users. The key to cutting down the number of servers—a goal for many companies—is to boost each server's utilization rate, so fewer systems can handle more work. To do that requires allocating and reallocating computing resources to the applications and business processes that need the most horsepower.

A server's hard disk is one of the network's primary commodities since it holds files for network-based programs. Personal files and data may also be stored on a network drive attached to the server, but the administrator must ensure that disk space is available at all times for legitimate users of the network. The file server is typically composed of a fast microcomputer, a fast hard drive, a CD-ROM, a RAID, and a tape backup. Here, the emphasis is on speed, so Small Computer System Interface (SCSI) disk subsystems are used. SCSI is a set of evolving ANSI standard electronic interfaces that allow computers to communicate with peripheral hardware such as disk drives, tape drives, CD-ROM drives, printers, and scanners faster and more flexibly than did previous interfaces. SCSI allows multiple devices to be connected to a single port.

An administrator can apply different types of RAID as shown in Figure 9–3 and Figure 9–4. RAID 0, also known as striping, increases the performance of a hard drive by writing stripes of data across multiple hard drives. The data is broken down into blocks and each block is written to a separate disk drive. Data is accessed more rapidly when written across multiple drives that can be read in parallel. Even though it increases performance, RAID 0 is less reliable than a single hard drive, because, if one of the drives fails, all data in an array will be corrupted. RAID 1, also referred to as mirroring, is fault tolerant because it keeps a copy of the contents written to one drive on the other drive. If one of the drives within a RAID 1 fails, simply replace the drive and rebuild the RAID with the contents of the other drive. The RAID 1 architecture can be designed such that a disk failure will cause the I/O subsystem to switch to one of the replicated disks with no service interruption. This is the most reliable kind of **RAID level** and is often used in servers. Where performance is also important, parity-checking RAID (level 3, 4 or 5) is better than RAID 0 because it combines striping with redundancy; the parity information allows recovery from the failure of any single drive.

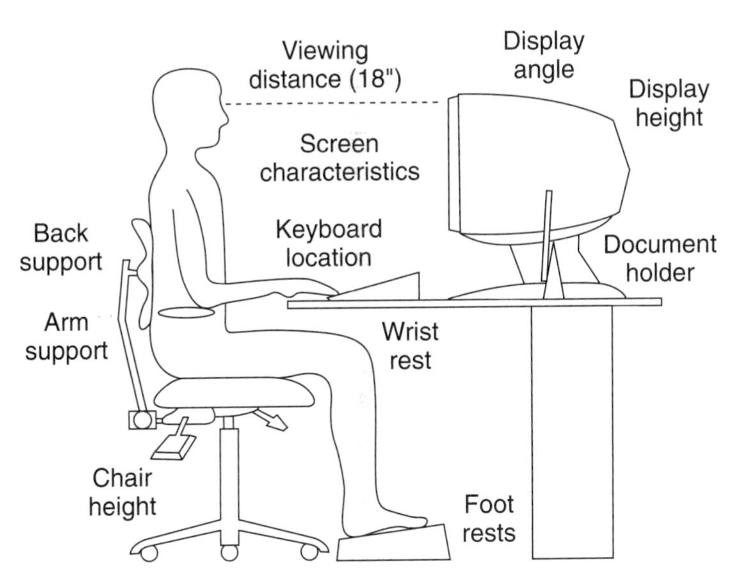

Viewing distance (18")
Display angle
Display height
Screen characteristics
Keyboard location
Back support
Arm support
Document holder
Wrist rest
Chair height
Foot rests
Considerations for Ergonomically Designed Workstations
1. The monitor should have a nonglare screen that can be tilted back 10 to 20 degrees.
2. The screen of the terminal should be about 18 inches from the operator's eyes.
3. The top of the screen should be no higher than eye level.
4. The chair should have a seat that adjusts up and down and a backrest that adjusts separately. When seated, the operator's feet should rest flat on the floor or on a footrest.
5. The keyboard should be attached to the rest of the terminal by a cable so it can be moved on the work surface. The keyboard should be positioned so that the operator's lower arms are parallel to the floor and the upper arms are perpendicular to it.
6. Nearby blinds should be closed and other sources of glare eliminated. Small adjustable task lights are usually preferable to overhead lights.

Figure 9–2 Ergonomically-designed workstation. (Courtesy *Telecommunications for Managers 4/E* by Rowe, S.H., copyright 1995. Reprinted by permission of Prentice-Hall, Inc., Upper Saddle River, N.J.)

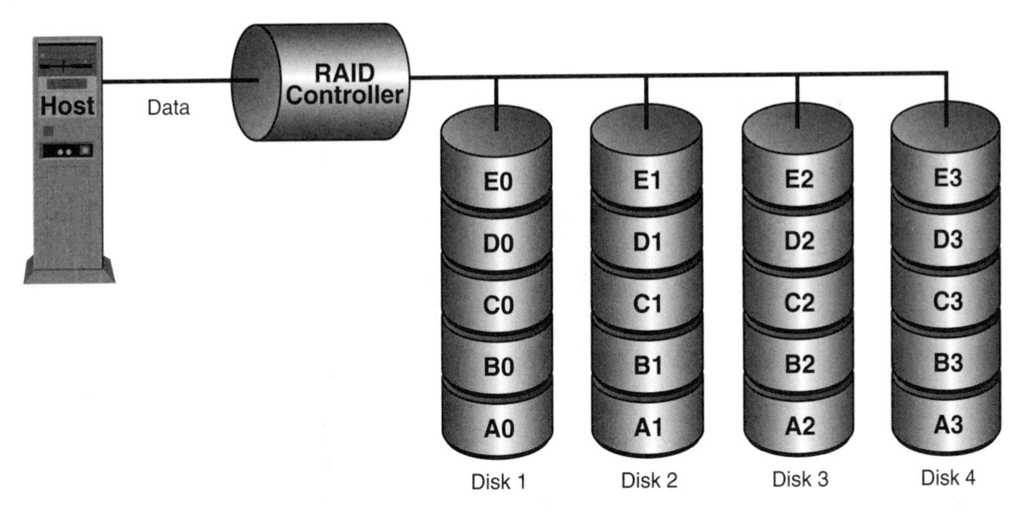

Host
Data
RAID Controller
E0
D0
C0
B0
A0
E1
D1
C1
B1
A1
E2
D2
C2
B2
A2
E3
D3
C3
B3
A3
Disk 1
Disk 2
Disk 3
Disk 4

Figure 9–3 RAID level zero (RAID 0 or striping).

A network consists not only of servers and workstations, but also of routers, switches, hubs, and shared input and output devices such as printers, copiers, and scanners. The network administrator must ensure device compatibility so that they can be interfaced seamlessly with the LAN protocols.

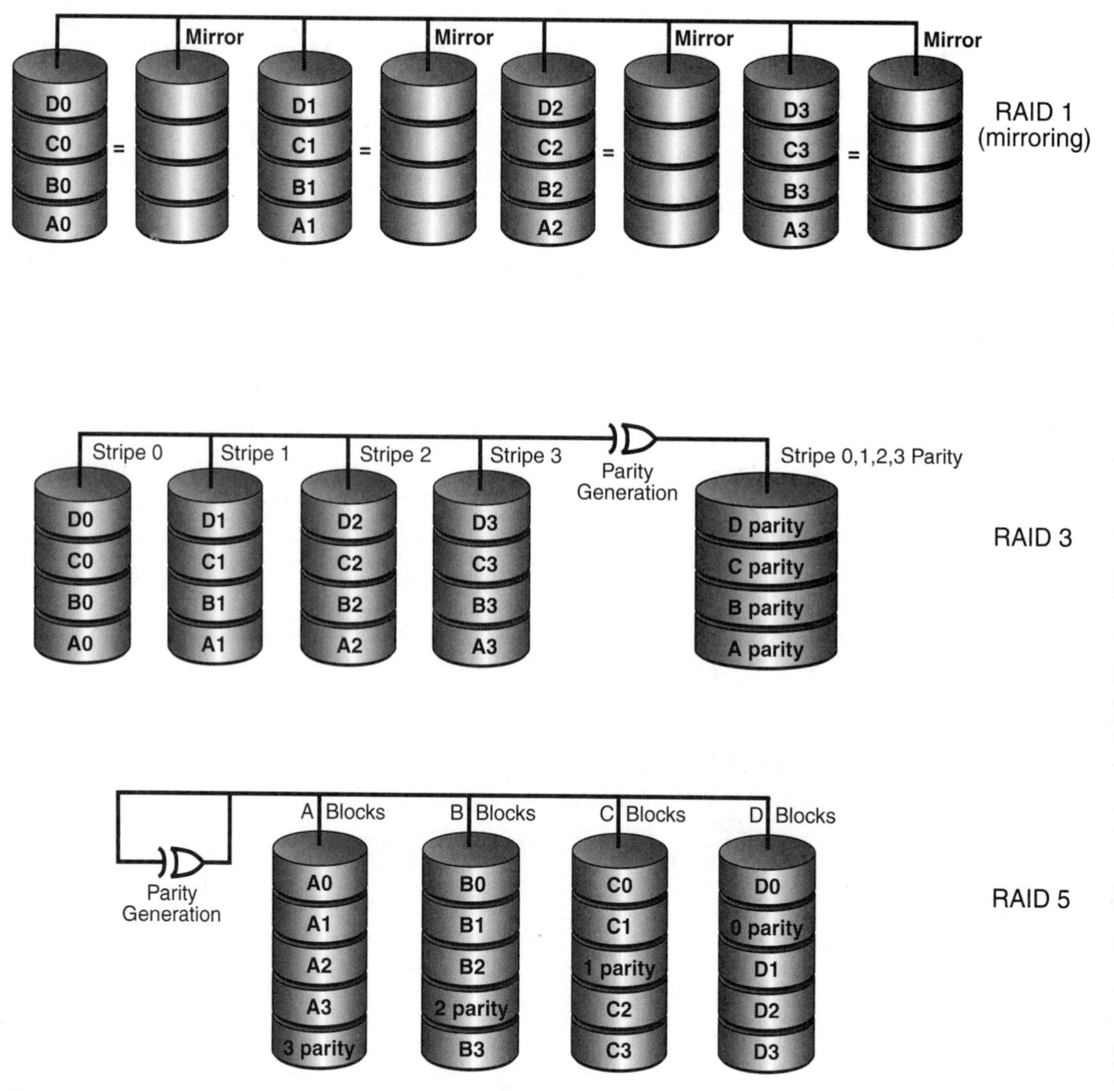

Figure 9–4 RAID levels with redundancy.

Software

With reference to computers, a **platform** is an underlying computer system on which application programs can run. It consists of an operating system and a microprocessor. The operating system must be designed to work with the instruction set of that particular microprocessor. As an example, Microsoft's Windows is built to work with a series of microprocessors from the Intel Corporation that share the same or similar sets of instructions. There are usually other implied parts in any computer platform such as a motherboard and a data bus, but these parts have increasingly become modularized and standardized. On personal computers, Microsoft's Windows, Solaris by Sun Microsystems, and Apple's Macintosh are examples of different platforms; Unix, Novell's Netware and IBM's Operating System/390 are representative of an enterprise server or mainframe platform.

Historically, most application programs were written to run on a particular platform providing a different application program interface for different system services. Thus, a desktop program would have to be written to run on the Windows platform and then again to run on the Macintosh platform, as represented in Figure 9–5. Although these platform differences continue to exist and there will probably always be proprietary differences between them, new open or standards-conforming interfaces now allow some programs to run on different platforms through mediating programs, commonly referred to as Application Programming Interface (API).

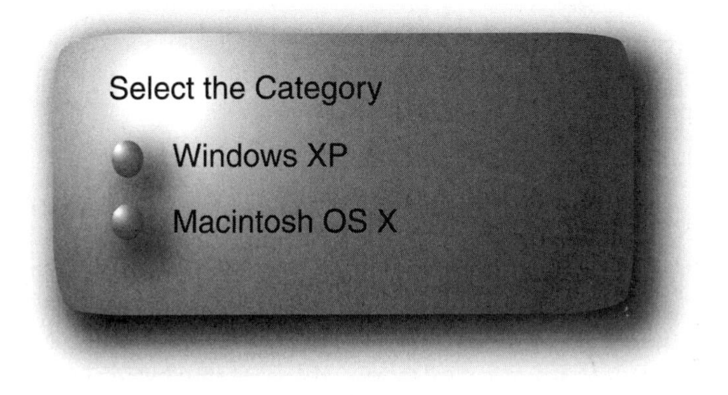

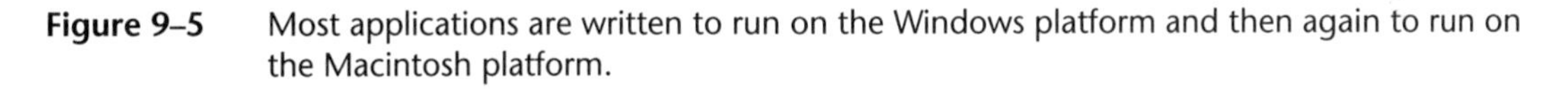

Figure 9–5 Most applications are written to run on the Windows platform and then again to run on the Macintosh platform.

The processes that take place on the hardware devices of a LAN must be controlled by software called the **Network Operating System (NOS)**. The NOS provides centralized administration of the entire network; it enables the sharing of printers, files, databases, and applications and provides directory services, security, and housekeeping aspects to manage a network. The NOS controls the operation of the servers and provides user-friendly access to network resources. It enables differential access to files on network disks; that is, it controls which files a user can access as well as how the user accesses those files. For example, a user may have access to a word processor file, but only to read it, not to modify it. At the same time, another user may have access to the same file and be able to modify it.

When selecting the NOS, its following characteristics should be carefully evaluated:

- Architecture addresses the ability to support multiple microprocessor architectures so as to accommodate a dynamic and growing workload.

- Functionality, Reliability, and Scalability pertains to the same NOS being able to support small, medium-sized, and large environments, which results in reduced management and system costs. Also, a NOS that supports an array of fault-tolerant hardware configurations minimizes network downtime.
- Broad Network Media and Client Support is most important because without this support, it may be difficult for a company to build the most optimal network infrastructure supporting different processors.
- Network Services and Applications available on a NOS platform include mail client, mail server, integrated office suites, support of RAID, file sharing and locking features, support of peripheral devices such as printers and fax machines, directory services, and security management.
- Network Protocols on a corporate network should be compatible with the NOS. Typical examples are TCP/IP, Telnet, SMTP, SNMP, SNA, FTP, IPX/SPX, NetWare, NetBIOS, and NetBEUI.
- Server Management reviews the tools available to manage NOS platforms, including file management, user account management, error reporting, security breaches, and server performance reporting.
- Application development tools include object-oriented development tools, distributed multitier application architectures, and tools for team programming on each NOS platform.

A networking software must be chosen based on needs—both present and future—and a careful comparison of the capabilities of existing products, as well as on the vendor's capabilities to deliver future enhancements. Users who implement open standards-based solutions will protect their investments, interoperate better with other competitive solutions, and avoid the pitfalls of proprietary systems, including forced upgrades.

NETWORK ADMINISTRATION AND MAINTENANCE

Network administration and maintenance is an infrastructure of techniques and procedures that assure the proper day-to-day operation of a system by critical network components to detect failures and degraded performance and to take corrective action before services are affected. A *Network Operations Center (NOC)* is usually a separate room from which a telecommunications network is managed, monitored, and maintained to ensure uninterrupted service for its users. Enterprises with large networks typically have computer simulations or visualizations of the networks to monitor its status and the necessary software for remote management. The NOC is the focal point for network troubleshooting, software distribution and update, performance monitoring, and coordination with affiliated networks.

Maintenance is a key element of network administration. It is a three-step process; the first step is the ability to detect and report problems, which is mostly automated. Second, the problem must be isolated to determine the cause, which may also be automated. Finally, the problem must be resolved, which is a human task. Maintenance management activities can be supported by a database of events and network characteristics and components and some automated capabilities for determining problem resolution.

A *network administrator* is a person who performs the following maintenance tasks on a day-to-day basis:

- Monitor and control hard disk space
- Monitor network workload and performance
- Add to and maintain user login information and workstation information
- Setup Internet and email access accounts
- Manage resource and file access
- Monitor and reset network devices
- Coordinate timely resolution of network troubles
- Provide timely, consistent and meaningful communications regarding the availability of network services
- Perform regular maintenance on software and data files stored in the servers
- Schedule backup programs to run at convenient times
- Update anti-virus and security software
- Install software upgrades for servers and workstations
- Maintain records of user accounting and billing
- Manage network changes and moves to minimize disruption
- Collaborate with service providers to improve network performance and management
- Keep abreast of emerging technologies and related management tools

Implementing Virtual LANs (VLAN)

The network management environment continues to change as solutions are developed to reduce complexity, improve overall network administration, and provide maximum return on IT investment. Organizations are moving toward networks featuring single user/port LAN-switching architectures. In a network using only routers for segmentation, segments and broadcast domains correspond on a one-to-one basis. Switches enable organizations to divide the network into smaller, layer 2–defined segments, providing increased bandwidth per segment. A **Virtual LAN (VLAN)** environment implements the corollary *switch when you can, route when you must.* Optimal VLAN deployment is predicated on reducing traffic traversing the router to a minimum, which minimizes the possibility of the router becoming a bottleneck. VLANs are created to provide the segmentation services traditionally provided by routers in LAN configurations. Routers play two specific roles in VLANs: to provide connectivity between VLANs and to provide broadcast filtering capabilities for WAN links, where VLANs are generally not appropriate.

Figure 9–6 compares a virtual LAN with a traditional LAN. A VLAN is a switched network that is logically segmented on an organizational basis, by functions, project teams, or applications rather than on a physical or geographical basis, as depicted in Figure 9–7.

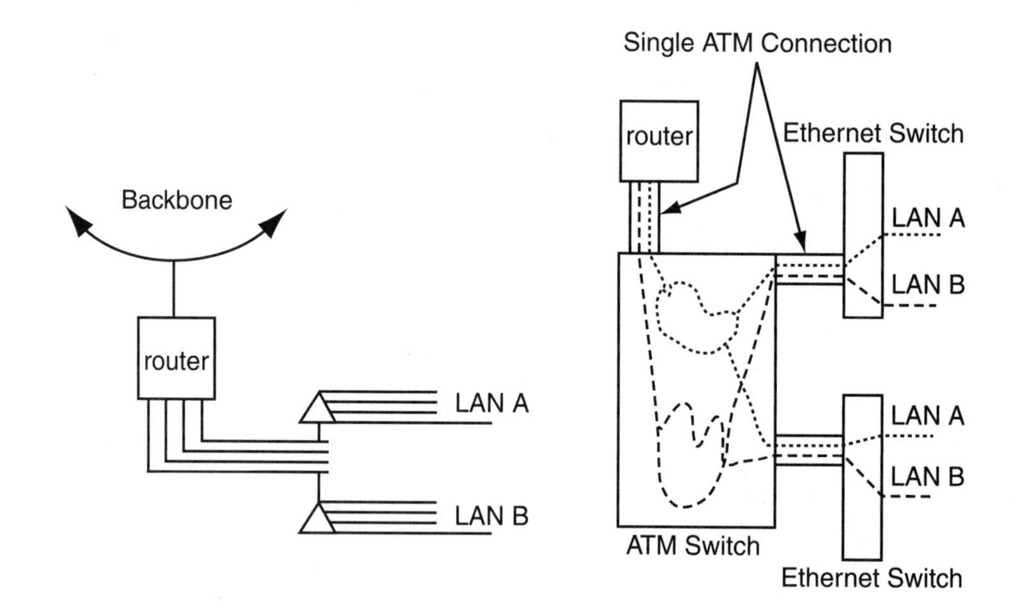

Figure 9–6 VLAN versus traditional LAN.

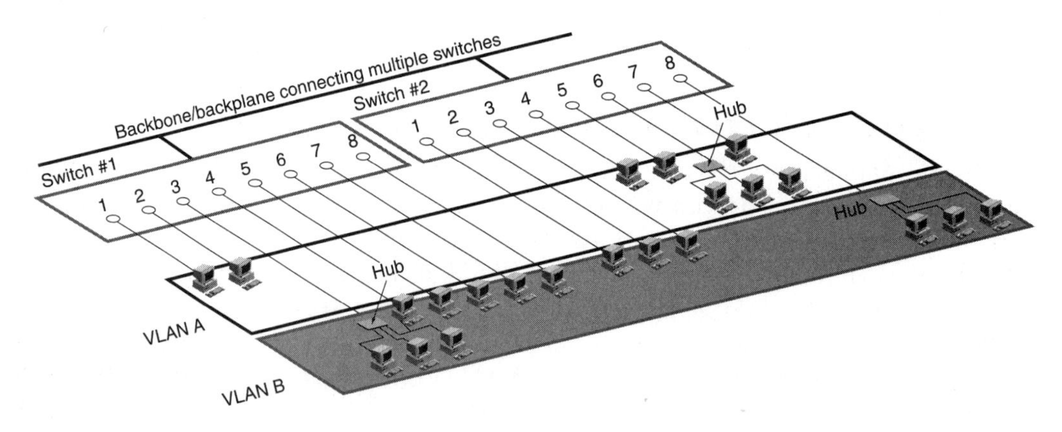

Figure 9–7 VLAN based on logical segmentation rather than on a physical basis.

On a VLAN, all workstations and servers used by a particular workgroup team can communicate as if they were on a common LAN, regardless of their physical connections to the network or the fact that they might be intermingled with other teams. Reconfiguration of the network can be done through software rather than by physically unplugging and moving devices or wires. VLANs have some drawbacks—they are mostly proprietary, single-vendor solutions and have their own administrative costs.

Network Performance Management

Performance management facilities provide the network manager with the ability to monitor and evaluate the performance of a network—to collect and disseminate data concerning the current level of performance of resources and to maintain and examine performance logs for purposes such as planning and analysis.

Three measures of LAN performance are commonly used: delay, throughput, and the total load on the LAN, including control packets like tokens, and collisions, which are destroyed packets that must be retransmitted. A graph of a typical total traffic load by hour of the day is illustrated in Figure 9–8. The total load determines **network utilization** which is defined as the ratio of total load to network capacity, as shown in Equation 9–1. Since utilization cannot exceed 100%, transmitted frames beyond network capacity are lost and must be repeated. From the user's perspective, the efficiency or utilization of the network may seem of minor importance but it is directly related to network costs.

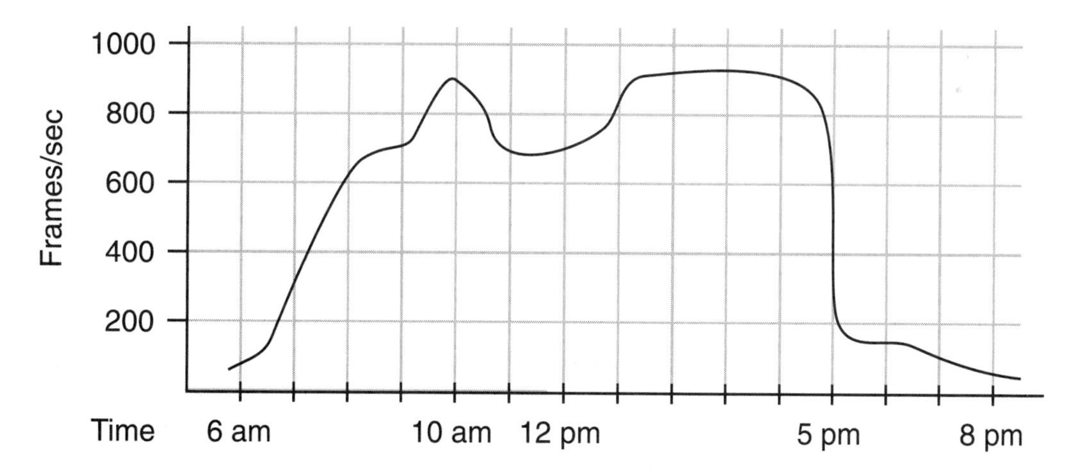

Figure 9–8 Total traffic load by hour of the day on a typical telecommunications network.

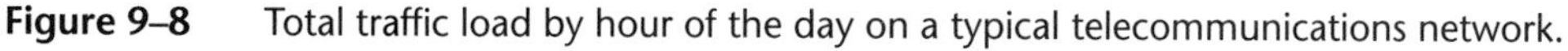

$$\text{Network Utilization} = \frac{\text{Total Load}}{\text{Network Capacity}} \times 100\% \qquad (9\text{–}1)$$

Example 9–1

Problem

A network has a capacity of 1000 frames per second and the input to the network is 500 frames per second. If it is assumed that 1% of all transmitted frames are lost and must be repeated, find the network utilization.

Solution

At an input of 500 frames/second, 5 frames per second will be repeated. Thus the total load = 505 frames per second.

$$\text{Network Utilization} = \frac{\text{Total Load}}{\text{Network Capacity}} \times 100\%$$

$$= \frac{505}{1000} \times 100\%$$

$$= 50.5\%$$

Factors that are exclusively under the control of the local network designer are:

- Capacity
- Propagation delay
- Number of bits per frame
- Local network protocols
- Offered load
- Number of stations

Of the local network protocols—physical, medium access, and link—the physical layer is not likely to be much of a factor. The link layer will add some overhead bits to each frame and some administrative overhead, but the medium access layer can have a significant effect on network performance.

The NOS contains several management tools intended to increase traffic efficiency and minimize operator intervention. It provides tools that show statistical data about the use of the network and outline potential problems. An experienced network administrator uses these statistics to ensure that the network operates at its peak level at all times. A bar graph of the historical trend analysis of network downtime in hours is shown in Figure 9–9; a representative example. However, many LAN administrators find that such monitoring and management utilities are not enough to get a complete picture of the network and where some of the problems may be located.

For this purpose, many third-party vendors offer management and monitoring equipment and software that enhance the software available with the LAN. Fault management

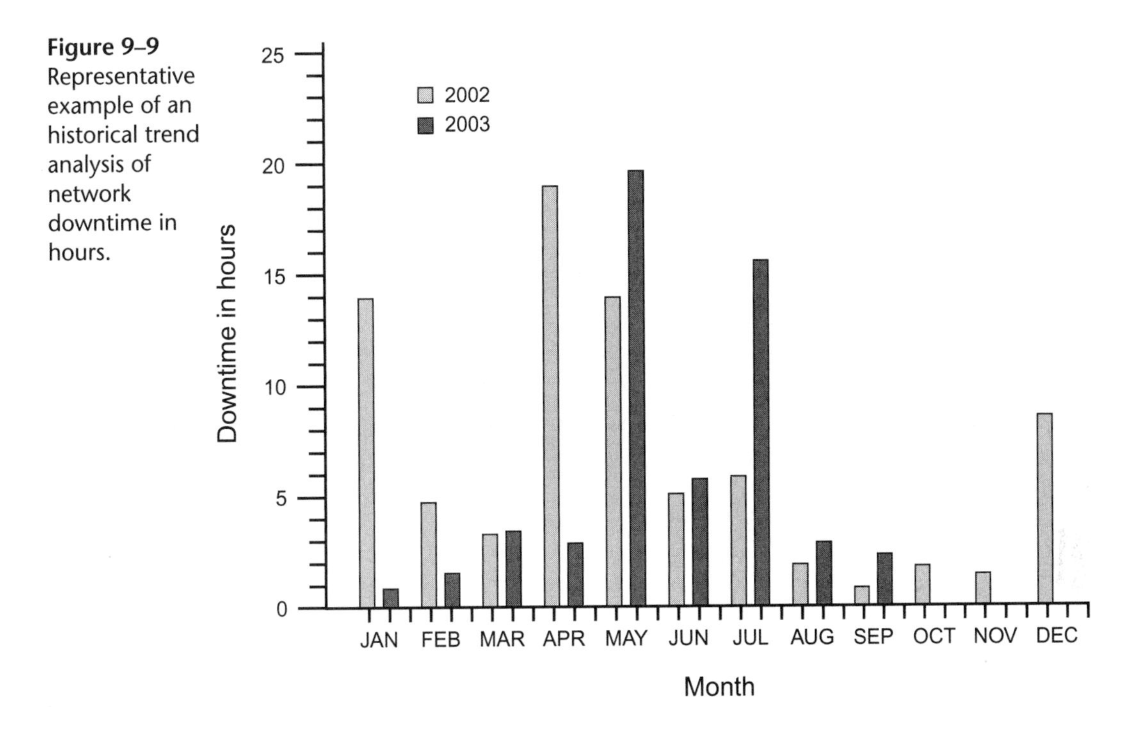

Figure 9–9 Representative example of an historical trend analysis of network downtime in hours.

facilities allow network managers to detect problems in the communications network and the OSI environment. The *headend* is a single point of failure, that is, loss of a headend means loss of an entire network. Since network availability is a major concern, a simple, effective measure is to partition the networks into sub-networks joined by bridges or switches so that each sub-network has its own headend.

Questions concerning possible errors or inefficiencies include:

- Is traffic evenly distributed among the network users or are there source-destination pairs with unusually heavy traffic?
- What is the percentage of each kind of packet? Are some packet types of unusually high frequency, indicating an error?
- What is the distribution of data packet sizes? Are variable-size data packets worth the additional overhead or would fixed-size packets suffice?
- What are the channel acquisition and communication delay distributions? Are these times excessive?
- What is the information utilization and throughput? How do the information statistics compare with the channel statistics?
- Are collisions a factor in getting packets transmitted, indicating possible faulty hardware or protocols?

NETWORK SECURITY

In recent years, enterprise network environments have become more complex, with an increasing reliance on digital assets to provide services that meet business demands. The greatest challenge is to overcome the implementation and management of security solutions to ensure network and data integrity.

In an enterprise network, the idea that the security of the entire network is only as strong as its weakest part is termed the "weakest-link axiom." **Computer and network security** can be defined as the protection of network-connected resources against unauthorized disclosure, modification, utilization, restriction, incapacitation, or destruction. The generic name of the collection of tools designed to protect stored data is computer security, while network security measures are needed to protect data during its transmission. Network security measures may be implemented at different layers of the OSI model, as shown in Figure 9–10. Security implies safety, including assurance of data integrity, freedom from unauthorized access, freedom from snooping or wiretapping, and freedom from disruption of service.

OSI Layer		
7	Application	Passwords, Key Management, Firewall
6		
5		
4	Transport	Transport Encryption
3	Network	Network Encryption, Screening router
2	Data link	Encryption
1	Physical	

Figure 9–10 Network security measures at different layers of the OSI model.

Security Threats

Hundreds of thousands of systems are now connected to the Internet. There is no accurate way of measuring the threat that may be launched by an inimical agent. Security risks vary from uploading files with embedded malicious code onto a network to stealing information. Each time a company deploys a new Internet gateway, LAN, or distributed client/server system, it risks leaving another virtual window open for cyber-prowlers, disgruntled employees, or unethical competitors to work through. Scripted attack methods automate the process of breaking into a network to the level of point-and-click; very little skill is required to compromise a network and disrupt business, steal proprietary data, or maliciously damage or modify information and data files. The need to deal with intrusions effectively has never been greater. Using a combination of automation and human expertise, attack patterns and trends can be discerned and tracked. Security threats can be divided into two major categories: *passive threats* and *active threats*.

Passive Threats

Passive threats involve monitoring the transmission data of an organization. The goal of the attacker is to obtain information that is being transmitted. These threats are difficult to detect because they do not involve alteration of the data, so the emphasis in dealing with passive threats is on prevention rather than detection. Although they can be directed at communication resources (routers and lines), they generally perpetrate at the host level.

Active Threats

Active threats involve some modification of the data stream or the creation of a false stream. It is difficult to prevent active attacks because this would require physical protection of all communications facilities at all times. Instead, the goal is to detect active attacks, quickly recover from disruption or delays caused by the attacks, and possibly pursue legal action against the hackers, all of which have a deterrent effect and may contribute to prevention. These threats are most successful when directed to what could be the weakest link in the overall system, namely, at the host level.

Viruses and Worms

Security threats to business-technology systems keep growing. The biggest threat is the growing sophistication of the viruses and worms attacking IT systems. Most often, attacks crash networks and systems; in a small number of cases, they result in unauthorized data access, financial loss, and identity theft. A **virus** is a program that can affect other programs by modifying them; the modified program includes a copy of the virus program, which can then go on to infect other programs. Whenever the infected computer comes into contact with an uninfected piece of software, a fresh copy of the virus passes into the new program. A **worm** is a program that makes use of networking software to replicate itself and move from system to system. The worm performs some activity on each system it gains access to, such as consuming processor resources or depositing viruses. In a LAN environment, the ability to access applications and system services on other computers provides a perfect culture for the spread of viruses and worms.

Although the best solution for these threats is prevention, this goal is almost impossible to achieve. The next best approach is to do the following:

- Detect
- Purge
- Recover

The first step is to locate the virus, preferably before or soon after the infection has occurred. Next, remove the virus from all infected systems so that the disease cannot spread further, and finally, recover any lost data or programs. The issue of network security has become increasingly important as more and more businesses and people go on the Internet. Teams of people have been formed to assist in solving hacker attacks and to disseminate information on security attacks and how to prevent them. Two such teams are Computer Emergency Response Team (CERT), and Forum of Incident Response and Security Teams (FIRST). CERT exists as a point of contact for suspected security problems related to the Internet. It can help determine the scope of the threat and recommend an appropriate response. FIRST is made up of a variety of computer emergency response teams from government, business, and academic sectors. It was designed to foster cooperation and coordination between teams in an attempt to decrease reaction time to security incidents and promote information sharing among team members.

Network Vulnerability Management and Security Policy

Traditional network vulnerability assessment is used to identify existing vulnerabilities and risk profiles to make recommendations on improving security practices. The two key areas are *internal assessment* and *external assessment*. An internal assessment consists of various audits that are conducted throughout the internal network, including all devices and network applications. The result is a security policy that outlines the procedures for business operations based on industry best practices or government regulations. An external assessment may require outsourcing security services to perform penetration tests and an audit of the company's network from the perspective of an attacker. These audits are usually performed via the Internet, from partner networks, and remote offices using hacking techniques or commercially available vulnerability tools.

Network exposure is defined as all information that can be gathered remotely about the network, including vulnerabilities; the process to identify these exposures is called *vulnerability assessment*. Examples of exposures include the way an IP-enabled device responds to network connection requests, whether specific ports are open, and how the applications on those ports respond. It is important to have multiple views of IP-enabled devices to fully understand their security posture.

As represented in Figure 9–11, **vulnerability management** is a cyclical process of identifying, measuring, prioritizing, monitoring, and remediating potential security risks. It determines weaknesses within the network proactively by probing IP-enabled devices for their susceptibility to known exposures, managing the process of addressing those exposures, and monitoring for attacks to still-vulnerable devices. In a time when networks are barraged by accelerating numbers of attacks, the only effective defense is efficient prevention. Because enterprise networks are large and constantly changing, with devices and applications being added, removed, activated, and deactivated all the time, the results from one point in time are likely to be outdated quickly.

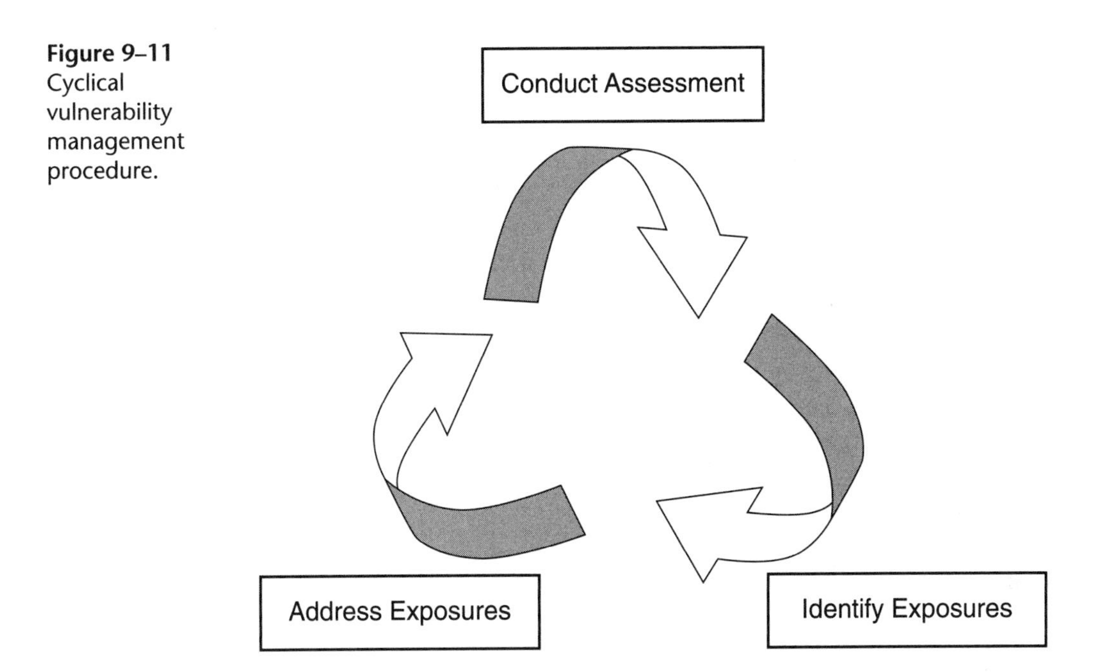

Figure 9–11 Cyclical vulnerability management procedure.

An organization must assess risks and develop an unambiguous **security policy** regarding access to each element of information, the rules an individual must follow in using and disseminating information, and a statement of how the organization will react to violations. Establishing a security policy and educating employees is critical because humans are usually the most susceptible point in any security scheme.

Security Measures

Some of the basic security strategies that can be utilized to combat the threats include:

- Authorization
- Encryption
- Authentication
- Intrusion Detection System (IDS)
- Firewall

These strategies are represented in Figure 9–12.

Authorization

The purpose of **authorization** is to ensure that only authorized users have access to a particular system and/or specific resources and that modification of a particular portion of data is limited to authorized individuals and programs. In most cases, access control

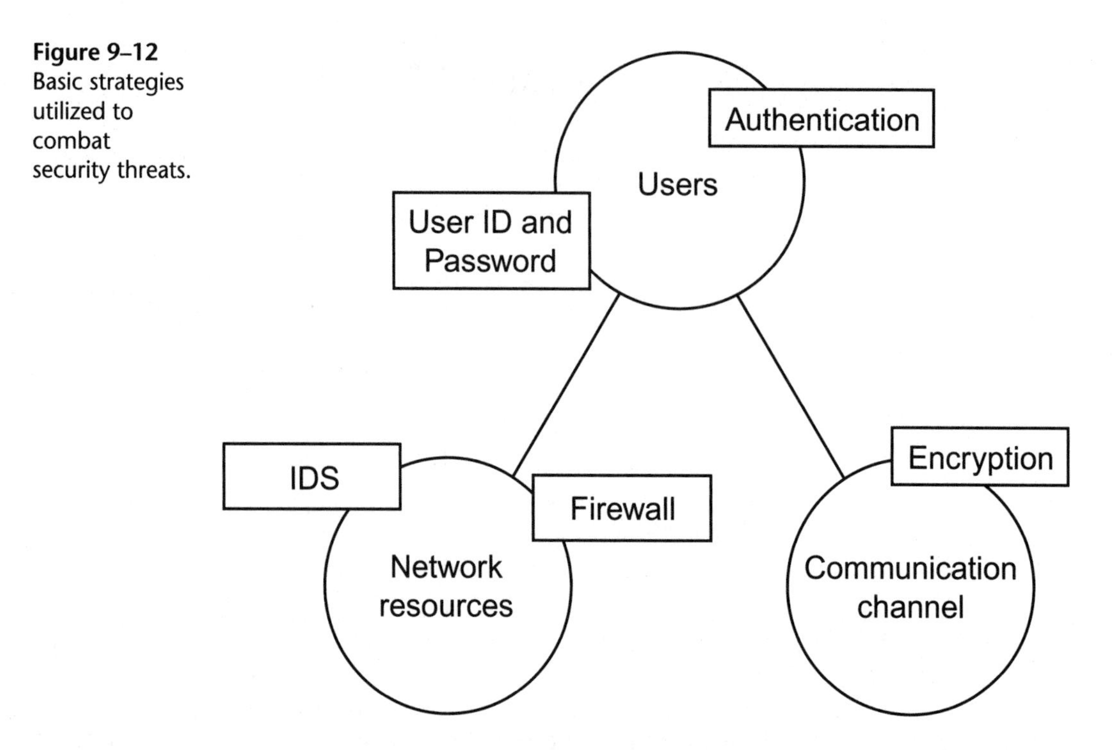

Figure 9–12 Basic strategies utilized to combat security threats.

mechanisms, essentially a form of authorization, are implemented in a single computer to control access to that computer. An example of user access control on a time-sharing system, such as a desktop in a computer lab on campus, is the user logon, which requires both a user identifier (user ID) and a password.

Encryption

The most effective way of securing the integrity of electronic data is by the use of **encryption**, which involves scrambling of data by use of a mathematical algorithm so that the scrambled information is undecipherable and meaningless. This can be done at various levels of the OSI model with software, hardware, or both. The function involves both an algorithm and a *key*, where the ease of figuring out the key depends on its length. A key is a secret code that the sender and recipient must share to encrypt and decrypt a message.

There are three kinds of encryption functions:

- Hash functions, which do not involve the use of keys
- Private-key functions, which involve the use of one key
- Public-key functions, which involve the use of two keys

Private-key encryption is compared with public-key encryption in Figure 9–13. The U.S. government's approved private-key encryption specification, *Data Encryption Standard (DES)*, was adopted in 1977 and is used extensively in the military, aerospace, intelligence,

Private-key Encryption	Public-key Encryption
✦ The same-length algorithm with the same key can be used for encryption and decryption	✦ One algorithm is used for encryption and decryption with a pair of keys: one for encryption, and one for decryption
✦ The sender and receiver must share the algorithm and the key	✦ The sender encrypts with the public key and the receiver decrypts with the private key
✦ The key must be kept secret	✦ One of the two keys must be kept secret
✦ It must be impossible or at least impractical to decipher a message if no other information is available	✦ It must be impossible or at least impractical to decipher a message if no other information is available
✦ Knowledge of the algorithm plus samples of ciphertext must be insufficient to determine the key	✦ Knowledge of the algorithm plus one of the keys plus a sample of the ciphertext must be insufficient to determine the other key

Figure 9–13 Encryption functions.

and financial institutions. In DES, data is encrypted in 64-bit blocks using a 56-bit key, while public-key encryption uses two keys, a private key and a public key, as illustrated in Figure 9–14. The public key is used to encrypt messages and the private key is used to decrypt. The two keys are mathematically related such that data encrypted with either key can only be decrypted using the other key. The public encryption key is made available to whoever wants to use it, but the private key is kept secret by the key owner.

When applications use public-key cryptography it leads to a requirement for a *Public Key Infrastructure (PKI)*, which is a comprehensive system that provides the public-key-based encryption and digital signature services on behalf of the applications. A major benefit of the PKI is that it enables the use of encryption and digital signature services in a consistent manner across a wide variety of applications, and it provides for authentication, data integrity, and nonrepudiation. Of course, there is a price to pay for a high level of security. Most encryption devices sit in the data path, which can lead to latencies that negatively effect performance.

Authentication

Authentication methodologies include one-way encryption to validate information transmitted, digital signatures that verify both the information sent and the sender, and digital certificates that are distributed by a *Certification Authority (CA)* to acknowledge the identity of the user. This process is depicted in Figure 9–15. CA is the trusted third party responsible for verifying the digital authentication keys and certificates issued to and used by the organization in relation to its documents. Various forms of individual

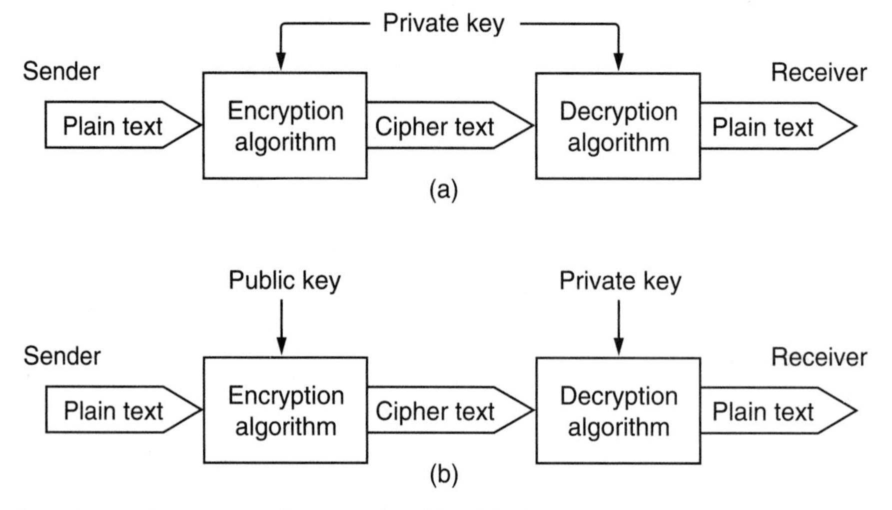

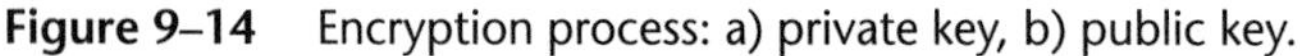

Figure 9–14 Encryption process: a) private key, b) public key.

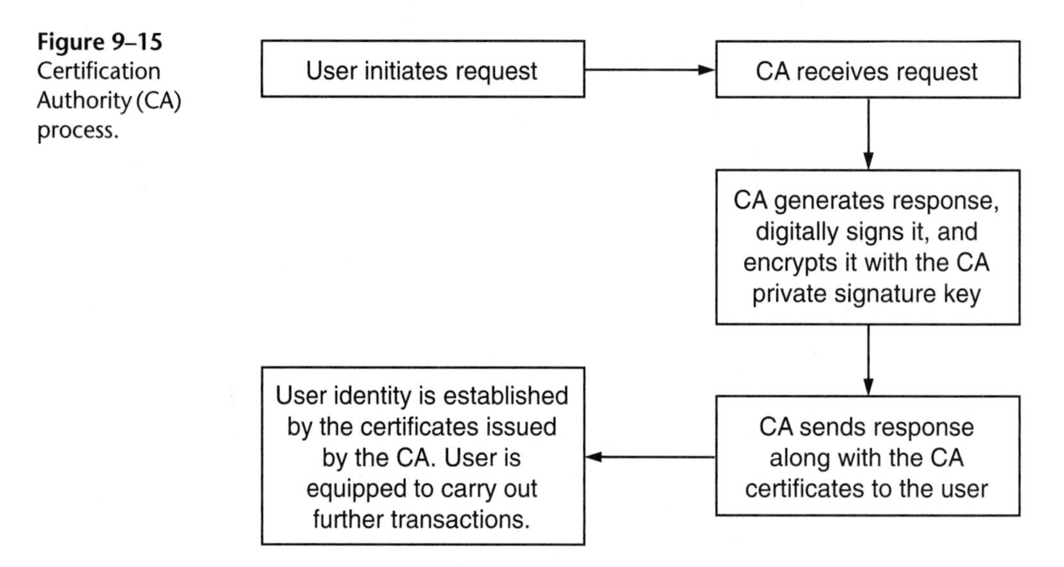

Figure 9–15 Certification Authority (CA) process.

authentication, such as one-time passwords, security tokens, Personal Identification Number (PIN), biometric devices, and single sign-on systems provide sophisticated methods of ensuring that the user is genuine.

Intrusion-Detection Systems (IDS)

Monitoring/analysis tools and **Intrusion Detection Systems (IDS)** are an important security measure because they let a network manager be aware of attempts to bypass security. Figure 9–18 shows how IDS are implemented. In *active monitoring,* IDS notify the administrator whenever an incident occurs; the major advantage here is speed—a poten-

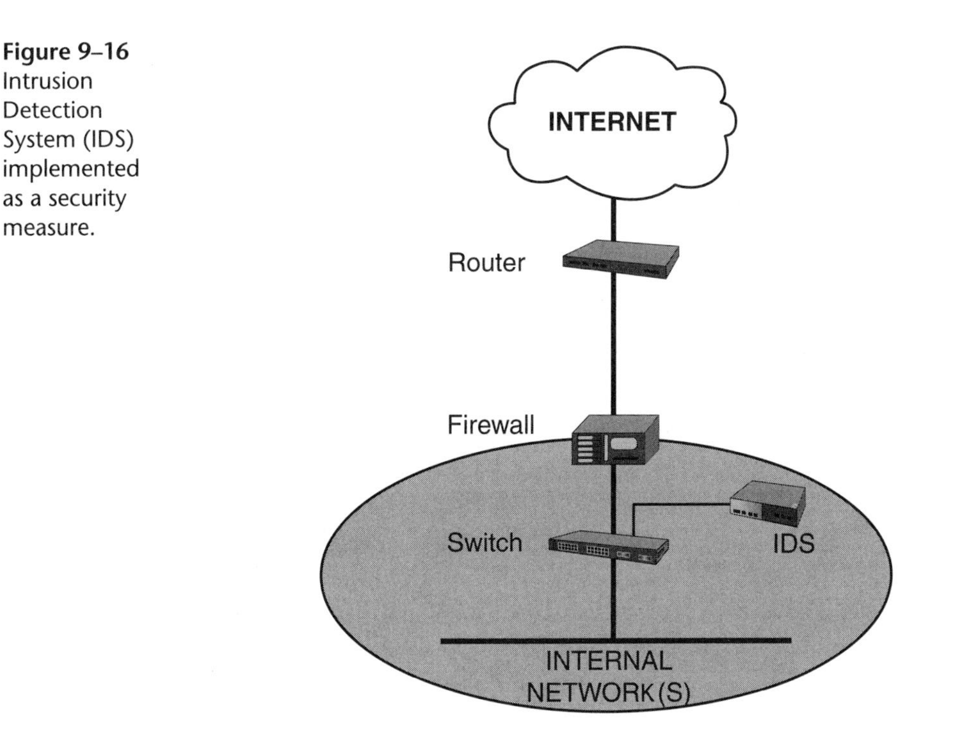

Figure 9–16 Intrusion Detection System (IDS) implemented as a security measure.

tial problem is identified immediately. Its disadvantage is that active monitors often produce so much information that a manager cannot comprehend it or notice problems. In *passive monitoring*, IDS log a record of each activity against it in a file, which can be accessed at any time. The major advantage of passive monitoring arises from its record of events: a manager can consult the log to observe a trend and, when a security problem does occur, review the history of events that led to the problem. Most managers prefer passive monitoring with a few high-risk incidents also reported by active monitoring.

IDS gather and analyze information about break-ins or break-in attempts via software by scanning security logs or auditing other information available on the network. The data is captured while minimizing impact on its flow. IDS are not designed to prevent attacks, only to alert administrators that there appears to be an attack taking place. Management of any IDS solution requires extensive time and effort because it can generate a flood of alerts and other kinds of data. So the increasingly important task is finding serious threats among thousands of minor alerts, which requires intelligent trend analysis to make better decisions. IDS will only allow you to monitor the traffic but do not let you block it; they provide a second layer of protection.

Network-based IDS solutions require constant changes, such as adding signatures when new hardware or software is added to a subnet or constantly updating the system with vendor-issued updates and new attack signatures. Host-based IDS solutions are also resource intensive and only protect the device on which it is installed. There are two primary methods of acquiring and reading data on switched enterprise networks: port mirroring (or spanning) and in-line network Tap (Test Access Port)—both with advantages and disadvantages.

In port mirroring, there is a designated mirror port to which an analyzer or probe is attached. Traffic from one or more network ports is switched through the backplane to its normal destination port and copied to the mirror port. Port mirroring is economical and easy to use, with no new equipment required, and allows viewing of traffic on all links of a switch, mirrored to one analyzer. However, multiple ports to one port can cause buffer overflow and dropped packets. In addition, most mirror ports filter anomalies—corrupt network packets, packets below minimum size, and layer 1 and 2 errors are usually dropped by the switch, thus making troubleshooting a challenge. Also, port mirroring puts a load on the switch's CPU/transfer logic, degrading its performance.

In-line Taps are inserted directly into a link, as illustrated in Figure 9–17. They split or copy the signals from both channels and retransmit the data streams back out to the link and to the probe. Taps can see 100 percent of the packets, as well as anomalies to support troubleshooting, but do not require configuration and are passive and fault tolerant. Taps are best applied for mission-critical or business-critical links, where IT managers can troubleshoot occasional anomalies and other problems without taking down the link to insert an analyzer. Taps are most useful in an environment where you cannot span all the ports off a large core switch.

Firewall

Firewall, a piece of hardware and software, is a security device that allows limited access into and out of one's network from the Internet. It operates at the application layer or at the network and transport layers of the protocol stack. In essence, it partitions an enterprise network into two areas that are informally referred to as the inside and outside. A firewall is only as effective as its configuration, which should be customized around the network profile. Companies should be as restrictive as possible about what data can come in or go out and then simply block everything else. In no case should companies use a firewall's default settings, and firewall rules should be periodically reviewed and assessed. Firewalls are classified into three main categories:

1. Packet filters
2. Application-level gateways
3. Proxy servers

Packet filtering at the network layer is used as a first defense and comes as part of most routers' software, illustrated in Figure 9–18. Each packet is either forwarded or dropped based on its source address, destination address, or a defined (TCP) port. Configuring a filter involves some determination of what services/addresses should or should not be permitted to access the network or server.

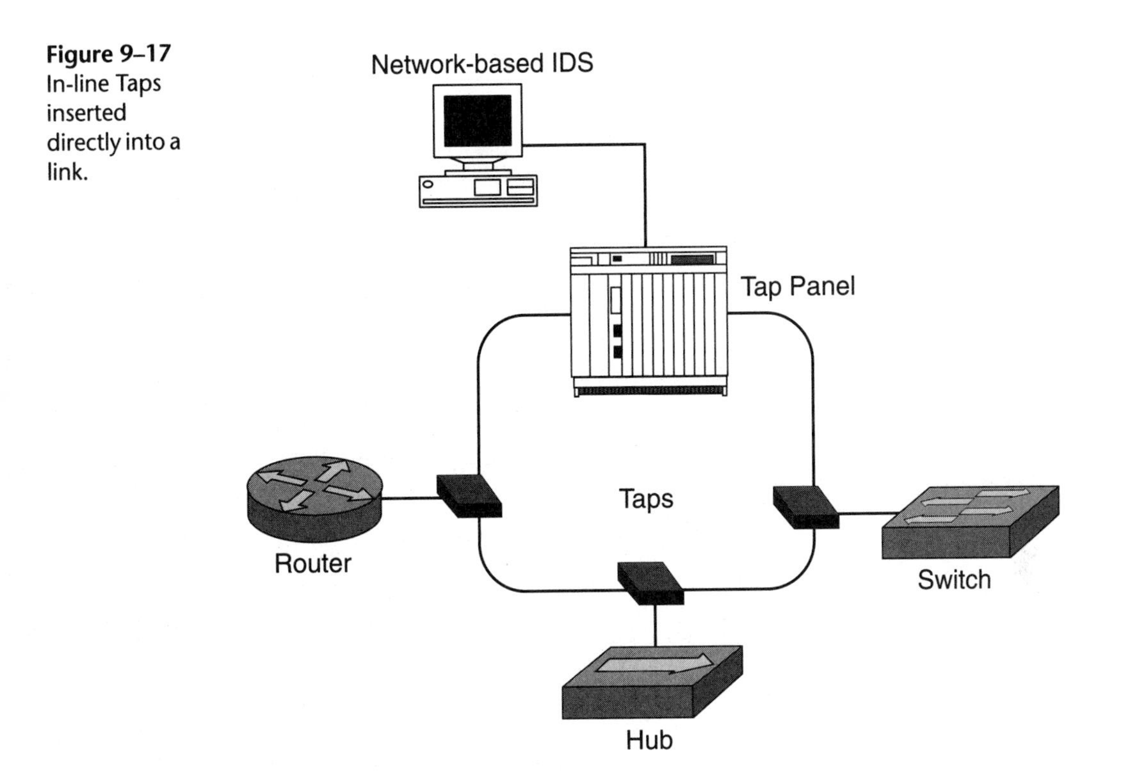

Figure 9–17 In-line Taps inserted directly into a link.

Application-level gateways provide a mechanism for filtering traffic for various applications. The network administrator defines and implements code specific to applications or services used by the user's site. The gateways have a number of advantages over routers, including logging, hiding of internal host names and IP addresses, robust authentication, and simpler filtering rules.

Proxy servers terminate a user's connection and set up a new connection to the ultimate destination on behalf of the user, in effect, substituting for the user. A user connects with a port on the proxy; the connection is routed through the gateway to the destination. The configuration is represented in Figure 9–19. A Web services proxy server is quite common.

Security Provisions in a VPN

VPNs share similar features at varying degrees of sophistication, cost, and ease of implementation. The main components integral to VPNs—encryption, authentication, and tunneling protocols—each provide a different way of ensuring privacy. Combined, they complement and strengthen each other, enhancing the integrity of the transmission and ensuring security. A VPN has three main components:

- Security gateways
- Security policy servers
- Certification Authorities (CAs)

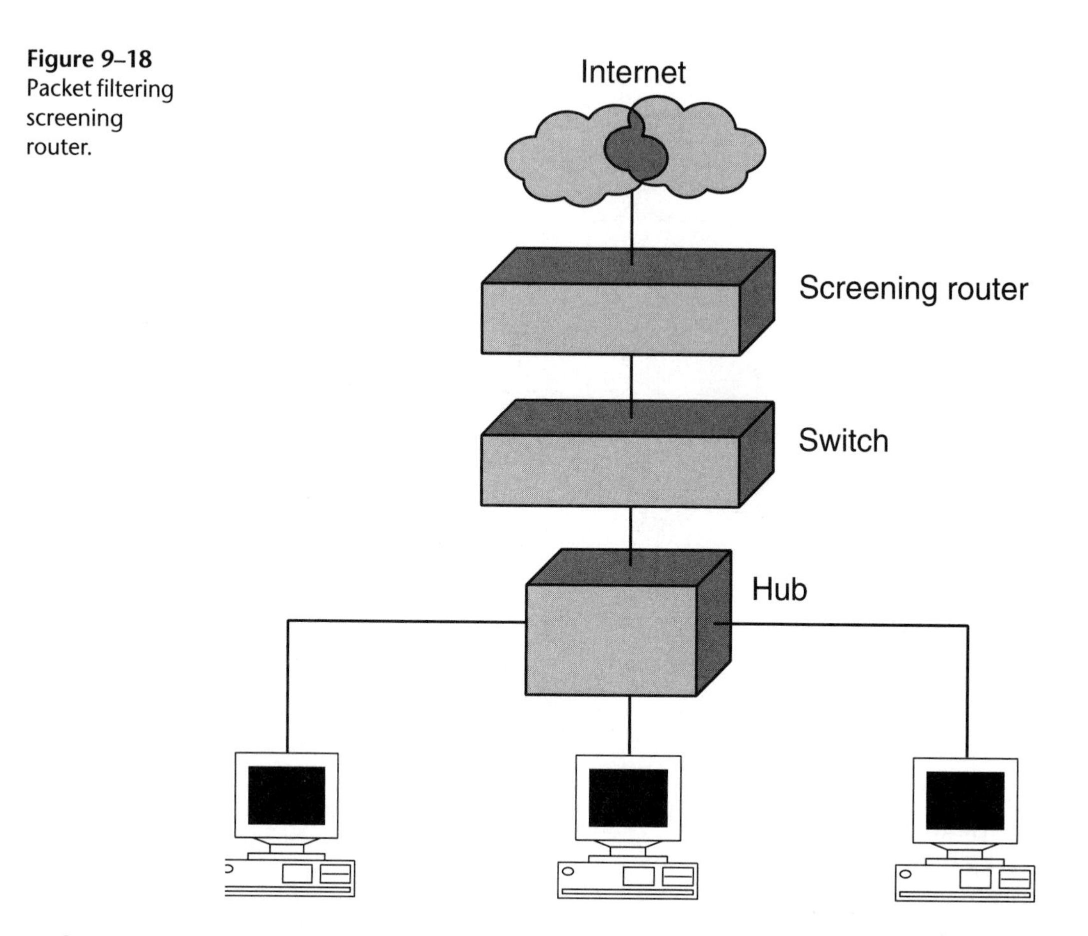

Figure 9–18 Packet filtering screening router.

Security gateways sit between public and private networks, preventing unauthorized intrusions into the private network. They provide tunneling capabilities and encrypt and decrypt private data before and after it is transmitted. User authentication is provided through a security policy server that maintains the access control lists and other user-related information that the security gateway uses to determine which traffic is authorized. The security policy server aggregates user authentication information from a variety of sources. Many companies are building VPNs for dial-in users so that they can reduce their remote access costs. The main difference between LAN-to-LAN VPNs and dial-in VPNs is that the security gateway for the main site of a dial-in VPN to which the remote users connect must be able to authenticate individual users rather than other gateways. A dial-in VPN also requires the remote user's computer to run VPN client software that handles tunneling and encryption. Digital certificates authorized by CAs are attached to transmissions to validate the sender of the information.

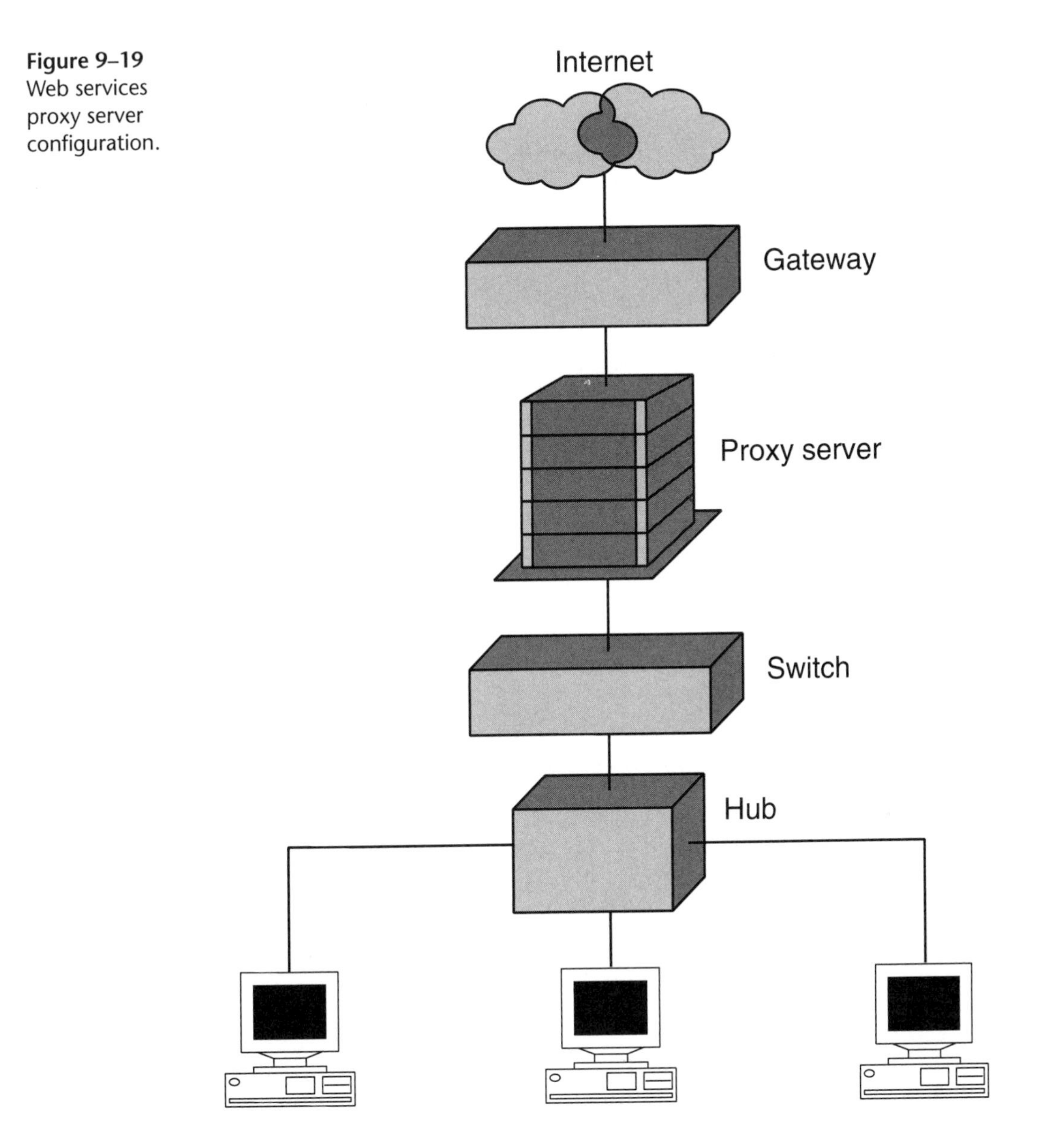

Figure 9–19 Web services proxy server configuration.

CONFIGURATION MANAGEMENT

Configuration management (CM) is the effective management of a network's life cycle and its evolving configuration by tracking each component through the system life cycle, documenting and controlling any changes, and ensuring that the overall system retains its integrity and conforms to requirements. CM identifies all the hardware and software components in the network and how they are connected. Equipment inventory is updated to keep track of all additions and deletions to the installed network hardware. This data will be useful if network operations personnel want to find the location of a control unit with a certain serial number or if company management needs to know the total value of all the terminals. Other documentation includes software listings, vendor manuals for all hardware and software, and routine operating procedures, as well as disaster recovery plans.

Network diagrams showing how the pieces of equipment and circuits are connected are another useful form of documentation. Figure 9–20 is a map of the network locations, controllers, and computers, as well as the circuits connecting them. Another level of detail, shown in Figure 9–21, is a listing of each circuit and the devices attached to it, with their model numbers and serial numbers. A third level of detail is the wiring diagram, which is drawn for a single building or floor of a building. This may show in detail the exact cable runs and type of wires with the pin numbers of each component.

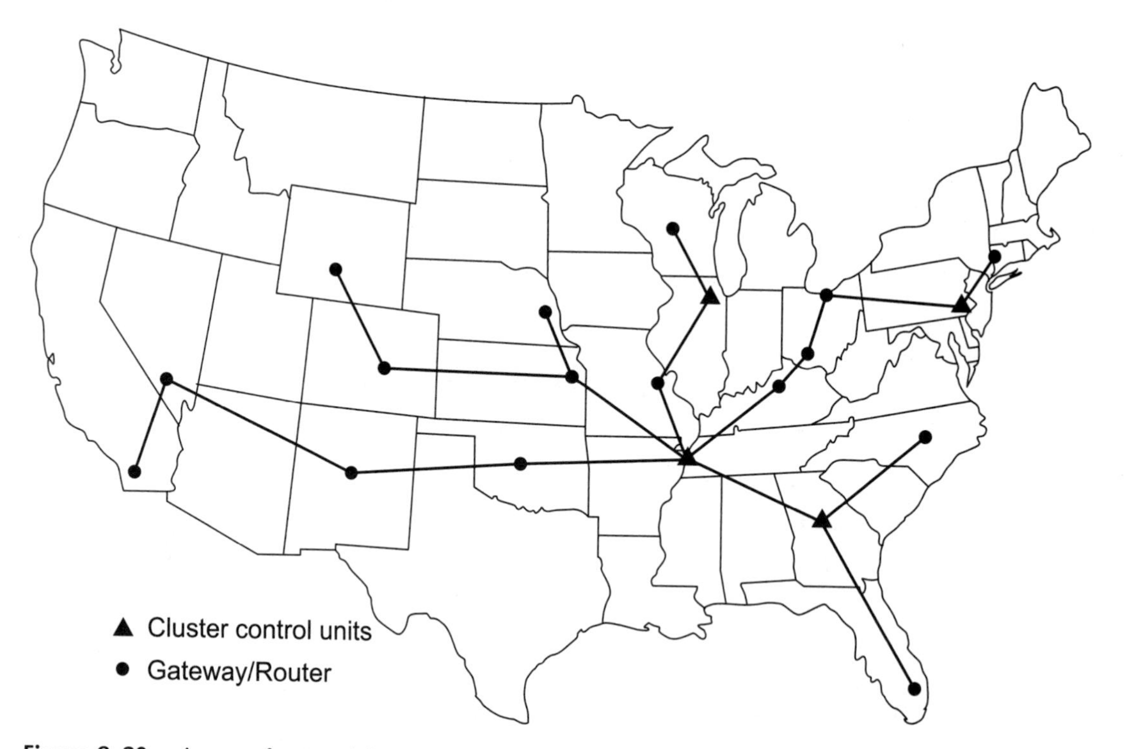

Figure 9–20 A map of network locations, controllers, and computers representing an implementation of configuration management.

Circuit Name	Circuit Number	Controller Type	Model No.	Serial No.	Network Address	Location
West	EUB3856B	3725	N.A.	135269	3E1C	Midland
		3174	3	4426578	3E9D	Chicago
		3174	2	890124	3E1A	Denver
		3274	41C	182930	3E4F	San Diego
East	EUB3472	3725	N.A.	135269	3E1C	Philadel-phia
		8130	30	42457	1A46	Toledo
		3174	2	675645	1A46	Atlanta
		3274	61C	276578	1A88	Mobile
		3174	2	764190	1A6A	Miami

Figure 9–21 A detailed implementation of configuration management consists of a listing of circuits and attached controllers.

NETWORK APPLICATIONS AND SERVICES

To most IT professionals, network availability means providing a level of application accessibility that is in line with business goals, management expectations and user demands. Unfortunately, there is no single figure that fits for every enterprise. For some, it is perfectly acceptable to bring systems down on a regular basis for disruptive backups or other maintenance tasks, but for an increasing number of businesses that provide global services on a true 24/7/365 basis, continuous availability is not an option, it is a business imperative.

Storage

Today, the value of the IT infrastructure can be described in financial terms, but the value of the data it stores is seldom known. Who knows how much the Human Genome Project's data is worth? If the cures for serious diseases are unlocked or genetic problems can be addressed, the value of this data is infinite. Businesses are beginning to realize that the real value of their IT function is not the infrastructure, but the *data* it contains. There are three distinct types of storage strategies: **Direct Attached Storage (DAS)**, **Network Attached Storage (NAS)**, and **Storage Area Network (SAN)**.

Direct-Attached Storage (DAS)

Enterprise storage primarily exists in a relatively fixed and controlled environment, usually attached directly to servers. A traditional DAS configuration, represented in Figure 9–22, includes servers with parallel SCSI-attached storage devices. Due to this parallel cabling

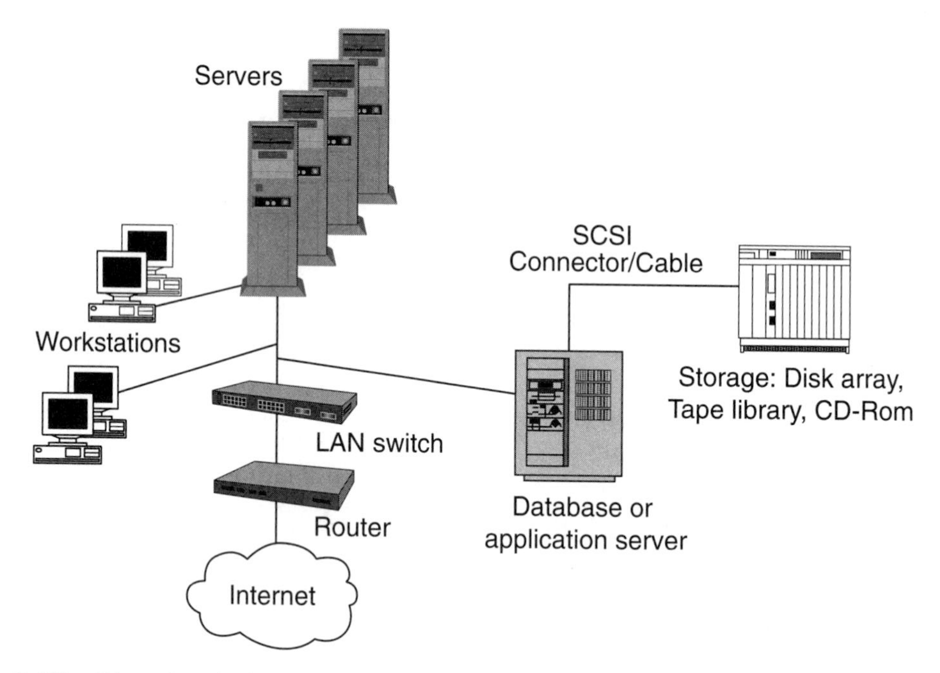

Figure 9–22 Direct Attached Storage (DAS).

scheme, each server is the exclusive of its own storage, which may result in loss of data access if the server fails. In addition, since each server is capable of supporting a limited amount of SCSI storage, adding more storage capacity may mean adding more servers, increasing administrative costs. In DAS, clients typically access file or application servers in a homogeneous operating system (OS) environment—Windows clients only have access to Windows servers; Solaris clients only have access to Sun servers.

Although the percentage of data centers using NAS and SANs is increasing, the majority of storage devices are still attached directly to servers via SCSI cables. Some DAS has been converted from SCSI to Fiber Channel (FC). Direct-attached RAID disk arrays can be backed up to other arrays and tape libraries. DAS has the advantage of being simple, but it places an added load on the servers and the LAN, because they handle the same data twice (once to write it and again to replicate it). In conjunction with other approaches, backup to tape remains an important part of a company's storage strategy.

Network-Attached Storage (NAS)

NAS is hard disk storage that is set up with its own network address rather than attached to a server as in DAS, as shown in Figure 9–23. It overcomes the OS-specific limitation of DAS by allowing cross-platform access to file systems, thus enabling storage capacity sharing heterogeneous computer platforms. For many customers, this is the main benefit of a NAS solution in addition to low cost and ease of installation. In many respects, a NAS device strongly resembles a traditional file server based on DAS. A client requests a file from the server, which in turn retrieves and assembles the data blocks that compose file from disk. What elevates a NAS device above a traditional file server component is that NAS has been optimized for efficient file access and stripped of auxiliary utilities common to most operating systems.

The lower cost of ownership and deployment times, combined with the ability of NAS to provide scalable, high-performance data sharing are proving to be very appealing for a large number of environments including collaborative development, engineering, e-mail, Web, and general file serving. Nevertheless, NAS boxes include multi-disk RAID systems that are deployed as islands of storage, where each island is managed individually. While administrators can see one view of all devices, they cannot manage the devices as one unit: for example, replication, permissions, and storage allocations are difficult because each device must be managed individually. Another basic NAS management problem is data migration. When an environment is scaled up, it is a time-consuming process to migrate data to a new filer, partition it, and bring it online.

Storage-Area Network (SAN)

A SAN is composed of servers and storage devices such as disk arrays or tape subsystems that are connected by a network infrastructure. It is isolated from the LAN and is maintained as a dedicated network. SAN topology, depicted in Figure 9–24, can support high-end server systems with Fiber Channel as well as mid-range servers with the Internet Small Computer Systems Interface (iSCSI). The servers and storage devices on a SAN send and receive block data. SAN technology is an expensive but efficient platform for building an information infrastructure because it is flexible and interoperable, highly available, and very manageable. SANs increase the availability of data by letting any server on the network access any storage device on the SAN. SANs also eliminate the DAS access bottleneck and other issues such as limited scalability.

The primary distinction between NAS and SAN is that NAS devices provide file access services to client computer systems, while SANs provide block data access. Since SANs implement a block-mode architecture, they do not support heterogeneous data sharing. A single enterprise may acquire both NAS and SAN solutions to address different end-user requirements. An engineering department, for example, may implement NAS to solve cross-platform issues. A human resources department may implement a SAN to support terabytes of employee information. DAS is more secure than SAN or NAS, which expose access points to data streams and repositories.

Figure 9–23 Network Attached Storage (NAS).

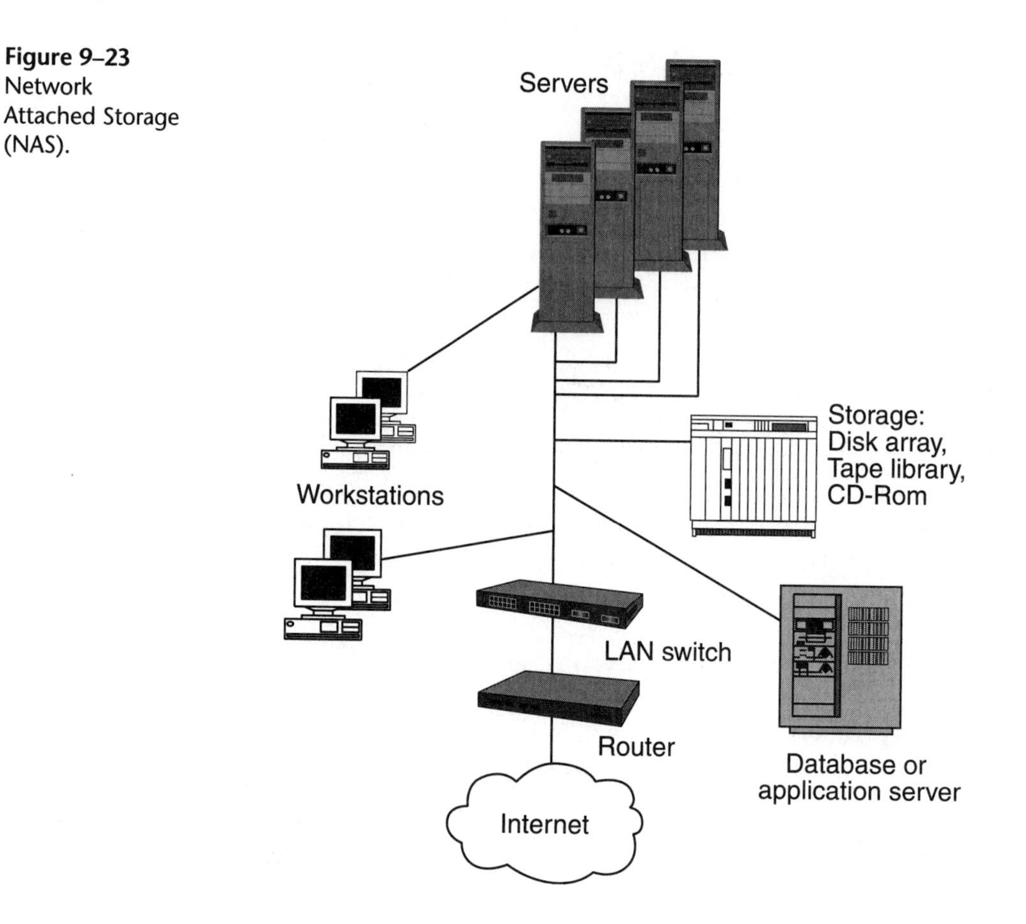

Network Application Software

Careful consideration must be given when selecting network application software to ensure that it is robust and scalable and incrementally upgradable. A three-tier application is an application program that is organized into three major parts, each of which is distributed to a different place or places in a network, as shown in Figure 9–25. The three parts are:

1. Workstation or presentation interface
2. Business logic
3. Database and programming related to managing it

In a typical three-tier application, represented in Figure 9–26, the application users' workstation contains the programming that provides the GUI and application-specific entry forms or interactive windows. Business logic, which is located on a LAN server or other shared computer, determines what data is needed and where it is located. The business logic itself acts as a client in relation to a third tier of programming that might be located on a mainframe computer. The third tier includes the database and a program to

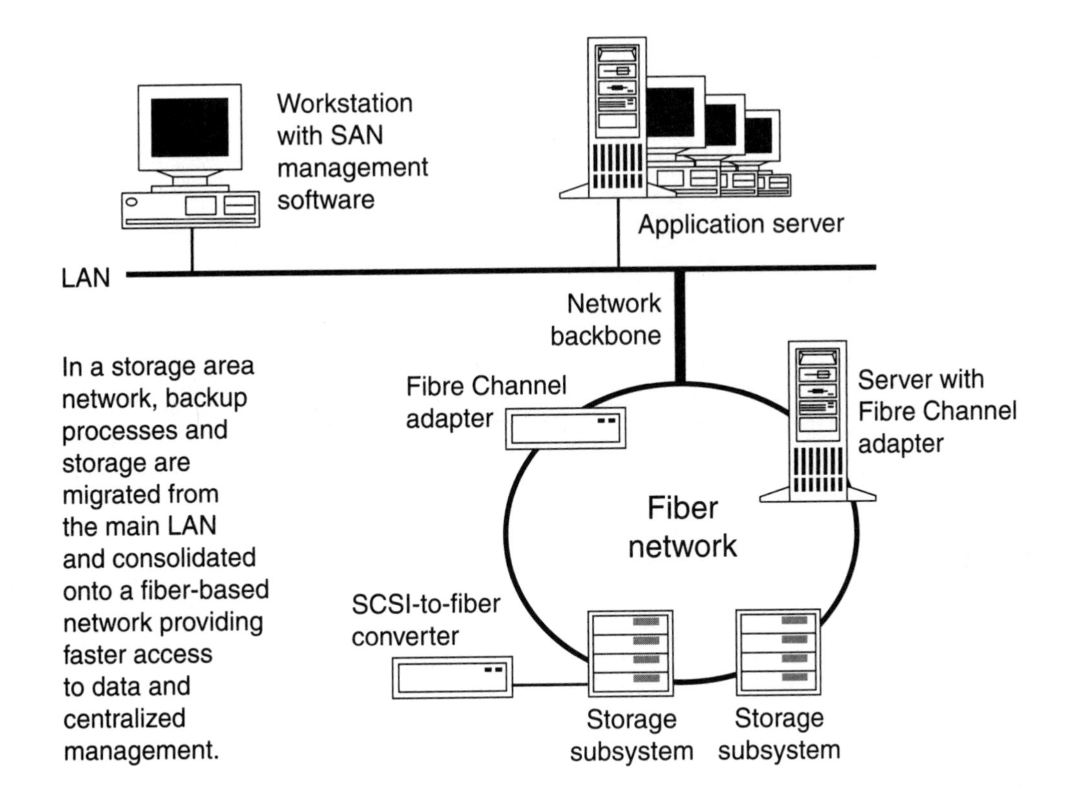

Figure 9–24 A SAN topology.

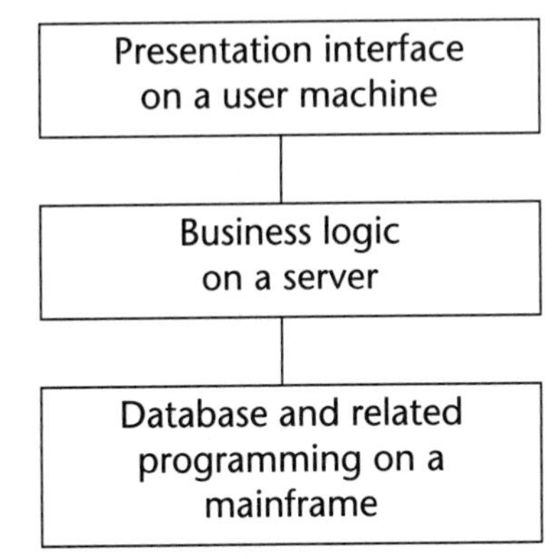

Figure 9–25 A three-tier application.

manage *read and write* access to it. While the organization of an application can be more complicated than this, the three-tier view is a convenient way to think about the parts in a large-scale program.

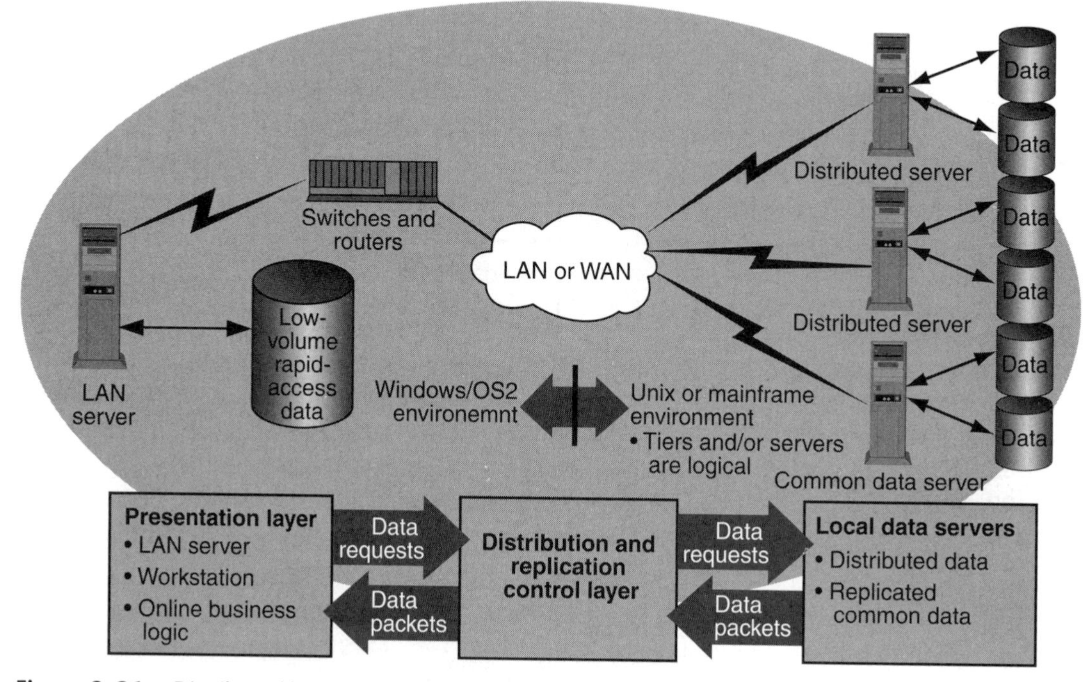

Figure 9–26 Distributed/scalable three-tier client-server architecture.

A three-tier application uses the client/server computing model where each tier can be developed concurrently by different teams of programmers coding in different languages. Because the programming for a single tier can be changed or relocated without affecting the other tiers, the three-tier model makes it easier for an enterprise or software packager to continually evolve an application as new needs and opportunities arise. In contrast to a **two-tier** application illustrated in Figure 9–27, the **three-tier** application architecture is consistent with the concept of distributed object-oriented programming.

Business Continuance

Business continuance is not a new concept; however, it has become a primary concern for corporate executives from all industry segments. In fact, implementation of a business continuance strategy has become a necessary cost of doing business today. **Business continuance** describes the processes and procedures an organization puts in place to ensure that essential functions can continue during and after a disaster. The planning seeks to prevent interruption of mission-critical services and to reestablish full functioning as swiftly and smoothly as possible. Besides obvious revenue impacts associated with network downtime and lost data, other serious considerations include lost productivity, brand dilution, legal consequences, and customer retention. Moreover, in some seg-

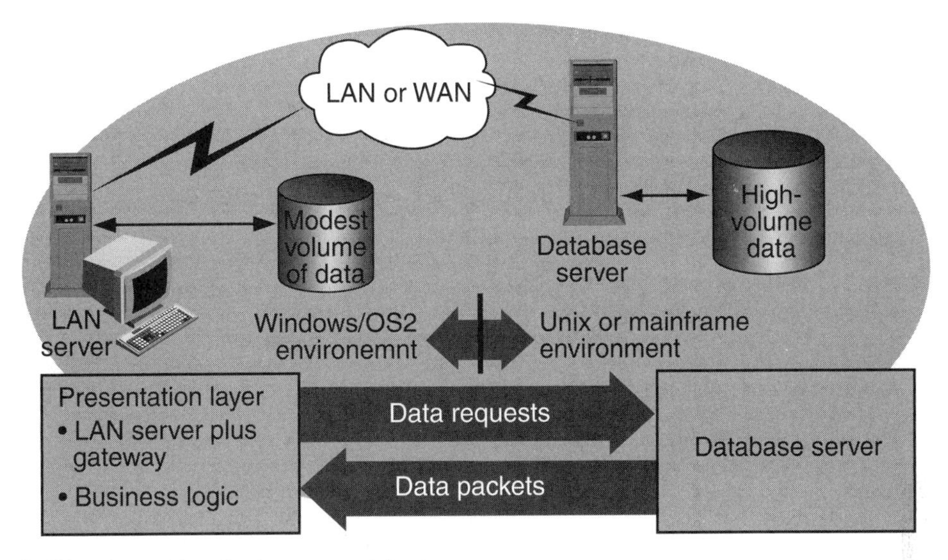

Figure 9–27 A two-tier client-server architecture.

ments, such as financial services and healthcare, business continuance is being regulated as a critical requirement with which these organizations are expected to comply.

The strategy depends on an appropriate storage solution that enables data to be replicated and backed up, systems to be rapidly redeployed, and users to access the most available systems regardless of location. It is essential that businesses objectively understand their critical applications, how long they can afford to be down, and what they can afford to lose. Some applications must never go down, some may tolerate minutes of interruption, and others may be measured in hours. This understanding will drive the solution—an end-user may have high performance storage for online applications, lower performance storage for migrate data, and tape backup solution for archive.

As shown in Figure 9–28, deploying redundant data centers, then interconnecting them over a wide area network, is a critical component of most sound plans. Industry analysts recommend that a secondary site for disaster recovery be located far enough away from the primary data center to reduce the likelihood of being affected by the same disaster. The storage network that connects these sites with high-performance, highly reliable and secure communications provides a critical foundation for enabling a business to continue as usual even while recovering form a disaster.

For those applications that are not as critical but require much faster recovery than can be expected from tape backup, organizations look to an *asynchronous replication* technique. Asynchronous replication software uses network bandwidth efficiently by transferring only changed data blocks or tracks to a remote backup system. This technique requires much less bandwidth than does the transfer of full-tape backups, and it enables faster data recovery in the case of data corruption or a major disruption at the production site.

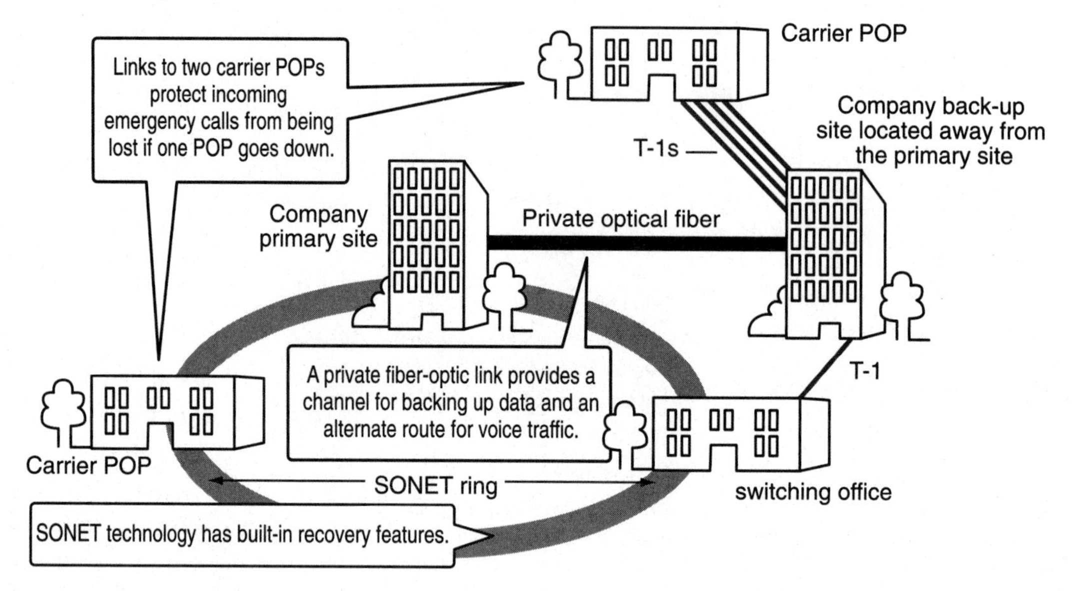
The company's call centers must be working at all times in order to provide emergency services to its customers, so its two call centers are designed for resiliency.
Links to two carrier POPs protect incoming emergency calls from being lost if one POP goes down.
Carrier POP
Company back-up site located away from the primary site
T-1s
Company primary site
Private optical fiber
A private fiber-optic link provides a channel for backing up data and an alternate route for voice traffic.
T-1
Carrier POP
SONET ring
switching office
SONET technology has built-in recovery features.

Figure 9–28 Deployment of redundant data centers ensures business continuance.

Synchronous replication or mirroring is ideal for applications that require complete data integrity and that cannot afford a single lost transaction. Mirroring synchronously replicates all disk writes to backup storage system located at a remote site. This requires a low-latency, high-speed, highly reliable storage network, since substantial latency between the two sites can negatively impact application performance. IT organizations can consider data center mirroring, where both production and backup data centers share workloads and are fully synchronized. This can be achieved with either of two designs. In the first and more complex design, the same application is load-balanced across two mirrored sites, and user sessions are directed to the most available site by an intelligent load balancing system. In the second and more common design, each data center supports a different set of applications. When a failure occurs in one site, the surviving site takes over supporting the affected applications.

TELECOMMUNICATIONS MANAGEMENT NETWORK

Telecommunications Management Network (TMN) offers a method for integrating network management across different networks, as shown in Figure 9–29. It is equally applicable to service provider networks or companies having their own private networks. It can be used to manage analog and digital networks, circuit- and packet-switched networks, sig-

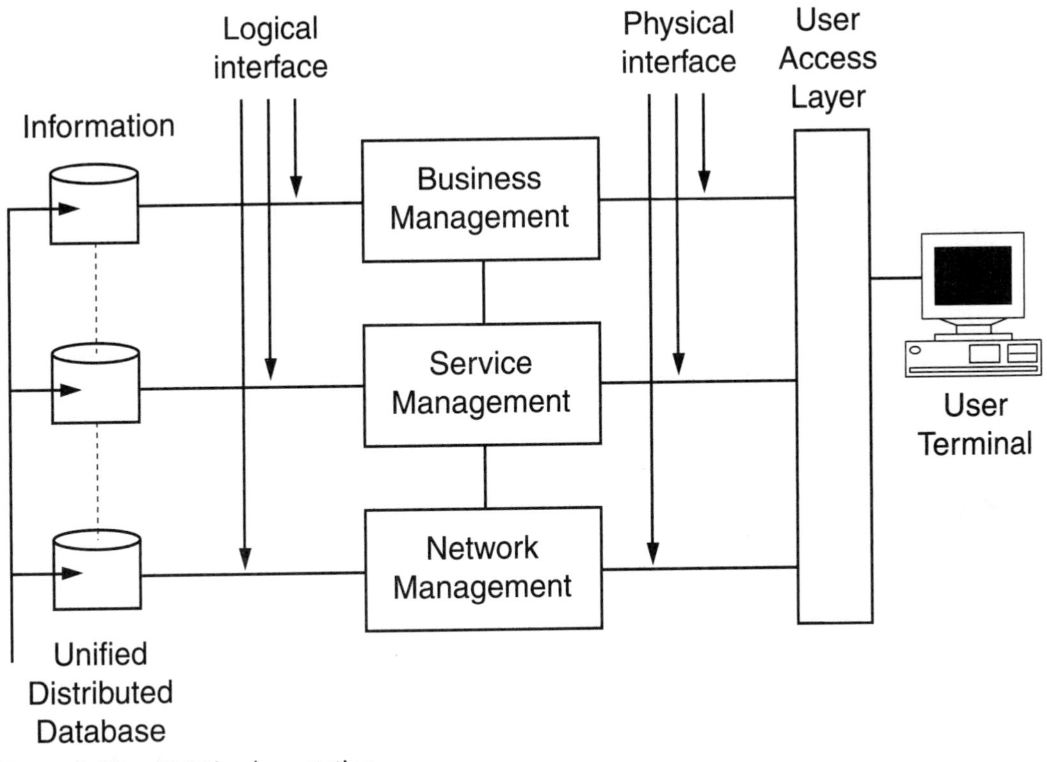

Figure 9–29 TMN implementation.

naling systems, customer terminal equipment, and telecommunications support equipment. The TMN architecture has three parts:

- Logical
- Physical
- Information

The *logical* part specifies the management functions and the reference points for data exchange between the functions; the *physical* part defines how these functions are implemented on real systems and the interfaces between them; and the *information* part defines the data structures.

There are two distinct approaches: preventive and reactive control. Preventive controls are designed into the network for the purpose of limiting the data coming into the network using intelligent routing and switching algorithms and buffers to absorb heavy traffic, within the limits of budget and size constraints. Reactive control is activated when the network senses that heavy traffic is threatening its efficiency. If there are focused points of overload, routing can be altered, buffer allocation schemes can be modified, or particular customers can be shut off. If these methods are unsuccessful, selective discarding allows specific packets to be dropped in specific services. For example, voice traffic could tolerate some dropped packets and still retain sufficient quality for intelligibility, whereas data traffic between computers would be a very poor candidate for this control method.

SUMMARY

The role of the network is to offer services to enhance user productivity and reduce costs, which is achieved by sharing hardware and software resources, documents, and equipment. The manager of a network configures the network, monitors its status, reacts to failures and overloads, and plans intelligently for future growth. Security is of vital concern among both users and network mangers; it is indicative of a mindset that believes no system is completely unhackable. An ideal network management system is an intelligent self-learning system that learns where a problem can arise and tries to resolve it automatically, and at the same time, alerts the network manager and calls attention to any special characteristics of the situation. Today, government and industry are more aware about network vulnerabilities and the necessity to have disaster recovery and business continuity plans.

As enterprises continue to migrate to networked storage to save costs, the threat to storage security is becoming a reality. With server consolidation and shared network resources, there is no longer just one point of entry to silos of stored data but, rather, multiple paths to a collection of consolidated storage data. If a virus infects a file on a NAS device, it can spread to the rest of the files in the NAS environment. Further, if the virus is not identified and eradicated immediately, every time NAS data is backed up the virus can begin a cycle of re-infection. In a local environment, servers and storage devices use parity and checksums to ensure the integrity of data. When storage is extended over WAN, additional components such as routers, gateways, and multiplexers handle the data blocks, which increase the risk of data corruption.

Business applications will continue to demand increasing amounts of storage. Applications driving capacity growth include e-mail, e-commerce, transactional databases, digital photography/audio/video, medical imaging, and regulation-required documentation. Not all storage solutions are created equal. Each application requires storage that has specific characteristics in the areas of capacity, performance, security, and availability. Purchasing the bigger, newest, and fastest storage components is not only economically unfeasible, but it is also impractical when such characteristics are unnecessary. Instead, legacy equipment should be fully leveraged for maximum return on investment. As we debate and implement different storage topologies—SAN, NAS, and DAS—the future value lies in eliminating this decision completely, as self-managed storage networks will decide where data is stored.

REVIEW QUESTIONS

1. Define the following terms:
   - A. Virus
   - B. Worm
   - C. Packet filtering
   - D. Application-level gateway
   - E. Proxy server
   - F. Network utilization
   - G. Encryption
   - H. Virtual LAN
2. Evaluate the importance of policy management.
3. Describe the characteristics of different RAID levels and specify their applications.
4. Analyze the specifications and characteristics that should be considered when evaluating a network operating system.
5. Outline the responsibilities of a network manager.
6. Discuss the benefits of configuring VLANs and provide examples.
7. Assess the need for a network operations center.
8. Differentiate between network administration and maintenance, and management tasks.
9. Compare and contrast:
   - A. Active threat and Passive threat
   - B. Active monitoring and Passive monitoring
   - C. DAS and NAS
   - D. NAS and SAN
   - E. Internal vulnerability assessment and External vulnerability assessment
   - F. Asynchronous replication and Synchronous replication
10. Assess different security threats and the need for a security policy.

11. Evaluate the following security measures:
    - A. Access Control
    - B. Encryption
    - C. Authentication
    - D. IDS
    - E. Firewall
12. Assess the significance of configuration management.
13. Evaluate different network services and describe how they are implemented:
    - A. Storage
    - B. Application software
14. Explain the term *business continuance* and discuss its significance.
15. Describe the Telecommunications Management Network concept.

10 TELECOM POLICY AND BUSINESS CONTRACTS

KEY TERMS

Telecommunications Act of 1996
Reciprocal Compensation
Unbundled Network Elements (UNEs)
Access Charge
Local Number Portability (LNP)
Location Routing Number (LRN)
Advanced or Enhanced Telecommunications Services
Intellectual Property
Copyright
Trademark
Patent
Service Level Agreement (SLA)
Minimum Annual Commitment (MAC)
SLA Clauses
Usage-based Model
SLA Monitoring Tools
Total Cost of Ownership (TCO)
Cost
Hard Benefits
Soft Benefits
Life Cycle
Return on Investment (ROI)
Net Value

OBJECTIVES

Upon completion of this chapter, you should be able to:

- Discuss the significance of the Telecommunications Act of 1996
- Analyze the implications of sharing Unbundled Network Elements (UNEs) with competitors and reciprocal compensation
- Evaluate the benefits of Local Number Portability (LNP)
- Describe the implications of regulation on electronic documents
- Identify the need for intellectual property and copyright laws
- Analyze the effects of lack of global intellectual property and copyright laws

- Discuss the impact of telecom policy on business
- Describe some of the impediments faced by U.S. vendors of telecom equipment in a global market economy
- Determine the components for effective implementation of electronic commerce
- Evaluate the components of a Service Level Agreement (SLA)
- Identify the significance of different SLA clauses
- Explain the term Total Cost of Ownership (TCO) and its value to business
- Describe cost/benefit analysis and its applications

INTRODUCTION

In the United States, there are general trends toward continuing deregulation and encouraging competition. On the other hand, international telecommunications environment often appears to favor institutionalized subsidies, centralized management, and pricing policies at the expense of efficiency. Less-developed countries tend to be more protective of their telecommunications infrastructure and thus, they tend to keep it highly regulated. In some countries, such as China, the government totally controls both the finances and the programming of telecommunications entities. In other nations, like Australia, Japan, and most countries in Europe, the government oversees and directs the overall philosophy and content.

In the United States, regulation of the telecommunications industry is based on a pro-competitive philosophy, with many entities involved in the process, both formally and informally. The regulatory groups, namely, the Congress and the FCC, intermingle partly because of the manner in which the nation's forefathers established the basic government. The legislative branch writes the laws, the executive branch administers them, and the courts adjudicate them. The courts of the United States have had a significant impact on telecommunications, primarily through the appeal process. FCC decisions can be appealed through any of the U.S. circuit courts.

The telecommunications industry is subject to federal, state, and local regulations. Starting at the most fundamental level, local governments have to ensure that their jurisdiction over rights of way, zoning, and taxation are implemented in ways that do not artificially hinder network deployment or inadvertently tip the direction of technology away from the most efficient choice. At the state and federal levels, there are a number of complex rules concerning universal service, interconnection, and access charges. The FCC and state Public Utility Commissions (PUCs) have been transitioning these rules to a more competitively neutral system.

Figure 10–1 summarizes key regulatory events in telecommunications history. The Communications Act of 1934, which created the FCC, and the Communications Satellite Act of 1962 established the basic broadcasting structure. Although the FCC has day-to-day regulatory authority, major changes in telecom policies must be approved by the

1934	*Communications Act of 1934* establishes Federal Communications Commission (FCC) to regulate interstate, international, and maritime communications, with universal service stated as the goal. Department of Justice begins major antitrust action against Bell System, which is delayed due to issues of national interest during WWII.
1935	First state Public Utility Commissions (PUCs) formed to assume intrastate regulatory authority from municipal and city governments.
1962	*Communications Act of 1962* places authority with FCC to assign commercial satellite frequencies. The Act establishes Communications Satellite Corporation (Comstat) to act as a carriers' carrier (wholesaler) for international satellite service and in conjunction with Intelsat (International Telecommunications Satellite Organization) established as an international financial cooperative that owns and operates satellites for international communications.
1996	*Telecommunications Act of 1996* passed by Congress and signed into law February 8, allowing full and open competition across all dimensions, including manufacturing, local service, and long distance. Conditions established for RBOC entry into long distance market within home states. FCC begins a process of establishing rules for implementation of the Act. Implementation specifics in process of full definition; ultimate impact remains unclear. *Communications Decency Act* enacted to hold both creators of content and service providers responsible for access of minors to indecent or offensive material over the Internet. The Act was ruled unconstitutional, in violation of free speech guaranteed by the First Amendment.

Figure 10–1 Key regulatory events in telecommunications history.

Congress. The Telecommunications Act of 1996 makes sweeping amendments to the Communications Act of 1934, which set up the regulatory structure that prevailed for over 60 years. Figure 10–2 graphically depicts the U.S. telecommunications policy-making process.

TELECOMMUNICATIONS ACT OF 1996

The **Telecommunications Act of 1996** was enacted by the 104th Congress of the United States of America and signed into law by the President. It promotes competition and reduces regulation in order to encourage rapid deployment of broadband communications, and it secures lower prices and higher-quality services for American consumers. While the Act covers a wide range of policies, the biggest changes are in the areas of two-way voice, data, and video services.

The Act has three primary objectives:

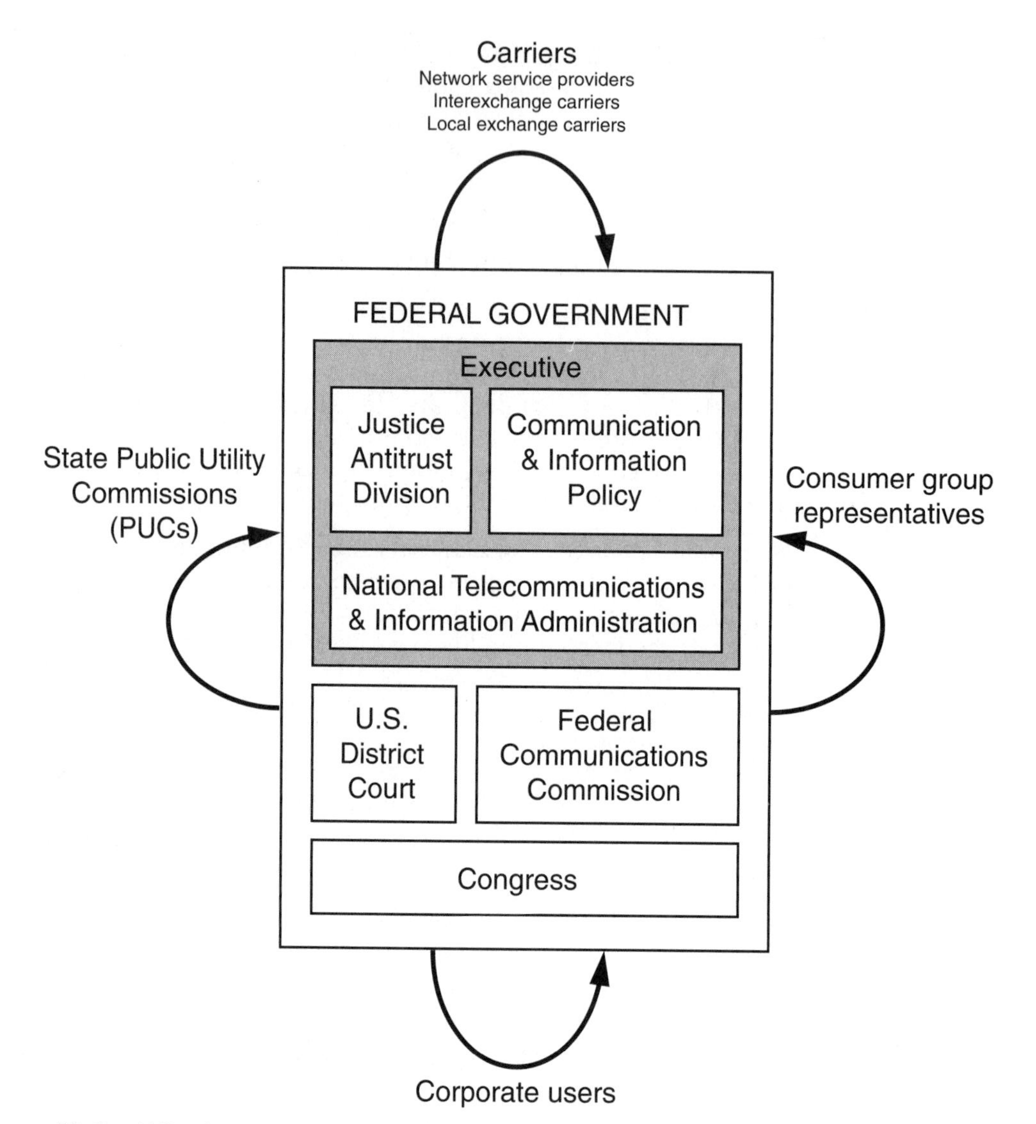

Figure 10–2 U.S. telecommunications policy structure.

1. To ensure both a timely deployment of advanced services and an underlying infrastructure necessary to support this deployment
2. To promote competition
3. To ensure universal service, that is, the widespread availability of basic and advanced services to consumers in all parts of the United States

Infrastructure for a Timely Deployment of Advanced Services

One of the most critical pieces of broadband deployment is the connection to the end user. In a network environment, these connections are not only dependent on the cost of connection, but also on the ability to interconnect with other users and other networks. Without these provisions, it would be difficult for an integrated provider of both voice and data communications to compete with an ILEC, formerly known as RBOC, given the huge advantage the incumbent has with respect to the local network infrastructure. Let us consider the following scenario.

Suppose that an entrant built an innovative network covering, at first, only a small number of subscribers. Users would be very reluctant to subscribe to this network instead of the incumbent's network if they could call only the subscribers of the entrant's network and not the vast majority of users who remain with the incumbent carrier. Competition would no doubt be thwarted if subscribers of a smaller network could not communicate with subscribers of a larger network. Therefore, the 1996 Act makes provisions for competing carriers to be able to rely on one another's facilities to complete calls, so that subscribers on one network can seamlessly connect with subscribers on another.

The Act requires all carriers, including incumbents, to offer such transport and termination service to rivals on a **reciprocal compensation** basis. Reciprocal compensation are fees paid to local phone companies for use of their networks to complete the calls; they ensure that local phone companies are compensated for providing the local infrastructure that other providers may use. The wholesale components of the ILEC's local infrastructure are known as **Unbundled Network Elements** (UNEs), which primarily include seven elements: local loops; network interface devices; switching capabilities; interoffice transmission facilities; signaling networks and call-related databases; SS7 functions and Operations Support Systems (OSSs); and operator and directory assistance; as shown in Figure 10–3.

Unbundled Network Elements
✦ local loop
✦ network interface devices
✦ switching capabilities (hardware and software)
✦ interoffice transmission facilities
✦ signaling networks and call-related databases
✦ SS7 functions and operations support systems
✦ operator and directory assistance

Figure 10–3 Components of Unbundled Network Elements (UNEs).

The 1996 Act mandates ILECs to make UNEs available to their competitors at any technically feasible point through a process of negotiation backed by the prospect of regulatory intervention to set prices if negotiations fail. If negotiation failed and state regulatory arbitration was invoked, the prices of UNEs and of transport and termination are based on forward-looking economic costs of providing those elements and services, with due allowance for risk and the faster economic depreciation of investments that will result if competition and innovation accelerate.

The FCC's unbundling rules forbid ILECs from separating individual network elements before leasing them to a CLEC. This rule is designed to prevent CLECs from having to incur the expense of re-combining these UNEs. The local loop, which is the connection between customer premises and the local switching exchange, is often regarded as the most essential UNE because it is the one element that cannot be readily duplicated on a mass scale. The Act clears the path for increasing numbers of businesses to become CLEC competitors: companies that offer local service by purchasing UNEs or reselling ILEC services.

The FCC has allowed ILECs to deploy infrastructure for advanced data services out of a CLEC subsidiary without being subject to the wholesale restrictions of the Telecom Act. This lets an ILEC subsidiary, the CLEC, to provide DSL service, for example, without having to share its infrastructure with competitors. Also, CLECs can bundle voice access with DSL, something the Act forbids ILECs to do. As a result, there is movement toward ILECs becoming CLECs by restructuring their organizations and facilities to divide themselves into two separate entities.

Regulation affects the tariff structure, and, in effect, the fees charged for end-user services and for carrier-to-carrier services. **Access charge** is a fee paid by a long distance carrier to a local exchange provider for use of its local facilities, as illustrated in Figure 10–4. Any time there is a price increase or a new service is introduced, it is subject to review by that state's PUC. Most services sold by LECs are subject to tariff, guidelines for which are incorporated into a voluminous set of written service descriptions and prices.

Reciprocal compensation has implications for both ILECs and CLECs. An ILEC is required to compensate a CLEC when the CLEC delivers ILEC-originated traffic. This situation arises, for example, when an ILEC customer dials up an ISP who is a customer of a CLEC or an ISP who is a CLEC competing with the ILEC in the same service area. The call leg between the customer and the ILEC switching exchange is handled by the ILEC; the completing call leg is handled by the CLEC. In such a case, the ILEC owes reciprocal compensation to the CLEC only if the call is deemed local, because long-distance calls are exempt from the reciprocal compensation scheme.

But what constitutes a local call? ILECs have taken some pains trying to prove that if somebody dials up an ISP's server in CLEC territory, then connects via that server to a distant Web site, the call is not really local, and hence is not subject to reciprocal compensation. Since the FCC ruled that calls to ISPs are long distance, ILECs do not pay any reciprocal compensation fees to the ISPs when their customers dial an ISP for Internet access. In the same token, ISPs rarely pay any reciprocal compensation to the ILEC because although ISPs receive many local phone calls, they place almost none. The calls

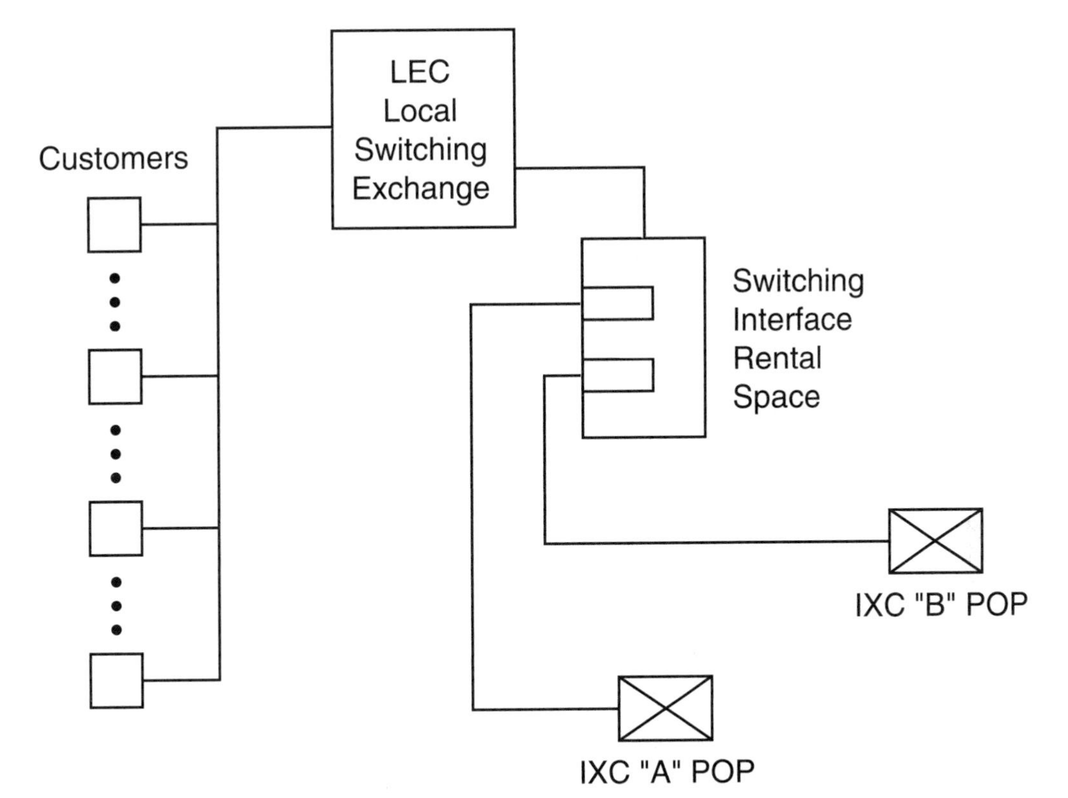

Figure 10–4 IXCs may rent closet space and pay an "access charge" only to provide a physical interface to the LEC facilities. From there, the IXC can lease lines or install its own lines to where its POP is located.

that ISPs receive travel over the ILEC infrastructure, but there is no exchange of fees, as depicted in Figure 10–5. This provides for a competitively neutral system.

The Promotion of Competition

In the past, the RBOCs or Baby Bells have resisted opening their networks to rival phone services. The 1996 Act takes several steps to promote competition in the provision of local access to voice, data, and video services. The two most significant ones are:

1. Encouraging the opening of local exchange to competition
2. Removing legal barriers to telephone companies entering the cable business, and vice versa

Competition among LECs

Section 251, which addresses interconnection, is most relevant with respect to the market-opening provisions of the Act. It stipulates that the local operating companies may enter

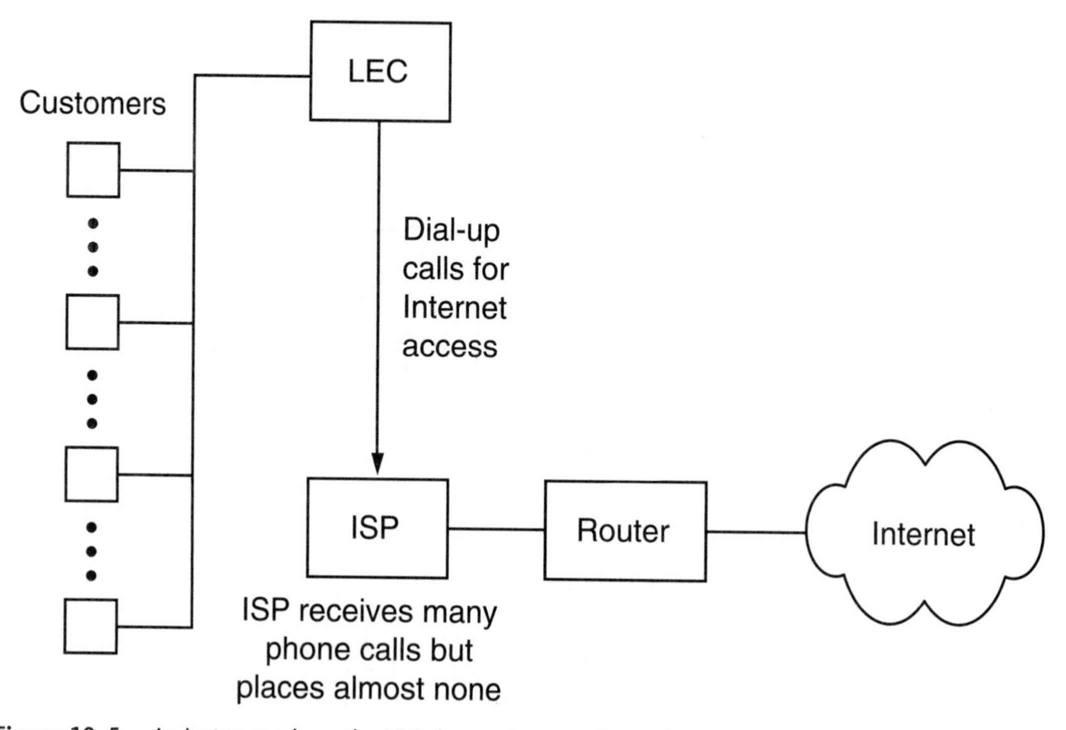

Figure 10–5 An instance where the LEC does not pay reciprocal compensation to the ISP.

the long-distance business as competitors to companies like AT&T, MCI Worldcom, and Sprint only if they first open their own networks to competing local services. Otherwise, ILECs may provide local broadband connections to the end user but would have to connect to other providers for the interLATA portions of the service. Section 251(c) specifies that the ILECs may introduce competition in several ways: by providing rivals with interconnection to their networks, by letting competitors use UNEs on a nondiscriminatory basis, or simply by letting them resell the services the ILECs themselves offer. Section 271 spells out a 14-point checklist that the incumbent carriers must satisfy before being allowed into long distance. The distinction between local and long-distance service is going away or becoming less relevant.

In an attempt to spur competition in the marketplace, the FCC required that LECs offer services allowing a business or residential phone user to change service providers while retaining the same phone number, as illustrated in Figure 10–6.This is referred to as **Local Number Portability (LNP)**, a program initiated by the FCC via the Telecom Act of 1996. Previously, switching local carriers meant having to switch phone numbers, which has been a barrier to local competition. Establishing a new phone number takes time, effort, and paperwork, which translate to a potential loss of revenue for businesses. LNP removes this barrier that would otherwise deter LEC customers, especially business customers, from changing providers. The ultimate goal is to have geographic portability,

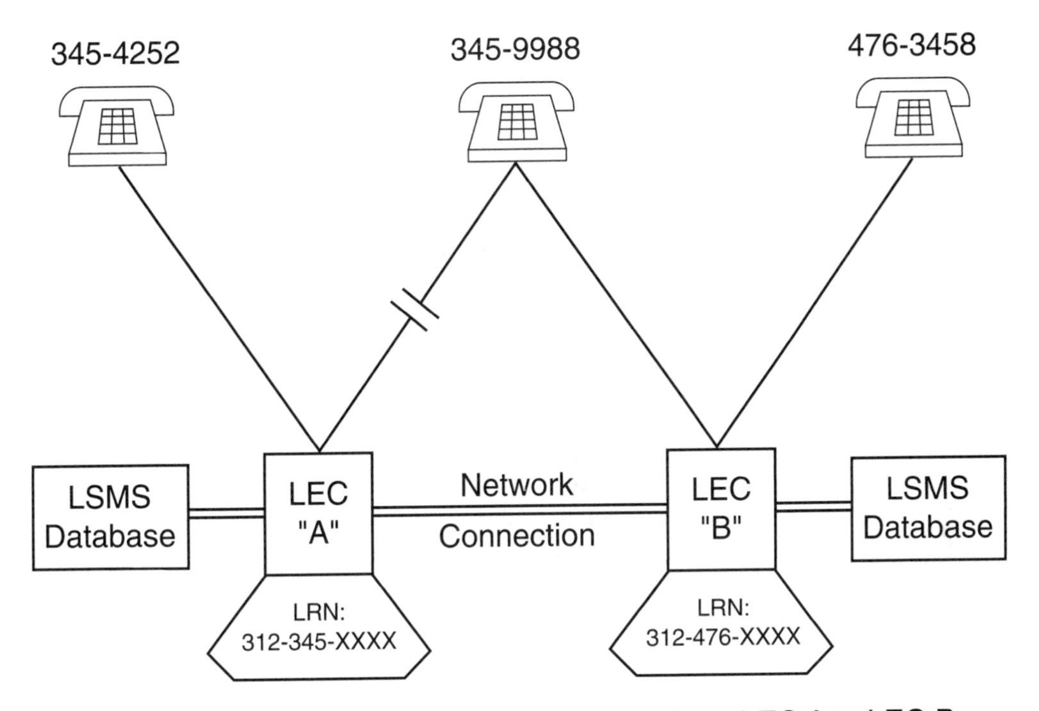

(312) 345-9988 changes service providers from LEC A to LEC B.

Scenario 1: 345-4252 calls 345-9988. Since 345-9988 cannot be found on LEC A's switch, query is launched to LEC A's LSMS database. The LRN returned is 312-476-XXXX. Call is then routed to LEC B's switch and terminated to 345-9988.

Scenario 2: 476-3458 calls 345-9988. The number is found on LEC B's switch and call is terminated.

Figure 10–6 Local Number Portability.

that is, a one-number-for-life scenario where the phone number travels with the person regardless of location and the type of service used: wireless or wireline.

Key to LNP success is the **Location Routing Number (LRN)** and Local Service Management System (LSMS) databases distributed among the exchange carriers. The LRN uniquely identifies the switch that is serving a particular number. When a customer moves the local service to another carrier, a new LRN is assigned to the telephone number being ported. The regional Number Portability Administration Center (NPAC) updates the LSMS databases with the newly assigned LRN. NPAC centralizes LNP service data, allowing for quick and relatively inexpensive administration changes by a neutral third party. The LSMS is linked to the SS7 SCP, which, when queried, identifies the LRN and the call is passed to the proper switch. All facilities-based telecommunications providers in the United States and Canada are required to have the LSMS functionality.

Competition among Phone and Cable Companies

Secondly, cable companies are becoming multiservice operators to increase revenue by offering new voice and data services over their existing video infrastructures. Time was you would buy TV service from your cable company, telephone service from your phone company, and that was that. But now cable companies are offering phone services at hard-to-pass-up prices, while phone companies are fighting back with plans for TV services delivered through brand-new agreements with leading satellite TV providers. The chief business beneficiaries of this competition are home office workers and very small companies.

Universal Availability of Advanced Services

The 1996 Act expresses strong concern that **advanced** (also called *enhanced*) **telecommunications services** be available in all regions of the United States on a reasonable and timely basis. Enhanced services are defined without regard to any transmission media or technology as high-speed, switched, or broadband telecommunications capabilities. As shown in Figure 10–7, these include a variety of computerized applications combined integrally with basic telecommunications offerings, such as voicemail, e-mail, protocol conversion, various Internet-based services, and telemessaging services—for example, sports scores, news updates, and stock quotes.

Advanced or Enhanced Services
✦ voicemail ✦ e-mail ✦ protocol conversion ✦ various Internet-based services ✦ telemessaging

Figure 10–7 Advanced services.

The pro-competitive philosophy of the 1996 Act affects investment incentives, which in turn affects universal availability of enhanced services. The FCC has ruled that new investments made by all ILECs must be given to their competitors at some variant of forward-looking costs, and that new services using such investments must be made available to competitors at sizable discounts. Fundamentally, companies do not invest in equipment when they cannot anticipate using the investment to secure a competitive advantage, and the ruling largely prevents any ILEC from securing any competitive advantage through its investments. Thus, the FCC's rules rob ILECs of the economic incentive to make new investments, including investments in data communications equipment. Although it calls for explicit subsidies for broadband services to schools, libraries, and rural health care providers, it does not specify how this rollout will be funded if the services are not commercially viable everywhere.

The FCC tried to distinguish between regulated common carrier telecommunications services and enhanced services, which, under its policies, are deregulated data processing services not subject to carrier regulation. The rules worked tolerably well as long as the FCC's definitions properly reflected a reality-based difference between the two types of services. Today, however, data telecommunications services consist of both basic telecommunications services and enhanced services, and the technologies are inextricably intertwined. A properly functioning data network contains a variety of functions that the FCC had previously defined as enhanced services, most notably the ability of the network to support more than one interface—an ability called protocol conversion by the FCC.

The FCC's rules required that when the companies offer data services, they split the basic and enhanced elements of the data network, although this is often cumbersome, uneconomical, and inefficient. The agency has been reexamining its policies, and one of its chief goals is to devise a regulatory structure that does not require the companies to split their data networks along the basic and enhanced lines, which currently impede efficiency in data service offerings.

REGULATION ON ELECTRONIC DOCUMENTS

The real challenge of the new electronic environment lies in the way in which we conduct and document our business. As electronic records become more widely accepted, regulations governing their handling and legality are being adapted. Document retention policies apply to electronic as well as paper-based records, but the accuracy and use of electronic documents and records has become a critical legal issue.

The Code of Fair Information Practices, which laid the foundation to the Fair Information Act of 1974, is based on five principles:

1. There must be no personal data record-keeping systems whose very existence is secret.
2. There must be a way for people to find out what information about themselves is in a record and how it is used.
3. There must be a way for people to prevent information about themselves that was obtained for one purpose from being used or made available for other purposes without their consent.
4. There must be a way for people to correct or amend a record of identifiable information about themselves.
5. Any organization creating, maintaining, using, or disseminating records of identifiable personal data must assure the reliability of the data for their intended use and must take precautions to prevent misuses of the data.

The challenge for organizations is to demonstrate consistent retention and preservation compliance when faced with the creation and storage of paper, magnetic, optical, and microfilmed documents. In addition to statutory and transactional documents, e-mail, voicemail, and Web data are considered corporate documents and may be produced for a regulatory or legal request. Legal experts stress the importance of recognizing electronic documents and e-mail as part of the corporate base. The Securities Exchange Commission (SEC) regulates that e-mail must be treated like other forms of written communication and meet compliance guidelines. Finance, pharmaceutical, healthcare, telecommunications, and government-related firms must observe strict electronic document retention requirements. For instance, the SEC insists that American securities firms retain their electronic documents for five years—and be sure they can search and restore specific messages and threads in a short turnaround.

The handling of electronic documents is a monumental task; the key is being able to set appropriate e-mail polices and then build the appropriate architecture to support those polices. The relative immaturity of the e-mail management products, varying user definitions for data-retention polices, and a number of federal and state regulations, for example the Health Insurance Portability and Accountability Act (HIPAA), dictating various storage-retention requirements complicate the matter. Companies must walk a narrow course between the expensive and risky extremes of e-mail management: neither keeping all messages (strains storage and management resources) nor deleting e-mail without a strict retention policy (violates regulatory laws). Establishing data-retention guidelines, specifically what e-mails are important, what terms should be tracked, and what files or client materials should be kept is a challenge. The opportunity for niche software has emerged in the form of e-mail compliance tools. As an example, these tools are used by brokerage houses in the financial services industry to monitor e-mail for things like illegal stock hyping, insider trading, and high-pressure sales tactics.

Global Perspective

In order to get a global perspective on regulation of electronic documents, let us look at one specific example. Following two years of work involving some 130 organizations, the British Standards Institution (BSI) issued a Technical Code of Best Practice for the storage of documents in electronic form. The purpose of the code is to set a standard of system and process controls, which instill confidence that records have been stored and retained according to best practice.

Code 1 addresses system planning and the use of the document storage system. Although documents are commonly considered text-based, the code applies to any type of data file.

Code 2 describes procedures and processes for transferring electronic documents between computers. If a document has been created and stored so that it can be presented as legally admissible evidence, any communication system that is used to move the document between locations must be similarly stringent. The code is applicable to networks, remote communication via carrier, circuit-switched or message-switched systems, and to any type of communications hardware.

Code 3 assesses the suitability, control and use of digital techniques providing copyright protection and signatures for legal admissibility.

Code 4 provides guidelines for using and acting as a Certification Authority.

Code 5 addresses the use of trusted remote archives: A legally compliant remote archive, operated by a trusted and independent third party, enables storage of documentation without threat to its confidentiality, integrity, and availability.

INTELLECTUAL PROPERTY

Intellectual property is a discipline that deals with issues in *copyright, trademark,* and *patent* law; its nature, scope, and purpose are depicted in Figure 10–8. **Copyright** ensures authors the right to protect their work and to benefit from new inventions by giving them exclusive rights to such works.**Trademarks** are awarded to any sign or symbol capable of distinguishing goods or services. **Patents** are awarded for inventions or non-obvious improvements to existing products or processes.

	Copyright	Trademark	Patent
Nature	Original literary and artistic works such as software, music, books, paintings, and movies	Commercial identifications such as words, designs, slogans, and symbols	Inventions or nonobvious improvements to existing products or processes
Scope	Protects against unauthorized use or copying	Protects against creating a likelihood of confusion or diluting a famous mark	Excludes others from making, selling, or using the invention
Purpose	Encourages and rewards creative expression	Protects owners from unfair competition	Encourages and rewards innovation

Figure 10–8 Intellectual property protection.

In the past, ideas or stories were passed to others through speaking. Only when print technology was invented did the idea of authorship and ownership come into play. In the United States, the primary objective of copyrights was established by Congress in the U.S. Constitution (Article 1, section 8), stating "Congress shall have power…to promote the progress of science and useful arts by securing for limited times to authors and inventors the exclusive right to their respective writings and discoveries."

Copyright attracts investments into the production and distribution of original, expressive information by promising authors and publishers a reward. It protects a wide range of materials including musical lyrics, literary works, computer programs, and broadcasts. The protection is automatic for any written work and circuit layouts; no formal applications have to be filed with the government. However, patents and trademarks must go through an application process where timing is critical because they are awarded by the United States Patent Office on a first-come-first-served-basis. Some well-known trademarks are the globe of AT&T, the star of Texaco, and the Coca-Cola logo. Intellectual property rights are tools that can assist in licensing and franchising; they are considered an asset to the company that owns them and can be sold or traded as if they were tangible items.

However, intellectual property has become a somewhat controversial topic in the age of electronic publishing. As there are more publications in the form of software on disks, on-line databases, electronic mail, and videodisks, issues such as copyright, trademark and patent infringement can easily arise. Consequently, these issues force lawmakers and the courts to make interpretations of existing laws and create new laws to protect inventors and prosecute individuals who undertake inappropriate acts. For instance, the development of computer software by teams of writers and interface designers complicates the concept of ownership.

Patents cover more areas than ever before. While they used to be intended to cover mainly gadgets, they have broadened to cover business processes. Most software patents today are business methods. For example, Web services include a way for accessing and using data; in many cases, this had been an internal software application or even a single function in an application, such as order tracking, inventory checking, or customer profiling. However, by making it available on the Web, a company is making available its own intellectual property (a business method), which can possibly be used by someone else. Therefore, a patent is necessary.

Technology threatens the very concept of intellectual property, which has a long history of legal protection. The advent of computers has quickened the pace of copying, transmitting, and disseminating information. Since it is difficult to own something so easily transferable, computers challenge the entire mechanism of knowledge, not to say ownership of knowledge. A working group on intellectual property rights recognized the need to review current copyright laws in light of the fact that the establishment of high-speed, high-capacity electronic information systems makes it possible for one individual to deliver perfect copies of digitized worked to scores of other individuals, or to upload a copy to a bulletin board where thousands of individuals can download it on disks or print unlimited hard copies on paper. The Information Infrastructure Task Force (IITF) is dealing with proposed changes to the Copyright Act. The ultimate goal is to establish laws that are forward-looking and flexible enough to adapt to incremental changes in technology without the need for frequent statutory amendment.

Global Intellectual Property Law

The U.S. government is still trying to achieve a global standard for intellectual property law. But laws are *territorial* and can only be applied over the same jurisdiction that the government has. Intellectual property rights were initially written to encourage progress, especially in the world of technology and art. Modern telecommunications have enabled countries to link together, but unfortunately, the Internet is being used for illegal copying of programs and documents. High prices can be attributed to this illegal copying of information, as U.S.-based corporations are losing an estimated $60 billion annually as the result of intellectual property infringement. The same advancements, that have enabled information to be shared all over the globe, have created a problem because of different intellectual property laws.

Let us consider the following scenario: users in the United States can access information held in foreign countries. Some of these foreign Web sites offer information that would violate U.S. intellectual property rights if the sites were within the boundaries of the United States. Since these foreign sites can give the same information at a lower price, usually at the cost of transmission, the U.S. user may be tempted to access the lower priced item in spite of the fact that obtaining copyrighted material from foreign sites involves a violation of the copyright law. However, litigation is not the preferred long-term solution for revenue protection. The preferred solution is for programmers and Web designers to get involved in the development of legislation.

TELECOM POLICY IMPACT ON BUSINESSES

Telecom-related regulations have significant impact on carriers that sell these services, as well as on their business customers. Since the past few years, the telecommunications industry has been experiencing mergers, acquisitions, alliances, and restructurings. Global markets are moving toward a free-market economy, and companies are positioning themselves to offer one-stop services for all of the customer's telecommunication needs. The imminent rise in high-speed data services, voice communications over the Internet, and multimedia applications translates into more research and development, competition, and consolidation.

Global Market Competitiveness

One of the biggest global challenges facing U.S. vendors is not technological, but rather, regulatory. In order to sell their products in many countries, vendors must go through a certification process. In some countries, the certification process is simple: if the equipment is approved by the FCC and meets certain minimum requirements, it is certified. In other countries, government regulations are used to exclude foreign manufacturers from their markets by requiring vendors to meet very exacting standards and to customize their hardware designs. While this is not usually technically demanding, it is a very time-consuming process. Unless there is a lot of money to be made by selling into a particular market, vendors bypass countries that impose many arbitrary requirements.

In Australia, for example, getting equipment approved for sale within the country is a very laborious process. In addition, the national telephone company fines businesses that use non-approved telephone systems. As a result, most vendors have chosen to ignore this market, and customers in Australia lose because of fewer choices and little free-market competition. Europe, as a whole, falls between two extremes. With the creation of the European Union, the certification process is being streamlined so that vendors can certify their products for use in the EU, whereas before they had to certify their products in each member country.

ELECTRONIC COMMERCE

Companies are using the Internet to conduct transactional business, to outsource, to bypass, and to partner. Business planners view technology not only as a support system, but also as a key strategic tool. An ongoing search for more efficient ways of doing business is a driving force behind *electronic commerce* (or *e-commerce*), which is any purchasing or selling through an electronic communications medium. Figure 10–9 identifies major components of electronic commerce. It allows business applications within different organizations to automatically exchange information related to the sale of goods and services through a symbiotic integration of three key elements:

- Communications
- Data management
- Security capabilities

Communications support the transfer of information from the originator to the recipient, while data management defines the exchange format for that information. Security mechanisms authenticate the source of information, guarantee the integrity of the information received, prevent disclosure of the information to inappropriate users, and document whether the information was received by the intended recipient.

Electronic Commerce Encompasses One or More of the Following
✦ EDI (Electronic Data Interchange) ✦ EDI on the Internet ✦ e-mail on the Internet ✦ shopping on the Web ✦ product sales and service sites on the Web ✦ electronic banking or funds transfer ✦ outsourced customer and employee care operations

Figure 10–9 Major components of electronic commerce (e-commerce).

E-commerce in general, and Web commerce in particular, differ from traditional commerce in the way information is exchanged and processed. Traditionally, information has been exchanged through direct, person-to-person contact or through the use of the telephone or mail systems. In e-commerce, information is conveyed via a communications network, a computer system, or some other electronic media. It pulls together a gamut of business support services, including inter-organizational e-mail; on-line directories; trading support systems for products and custom-built goods and services; ordering and logistic support systems; settlement support systems; and management information and statistical reporting systems.

E-commerce clearly depends on the availability of reliable, secure, inexpensive and ubiquitous connectivity of key networks, including:

- Organizations' own enterprise networks, which house appropriate information
- PSTN
- Internet
- ISP's networks that can be accessed by dial-up or private lines

All these networks can actually be seen as overlays, where one network may depend on facilities of another network. Figure 10–10 provides an encapsulated view of electronic commerce. The private enterprise network is usually much more extensive than seen in this figure and can have significant geographic scope using various communication facilities available in the public network, for example, dedicated lines, ATM, and Frame Relay.

The benefits of e-commerce include:

- Reduced costs to buyers from increased competition in procurement as more suppliers are able to compete in an electronically open marketplace
- Reduced costs to suppliers by electronically accessing on-line abilities to submit bids and by on-line review of awards
- Reduced errors, time, and overhead costs in information processing by eliminating requirements for reentering data
- Reduced inventories, as the demand for goods and services are electronically linked through just-in-time-inventory and integrated manufacturing techniques
- Increased access to real-time inventory information, faster fulfillment of orders, and lower costs due to the elimination of paperwork
- Increased access to a client base
- Improved market analysis as the large and increasing base of Internet users can be targeted for the distribution of surveys for an analysis of the marketability of a new product or service idea
- Rapid information access over the Internet
- Rapid interpersonal communications through e-mail
- Wide-scale information dissemination since one can place documents on Web servers and make them accessible to millions of users
- Increased productivity as a result of enhanced communications capabilities

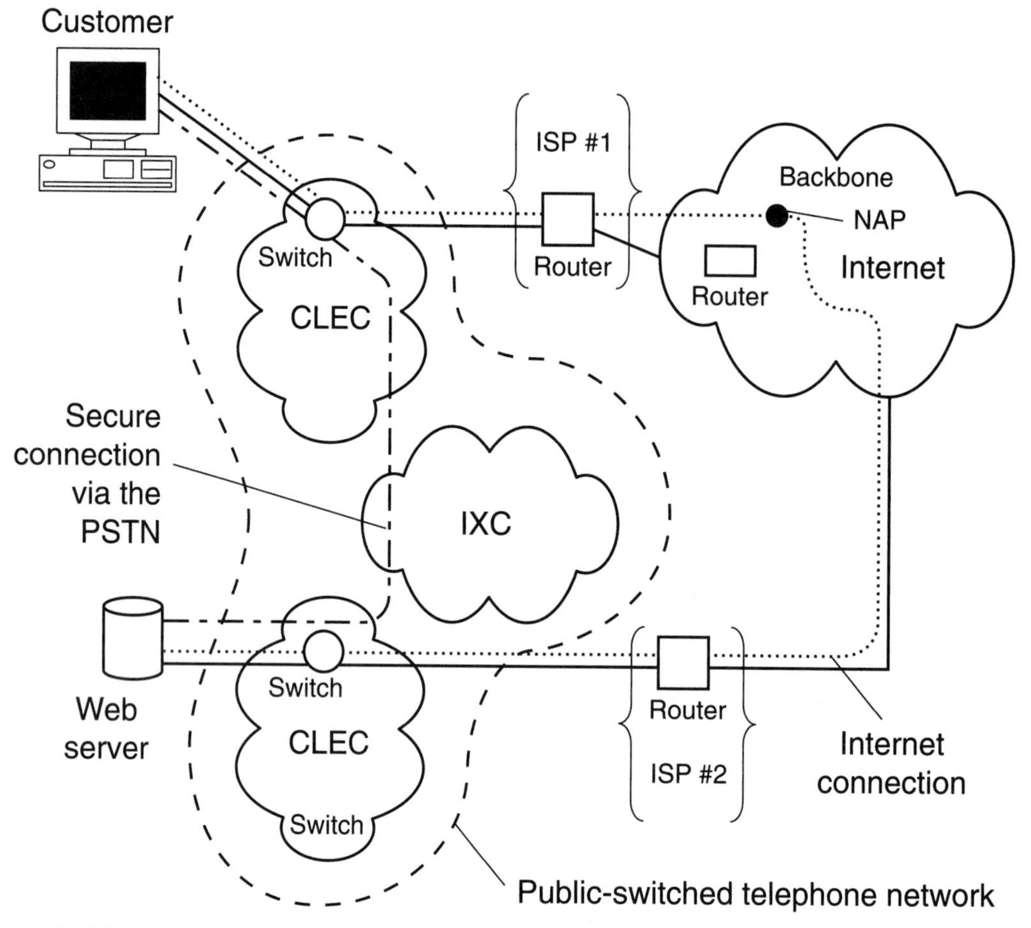

Figure 10–10 An encapsulated view of electronic commerce.

Network availability, security, and performance of user applications are some of the most significant enabling factors for successful implementation of e-commerce technologies, which can be ensured by designing a sound agreement between the user organization and the network service provider. Figure 10–11 provides a comprehensive look at different elements involved in the e-commerce process.

SERVICE LEVEL AGREEMENT (SLA)

A **Service Level Agreement (SLA)** represents a contract between a network service provider and a customer that specifies, usually in measurable terms, certain levels of network and application performance such as throughput, delay, and availability, and a promise of rebates if those parameters are not met by the provider, in return for **Minimum**

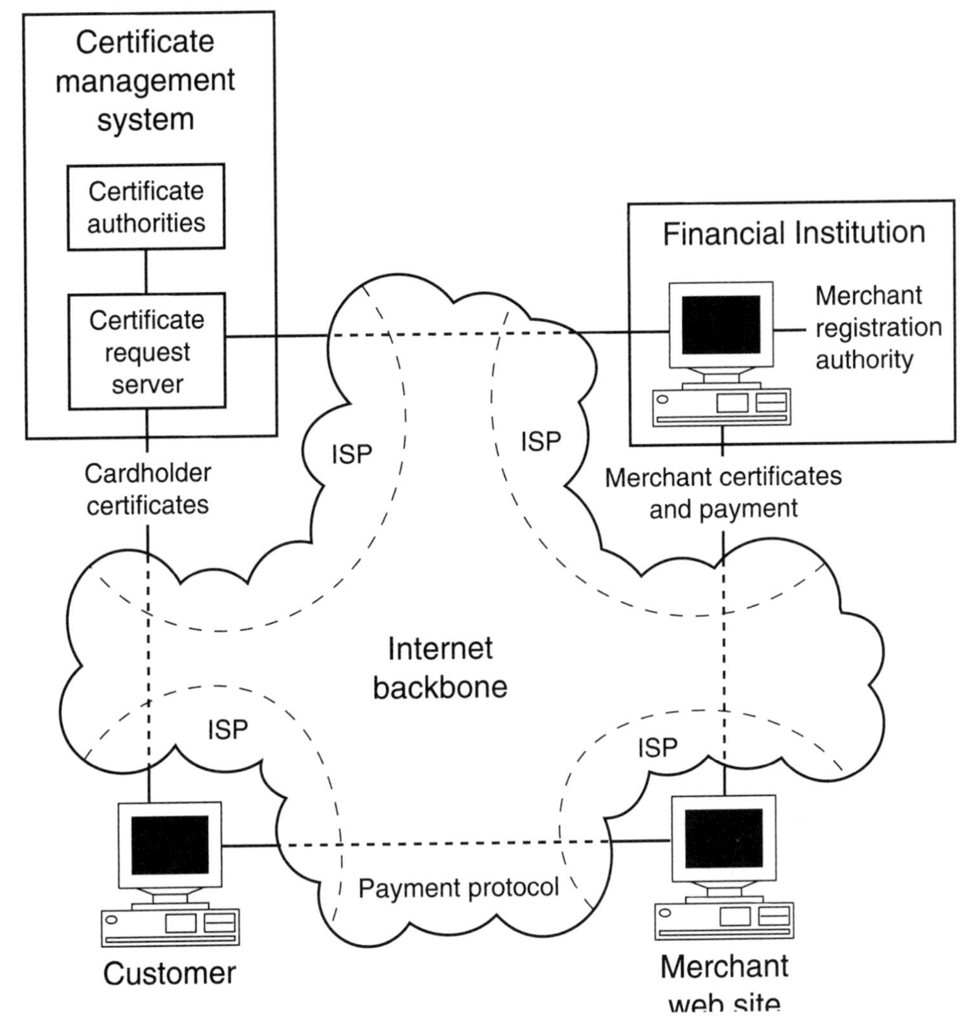

Figure 10–11 Essential elements of the e-commerce process.

Annual Commitment (MAC) from the customer. MAC is the amount of money a user organization agrees to pay a carrier each year of a multiyear contract in exchange for negotiated discounts. The penalty for not meeting the MAC is usually either the entire amount of the shortfall or a pre-arranged alternative.

Need for a SLA

A dramatic increase in network outsourcing has brought more attention to the value of SLAs as a means to ensure optimal performance of the enterprise network and the mission-critical applications, which in many cases now reside on the service provider's servers.

Since network downtime translates to dollars lost, organizations expect and demand more from their service providers. A well-thought-out SLA is valuable to both the service provider and the client. The service provider gets the benefit of avoiding unrealistic expectations that would be technically impossible or too expensive to deliver, while the client gets a guaranteed level of service. With the advent of more interoperable methods for delivering QoS, customers are demanding guarantees for their applications and end users, and many service providers are making network performance guarantees a competitive advantage.

Selecting a Service Provider

Selecting a carrier is a business-critical decision that is primarily a process of measuring technology, product quality, and level of customer service against the price. It is a process of analyzing and comparing various offerings by carriers with the organization's unique needs. The carrier's business requirements and operations, service management, and ability to stay competitive must all be taken into account. The main issues to evaluate should be accessibility, latency, response time, security, dial-in access availability for remote users, ease of use, support, redundancy measures, and network monitoring and reporting capabilities.

A new form of service provider, the *Storage Service Provider (SSP)*, has emerged to provide outsourced storage solutions. SSPs address the explosive growth of storage and the current lack of skilled storage expertise. They deliver flexible, cost-effective storage networking solutions and provide managed services such as data protection and storage-on-demand to customers located either within their data center or geographically dispersed over an optical network infrastructure. SSPs implement metered on-demand storage programs that automatically charge customers only for actual usage. The storage company places a server or workstation at a customer location to monitor usage and performance information. This server polls storage devices, servers, applications, and databases on the network and generates invoices based on the usage pattern. The software reads the system and tells the company how much storage capacity is being used, just like an electric meter. These storage companies are appealing because they provide businesses the flexibility to increase resources without having to buy a lot of equipment up front or to avoid having to predict future needs precisely.

Application Performance Evaluation

SLAs date back to the mainframe era when it became necessary to spell out acceptable degrees of application performance. In those days, since interactive applications ran from terminals, *keystroke response*—the time it took for the screen to redraw after the Enter key was depressed—was the most common performance metric. Today, application performance requirements of traditional applications are relatively well understood: three-second response time for mainframe applications; low latency for telephony and video conferencing; and sufficient bandwidth for FTP, e-mail, and HTTP. A one-day graph of link utilization and application response time is shown in Figure 10–12. It is clear that greater utilization yields higher response time. New applications, many of which are critical to business success, include e-commerce and Internet-enabled call centers, distance training, integrated messaging, IP telephony, enterprise resource planning, and multimedia collaboration. Instrumenting the application layer to gauge QoS metrics is not as simple because the underlying layers complicate matters. Next-generation tools are being developed to deliver an end-to-end view of services, many of which may cross domains, that is, between enterprise and service provider, or between one service provider and another.

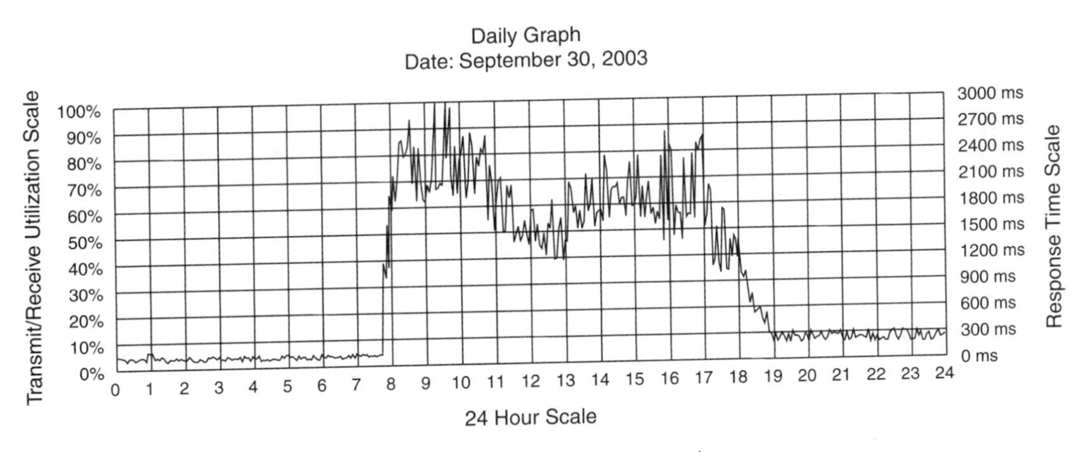

Figure 10–12 Link utilization and application response time graph.

Network Performance Evaluation

Instrumenting just the data link and physical layers, which involves measuring throughput and latency of WAN link providers, is substantially easier than application performance management. All SLAs specify the bottom three layers of the OSI model. With TDM-based leased lines, such as T-1 and 56 kbps services, customers have a pretty good idea of the QoS they will receive. The circuit-switched link is either alive or it is not, and it either transfers bits at the designated rate or it does not. In this case, a SLA would be principally concerned with availability. On the other hand, virtual circuit services such as Frame Relay and ATM do not hold open specific time slots for a particular customer as TDM services do. As a result, even with the most scrupulous service provider, the price of shared facilities is the potential for periodic congestion.

A network should have enough capacity to handle critical applications at peak load. The cost of network unavailability goes beyond the immediate impact of lost revenue and lost productivity. From a customer's or business partner's perspective, not being "open for business" can do incalculable damage to a company's customer satisfaction, reputation, and identity. From an internal perspective, employee morale can suffer. These indirect consequences can also have significant future costs associated with them, including loss of market share. Network downtime is caused by a wide variety of factors including loss of power, loss of facilities, network overload, service provider problems, software bugs, or hardware component failures. Network resource outages are related to network overload, that is, performance overload attributable to insufficient infrastructure, as well as the lack of appropriate enterprise management tools to detect overload conditions proactively, before they become a source of contention within the enterprise.

SLA Clauses

Service Level Agreements can be external or internal. The objective of an external SLA is to define and monitor a provider's service levels, while the objective of an internal SLA is to set and manage user or customer expectations. A common thread through both internal and external **SLA clauses** is a measurement of network availability and performance, and application availability and performance. Most SLAs include an abundance of very specific clauses, stipulations, exceptions, and fine print. Although every SLA is different, here are a few basic elements that should be a part of each one:

- **Parties to the Agreement:** All parties involved in the agreement should be listed, especially when there are multiple service providers and/or client groups.
- **Terms of the Agreement:** The period of time that the agreement will be in place should be specified; a typical length is two years.
- **Services Included:** An SLA should identify and define each service covered, identify service level indicators, describe how the indicator will be measured, and specify who is responsible for performing the measurement. Typically, network metrics include availability, latency, and error rates, while application metrics include availability and response time.
- **Nonperformance:** An SLA should specify the response to nonperformance and the action to be taken when the indicators do not meet the levels specified. However, some consideration has to be given to the amount of deviation. For example, instead of stating that none of the transactions will endure more than a 2 second response time, an alternative statement might be that 95 percent of the transactions will have less than a 2 second response time, and the remaining 5 percent may have a response time of 2 to 5 seconds.
- **Optional Services:** This is a list of optional services that the service provider is willing to supply on request, in addition to those listed in the current agreement.
- **Monitoring and Reporting:** This provides a description of the reports created using the monitoring tools, the frequency of reporting, access to reports, and availability of real-time reports.
- **Other provisions:** These may include unique provisions that reflect the organization's business or regulatory environment.

Figure 10–13 illustrates the depth and quality of an SLA. An SLA should define a process for modifying the agreement. For instance, a new piece of equipment may be added, and the client may therefore have heightened expectations of performance, or the introduction of new applications may change QoS and the cost of delivering it. For instance, an organization's toll-free number applications may be replaced by e-mail, or voice and legacy data traffic may be moved to new IP-based transport schemes, resulting in MAC shortage even when the traffic volume stays the same. If the traffic falls short of the MAC, some SLAs may impose a hefty financial penalty. Therefore, users should consider including some additional provisions in a multiyear SLA to protect against the unpredictable, such as a:

- **Technology displacement clause,** which typically states that the MAC will be reduced or disregarded if a major technology shift causes a money-saving WAN service to become a viable option.
- **Business change clause,** which provides for reduction in, or addition to MAC in case of corporate divestitures, acquisitions, or sellouts. For example, network expenses will be consolidated or reduced if a company is sold or business units are purchased.
- **Competitive termination clause,** which provides for an early end to contract if the market rates drop dramatically and a competitor offers an alternative. Usually, existing carrier is given the right to make a new offer within 5% of the competitor's rate.
- **International clause,** which provides for automatic rate reopening each year for international traffic. This is important because rates are expected to come down significantly as the FCC's policy of forcing down international settlement rates is working and foreign competition is taking hold.
- **Carry forward/carry back clause** lets the customer use excessive minutes as a credit to cover a previous or future shortfall, in order to satisfy the MAC.

SLA Negotiation

For most purchases, vendors use a one-third rule. One-third of the quoted price is for the product, one-third is for the service, and one-third is for profit margin. The final third is negotiable. The goal of the carriers is to get its customers to sign long-term agreements in return for a lower price. Some carriers will offer low prices but charge in one-minute increments. This means for a short call the customer will pay for the whole minute. The industry standard is to use six-second increments but most use somewhere between 18 and 6 seconds as minimum.

One must be careful not to be tempted to accept a yearly credit from the carrier in lieu of a straight rate discount because it may not be in the best interest of the customer. For example, consider a million minutes per month user who has a 7¢ per minute rate. If the vendor offers a $50,000 credit per year for three years, it brings the rate to 6.58¢ per minute.

No.	Question	Yes	No	Requires Action
5.3	Does the SLA cover payment terms for charges?	❐	❐	❐
5.4	Does the SLA include statements concerning payment of taxes arising out of the agreement?	❐	❐	❐
5.5	Does the SLA include notification of penalty interest for life of payment?	❐	❐	❐
6.0	Customer Duties and Responsibilities	❐	❐	❐
6.1	Does the SLA include information on the clients responsibilities for providing access, facilities and resources?	❐	❐	❐
6.2	Does the SLA cover Client responsibilities for providing training to their personnel on operating technical or specialized equipment?	❐	❐	❐
7.0	Warranties and Remedies	❐	❐	❐
7.1	Does the SLA include a warranty in respect of quality of service?	❐	❐	❐
7.2	Does the SLA include identification in respect of supplier negligence?	❐	❐	❐
7.3	Does the SLA include a warranty in respect of copyrights, patents, and trade secrets?	❐	❐	❐
7.4	Does the SLA exclude responsibility for client errors contributing to such infringements to third party copyrights, patents, and trade secrets.	❐	❐	❐
7.5	Does the SLA include information in respect of remedies for breaches?	❐	❐	❐
7.6	Does the SLA contain a Force Majeure clause?	❐	❐	❐
8.0	Security	❐	❐	❐
8.1	Does the SLA allow for reasonable physical access to be provided to the suppliers representative?	❐	❐	❐

Figure 10–13 Sample page from an SLA Audit and Review illustrates the depth and quality of an SLA.

12 million minutes per year x .07¢ = $840,000 annual expenditure

and

$$\frac{(\$840{,}000 - \$50{,}000)\text{ credit}}{12\text{ million minutes}} = (6.58¢\text{ per minute) rate}$$

But when the company's traffic doubles to 2 million minutes/month, the rate becomes 6.79¢ because the credit is being spread over a larger volume. Accepting a credit is appropriate if there is an immediate budget need; for instance, the company may need a new switch but does not have the cash to buy it.

For years, the norm has been flat fee for bandwidth even though one sometimes paid for more than one used. Now the ISPs are pushing for a **usage-based model**, or pay-as-you-use system, largely to set up an equitable way to charge for high-bandwidth services such as videoconferencing. Under a usage-based system, business users benefit because they no longer pay for a T-1 line when all they really need is 256 kbps; on the supplier side, ISPs have a way to charge for high-end premium services. The underlying concept in usage-based pricing is metering IP packets, where the charges are based on the number of IP packets that are transmitted. Pricing schemes for usage-based systems include charges by the bit, per event, per broadcast, or per audio stream using a fixed period of time for a certain rate of transmission. Some carriers charge by the amount of port bandwidth or the clock speed at the entry port going to the Internet. However, the biggest technical hurdle is collecting detailed packet-by-packet information for billing and customer reports, which requires expensive back-office overhead. Usage-based pricing will co-exist with the flat-rate-fee model for some time.

SLA Monitoring Tools

Most notably, the performance levels specified in an SLA must be attainable and measurable. In addition, an SLA ought to spell out the process for measuring them; otherwise, that may leave some loopholes. For example, if the service provider's downtime timer does not begin until the customer notifies the provider of an outage, it may be worth the expense to install automated alarms at the customer's premises. In situations where service providers are unable or unwilling to provide performance information to customers, or when the customer has reason to mistrust a provider's figures, customers may want to install their own measuring systems. Network utilization reports can show customers whether they are buying too much or too little capacity, which may result in savings. In many real-world cases, an organization that has a WAN with few dozen nodes can reasonably expect to recoup the investment in network monitoring hardware and software in as few as six months.

SLA monitoring tools typically offer two perspectives: *real-time* and *historical*. Real-time monitoring is particularly useful for spotting problems before they escalate, which gives an opportunity to address a problem before the terms of the agreement are broken.

It also helps service providers capture data to report service quality statistics to customers. Historical trend analysis lets the parties spot trends that may lead to future problems as well as verify service levels over time, as depicted in Figure 10–14.

Monitoring tools can be broadly divided into two major categories: hardware and software. An example of a hardware option is a passive WAN probe that is inserted between a CSU/DSU and the service provider demarcation point. A CSU/DSU with SNMP capabilities may be classified as a software tool. Monitoring specific types of networks, such as ATM, Frame Relay, and IP WANs, is often done with hardware probes in conjunction with software, which are usually proprietary.

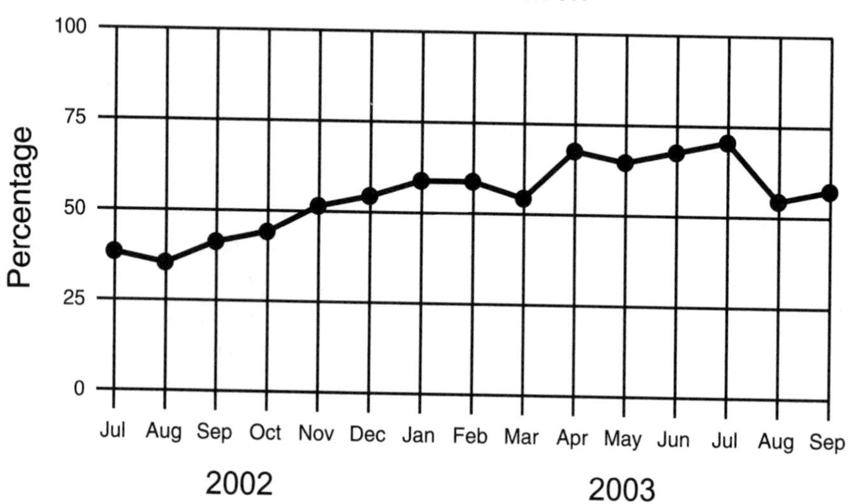

Figure 10–14 Historical trend analysis of circuit utilization.

Implementing an SLA

Implementing a successful SLA requires a clear understanding of the performance metrics and a data collection mechanism that can generate enterprise-wide information on a real-time basis over an extended period of time. Because customer satisfaction typically ranks at the top among business objectives, application-layer SLAs, which relate to customer service and help desk operations based on call centers, are important. But they are difficult to implement because of the complexities of application performance measurement. Lower-layer SLAs are worthwhile because high-speed data services are expensive. They are easy to implement because there is a well-defined demarcation point between a service provider and a customer.

TOTAL COST OF OWNERSHIP (TCO)

Total Cost of Ownership (TCO) consists of the cost of equipment, bandwidth, network, and operations. Using TCO as a strategic tool has value beyond the usual standards of measurement such as return on investment, return on assets, or profit and revenue. TCO analysis captures the current cost associated with delivering network and desktop support in a distributed environment. A holistic approach to TCO performance measurement includes client satisfaction, service levels, and business risk. When analyzing the TCO, it is important to look across the enterprise at four major components:

- Administrative activities
- Operational activities
- Technical activities
- Capital

Administrative activities include procurement, end-to-end asset management, and invoice reconciliation; operational activities include levels of support provided by help desks, LAN administrators, and vendors, as well as overall productivity levels and end-user training; technical activities include installations, moves, additions, and changes; and capital relates to hardware, software, and LAN connections. A generic telecom company expense model is shown in Figure 10–15. A TCO analysis can provide the best methodology for justifying information technology decisions, and a cost/benefit analysis may be used to demonstrate its true value.

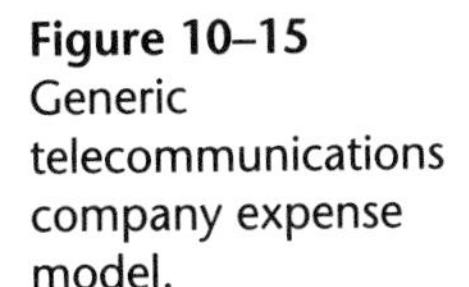
Figure 10–15 Generic telecommunications company expense model.

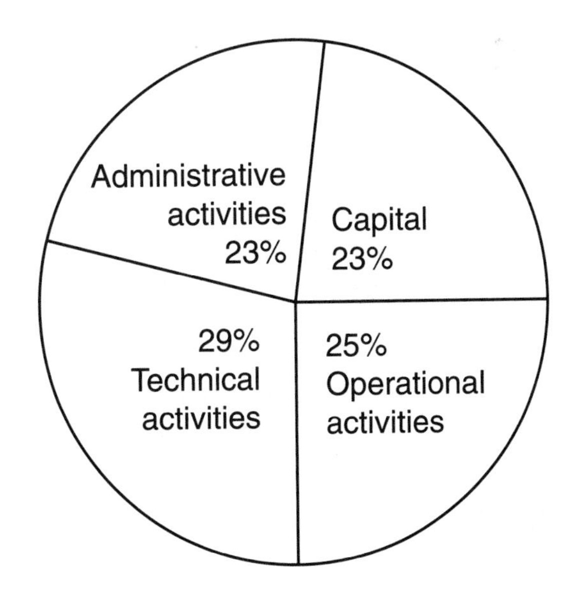

Cost/Benefit Analysis

Cost is a relative term. We may think of it as an outlay of dollars yet, true costs include both the long-term investments and support, coupled with the returns on investment. For example, systems in raw dollars may cost approximately $400 per station when dealing with PC-based PBXs. The long-term dollars over a five-year period can be three times that amount. However, if the PC is used to produce revenue of $50,000, what is the cost? Management regards most telecommunication, which includes both data and voice, as costs of doing business, an overhead function. This image must be dispelled because as an overhead function, it is a prime candidate for cost control and optimization, especially with the increasing number of ways to satisfy any given communications requirement. Telecommunications constitutes a significant expenditure of funds, which tends to increase steadily, and cost is always an issue no matter what financial position an organization is in. Management must be convinced of the use of telecommunications as a strategic resource to increase productivity, reduce costs of doing business, and improve the bottom line.

Companies are increasingly becoming aware that telecommunications can be used as a competitive weapon in the marketplace as a method of improving their attractiveness to customers, and in some cases, as a way of directly generating additional revenue. However, when telecommunications and telephony people describe technologies and applications concepts to their management, the jargon and use of technical terms can turn management off. Therefore, you must learn the language of business. If you apply business principles to your purchasing decisions, you will stand a better chance of getting the funds needed to support the organization.

When considering any expenditure, the first step is to identify expected benefits, which are normally categorized into *hard benefits* and *soft benefits*. **Hard benefits** are those that can be easily assigned objective and predictable dollar figures. Examples include: reduce dial-out voice communications costs by $3,000 per month, decrease IP-PBX maintenance costs by $500 per month, reduce travel costs (airline, hotel, meals) by $8,000 per month, and control overnight package shipment costs by avoiding $2,500 in expected monthly bills. Hard benefits share a few key elements: they specify a particular type of expenditure, a dollar amount to be saved, and a time period. **Soft benefits** are results that are expected to be good for the company but are more difficult to quantify and sometimes even to identify. Examples include improving productivity of call-handlers in the customer service area, or reducing the cycle time required to develop a new product. The goal is to make the soft benefits as hard as possible.

Once it has been determined that a true need exists for the purchase of any communications system, the next step is to evaluate everything on both a financial long-term basis and a functional basis. **Life Cycle** is a key concept in the analysis of acquisitions of both communications and data-processing equipment. It is the length of time that an organization can realistically expect to use the item in its planned role before discarding or replacing it. The term is not applicable to leased items paid for on a monthly basis but rather only to purchased equipment with an expected useful lifetime of several years.

Return On Investment (ROI)

In an enterprise, the **Return on Investment (ROI)** is how much profit or cost saving is realized for a given use of money. An ROI calculation is sometimes used along with other approaches to develop a business case for a given proposal. The overall ROI for an enterprise is sometimes used as a way to grade how well a company is managed. The focus point is: how to calculate the ROI of any fault-resilient technology?

The process of calculating ROI involves three steps: identify a business problem, assess technology alternatives, and then make a business decision as to the best solution. Any solution that costs less than the problem provides a good return on investment. So first, let us quantify a business problem. This means finding a dollar value per unit of lost time/data. The validity of the data is based on consensus between you and your management team. Although it can vary by industry, a general formula is as follows:

$$\text{Cost of Outage} = (T_o + T_d) \times (H_r + P_r)$$

where T_o is the Time (or length) of Outage

T_d is the Time (or length) of Data Loss

H_r is the Rate for Human costs

P_r is the Rate of Profitability

To determine T_o, ask questions like: How long are you down? How long are your users idle? How long is the business function unavailable? For T_d ask: How much data was lost? How long since the last backup? How much work will need to be repeated? The total time is $T_o + T_d$. Your situation may have different variables, but the key is to recognize the entire window of exposure. Consider the tape backup. If the outage occurred at 4 p.m. on Tuesday, the window from exposure is from the last recoverable backup occurred until the system is fully restored. T_d of data loss was from Monday night's back up until the point of outage. T_o or outage time was from outage to when the entire system is restored.

To determine H_r, questions to ask are: What does it cost (per user or department) when the personnel are sitting idle? The generic version of this is to look at the average salary per job description and divide it by a unit of time. As an example, the average white-collar American costs $36 per hour per person. If a given department were all engineers, with an industry average salary of $100,000 plus an assumed additional 20% for benefits, an engineer would cost $120,000 annually or $60 per hour. If 30 engineers relied on a server, that outage has a fixed human cost of $1,800 per hour. Your payroll department can provide the best numbers to use.

For P_r ask: How many dollars does a department generate (e.g., telesales or eCommerce website)? How many dollars are lost in compliance (e.g., regulatory issues)? How many dollars are lost when functions stop (e.g., shipping)? Combined, $H_r + P_r$ is Dollars Lost during Outage. This formula can vary dramatically. For example, some business units may have a huge financial impact immediately upon outage but only a trickle after.

Other units may not start feeling a financial burden for the first several hours after the outage. The point is for you and your management to agree on a method by which to quantify the cost of outage.

After identifying the cost per outage, one should now look at the frequency of outage. Multiplying annual frequency times the cost of the outage results in the business impact of the problem. In other words, how much does the company lose annually as a result of this problem? Since you are assessing a problem against technology alternatives, the problem can be quantified over the lifecycle of the technology. For example, if tape backup hardware/software has a three-year useful expectancy, then how many outages do you have over three years? Multiply that by the cost per outage and you now have a dollar figure that depicts the business impact of the problem.

Now, we simply have to compare each technology alternative (and its associated costs) to the problem. It always comes down to "how much does my problem cost? And how does that compare with the cost of a solution?"

Net Value

To analyze the tradeoff between benefits of the network and its costs, a concept of *Network Value*, or **Net Value**, is useful. Net value can be thought of as a ratio: the benefits of a network divided by the costs of a network. More explicitly, the benefits of a network are the quantifiable business results attributable to the network less the cost of network downtime; that is, the lost business and lost productivity that result when the network is not available. Regardless of the type of network and the purposes to which that network is dedicated, net value focuses on three key strategies: maximizing application performance, minimizing network downtime, and minimizing the lifecycle costs of operation.

A unifying enterprise networking strategy that maximizes net value can help the business improve its bottom line. Key approaches are to:

- Optimize the LAN/WAN infrastructure for maximum price/performance ratio
- Extend security across VPNs and Internet applications for improved business reach
- Combine telephone and IP to leverage connectivity and new applications
- Unify network management across multiple domains to simplify operations

TCO Applications

Total Cost of Ownership analysis can help a company evaluate the buy-versus-lease question. Many businesses set up VPNs as a way to save money when they cannot afford the cost of building a WAN through leased lines or direct dial-up. In most cases, it is cheaper to outsource. For instance, setting up digital certificates and a certificate authority to authenticate users is a complex task. In addition, putting together the various pieces of a VPN—access control, encryption, authentication, and firewall—requires expertise many enterprises do not have. A company would therefore have to divert peo-

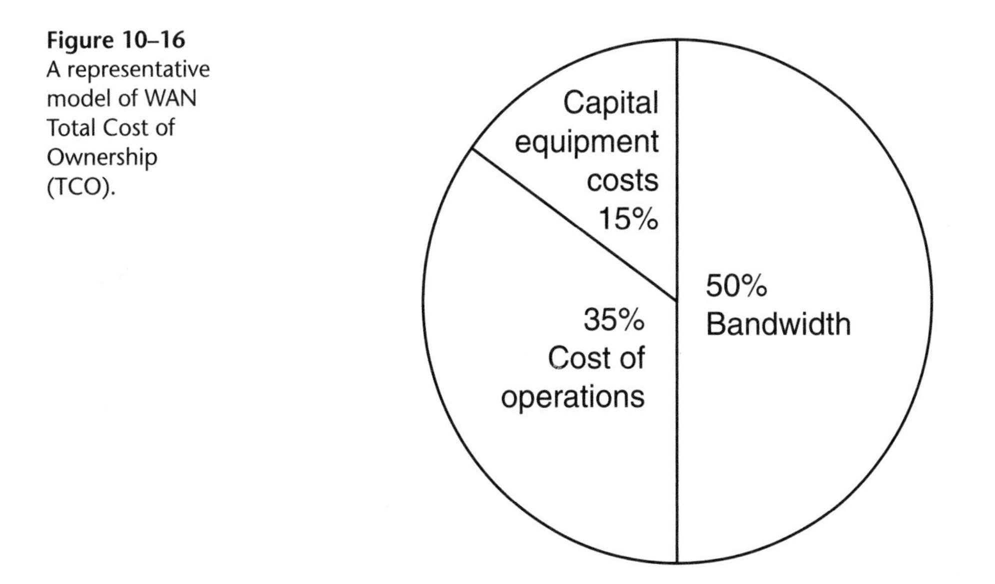

Figure 10–16 A representative model of WAN Total Cost of Ownership (TCO).

ple from whatever work they were doing to train them, and that is a cost. A simple way is for a service provider to do the job, although customers sometimes have to trade off between cost and features. With no outsourcing, enterprises maintain control of their networks and do not feel locked into a single service provider.

Technology optimization profiles, which are part of TCO analysis, can be used to align end-user needs with the most appropriate assets, allowing for a cascading of equipment and enhanced overall savings. For example, high-end users will have the greatest need for speed and bulk, while others in the organization can get by with more basic desktops. As the high-end computers become obsolete for the aggressive users, they can find provide for upgrades further down the line.

A 2003 study of WAN TCO, represented in Figure 10–16, indicated that bandwidth is the biggest cost area, accounting for about 50 percent of WAN TCO, and the cost of operations accounts for approximately 35 percent. Capital equipment costs, which make up the balance, are the least significant aspect of TCO. However, the choice of equipment does influence how effectively bandwidth is managed. Operations costs include installation, training, troubleshooting, maintenance, and other costs associated with managing users, applications, and equipment on the network.

In the LAN, bandwidth is used to eliminate complexities in the network environment since it is inexpensive relative to engineering. On the other hand, a major requirement of WAN infrastructure is to minimize the bandwidth required across the wide area. Bandwidth effectiveness is simply having enough bandwidth to support optimal application performance and doing so at the lowest possible cost. Increased bandwidth effectiveness is achieved through high-performance switching in the LAN and through consolidation of multiple traffic types across the WAN.

SUMMARY

The computer industry, through reliance on ingenuity, entrepreneurship, and most importantly, competition, has advanced rapidly to meet consumers' demands. The government has judiciously intervened only in limited circumstances. The Telecommunications Act of 1996 is an important piece of legislation that created the framework for competitive delivery of telecommunications services to the last mile. The Act is potentially one of the most far-reaching of any U.S. law adopted during the 1990s. Consumers have started to realize the benefits of this new legislation, which facilitates a communications infrastructure in which market forces, not regulations, drive technology, service, and price.

The telecommunications industry is currently undergoing a massive transformation because of larger industry forces, namely deregulation and globalization, resulting in increased competition and rapidly dropping rates. Mergers, acquisitions, and partnerships continue in the telecommunications industry, with most companies geared to acquire a piece of technology that will provide any advantage in new product development and time to market. To compete in this new market, carriers need lower cost, lower maintenance, and more efficient networks. From the users' viewpoint, the application that has realized enormous growth in recent years is electronic commerce, which increases the speed, accuracy, and efficiency of business and personal transactions.

Case Study

Physical layer network management systems can offer return on investment benefits that reduce the total cost of network ownership. By bringing real-time visibility to the network, physical layer network management systems encourage proactive maintenance. This technology is non-intrusive and in parallel to the data transfer path, ensuring that any faults in the management system will not affect the data transfer continuity. A large enterprise installed a physical layer network management system that resulted in decreased downtime, reduced network maintenance costs, increased asset utilization and improved planning capabilities. By reducing network cost of ownership and increasing physical layer performance, the return on investment of this system was less than a year.

Question

1. Specify a problem with regards to your campus network. Identify different solutions and conduct:
    - A. ROI for each alternative
    - B. Cost/benefit analysis for each alternative
2. Recommend the best solution to resolve the problem.

REVIEW QUESTIONS

1. Define each of the following terms:
   - A. Reciprocal Compensation
   - B. Unbundled Network Elements
   - C. Access Charge
   - D. Local Number Portability
   - E. Location Routing Number
   - F. Enhanced Services
   - G. Intellectual Property
   - H. Copyright
   - I. Trademark
   - J. Patent
   - K. Service Level Agreement
   - L. Minimum Annual Commitment
   - M. Usage-based model
   - N. Real-time monitoring
   - O. Cost
   - P. Hard Benefits
   - Q. Soft Benefits
   - R. Life Cycle
   - S. Net Value
2. Discuss the significance of the Telecommunications Act of 1996.
3. Analyze the implications of sharing Unbundled Network Elements with competitors.
4. Develop a scenario to depict the application of reciprocal compensation.
5. Evaluate the benefits of Local Number Portability from two viewpoints: customer and carrier.
6. Describe the implications of regulation on electronic documents.
7. Identify the need for intellectual property and copyright laws.
8. Analyze the effects of lack of global intellectual property and copyright laws.
9. Discuss the impact of telecom policy on business.
10. Describe some of the impediments faced by U.S. vendors of telecom equipment in a global market economy, giving examples.
11. Evaluate the components of a Service Level Agreement.
12. Identify three clauses that are part of every SLA, and three clauses that should be included to protect against the unpredictable.
13. Explain the term Total Cost of Ownership and its application in business.

ACRONYMS

B825	Bipolar with 8 Zero Substitution
2B1Q	2 Binary 1 Quaternary
2G	Second Generation
3G	Third Generation
4-PSK	Quadrature Phase Shift Keying
A2LA	American Association of Laboratory Accreditation
AA	Automated Attendant
AAL	Adaptation Layer
ABR	Available Bit Rate
ACD	Automatic Call Distribution
ACE	Access Control Entry
ACK	Acknowledgment
ACLs	Access Control Lists
ACP	Action Control Point
ACTS	Advanced Communications Technology Satellite
ADSL	Asymmetric DSL
AIEE	American Institute of Electrical Engineers
AIN	Advanced Intelligent Network
AM	Amplitude Modulation
AMI	Bipolar Alternate Mark Inversions
AMPS	Advanced Mobile Phone Service
ANI	Automatic Number Identification
ANSI	Anerican National Standards Institute
AO/DI	Always On/Dynamic
APD	Avalanche Photodiode
APPN	Advanced Peer-to-Peer Networking
API	Application Programming Interface
APS	Automatic Protection Switching
AQR	Assured Quality Routing
ARP	Address Resolution Protocol
ASCII	American Standard Code for Information Interchange
AT&T	American Telephone and Telegraph
ATM	Asynchronous Transfer Mode
ATM	Cell-Relay or Asynchronous Transfer Mode
ATM-PON	ATM-Passive Optical Network
AWG	American Wire Gauge
BBN	Bolt, Beranek, and Newman
BCC	Block Check Character
BER	Bit Error Rate
BHCA	Busy Hour Call Attempts
BICI	Broadband InterCarrier Interface

B-ISDN	Broadband ISDN
BISYNC	Binary Synchronous Communications
BITNET	Because It's Time Network
bps	bits per second
BPSK	Binary Phase Shift Keying
BRI	Basic Rate Interface
BSC	Binary Synchronous Communications
BSI	British Standards Institution
BW	Bandwidth
CA	Certification Authority
CAP	Carrierless Amplitude Modulation/Phase Modulation
CAT 3	Category 3
CAT 5	Category 5
CATV	Cable TV
CBR	Constant Bit Rate
CCIR	International Radio Consultative Committee
CCIS	Common Channel Interoffice Signaling
CCITT	International Telephone and Telegraph Consultative Committee
CCS	Centi-Call Seconds
CDMA	Code Division Multiple Access
CDPD	Cellular Digital Packet Data
CE	Conformitè Europèenne
CENTREX	Central Office Exchange Service
CEPT	European Conference of Postal and Telecommunications Administrations
CERT	Computer Emergency Response Team
CIP	Classical IP
CIR	Committed Information Rate
CITEL	Inter-American Telecommunication Commission
CLASS	Custom Local Area Signaling Services
CLECs	Competitive Local Exchange Carriers
CLNP	Connectionless Network Protocol
CM	Cable Modem
CM	Configuration Management
CMTS	Cable Modem Termination System
CORBA	Common Object Request Broker Architecture
CoS	Class of Service
CPE	Customer Premise Equipment
CPU	Central Processing Unit
CRC	Cyclic Redundancy Check
CREN	Corporation for Research and Education Networking
CSA	Canadian Standards Association
CSMA/CD	Carrier Sense Multiple Access with Collision Detection
CSNET	Computer Science Network
CSU	Channel Service Unit
DARPA	Defense Advanced Research Projects Agency
DAS	Direct Attached Storage
DCCH	Digital Control Channel
DCE	Data Communications Equipment
DEC	Digital Equipment Corporation
DEMUX	Demultiplexer
DES	Data Encryption Standard
DFS	Distributed File System
DHCP	Dynamic Host Configuration Protocol
DID	Direct Inward Dial
DiffServ	Differential Services

DIX	DEC-Intel-Xerox
DMD	Differential Mode Delay
DMT	Discrete Multitone Technology
DNA	Digital Network Architecture
DNS	Domain Name System
DoD	Department of Defense
DOD	Direct Outward Dial
DQDB	Distributed Queue on a Dual Bus
DSI	Digital Speech Interpolation
DSL	Digital Subscriber Line
DSSS	Direct Sequence Spread Spectrum
DSU	Data Service Unit
DTE	Data Terminal Equipment
DTM	Dynamic Synchronous Tranfer Mode
DTMF	Dual Tone Multiple Frequency
DVC	Desktop Videoconferencing
E&M	recEive and transMit
E/M	Electromagnetic
EBCDIC	Extended Binary Coded Decimal Interchange Code
ECTRA	European Committee for Regulatory Telecommunications Affairs
ELFEXT	Equal-Level Far-End Crosstalk
EMC	Electromagnetic Compatibility
EMI	Electromagnetic Interference
EPABX	Electronic Private Automatic Branch Exchange
ERC	European Radio-communication Committee
ERP	Enterprise Resource Planning
ES	Enhanced Services
E-SMR	Enhanced Switched Mobile Radio
ETSI	European Telecommunications Standards Institute
ETX	Special Code
FAT	File Allocation Table
FC	Fiber Channel
FCC	Federal Communications Commission
FCS	Frame Check Sequence
FCS	Frame Check Sequence
FDDI	Fiber Distributed Data Interface
FDM	Frequency Division Multiplexing
FDMA	Frequency Division Multiple Access
FEP	Front-End Processor
FEXT	Far-End Crosstalk
FHSS	Frequency Hopping Spread Spectrum
FIRST	Forum of Incident Response and Security Teams
FM	Frequency Modulation
FPLMTS	Future Public Land Mobile Telecommunication System
FRADs	Frame Relay Access Devices
FSAN	Full-Service Access Network
FSK	Frequency Shift Keying
FTP	File Transfer Protocol
FTP	Foil Twisted Pair
FX	Foreign Exchange
GFR	Guaranteed Frame Rate
GoS	Grade of Service
GPO	Geosynchronous or Geostationary Orbit
GPS	Global Positioning System
GSM	Global System for Mobile Communications

GUI	Graphical User Interface
HDLC	High-level Data Link Control
HDSL	High-bit Rate Digital Subscriber Line
HFC	Hybrid Fiber/Coaxial
HIPAA	Health Insurance Portability and Accountability Act
HPFS	High Performance File System
HTTP	HyperText Transmission Protocol
I2	Internet2
IANA	Internet Assigned Numbers Authority
IBM	International Business Machines
IC	Integrated Circuit
ICANN	Internet Corporation for Assigned Names and Numbers
ICLID	Incoming Caller Line ID
ID	Identification
IDS	Intrusion Detection System
IEC	International Electrotechnical Commission
IECs or IXCs	Inter-Exchange Carriers
IETF	Internet Engineering Task Force
IITF	Information Infrastructure Task Force
ILECs	Incumbent Local Exchange Carriers
IMAP	Internet Message Access Protocol
IMT-2000	International Mobile Telecommunication 2000
IMTS	Improved Mobile Telephone System
IN	Intelligent Network
INTELSAT	International Telecommunication Satellite Organization
IP	Internet Protocol
IPSec	Internet Protocol Security
IPv4	IP version 4
IPv6	IP version 6
IPX	Internetwork Packet Exchange
IR	Infrared
IRB	Integrated Routing and Bridging
IrDA	Infrared Data Association
IRE	Institute of Radio Engineers
iSCSI	Internet Small Computer Systems Interface
ISDN	Integrated Services Digital Network
ISM	Industrial, Scientific and Medical
ISO	International Standards Organization
ISP	Internet Service Provider
IT	Information Technology
ITU	International Telecommunication Union
IVR	Interactive Voice Response
IXCs	Inter Exchange Carriers
JPEG	Joint Photographic Experts Group
L2TP	Layer 2 Transport Protocol
LANE	LAN Emulation
LATA	Local Access and Transport Area
LCD	Liquid Crystal Display
LDAP	Lightweight Directory Access Protocol
LEC	Local Exchange Carrier
LEDs	Light Emitting Diodes
LEO	Low Earth Orbit
LLC	Logical Link Control
LMDS	Local Multipoint Distribution System
LNA	Low-Noise Amplifier
LNP	Local Number Portability

LRC Longitudinal Redundancy Check
LRN Location Routing Number
LSMS Local Service Management System
MTP Message Transfer Part
MAC Media Access Control
MAC Medium Access Control
MAC Minimum Annual Commitment
MAU Multistation Access Unit
MCI Microwave Communications, Inc.
MDC Main Distribution Center
MEO Medium Earth Orbit
MF Multi-Frequency
MGCP Media Gateway Control Protocol
MIME Multipurpose Internet Mail Extensions
MMDS Multichannel Multipoint Distribution System
MPEG Motion Picture Experts Group
MPLS MultiProtocol Label Switching
MPOA Multiprotocol
MRA Mutual Recognition Agreement
MSC Mobile Switching Center
MTS Mobile Telephone System
MUO Multi-User Outlet
MUX Multiplexer
NA Numerical Aperture
NAP Network Access Point
NAS Network Attached Storage
NASA National Aeronautics and Space Administration
NCP Network Control Point
NCP Network Control Protocol
NDIS Network Distributed Interface Standard
NDS Novell Directory Services
NEC National Electrical Code
NETBEUI NetBIOS Extended User Interface
NEXT Near-End Crosstalk
NF Noise Figure
NIC Network Interface Card
NIST National Institute of Science and Technology
NMT Nordic Mobile Telephone
NOAA National Oceanic and Atmospheric Administration
NOC Network Operations Center
NOS Network Operating System
NPAC Number Portability Administration Center
NR Noise Ratio
NREN National Research and Education Network
NRZ-I Non-Return to Zero Inverted
NRZ-L Non-Return to Zero Level
NSF National Science Foundation
NSFNET NSF Network
NTFS NT File System
OAM Operations, Administration, and Maintenance
OC-1 Optical Carrier, Level 1
ODI Open Data Interface
OLT Optical Line Terminal
ONU Optical Network Unit
OSI Open Systems Interconnect
OSI Open Systems Interconnection
OSP Open Settlement Protocol
OSS Operations Support System
OUI Organizationally Unique Identifier

PABX Private Automatic Branch Exchange
PAD Packet Assembler/Disassembler
PAM Pulse Amplitude Modulation
PBX Private Branch Exchange
PC Personal Computer
PCL Packet Loss Concealment
PCM Pulse Code Modulation
PCMCIA Personal Computer Memory Card Industry Association
PCN Personal Communications Network
PCS Personal Communications System
PDA Personal Digital Assistant
PDU Protocol Data Unit
PHY Physical Layer Protocol
PIC Primary Interexchange Carrier
PIN Personal Identification Number
PIN Positive-Intrinsic-Negative [Photodiode]
PINT IETF PSTN/Internet Inter-Networking Working Group
PKI Public Key Infrastructure
PLC Packet Loss Concealment
PM Phase Modulation
PMD Physical Media Dependent
PNC Personal Communications Networks
PNNI Public/Private Network-to-Network or Node-to-Node Interface
POH Path Overhead
POI Point of Interface
POP Point of Presence
PoS Packet over SONET
POTS Plain Old Telephone Services
PPM Pulse Position Modulation
PPP Point-to-Point Protocol
PPS Pulses Per Second
PPTP Point-to-Point Tunneling Protocol
PRI Primary Rate Interface
PSE Packet Switching Exchange
PSK Phase Shift Keying
PSTN Public Switched Telephone Network
PUCs Public Utility Commissions
PVC Permanent Virtual Circuit
PVC Polyvinyl-Chloride
PWM Pulse Width Modulation
QAM Quadrature Amplitude Modulation
QoS Quality of Service
QPSK Quaternary Phase Shift Keying
RADSL Adaptive DSL
RAID Redundant Array of Independent Disks
RAID Redundant Array of Inexpensive Disks
RAS Remote Access Server
RBOCs Regional Bell Operating Companies
REN Ringer Equivalence Number
RF Radio Frequency
RL Return Loss
RLE Run-Length Encoding
RLOR Receive Loudness Objective Rating
RMOA Real-time Multimedia Over ATM
ROI Return on Investment
ROM Read Only Memory
RSVP Resource Reservation Protocol
RTP Real-time Transport Protocol
SAN Storage Area Network

SCCP	Signaling Connection Control Part
SCP	Service Control Point
SCSI	Small Computer System Interface
SDH	Synchronous Digital Hierarchy
SDLC	Synchronous Data Link Control
SDSL	Symmetric Digital Subscriber Line
SEC	Securities Exchange Commission
SF	Single Frequency
SIP	Session Initiation Protocol
SIR	Sustained Information Rate
SLA	Service Level Agreement
SLIP	Serial Line Internet Protocol
SMDI	Station Message Detail Interface
SMDS	Switched MultiMegabit Data Service
SMF	Single Mode Fibers
SMTP	Simple Mail Transfer Protocol
SNA	Systems Network Architecture
SNMP	Simple Network Management Protocol
SNR	Signal-to-Noise Ratio
SONET	Synchronous Optical Network
SPE	Synchronous Payload Envelope
SS7	Signaling System Seven
SSP	Storage Service Provider
SSP	Service Switching Point
STAT MUX	Statistical Multiplexer
STDM	Statistical TDM
STM-1	Synchronous Transport Module, Level 1
STP	Screened Twisted Pair
STP	Shielded Twisted Pair
STP	Signal Transfer Point
STS-1	Synchronous Transport Signal, Level 1
STTs	Satellite Tie Trunks
SVC	Switched Virtual Circuits
SYN	Synchronization
Syncom	Synchronous Communications
TACS	Total Access Communications System
TASI	Time Assignment Speech Interpolation
TC	Telecommunications Closet
TCF	Technical Construction File
TCO	Total Cost of Ownership
TCP/IP	Transmission Control Protocol/ Internet Protocol
TDM	Time Division Multiplexing
TDMA	Time Division Multiple Access
TDR	Time Domain Reflectometer
TELNET	Terminal Emulation
TLOR	Transmit Loudness Objective Rating
TMN	Telecommunications Management Network
TR	Telecommunications Room
TSB	Telecommunication Standardization Bureau
UART	Universal Asynchronous Receiver Transmitter
UBR	Unspecified Bit Rate
UCAID	University Corporation for Advanced Internet Development
UDP	User Datagram Protocol
UL	Underwriters Laboratories, Inc.
UMTS	Universal Mobile Telecommunications System
UNEs	Unbundled Network Elements
UNI	User-to-Network Interface

UPS	Uninterruptible Power Supply
USENET	User's Network
USNC	U.S. National Committee
USOC	Uniform Service Ordering Code
UTP	Unshielded Twisted Pair
VBR-nrt	VBR non-real-time
VBR-rt	Variable Bit Rate real-time
VLAN	Virtual LAN
VoFR	Voice over Frame Relay
VoIP	Voice over IP
VPIM	Voice Profile For Internet Messaging
VPLS	Virtual Private LAN Service
VPN	Virtual Private Network
VSAT	Very Small Aperture Terminal
WAP	Wireless Application Protocol
W-CDMA	Wideband Code Division Multiple Access
WDM	Wavelength Division Multiplexing
WINS	Windows Internet Naming Service
WLANs	Wireless LANs
WLL	Wireless Local Loop
WML	Wireless Markup Language
WWW	World Wide Web
XOR	Exclusive-OR

GLOSSARY

10BaseT A popular Ethernet standard in a LAN environment where the *10* refers to the transmission speed, the *Base* refers to baseband signaling, and the *T* stands for twisted-pair wire.

10Base2 Thin coaxial cable where the *10* refers to the transmission speed, the *Base* refers to baseband signaling, and the *2* refers to the coaxial cable maximum segment length in meters (200 m). Also referred to as **thinnet.**

10Base5 Thick coaxial cable where the *10* refers to the transmission speed, the *Base* refers to baseband signaling and the *5* refers to the coaxial cable maximum segment length in meters (500 m). Also referred to as **thicknet.**

100BaseT Commonly referred to as Fast Ethernet where the *100* refers to the transmission speed, and the *T* stands for twisted-pair wire.

1000BaseT Also known as Gigabit Ethernet where the *1000* refers to the transmission speed, and the *T* stands for twisted-pair wire.

absorption A material property that is the result of atomic resonance in the glass structure of a fiber-optic cable.

acceptance angle The greatest possible angle at which light can be launched into the core of a fiber and still be guided through total internal reflection.

acceptance cone The range of the angle of incidence at which light can be launched into the core of a fiber and still be guided through total internal reflection.

access charge The cost incurred for space rented in the local exchange switch by each IXC.

access control The process of assuring that only authorized users have access to a particular system and/or specific resources through the use of time-sharing systems.

access point A point that bridges a wireless LAN to an existing wired LAN. It serves as a connection point for users of wireless devices to connect to wired LANs.

Action Control Point (ACP) See **Service Switching Point (SSP)**.

active monitoring An aspect of firewall design where the firewall notifies the network manager of attempts to bypass security.

active threat A network security threat that involves some modification of the data stream or the creation of a false stream by an unauthorized person.

Address Resolution Protocol (ARP) A mechanism that enables IP data to reach a LAN destination by providing address conversion.

addressing signals The destination information such as the actual telephone number being called.

Advanced Mobile Phone Service (AMPS) A high-capacity analog land mobile communication system that employs a cellular concept.

advanced telecommunications services See **enhanced services.**

aliasing A low sampling rate produces a form of distortion in which the reconstructed original signal results in a lower-frequency signal.

American National Standards Institute (ANSI) A private, nonprofit, voluntary standards organization.

analog Continuously varying signals.

analog-to-digital conversion A three-stage process for converting an analog signal into a train of pulses.

application-level gateways A device that provides a mechanism for filtering traffic for various applications.

ASCII character code A general-purpose seven-bit binary data code developed by the ANSI, originally defined for use by the U.S. government.

Assured Quality Routing (AQR) A set of processes and procedures that marries packet and circuit switching to automatically reroute IP calls to the PSTN when parameters on a given call exceed developed acceptable ranges.

asynchronous A type of transmission that uses framing bits to signal the beginning and end of each data character.

Asynchronous Transfer Mode (ATM) A cell-based transport mechanism for the transport of a broad range of information—voice, data, and video communication.

attenuation The loss of power that occurs in a signal as it travels down a cable.

authentication Methods used to verify both the information sent and the sender.

authorization The purpose of authorization is to ensure that only authorized users have access to a particular system and/or specific resources and that modification of a particular portion of data is limited to authorized individuals and programs.

Automatic Number Identification (ANI) The capability of a local switching office to automatically identify the calling station.

Automatic Protection Switching (APS) The built-in fault tolerance for SONET.

Avalanche Photodiode (APD) The more expensive, but more sensitive and accurate of the two types of widely-deployed photodiodes. APD photodiodes provide gain through an amplification process in which one photon releases many electrons.

Baby Bells Regional Bell Operating Companies (RBOCs) that were sole providers of practically all the local exchange telephone services.

backbone A larger network transmission line that carries data gathered from smaller lines that interconnect with the network.

backbone wiring The connection between the communication closet and the equipment room within a building, and the connection between buildings.

backoff The process that a network interface uses to attempt to transmit data without a collision.

bandwidth It is the rate of information transfer. It is also a measure of the transmission capacity of a communications medium.

base station Part of a wireless network, consisting of a transmitter, receiver, controller, and antenna system.

baseband The original frequency range of a signal before it is modulated into a higher and more efficient frequency range.

Basic Rate Interface (BRI) A categories of ISDN User-to-Network Interface with two channels that can carry voice, data, or video, and one channel that provides intelligent line management.

bend radius The radius of the loop when there are bends or angles in a cable route.

bending losses Losses that occur in fiber-optic cable because of improper installation.

Bit Error Rate (BER) The number of bits in error expressed as a portion of transmitted bits.

bit-interleaved TDM A single data bit from an I/O port is output to the aggregate, followed by a data bit from another I/O port, and so on with the process repeating itself.

Block Check Character (BCC) Produced by a Longitudinal Redundancy Check (LRC) to provide extra error-detection capabilities for a block of data.

blocking networks Older electromechanical telephone networks that contain fewer paths than terminations.

Bluetooth A uniting technology that allows any sort of electronic equipment, from comptuers to cell phones, to make its own connection without any wires.

bridge This device operates at Layer 2, and is used to interconnect LANs of the same type.

broadband The simultaneous transmission of multiple channels over a single line.

broadcast A type of network where all stations see all frames, regardless of whether they represent an intended destination.

buffered mode A process, which downloads an entire file to a client and plays the audio/video in the client environment rather than across the network.

bursty traffic Peak periods of data transmission followed by periods in which no transmission takes place.

bus Used in two different contexts: in computers, the data path on the motherboard that interconnects the microprocessor with attachments to the motherboard in expansion slots; and in a network, a topology in which all devices are directly attached to a line and all signals pass through each of the devices.

business continuance Business continuance describes the processes and procedures an organization puts in place to ensure that essential functions can continue during and after a disaster.

busy forwarding A telephone feature used to forward a call to another extension if the station is busy.

Busy Hour Call Attempts (BHCA) Describes the number of calls a telephone system can handle during the peak hour of the day.

byte The shortest string of bits that a computer will manipulate as a unit.

byte-interleaved TDM A data byte from an I/O port is output to the aggregate, followed by a byte from another I/O port, and so on with the process repeating itself.

Cable Modems (CMs) A high-speed data connection to the Internet through the coaxial cables that bring TV channels.

call Any demand to set up a connection on a telephone system.

call alert signals Establish the process for incoming and outgoing calls.

call attempts All accesses that require attention by the central processor including telephone call originations and features such as call pickup, call transfer, and call waiting.

call forwarding A telephone feature which is used to specify the extension to which a call will forward if the extension rings with no answer for longer than a specified period of time.

call hold A telephone feature that, when activated, will disengage both the transmitter and the receiver for a period of time.

call progress signals Establish the process for incoming and outgoing calls.

call transfer A telephone feature that is used to transfer a call to another line.

carrier sensing This protocol prevents transmissions that otherwise would result in collisions.

Category 3 A twisted-pair cable used in implementing a 10Base-T interface.

Category 5 Cabling consisting of four pairs wrapped in a thermal plastic insulator that are twisted around one another and encased in a flame-retardant polymer.

cell Used in two different contexts: A span of coverage in a wireless network, and a block of data in data transmission.

cell switching Combines some aspects of both circuit- and packet-switching; fixed-length cells are transmitted at a constant rate to produce networks with low latency and high throughput.

Cellular Digital Packet Data (CDPD) In a wireless network, it allows for a packet of information to be transmitted in between voice telephone calls.

cellular network Any mobile communications network with a series of overlapping hexagonal cells in a honeycomb pattern.

center frequency A carrier frequency.

central office See **local exchange.**

centralized cabling The highest functionality networking components reside in the main distribution center interconnected to intermediate distribution centers or to a Telecommunications Closet (TC).

centralized computing A central computer called a mainframe does all the processing associated with most tasks.

Centrex A brand name for a set of services offered by local telephone companies on business telephone lines.

Certification Authority (CA) A trusted third party responsible for verifying the digital authentication keys and certificates issued to and used by an organization in relation to its documents.

Channel Service Unit (CSU)/Digital Service Unit (DSU) Devices which come in either standalone units or combination CSU/DSUs. At least one of these devices is required for any digital line termination, but combination CSU/DSU is most common.

character A specific symbol that cannot be subdivided into anything smaller that retains its identity.

character code A common pattern for coding transmitted information.

circuit switching Connection-oriented network in which a communications channel remains dedicated to two users regardless of whether they actually need the full channel capacity for the entire time.

cladding The outer layer surrounding the core of an optical fiber that contains the light within the core.

classless addressing A mechanism in which an organization is assigned a number of bits to use as the local part of its address corresponding to its needs.

client/server In a distributed network, the client is any network device or process that makes requests to use server resources and services.

coaxial cable A two-conductor cable in which one conductor forms an electromagnetic shield around the other.

Code Division Multiple Access (CDMA) A wireless digital technology that allows multiple users to share a single frequency by encrypting each signal with a different code.

codec A coder-decoder device that accomplishes analog-to-digital conversion.

coherent Monochromatic light.

collision detection This protocol immediately stops the transmission of a frame upon detecting a collision.

Committed Information Rate (CIR) A guaranteed rate of throughput when using Frame Relay.

common channel signaling The most prominent out-of-band signaling system.

common control Switches on a telephone system that automatically route the call to the correct destination along the most economical route.

computer and network security The protection of network-connected resources against unauthorized disclosure, modification, utilization, restriction, incapacitation, or destruction.

computer A machine that can be programmed to manipulate characters.

conduit A pre-installed plastic or metal pipe that runs between or through buildings to ease the installation of cable.

conference call A telephone feature which enables three or more people to converse simultaneously on the telephone.

connection-oriented A connection is established prior to actual data transfer occurring.

connectionless Transmission occurs on a best-effort basis with an acknowledgment flowing back only after transmission is initiated.

connectors Nonpermanent joints used to connect optical fibers to transmitters and receivers or panels and mounts.

contention Two network interfaces try to transmit data at the same time.

conventional TDM A type of Time Division Multiplexing (TDM) scheme which employs either a bit-interleaved or byte-interleaved multiplexing.

converged data/voice networks The application of voice digitization and compression schemes through a variety of hardware and software products to enable voice to be transported on public and private networks.

convergence The merging of computers and communications, or packet-switched data and circuit-switched voice.

copyright Ensures authors the right to protect their work and to benefit from new inventions by giving them exclusive rights to such works.

core The inner layer of an optical fiber through which light travels.

cost Both long-term investments and support, coupled with the returns on investment.

couplers Used to split information in many directions.

critical angle The angle of incidence that occurs when the angle of refraction is 90° and light does not enter the second medium but is reflected along the interface.

crossover A cable in which the transmit and receive pairs are crossed. These cables are designed to connect two ethernet devices together without the use of a hub or two hubs via standard Ethernet ports in the hubs.

crosstalk Bleedover between adjacent channels.

cryogenics Artificially cooling an amplifier in order to reduce noise.

Cyclic Redundancy Check (CRC) A widely-used, reliable, and efficient scheme that detects multiple errors within any length of message, and requires far less transmission overhead as compared to other error-detection methods.

data communications Transmission of digital information.

Data Communications Equipment Devices that move information by implementing communications facilities.

data compression A technique applied to data prior to its transmission used to reduce the number of bits that must pass over the communications medium in order to reduce transmission time.

Data Encryption Standard (DES) The U.S. government's approved private-key encryption specification.

data network A collection of devices that can store and manipulate electronic data, interconnected in such a way that network users can store, retrieve, and share information with each other regardless of their physical location.

data rate Raw radio transmission speed in a wireless network.

Data Terminal Equipment The devices in a data network that transmit and receive data.

datagram One of the conceptual models of the network layer where packets are delivered individually so that they can take a different route and arrive at the destination in no particular order.

dedicated server A server dedicated to a single application.

demodulation A process that detects and restores the original signal, an inverse of modulation. See **modulation.**

Digital Control Channel (DCCH) In TDMA based IS-136, this channel handles signaling for administrative work and providing services.

dial tone An audible "ready" signal provided by a local exchange.

Direct Inward Dial (DID) lines A special type of telephone line used for incoming calls only.

directory services They allow users to access network services without having to know the network address.

Differential Mode Delay (DMD) Different modes of multimode fiber propagating at different speeds. Also called **modal dispersion.**

Differential Services (DiffServ) A protocol which establishes mechanisms to prioritize certain data streams.

digital Signals consisting of discrete quantities, most commonly binary.

Digital Subscriber Line (DSL) A platform for delivering broadband services to homes and small businesses over the existing copper infrastructure for telephone connections.

Direct Area Storage (DAS) Parallel SCSI-storage devices attached to each server in an enterprise.

direct control Older-type switches on telephone systems which lack alternate routing and digit translation capabilities.

Direct Sequence Spread Spectrum (DSSS) A spread spectrum technique which resists interference by mixing in a series of pseudo-random bits with the actual data.

dispersion Spreading of the light as it travels down the optical fiber.

distributed computing Spreads users across several smaller systems and limits the disruption caused if a system goes down.

Distributed Queue on a Dual Bus (DQDB) Distributes the network service requests of users into queues to handle the transfer of information on unidirectional buses.

Direct Outward Dial (DOD) lines Telephone lines that allow the extension user to dial directly outside the system.

dotted decimal notation A technique used to express IP addresses via the use of four decimal numbers separated from one another by decimal points.

Dual Tone Multiple Frequency (DTMF) The most commonly used method for address signaling for an outgoing telephone call.

echo A reflection that occurs when an electrical signal encounters an impedance irregularity. Also called **return loss.**

echo cancellers Used to control echo in digital circuits.

echo suppressors Used to control echo in analog circuits.

echo cancellation Process by which modem technology allows for full-duplex transmission to occur over only two wires (one set).

E&M signaling A signaling scheme used by trunks to communicate their status to the attached local exchange office equipment.

efficiency of transmission The ratio of the actual message bits to the total number of bits.

Electromagnetic Interference (EMI) A result of electromagnetic (E/M) emissions.

electromagnetic (E/M) spectrum All oscillating signals from 30 Hz to several hundred GHz.

electronic commerce Any purchasing or selling through an electronic communications medium.

encryption Scrambling of data by use of a mathematical algorithm so that the scrambled information is undecipherable and meaningless.

end office See **local exchange.**

enhanced services High-speed, switched, broadband telecommunications capabilities and services such as voice mail, e-mail, data processing, and gateways to online databases. Also called **advanced telecommunications services.**

equal access All IXCs have connections that are identical to AT&T's connection to the local telephone network.

European Conference of Postal and Telecommunications Administrations (CEPT) A European organization focusing on cooperation on commercial, operational, regulatory and technical standardization issues.

European Telecommunications Standards Institute (ETSI) A non-profit organization whose mission is to determine and produce the telecommunications standards in Europe.

even parity checking A type of parity checking to detect errors.

exchange The process of changing or exchanging wire pairs.

excess noise See **flicker noise**.

external assessment An external assessment may require outsourcing security services to perform penetration tests and an audit of the company's network from the perspective of an attacker.

external noise Noise that is a property of the channel.

Extranet An Internet that allows controlled access by authenticated outside parties to enable collaborative business applications across multiple organizations.

Federal Communications Commission (FCC) An independent United States government agency charged with regulating interstate and international communications by radio, television, wire, satellite, and cable.

feeder Cable that connects the main trunk line to the drop cable.

fiber-optic cable A transmission media designed to transmit digital signals in the form of pulses of light.

fiber zone A combination of collapsed backbone and a centralized cabling scheme.

figure of merit The ratio of a satellite's receiver antenna gain to the system noise temperature.

filter A tuned device that passes certain desirable frequencies and rejects the other.

firewall Hardware and/or software that acts as a security device and allows limited access out of and into one's network from the Internet.

fixed wireless Fixed wireless broadband gets its name from the antennas that need to be "fixed" high above the ground so they can provide broadband access to homes or businesses within a 35-mile radius.

flicker noise Excess noise that declines with increasing frequency and is rarely a problem in communication circuits. Also called **pink noise** or **excess noise**.

forward error correction The process by which an error is corrected by the receiving device in a network.

four-wire terminating sets Devices that convert the transmission circuit from four-wire to two-wire. Also called **hybrids**.

Fourier analysis	See **spectrum analysis.**
Fourier synthesis	A process of adding together the sine waves to recreate the complex waveform. The inverse of Fourier analysis.
Fourier theorem	States that any periodic function or waveform can be expressed as the sum of sine waves with frequencies at integer or harmonic multiples of the fundamental frequency of the waveform.
frame	A block of data.
Frame Relay	A fast packet-switching technique that provides a cost-efficient means of connecting an organization's multiple LANs.
Frame Relay Access Devices (FRADs)	Processes frames by traffic priority and maximum elapsed time in queue.
Frame Relay segmentation	Segments all voice and data transmission to a fixed frame or cell size.
frequency deviation	The process whereby the center frequency is made to deviate by an audio baseband signal.
Frequency Division Multiple Access (FDMA)	The division of the frequency band into channels allocated for wireless cellular communication.
frequency domain	A method of representing signals, where amplitude or power is shown on one axis and frequency is displayed on the other.
Frequency Hopping Spread Spectrum (FHSS)	A spread spectrum technique which resists interference by moving the signal rapidly from frequency to frequency in a pseudo-random way.
Frequency Shift Keying (FSK)	A popular implementation of Frequency Modulation (FM) for data applications where a carrier is switched between two frequencies.
front-end processor (FEP)	Handles all the routine communications procedures for the host computer.
full-duplex	Simultaneous two-way transmission.
Foreign Exchange (FX) Circuit	A leased-line service that allows a customer to draw dial tone from a remote local exchange in the LEC's service area.
gateway	A device which operates at Layer 7, or in other words, spans all seven layers of the OSI model.
geostationary orbit (GEO) satellite	A satellite that circles the equator without inclination and whose rotational period matches the Earth's own.
glare	Occurs when both the local end and the remote end of a telephone connection attempt to access the circuit at the same time.
Global System for Mobile Communications (GSM)	A second-generation global wireless system.

Grade of Service (GoS) The probability of blockage determined by the ratio of the number of telephone calls that cannot be completed or lost calls, to the total number of attempted calls during the busiest hours of the day.

Ground Start A DC line signaling method that is usually incorporated only on trunks and PBXs.

group A combination of twelve voice channels.

half-duplex Two-way communications, but in only one direction at a time.

handshaking The exchange of predetermined signals when a connection is established between two dataset devices.

hard benefits Benefits that can be easily assigned objective and predictable dollar figures.

Hartley's law The amount of information that can be transmitted in a given time is directly proportional to bandwidth.

headend A single point of failure in a network.

high-density or dense WDM The ability to optically multiplex each wavelength.

historical monitoring A type of monitoring used by a SLA that allows the parties to spot trends that may lead to future problems as well as verify service levels over time.

horizontal wiring The connection between the work area and the termination in the telecommunication closet.

host computer The processor that performs a variety of applications.

hub A wiring concentrator that is the focal point of adds, moves, or changes to the network.

Huffman code A lossless data compression technique in which the most common data is noted with the least number of bits, while the least common data is noted with the most number of bits, using a variable-length code.

hybrid A combination of two or more basic network topologies.

impedance Opposition to alternating current as a result of resistance, capacitance, and inductance in a component.

impedance matching The maximum transfer of power from an input to an output when the impedance of the input equals that of the output.

Improved Mobile Telephone System (IMTS) A full-duplex automatic switching system.

in-band signaling A signaling system that carries call setup, charging, and supervision signals over the same circuit that carries information.

index of refraction See **refractive index.**

interface Provides handshaking from one layer to the next in a network.

internal assessment Various audits conducted throughout the internal network, including all devices and network applications.

internetworking The equipment and technologies involved in connecting either LANs to LANs, WANs to WANs, or LANs to WANs.

Institute of Electrical and Electronics Engineers (IEEE) A worldwide technical, professional, and educational organization that promotes networking, information sharing and leadership through its technical publishing, conferences and consensus-based standards activities.

Integrated Services Digital Network (ISDN) Provides end-to-end digital connectivity, and distinguishes itself from other services by guaranteeing bandwidth and allowing users to simultaneously utilize voice and data applications over a single line.

intellectual property The discipline that deals with issues in copyright, trademark, and patent law.

Inter Exchange Carrier (IXC) Carrier that handles long-distance telephone calls.

internal noise Noise that originates within the communication equipment.

International Organization for Standardization (ISO) A non-governmental organization established in 1947 that promotes the development of standardization and related activities in the world with a view to facilitating the international exchange of goods and services, and to developing cooperation in the spheres of intellectual, scientific, technological and economic activity.

International Telecommunication Union (ITU) An international organization within which governments and the private sector coordinate global telecommunication networks and services.

Internet Message Access Protocol (IMAP) A protocol that permits true mail-to-client/server interaction by allowing the mail to stay at the server and be accessed and managed by the clients.

Internet Protocol (IP) A Layer 3 protocol which segments and packets data for transmission, and then places a header for delivery.

Intranet A private network that uses TCP/IP, HTTP, and other Internet protocols, but is contained within an enterprise.

Intrusion Detection System (IDS) Intrusion Detection Systems are an important security measure because they let a network manager be aware of attempts to bypass security.

inverse multiplexing The process of distributing a serial data stream, bit by bit, onto multiple T-1s, then reassembling the original data stream at the receiving end.

IP Version 4 (IPv4) – addressing A standardized scheme in which a unique, 32-bit address is assigned to each host connected to an IP-based network.

IP Version 6 (Ipv6) – addressing It has been developed to extend source and destination addresses, and provide a mechanism to add new operations with built-in security.

ISDN line All-digital transmission line that allows the transport of voice and non-voice services on a call-by-call basis.

jamming A process where a NIC remains in the transmit mode and adversely affects the attached device(s) as it renders the channel(s) useless.

jitter The variability of the latency in a network.

junctions See **trunks.**

key A secret code that a sender and recipient must share to encrypt and decrypt messages.

keystroke response The time it takes for a screen to redraw after the Enter key is depressed.

last number redial A telephone feature that enables the caller to automatically dial the last number called.

latency The amount of time it takes for a packet of data to get from one designated point to another.

learning bridge Examines the source field of every packet it sees on each port and builds up a picture of which addresses are connected to which ports.

leased lines Private lines that provide a permanent pathway between two communicating stations. Also called **dedicated lines**.

Lightweight Directory Access Protocol (LDAP) It is an access protocol that allows a collection of directories to function as a single integrated directory service even when different directories reside on different servers.

life cycle The length of time that an organization can realistically expect to use an item in its planned role before discarding or replacing it.

line-of-sight An infrared system for a wireless network which offers point-to-point, high-speed connectivity between stations located within 100 ft.

lines Circuits or paths that connect stations to the nodes.

link layer The layer of the SS7 and OSI protocol architecture that is responsible for flow control and error correction.

Local Access and Transport Area (LATA) A pre-determined area used to govern who could carry calls in what area.

local area The geographical area within which subscribers can call each other without incurring tolls.

Local Area Network (LAN) The most common type of data network that allows users to share computer related resources within an organization.

local exchange A class-5 office that serves every telephone subscriber in a localized area and also provides dial tone services to the user. See also **central office** and **end office.**

Local Exchange Carrier (LEC) Carrier that handles local telephone calls.

local loop A pair of copper wires that runs from a telephone to a local switching station.

Local Multipoint Distribution System (LMDS) A type of Wireless Local Loop system that supports transmission of signals over short distances using frequencies of about 30 GHz.

Local Number Portability (LNP) A program initiated by the FCC that will make it possible for a business or residential phone user to change service providers while retaining the same phone number.

Location Routing Number (LRN) Uniquely identifies the switch that is serving a particular number.

Logical Link Control (LLC) The part of the data link layer in a network that brings various topologies together in a common format.

Loop Start A DC line signaling method provided by the local exchange to communicate with a single-line telephone set.

loose buffer Allows the fiber in a fiber-optic cable to move which relieves the cable from stresses occurring during installation and frequent handling.

lost speech interpolation A technique that blends speech samples that precede and follow a lost frame to mask the lost speech elements.

material dispersion This property depends on the dopants of the core glass.

macrobending Stresses on fiber-optic cables which occur when optical fibers are wound on reels for transportation and during the installation process.

maintenance The ability to detect, report, isolate, and resolve problems in a network.

mastergroup Ten supergroups. See **supergroups** and **group.**

Media Access Control (MAC) Part of the data link layer in a network that specifies the access methods used.

mesh A network topology that is a hierarchical structure except that there are more interconnections between nodes at different levels, or even at the same level.

message switching A store-and-forward system that accepts a message from a user, stores it, and forwards it to its destination according to the priority set by the sender.

Message Transfer Part (MTP) The first three layers of the SS7 protocol architecture.

metaframing Keeps T-1s aligned during inverse multiplexing.

Metropolitan Area Network (MAN) Covers an area of between 5 and 50 km diameter and acts as a high-speed network to allow sharing of regional resources.

middleware The software layer that interacts between an application and a network.

Minimum Annual Commitment (MAC) The amount of money a user organization agrees to pay a carrier each year of a multiyear contract in exchange for negotiated discounts.

Mobile Switching Center (MSC) A control center for wireless services that is connected to the local exchange for wired telephones.

modal dispersion See **Differential Mode Delay (DMD).**

modem A modulator-demodulator device that converts digital signals received from a serial interface of a computer into analog signals for transmission over the telephone local loop, and vice versa.

modulation A means of controlling the characteristics of a signal in a desired way.

multi-frequency A method of address signaling for an outgoing telephone call.

multicast An addressing technique that allows a source to send a single copy of a packet to a specific group through the use of a multicast address.

Multichannel Multipoint Distribution System (MMDS) A type of Wireless Local Loop system which transmits microwave signals over the 2 GHz band of the radio spectrum.

multimode A type of optical fiber that is designed to carry multiple light rays or modes concurrently, each at a slightly different reflection angle within the fiber core.

multiplexing A protocol where two or more signals are combined for transmission over a single communication path.

MultiProtocol Label Switching (MPLS) A protocol that sets up a specific path for a given sequence of packets identified by a label put in each packet.

Multipurpose Internet Mail Extensions (MIME) Permits mail clients to send attachments of any kind, marking them with a label that indicates how the information was encoded and what application created the attachment.

Multistation Access Unit (MAU) Two pairs of wires located in the center of the network that routes the token from port to port.

multiple access An access method for a network in which any of the network devices can transmit data onto the network at will.

network A series of points or nodes interconnected by communication paths.

Network Access Point (NAP) One of several major Internet interconnection points that serve to tie all the Internet access providers together.

network administration An infrastructure of techniques and procedures that assure the proper day-to-day operation of a system.

network administrator The person who performs daily maintenance tasks on a network.

network architecture A coordinated set of guidelines that together constitute a complete description of one approach to building a communications environment.

Network Attached Storage (NAS) Disk storage that is set up with its own network address rather than being attached to the server that is serving applications to network users.

Network Control Point (NCP) See **Service Control Point**.

network downtime When the entire network comes to a halt.

Network Interface Card (NIC) A computer circuit board or card that is installed in a computer to provide a dedicated, full-time connection to a network.

network layer The layer of the SS7 and OSI protocol architecture which routes messages from source to destination, and provides an interface between lower layers and upper layers.

network management Human and automated tasks that support the creation, operation, and evolution of a network.

network manager	In a network, the person who is responsible for policy management, evaluation of hardware and software, administration and maintenance, security, and configuration management.
Network Operating System (NOS)	Software that operates between the user application and the data link layer of the OSI model developed by the ISO.
Network Operations Center (NOC)	A separate room from which a telecommunications network is managed, monitored, and maintained to ensure uninterrupted service for its users.
network utilization	The ratio of total load to network capacity.
network value	The benefits of a network divided by the costs of a network. Also called **network value**.
nodes	Exchange or switching points where two or more paths meet, enabling users to share transmission paths.
noise	Undesired and usually random variations that interfere with the desired signals and inhibit communication.
noise blanking	A technique used to improve communications by disabling the receiver for the duration of a short intense burst.
noise factor	See **noise figure.**
noise figure (NF)	The log of noise ratio; a quantity used to determine the signal quality. Also called **noise factor.**
noise ratio (NR)	The value of the signal-to-noise ratio at the input over the signal-to-noise ratio at the output.
non-blocking networks	Digital telephone networks that enable a connection to be made between any two ports independently of the amount of traffic.
noncontention	Two network interfaces cannot transmit data at the same time.
Numerical Aperture (NA)	The sine of the acceptance angle.
Nyquist sampling theorem	A theorem that if a waveform is sampled at a rate at least twice the maximum frequency component in the waveform, it is possible to reconstruct that waveform from the periodic samples without any distortion.
odd parity	A type of parity checking to detect errors.
open operating system	A system that adheres to a publicly known set of interfaces so that anyone using it can also use any other system that adheres to the standard.
OSI reference model	An architectural model for how digital information should be transmitted between any two points in a data communication network.
out-of-band signaling	A signaling system that uses a separate network to pass call setup, charging, and supervision information.
overheads	In a transmission, the synchronization, error detection, or any other bits that are not messages.
packet	Block of data delimited by header and trailer records.

Packet Assembler/ Disassembler (PAD) A software package or a piece of hardware that receives data and breaks it down into packets.

packet filtering A security feature of a network that is used as a first defense and comes as part of most routers' software.

packet switching Permits data or digital information to proceed over virtual telecommunications paths that use shared facilities and are in use only when information is actually being sent.

Packet Switching Exchange (PSE) A computer that analyzes each packet to verify that it was received without errors and routes the packet to the next appropriate port.

packetization delay The time it takes to fill a cell.

packets Blocks of data characters delimited by header and trailer records.

parallel Transmission of a group of bits at a single instant in time, requiring multiple paths.

parity checking One of the simplest error-detection schemes and is appropriate for use in asynchronous transmission because it checks one character at a time.

passive monitoring An aspect of firewall design where the firewall logs a record of each activity against it in a file, which can be accessed at any time.

passive threat A security threat to a network that involves monitoring of the transmission data of an organization by an unauthorized person.

patch panel A piece of cable termination equipment that connects raw cables to standard ports or connectors.

patent Awarded for inventions or nonobvious improvements to existing products or processes.

payload The actual data in a frame.

peer-to-peer network A LAN that does not make a distinction between a user workstation and a server.

performance management Facilities which provide the network manager with the ability to monitor and evaluate the performance of the network.

Permanent Virtual Circuit (PVC) Routing between stations is fixed and packets always take the same route.

Personal Computer Memory Card Industry Association (PCMCIA) Combines data connectivity with user mobility, and through simplified configuration enables movable LANs.

Personal Communications Networks (PCN) Also called Personal Communications Systems (PCS), they have expanded the horizon of wireless communications to provide integrated (such as voice, data, and video) services and near-universal access irrespective of time, location, and mobility patterns.

phase shift A time difference between two sine waves of the same frequency.

Phase Shift Keying (PSK) The most popular implementation of phase modulation (PM) for data applications where a carrier is shifted in phase to represent a particular symbol.

physical layer The bottom layer of the SS7 and OSI protocol architecture, which provides a physical connection between network nodes.

PING A command which lets one determine whether a workstation, server, or another TCP/IP network device recently connected to a network is correctly cabled to the network and whether the protocol stack is operational and correctly configured.

pink noise See **flicker noise.**

platform An underlying computer system on which application programs can run.

Point of Interface (POI) Same as **Point of Presence (POP)**.

Point of Presence (POP) A point where the Local Exchange Carrier (LEC) and the Inter Exchange Carrier (IXC) are interconnected.

Point-to-Point Protocol (PPP) A protocol that routes multiple protocols over dial-up and dedicated point-to-point links.

policy management An implementation of a set of rules or policies to dictate user connectivity and network resource priorities.

polling The process of a host computer asking an intelligent terminal if it has any data to send to the host computer.

port The place on a device where it connects to a cable.

Post Office Protocol A mechanism that provides a store-and-forward feature that allows for clients to be disconnected from the network and still receive mail.

Positive-Intrinsic-Negative Photodiode (PIN) The lower cost but less efficient of the two types of widely-deployed photodiodes. In the PIN photodiode, light is absorbed and photons are converted in a 1:1 relationship.

price to performance ratio The ratio of price to performance of a product, which is an important measure because the user always wants the most performance at the least cost.

Primary Rate Interface (PRI) A category of User-to-Network Interface (UNI) that is appropriate for a business that utilizes a T-1 line.

prioritization The means by which certain frames are given preferential treatment over others.

Private Branch Exchange (PBX) An on-premises telephone exchange that provides digital data services, as well as analog telephone service, along with a dial tone to the telephones.

private network A network built for exclusive use by a single organization.

probability of blockage See **Grade of Service**.

propagation The various ways by which an electromagnetic wave travels from the transmitter to the receiver.

protocol A set of rules that allow information to flow horizontally on a computer network.

protocol converter Connects two systems at the same layer and enables two different electronic machines to communicate with one another.

proxy servers Terminate a user's connection and set up a new connection to the ultimate destination.

Public Key Infrastructure (PKI) A comprehensive system that provides the public-key-based encryption and digital signature services on behalf of the applications.

Public Switched Telephone Network (PSTN) A mesh network that connects telephone lines with one another through multiple interconnections.

public network A network owned by a common carrier for use by its customers.

Pulse Amplitude Modulation (PAM) A technique that generates pulses whose amplitude varies with that of the modulating signal.

Pulse Code Modulation (PCM) A method of coding digital signals so that telephone (analog) signals can be transmitted digitally.

pulse dial A method of address signaling for an outgoing telephone call.

Pulse Position Modulation (PPM) A technique similar to Pulse Width Modulation except that the timing of the pulses varies with the amplitude of the modulating signal.

Pulse Width Modulation (PWM) A technique that generates pulses at a regular rate where their width is controlled by the amplitude of the modulating signal.

Quadrature Amplitude Modulation (QAM) A technique that uses two amplitude-modulated carriers with a 90° phase angle between them to produce a signal with an amplitude and phase angle that can vary continuously.

Quality of Service (QoS) A set of characteristics that define the delivery behavior of different types of network traffic and provides certain guarantees.

quantizer A circuit/device used at the receive end to determine whether the incoming digital signal has a voltage level corresponding to binary 0 or binary 1.

radio A wireless voice-transmitting system utilizing electromagnetic waves.

read and write Refers to a user's right to modify a file.

real When referring to a printed page, real data is that which is not white space.

real-time monitoring A system that allows an SLA to spot problems before they escalate.

reassembly The process by which the ATM Adaptation Layer reconstructs cells into high-level data and transmits the data to the destination devices.

reciprocal compensation Fees paid to local phone companies for use of their networks to complete the calls.

Redundant Array of Independent Disks (RAID) A way of storing the same data in different places, thus, redundantly on multiple hard disks.

reflection Occurs when a light ray traveling from one medium to another bounces back in the same medium.

reflective An infrared system in a wireless network which gets around the line-of-sight problem by bouncing the signal off walls, ceilings, and floors.

refraction Occurs when a light ray traveling from one medium to another changes speed as it travels in the second medium and is bent or refracted.

refractive index The ratio between the speed of light in free space and the speed of light in a medium. Also called **index of refraction.**

relay rack A metal frame used to secure and support networking equipment.

reliability The ability of a packet to reach its destination on a network.

repeater A receiver-transmitter combination that removes the effects of noise and distortion, amplifies the signal, and transmits it further down the channel.

resistance Opposition to current.

resistivity See **specific resistance.**

return loss See **echo.**

ring A network topology in which each device is attached along the same signal path to two other devices, forming a path in the shape of a ring.

router This device operates on Layer 3, the network layer that routes data to different networks.

Run-Length Encoding A method of compacting redundant data.

sampling A snapshot (sample) of the waveform taken for a brief instant of time, but at regular intervals.

sampling interval The time interval between each sample, which is also a reciprocal of the sampling rate.

sampling rate The rate at which a signal is sampled and expressed as the number of samples per second.

satellite A specialized wireless receiver/transmitter that is launched by a rocket and placed in orbit around the earth.

satellite earth station Establishes and maintains continuous communication links with all other earth stations in the system through a satellite repeater.

scalability The ability to smoothly increase the power and/or number of users in a network environment without major redesigns.

scatter An infrared system in a wireless network that uses diffused signals similar to the manner in which light scatters.

scattering Imperfections in the glass fiber as it is heated in the forming process, leading to attenuation in fiber-optic cables.

security policy An unambiguous policy enforced by an organization regarding access to each element of information, the rules an individual must follow in disseminating information, and a statement of how the organization will react to violations.

segmentation The process by which the ATM Adaptation Layer formats data into 48-byte cell payloads.

selecting The process that occurs when a host computer or a FEP sends data to a terminal after the terminal indicates that it is ready to accept data.

serial A transmission of bits one after another along a single path.

Serial Line Internet Protocol (SLIP) Can be used for carrying IP over an asynchronous dial-up or leased line.

Service Control Point (SCP) Data processing CPUs connected to databases of circuit parameters, routing, and customer profiles. Also called **Network Control Point (NCP)**.

Service Level Agreement (SLA) A contract between a network service provider and a customer that specifies certain levels of network and application performance and with a promise of rebates if those parameters are not met by the provider.

Service Switching Point (SSP) A tandem switch in an IXC network or a local exchange switch in a LEC network that provides switching and routing functions.

Shannon's Channel Coding Theorem Shannon's Channel Coding Theorem states that if the transmission rate is equal to or less than the channel capacity, then there exists a coding technique which enables transmission over the channel with an arbitrarily small frequency of errors.

Shielded Twisted Pair (STP) cable A 150 ohms cable made up of two copper pairs, where each copper pair is wrapped in metal foil and then sheathed in an additional braided metal shield and an outer PVC jacket.

shot noise Noise that has equal energy in every hertz of bandwidth because it is created by random variations of current flow in active devices such as transistors and semiconductor diodes.

Signal-to-Noise Ratio (SNR) The ratio of signal to noise power.

Signal Transfer Point (STP) Packet-switching nodes facilitating highly reliable and cost-effective network architectures.

Signaling Connection Control Part (SCCP) A layer of the SS7 protocol architecture that is responsible for addressing requests to the appropriate application and determining its status.

Signaling System Seven (SS7) An out-of-band signal control system that implements a layered protocol that is independent of the telephone network hardware.

silver satin Regular phone cable.

Simple Mail Transfer Protocol (SMTP) A mechanism for sending standard, interoperable text messages from one computer system to another.

simplex Communications in only one direction from the transmitter to the receiver.

single access A satellite system in which a single station is able to access and utilize the entire transponder bandwidth.

single mode A type of optical fiber that has a very small core and is designed to carry only a single light ray.

SLA clauses These define a provider's service levels and user or customer's expectations regarding network availability and performance, and applications availability and performance.

SLA monitoring tools These techniques measure the performance levels of networks and their applications.

slot The amount of time each user in a wireless network occupies the whole channel bandwidth.

slow start A transmission process used in TCP that determines the available bandwidth for a connection by starting with an initial window size of one segment and increasing the window size only when packets are delivered successfully.

Snell's Law States that a relationship exists between the refractive indices of the two media n_1 and n_2, and the angle of incidence and refraction, u_1 and u_{refr}.

soft benefits Outcomes that are expected to be good for the company but are more difficult to identify and quantify.

soft handoff Allows a handset to communicate with multiple base stations simultaneously in a wireless network.

spanning tree protocol A link management protocol used by Layer 2 devices.

speaker phone A telephone feature that enables the user to have hands-free conversation.

special code (ETX) The special block at the end of a signal indicating the end of the transmission.

specific resistance The comparison of the resistance of different materials according to their nature regardless of different areas or lengths.

spectrum analysis An amplitude spectrum of a signal represented in the frequency domain. See also **Fourier analysis.**

spectrum analyzer An instrument that provides an amplitude spectrum of a signal.

speed dialing A telephone feature which allows a user to preprogram telephone numbers into the memory of the telephone.

splice The point at which two ends of a fiber have been welded, glued, or fused together.

spoofing The process of simulating a communications protocol by a program that is interjected into a normal sequence of processes for the purpose of adding some useful function.

spread spectrum In a wireless network, users are separated by assigning them digital codes within a broad range of the radio frequency.

standards Documented agreements containing technical specifications or other precise criteria to be used consistently as rules, guidelines, or definitions of characteristics to ensure that the products, processes and services are fit for their purpose.

star A network topology in which all peripheral nodes are connected to a central node.

static A term typically used for atmospheric noise.

Statistical Time Division Multiplexing (STDM) A type of TDM scheme which assigns variable time slots on demand, depending upon the number of users and the data transmitted by each.

station drop The end of the telephone cable to the customer location.

stations Terminal points in a network; for example, telephone sets, data terminals, facsimile machines and computers.

Storage Area Network (SAN) A high-speed special-purpose network that interconnects different kinds of data storage devices with associated data servers on behalf of a larger network of users.

Storage Service Provider (SSP) A business that delivers flexible, cost-effective storage networking solutions and provides managed services such as data protection and storage-on-demand to customers.

store-and-forward See **Message Switching.**

straight-through Patch cables which are pinned so that pin one of one end connects to pin one of the other end.

streaming mode The process where information is sent in compressed form over the Internet in a continuous stream and is played as it arrives.

structured wiring A cabling system that meets stringent installation standards to protect the integrity of the cabling system and to eliminate the need for constant recabling with the addition of each new application.

subnetting A process in which the two-level hierarchy of Class A, B, and C networks is turned into a three-level hierarchy with the host portion of an IP address divided into a subnet portion and a host portion.

subscriber lines The telephone lines that connect users to the local exchange. Also called **local loops.**

supergroup Five groups. See **group.**

supervising signals Monitor the status of a telephone line or circuit to determine if it is busy, idle, or requesting service.

Sustained Information Rate (SIR) Provides control over how much information a node can place on the network, which in turn controls network congestion.

switch Sets up a communication path on demand and takes it down when the path is no longer needed.

Switched Multimegabit Data Service (SMDS) A public, packet-switched service aimed at enterprises that do not want to commit to predefined permanent virtual circuits, but need to exchange large amounts of data with other enterprises over a wide area network.

Switched Virtual Circuit (SVC) Routing between stations is determined with each packet.

switching The process of routing communications to different parties.

SyncML	An extensible and transport-independent technology that allows the device to support a single synchronization standard.
synchronous	A type of transmission that transfers a large block of data but requires a coherent clock signal between the transmitter and the receiver.
Synchronous replication	Synchronously replicates all disk writes to backup storage system located at a remote site. Also called mirroring.
Synchronous Optical Network (SONET)	A physical layer or Layer 1 technology that transmits data in frames over a wide area network.
T-1 line	A popular leased line option for businesses connecting to the Internet backbone, consisting of two pairs of UTP 19 AWG wire. Also referred to as a DS-1 line.
tandem office	See **toll center.**
telecommunication	The distant transfer of meaningful information from one location to a second location.
Telecommunications Act of 1996	The first major reform to the 1936 telecommunications legislation that established the Federal Communications Commission (FCC).
Telecommunications Closet (TC)	An enclosure in which wiring is terminated. Replaced by Telecommunications Room (TR).
Telecommunications Industry Association (TIA)	Accredited by the American National Standards Institute (ANSI) to develop voluntary industry standards for a wide variety of telecommunications products.
Telecommunications Management Network (TMN)	A method for integrating network management across different networks and is equally applicable to service provider networks or companies having their own private networks.
Telecommunications Room (TR)	A room in which wiring is terminated. Generally the connection point between the building backbone cable and the horizontal cable. This term replaces the older term "Telecommunications Closet."
telegraph	The earliest form of electrical communication designed to print patterns at a distance.
telephone	A device that converts voice into electrical signals, and vice versa.
telephony	The science of translating sound into electrical signals, transmitting them and then converting them back to sound.
thermal noise	An equal mixture of noise of all frequencies produced by the random motion of electrons in a conductor because of heat. Also called **white noise.**
thin-client architecture	A newer implementation of centralized computing where the level of computing power on each desktop may vary between end users.
throughput	The capacity available to the user on a wireless network.
tie lines	Point-to-point lines connecting two voice facilities, typically PBXs or Centrex, so that the users can communicate without dialing an outside call. Also called **tie trunks.**

tight buffer Layers of plastic and yarn material applied over the fiber in a fiber-optic cable.

Time Assignment Speech Interpolation (TASI) An analog Statistical Time Division Multiplexing (STDM) scheme for sharing voice circuits.

Time Division Multiple Access (TDMA) A second-generation wireless digital technology that allows users to access the assigned bandwidth on a time-sharing basis.

time domain A method of representing signals, where amplitude is shown on one axis and time is displayed on the other.

Tip and Ring A standard RJ-11 jack wired with a pair of conductors that connects the telephone to the plug in the wall. The Tip is the transmit wire connected to the transmitter of the telephone and the Ring is the receive wire connected to the receiver of the telephone.

thick coax A category of coaxial cable. See **10Base5**.

thicknet See **10Base5**.

thin coax A category of coaxial cable. See **10Base2**.

thinnet See **10Base2**.

three-tier architecture A layout in which a user's workstation contains the GUI interface, business logic is located on a LAN server, and the database is located on a mainframe computer.

toll center A class-4 office to which every local exchange or class-5 office is connected. Also called **tandem office.**

toll free lines Widely-used service, which is a reverse-billing service where the callers can call at any time without worrying about the cost of the call.

topology A pictorial description of the physical layout of a network including its nodes and connecting lines.

Total Cost of Ownership (TCO) The cost of equipment, bandwidth, network, and operations.

total internal reflection In a fiber, light striking the core cladding interface is totally reflected back into the core without any light entering the cladding.

total traffic intensity Measures usage of a telephone network in centi-call seconds.

TRACEROUTE A command that invokes a program that provides information about the route that packets take from source to destination.

trademark Any sign or symbol capable of distinguishing goods or services.

traffic Quantifies usage of a telephone network.

transceiver A device which sends and receives a signal usually over two separate fiber cables.

Transmission Control Protocol (TCP) A reliable, connection-oriented, unicast (point-to-point) guaranteed-delivery protocol, which performs end-to-end error checking, correction, and acknowledgement.

transmission paths Transmission lines.

transponder The device on a satellite that receives signals from ground station dishes.

tree A network topology that is a hierarchical structure resembling an interconnection of star networks.

trunk port One big logical link in a network.

trunking A technique that allows a networking device to bond together multiple physical links into a group that works like one logical link.

trunks Circuits or links that interconnect exchanges and switching centers. Also called **junctions** in Europe.

tunneling The process where the source end encrypts its outgoing packets and encapsulates them in IP packets for transit across the Internet, while at the receiving end, a gateway device removes and decrypts the packets, forwarding the original packets to their destination.

twisted pair A pair of copper wires twisted together and protected by a thin polyvinylchloride (PVC) or Teflon jacket.

Type I Signaling The most commonly used four-wire trunk signaling interface in North America.

Type II Signaling Used occasionally in North America, it is the least likely to cause interference problems in sensitive environment.

Type V Signaling The most popular interface outside North America.

Unbundled Network Elements (UNEs) Wholesale components of the ILEC's local infrastructure, including local loops, network interface devices, switching capabilities, interoffice transmission facilities, signaling networks and call-related databases, SS7 functions and Operations Support Systems (OSSs), and operator and directory assistance.

Universal Asynchronous Receiver Transmitter (UART) Converts data transmissions from parallel to serial and vice versa.

Uninterruptible Power Supply (UPS) A device whose battery kicks in after sensing a loss of power from the primary source and allows a computer to keep running for at least a short time.

Unshielded Twisted Pair (UTP) cable A low-cost copper media inherited from telephony, which is being used for increasingly higher data rates.

usage-based model For an ISP, a pay-as-you-use system.

User Datagram Protocol (UDP) An unreliable, connectionless protocol, but with less overhead as compared to TCP.

video communications One-way transmissions such as Cable TV (CATV) and two-way transmissions such as videoconferencing.

virtual circuit One of the conceptual models of the network layer where a connection from sender to receiver is set up only on demand.

Virtual LAN (VLAN) A VLAN is a switched network that is logically segmented on an organizational basis, by functions, project teams, or applications rather than on a physical or geographical basis.

Virtual Private Network (VPN) Encrypted tunnels through a shared private or public network that forward data over the shared media rather than over dedicated leased lines.

virtually non-blocking Telephone networks that are not designed to be totally non-blocking but provide enough paths that users rarely find themselves blocked by the network.

virus A program that can affect other programs on a network by modifying them with a copy of the virus program, which then goes on to infect other programs.

visual indicators Blinking lights or LEDs on a telephone used to convey a message to the user.

voice communications Telephone communications.

Voice over ATM Transmission of voice over an ATM network.

Voice over Frame Relay Transmission of voice over a Frame Relay network.

Voice over IP (VoIP) The process of transmitting telephone calls over the Internet rather than through the traditional telephone system.

voice processing systems Include Interactive Voice Response (IVR), Automated Attendant (AA), and voice mail, installed to reduced the number of agents required to handle customer calls.

vulnerability assessment Network exposure is defined as all information that can be gathered remotely about the network, including vulnerabilities; the process to identify these exposures is called vulnerability assessment.

vulnerability management Vulnerability management is a cyclical process of identifying, measuring, prioritizing, monitoring, and remediating potential security risks.

white noise See **thermal noise.**

white space Usually refers to blank area on a sheet of paper.

Wide Area Network (WAN) Usually refers to a network that covers a large geographical area and uses common carrier circuits to connect intermediate nodes.

wireless A communications system in which electromagnetic waves carry a signal through atmospheric space rather than along a wire.

Wireless Application Protocol (WAP) A standard for wireless data delivery and communications over the Internet.

Wireless LAN (WLAN) Transmits and receives data over the air, minimizing the need for wired connections.

Wireless Local Loop (WLL) A broadband wireless system that involves a low-power digital transceiver capable of supporting bi-directional communications in a small geographic area.

work area wiring The connection between a user station and the outlet.

workstation 1) Advanced machines typically used by engineers, architects, graphic designers and other individuals that require a faster microprocessor, a large amount of RAM, special features, and a relatively fail-safe system.
2) The place where a person sits or stands to do work, including the working surface, terminal, chair, and any other equipment needed to do the job.

workstation ergonomics The science of designing a workstation to better accommodate workers in a physical sense, allowing them to complete their jobs without the risk of injury.

worm A program that replicates itself and moves from system to system, performing some activity on each system it gains access to, such as consuming processor resources or depositing viruses.

X.25 One of the first packet-switching technologies that was built on the OSI model.

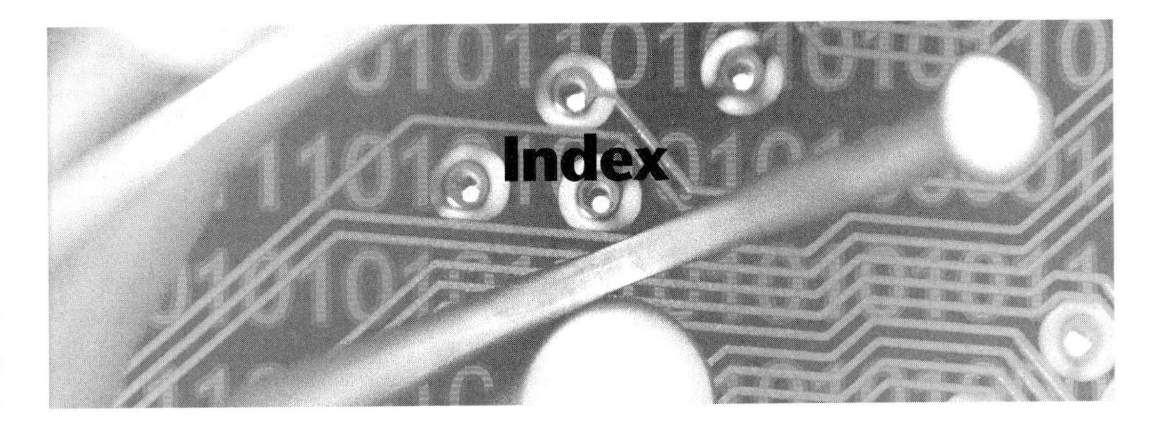

Numerics

A

G

H

N

T

U

V